The Theory of
TOROIDALLY CONFINED PLASMAS

Fourth Edition

The Theory of
TOROIDALLY CONFINED PLASMAS

Fourth Edition

Roscoe B. White

Princeton University

World Scientific

NEW JERSEY · LONDON · SINGAPORE · GENEVA · BEIJING · SHANGHAI · TAIPEI · CHENNAI

Published by

World Scientific Publishing Co. Pte. Ltd.

5 Toh Tuck Link, Singapore 596224

USA office: 27 Warren Street, Suite 401-402, Hackensack, NJ 07601

UK office: 57 Shelton Street, Covent Garden, London WC2H 9HE

British Library Cataloguing-in-Publication Data
A catalogue record for this book is available from the British Library.

THE THEORY OF TOROIDALLY CONFINED PLASMAS
Fourth Edition

ISBN 978-981-98-0917-2 (hardcover)
ISBN 978-981-98-0918-9 (ebook for institutions)
ISBN 978-981-98-0919-6 (ebook for individuals)

For any available supplementary material, please visit
https://www.worldscientific.com/worldscibooks/10.1142/14211#t=suppl

Desk Editor: Rhaimie B Wahap

Typeset by Stallion Press
Email: enquiries@stallionpress.com

兵の法を學ばんと思はば、此書を思案して、師は針弟子は糸と成て、絶えず稽古あるべき事也。

"If you want to learn the art search beneath the surface. The teacher is as a needle, the student is as thread. You must practice constantly."

A Book of Five Rings
Miyamoto Musashi, 1645

Preface to the Fourth Edition

Since publishing the third edition of this book I have spent much time on types of events that can produce problems in high energy confinement, both in tokamaks and stellarators, including mode avalanche, chirping, and the effects of particle resonances, which surprisingly can give large resonance islands in stellarators also in the absence of unstable modes. At low energy the resonances are located at surfaces where the magnetic field helicity is a low order rational, but orbit helicity is a function of particle energy, so there are interesting surprises to be found for fusion produced alpha particles in a reactor. Only with a complete understanding of these phenomena can a device design and an operation procedure be found which will reliably give a stable useful discharge.

Since I have no longer been teaching I have not provided problem sets for these subjects but I think the previously given problems can provide sufficient exercise to give a student confidence in the subject matter. Collaborators have been primarily Vinicius Duarte, Nikolai Gorelenkov, Erik Fredrickson and Mario Podesta in Princeton, Herb Berk in Texas, Bill Heidbrink in California and Andy Bierwage and Kunihiro Ogawa in Japan.

In the intervening years the effects of global warming have become much more evident, and the need for energy sources which are CO_2 free has become critical. Reactor design has advanced sufficiently that there are now several independent companies involved in trying to produce a commercially viable fusion power source, as well as strong national efforts in Japan, Korea, China and Europe as well as the U.S. Nevertheless, technical problems still make an early solution improbable, given the fact that even after an acceptable design is produced it takes ten years to construct it, and such a design is still not at hand.

Roscoe White
Princeton Plasma Physics Lab
Princeton University, 2024

Preface to the Third Edition

Since publishing the second edition of this book, I have been very much involved with examining the nature of mode particle interactions in toroidal confinement devices. This work has resulted in a large extension of the chapter on the theory of mode-particle interactions. There is also a full chapter devoted to the use of lithium as a particle sink either on the plasma wall or on the divertor. If the use of lithium turns out to be worth even a small fraction of its theoreticaly possible utility, it will greatly help the program. The development of a liquid lithium divertor is presently being carried out in China. Research on its use is also being carried out at Princeton, Frascati, and Padova.

Recent good news is the confirmation of the plans to proceed with the construction of IGNITOR, partly constructed in Italy and planned to be built at the trinity site at Troitsk, near Moscow. An important missing part of the experimental data to date is information obtained from the observation of an ignited burning plasma, or at least one in which the fusion energy is an important part of the plasma heating.

A section on nonlocal transport has been added, since it has increasingly become apparent that simple local descriptions of transport using locally derived diffusion operators are not sufficient to describe transport in toroidal devices. This is particulary apparent in the reversed field pinch at Padova, since many discharges consist of magnetic field structures that are above the stochastic threshold.

There is as well a new chapter deriving equations for full cyclotron motion in a toroidal system and the comparison of some results with guiding center simulations. In some present devices, in particular in NSTX, the cyclotron radius is very large, leading to some question concerning the use of guiding center equations to describe particle motion.

The presentation is necessarily not uniform; the sections treating the derivation of basic magnetohydrodynamic equilibrium and stability properties, and of particle orbits, are more mathematical, whereas other sections are more approximate and descriptive in nature. This is to provide a basic foundation and at the same time arrive at a description of current phenomena without becoming overwhelmed by minute detail.

Roscoe White
Princeton Plasma Physics Lab
Princeton University, 2013

Preface to the Second Edition

In the intervening years since the first edition of this book was published the field of magnetically confined fusion research has made great advances, and the goal of an inexpensive, abundant, clean, and inexhaustable energy source is tantalizingly close. The record value of fusion power produced in research reactors has increased even faster than the memory capacity of semiconductor chips, doubling approximately every year. The large tokamak at Princeton, TFTR, reached fusion power output greater than one fourth of the energy input, producing 11 Megawatts of fusion power, and the joint European tokamak in England, JET, passed the 60 percent mark, producing 16 megawatts of fusion power. Both JET and the Japanese reactor JT-60, operating without tritium, approached the equivalent break even point, *i.e.* in the same discharge; with tritium present, they would have produced as much power as used to heat the discharge. Research tokamaks regularly produce temperatures ten times hotter than the interior of the sun, and sustained stable operation at lower temperatures has been maintained for over an hour. In specially configured discharges in DIII-D at General Atomics the transport was reduced to the theoretical minimum neoclassical limit by using sheared rotation to stabilize the turbulence. Most of the rapid large scale instabilities impeding confinement have been understood and dominated, and research focuses on improving and extending the operating domain by the control of fine-scale instabilities.

At the same time the field experienced an often rancorous debate concerning the wisdom of constructing a large multibillion dollar demonstration reactor capable of igniting, producing net power, and testing associated engineering components. The reactor, ITER, was designed as a joint effort by the four partners, Europe, the Soviet Union, Japan, and the United States. The debate was complicated by the collapse of the Soviet Union and the withdrawal of the United States from all but token theoretical analysis due to congressional pressure to balance the US budget. ITER is now being built in Cadarache, France, and China, India and South Korea are also participating. In addition, large independent programs are underway in Japan, China and South Korea.

Many US physicists feel that a reactor produced following the design of ITER cannot be economically successful, and the US program, aside from an official but small support for ITER, has retreated to a search for a "smaller, cheaper, smarter" solution. One such approach was guided by the discovery of a large class of stellarators, the so called quasi-symmetric stellarators, which could offer distinct advantages over tokamaks provided they succeed in obtaining high beta and reasonable confinement. The parameter space of aspect ratio, elongation, and cross-sectional shape is being explored, as is the use of various means of profile control to produce regimes of very good confinement. The idea of covering the plasma-facing toroidal walls with flowing liquid metal, to eliminate the problems associated with wall erosion and the design of divertors. An important missing experimental step is the production of a burning, ignited plasma, and most physicists agree that such an experiment is crucial. Aside from ITER, the only design for an ignited plasma device to proceed beyond paper studies is the IGNI-TOR, partly constructed in Italy, but its future completion is uncertain.

In spite of the discord concerning the research strategy, the field is rather optimistic. The impressive list of successes obtained in the last decades leads many research fusion scientists to believe that a good solution is "out there" waiting to be found, and that it will be found. In the meantime, relatively cheap petroleum and natural gas, along with the development of renewable energy sources continue to suppress the urgency for arriving quickly at an economical design. The minimum commitment allowing preservation of the results established so far is to continue to train and maintain a cadre of professionals capable of carrying out the search, and of rapidly producing a good reactor when such a design is found.

The book has been extended in several ways. Many of the derivations have been improved. Wherever possible the analysis has been extended to include other toroidal devices as well as tokamaks, including the analysis of equilibrium, particle orbits and stability. The area of field stochasticity and the onset of chaos has been improved. The treatments of bootstrap current, stochastic ripple loss, and phase integral methods have all been rewritten and extended, and new sections added treating the classification of particle orbits in tokamaks, TAE modes and their saturation, neoclassical tearing, and general wave-particle interactions.

The second edition has bendfited from comments and corrections by Mike Beer, Alain Brizard, Yanlin Wu, Wonchull Park, Don Monticello, Allan Reiman, Wei Li Lee, Zhihong Lin, Rob Goldston, Paul Parks, Stuart Zweben, Leonid Zakharov, and many others, as well as the availability of LATEX word processing. Many of my students have contributed by posing good questions and by showing interest in the further development of some topics.

Roscoe White
Princeton Plasma Physics Lab
Princeton University, 2001

Preface to the First Edition

These notes accompany a graduate course taught at Princeton, designed to provide a basic introduction to plasma equilibrium, particle orbits, transport, and those ideal and resistive magnetohydrodynamic instabilities which dominate the behavior of a tokamak discharge, and to develop the mathematical methods necessary for their theoretical analysis. Primarily the book covers the consequences of ideal and resistive magnetohydrodynamics, these theories being responsible for most of what is well understood regarding the physics of tokamak discharges. No attempt is made to discuss the derivation of this formalism, this being a topic better left to a course devoted to kinetic theory. The focus is rather on the description of equilibria, the linear and nonlinear theory of large scale modes, and single particle guiding center motion, including simple neoclassical effects. Modern methods of general magnetic coordinates are used, and the student is introduced to the onset of chaos in Hamiltonian systems in the discussion of destruction of magnetic surfaces. The Hamiltonian formulation of guiding center motion provides another glimpse of modern methods in particle dynamics, preparing the way for the study of gyrokinetics. The interaction of a high energy particle component with a background magnetohydrodynamic plasma, of interest for the description of intensely heated and ignited plasmas, is also treated.

Much of the book is devoted to the description of the limitations placed on operating parameters given by ideal and resistive modes, and current ideas about how to extend and optimize these parameters. This permits the student to quickly arrive at the research level of a topic which is reasonably well developed and plays an important role in our current understanding of the dynamics of basic confinement behavior. The part of the book dealing with transport consists of an elementary introduction to the principle neoclassical mechanisms, and examples of the perturbation methods employed to deduce transport rates from the drift kinetic equation. There is a brief introduction to some of the primitive theories and phenomenological descriptions of anomalous transport. Particle loss due to symmetry breaking perturbations such as toroidal field ripple is treated. The last chapter is a treatment of the method of phase integrals, which has proved very valuable in the study of ballooning modes, parametric instabilities, and microinstabilities. In spite of its elegance and utility there is no treatment in the literature easily accessible to the student.

To a large degree the book is self contained, and most chapters depend logically on material developed previously. The exceptions are the chapter on phase integral methods, which is independent, and the chapter on transport, which requires only Chap. 1 through 3. The course is aimed at second year graduate students, but by changing emphasis on the subjects presented, it could also be given to first year students, or even to undergraduates with some previous plasma physics training. At the same time, the material in each topic can be easily expanded to make contact with current research, and the course could be taught at a more advanced level by using the text as a base. Notation is uniform and corresponds to that most commonly used among working plasma theorists.

Roscoe B. White
Princeton 1989

Acknowledgement

The material for these lectures has come into being during interaction with many colleagues, both at Princeton and elsewhere. The subject matter necessarily reflects my interests and the interests of the people with whom I have worked. The major credit for my career goes to Marshall Rosenbluth, who invited me to the Institute for Advanced Studies to work with him for two years, and had me spend one year working on laser fusion problems and one on tokamaks. These years saw the simulation of vacuum bubbles, prediced to be possible by Kadomtsov, and the beginning of my long time infatuation with Phase Integral methods. Particularly important were Paul Rutherford, who suggested the original outline of topics when I first taught the course and provided notes for some sections, Harry Mynick and Allen Boozer in the fields of guiding center motion and magnetic coordinates, Morrell Chance and Russell Kulsrud for magnetic coordinates and ideal MHD, Liu Chen, Jim Strachan, and Rob Goldston on the fishbone mode, and Marshall Rosenbluth, Don Monticello, Wonchull Park, and Kevin McGuire on resistive instabilities. I am indebted to Francis Troyon, James Van Dam, and Burton Fried for valuable suggestions regarding the text.

The second and third editions benefited from comments and corrections by Mike Beer, Alain Brizard, Yanlin Wu, Wonchull Park, Don Monticello, Allan Reiman, Wei Li Lee, Zhihong Lin, Rob Goldston, Paul Parks, Stuart Zweben, Leonid Zakharov, and many others, as well as by the availability of Latex word processing. The third edition has also benefited from my longtime interaction with Gianluca Spizzo and other physicists at the Reversed Field Pinch experiment in Padova, and close collaboration with Leonid Zakharov, who has carried out most of the theoretical analysis of the effect of lithium-coated walls and divertors, and many discussions with Stuart Hudson. Many of my students and also students of Zhihong Lin at UC

Acknowledgement

Irvine have contributed by posing good questions and by showing interest in the further development of some topics.

The fourth edition focuses on the understanding of further problems with high energy particles in tokamaks and stellarators, including mode chirping and avalanche, the presence of high energy particle resonances, and the effect of particle collisions on resonances.

Thanks are due to Barbara Sarfaty and Terry Greenberg for their care and perseverence in typing the first draft of the text, and finally much gratitude to Laura and Veronica for their patience while the book was being written.

Contents

Contents

List of Figures

Chapter 1

Toroidal Configuration

1.1 INTRODUCTION

Because of the divergence-free nature of the magnetic field, the simplest topological configuration it can assume with no field lines exiting from a fixed volume is toroidal. Since particle trajectories, to lowest order in gyro radius, follow field lines, this is also the simplest configuration which can give complete confinement within this approximation.

In addition, because of the very high particle velocity along the field lines, the field must, at least approximately, possess magnetic surfaces, *i.e.* the field lines trace out a nested set of toroidal surfaces, to prevent rapid particle transport from the inside of the torus to the exterior, and thus allow the existence of pressure and temperature gradients. The temperature must be sufficiently high so that ion collisions are strong enough to overcome the repulsive Coulomb potential and permit nuclear fusion by means of the strong short-range nuclear forces. The fuel which is most easily brought into such conditions is a mixture of deuterium and tritium, which fuses through the reaction

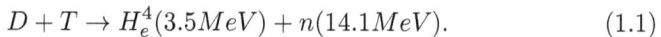

$$D + T \to H_e^4(3.5 MeV) + n(14.1 MeV). \tag{1.1}$$

The cross section for this reaction has a peak near 50 keV, and the plasma must be sustained at a temperature on the order of 15 keV, or around 100 million degrees centigrade, allowing the higher energy particles in the distribution to collide and fuse. For a plasma containing an optimum mixture of half deuterium and half tritium the ignition condition is (Lawson, 1957) given by

$$nT\tau_E > 3 \times 10^{21} keV s/m^3 \qquad (1.2)$$

where n is the ion density, T the ion temperature, and τ_E the energy confinement time, defined by

$$\tau_E = \frac{3}{2}\frac{\int n(T_i + T_e)dV}{P} \qquad (1.3)$$

with P the total power input and subscripts refer to ions and electrons.

Heating of a confined plasma to the necessary temperature is routinely obtained in fusion laboratories using neutral beam injection and radio-frequency heating, and densities $n > 10^{20}/m^3$ as well as plasmas with sufficient confinement times are also produced, but obtaining and sustaining a plasma with all three parameters sufficiently large requires, at least with present designs, a device approximately three times larger than has been constructed. Present research consists of extending the limits of maximum plasma temperature, density, length of discharge, and fusion power produced, and well as improving understanding of confined plasmas, to facilitate the search for design modifications permitting a reduction in the size and complexity of a device confining a burning plasma.

The first conceptual design of toroidal devices with controlled thermonuclear fusion in mind appeared in the early 1950s. In the United States, L. Spitzer initiated a research effort to build a stellarator, a device with helically wound external coils. In the Soviet Union, I. Tamm and A. Sakharov sketched a design for a tokamak, a device with a toroidal plasma current. The difficulty and complexity of the task of constructing a practical fusion device was vastly underestimated, and the ensuing 70 years of theoretical and experimental investigation by thousands of scientists of many nations revealed layer upon layer of sophisticated means by which a high temperature plasma will attempt to avoid confinement.

The story is by no means finished, and it still cannot be said with complete confidence that the goal of economic fusion power is within reach. Nevertheless, a large body of knowledge concerning the physics of plasma confinement now exists, and enormous progress has been made. In Fig.1.1 (Jassby and Meade, 2000) is shown the achieved fusion power in kilowatts as a function of year. The points are a composite of data from tokamak devices in the United States, Europe, and Japan. In recent history tokamak reactors have the only significant record for fusion power production, but

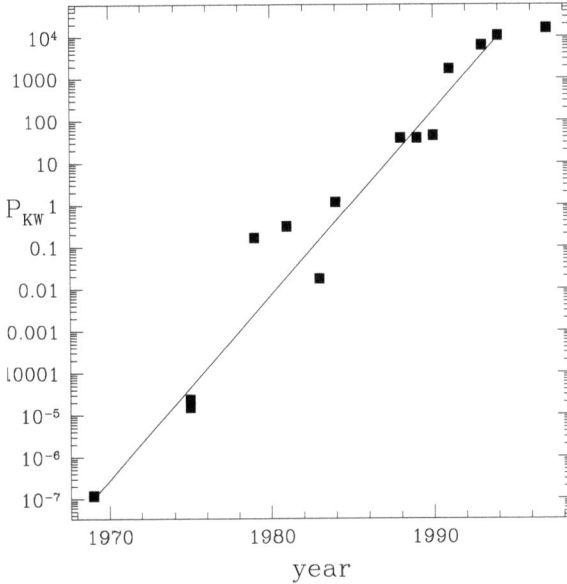

Fig. 1.1   Fusion power output history, kilowatts.

this may be only a historical accident; it is possible that stellarators or other devices may be just as successful. At present there are no serious attempts to increase the record of achieved fusion power. Present research is focused on solving problems to be anticipated in the operation of ITER, namely disruption control, wall erosion, loss of alpha particles due to magnetohydrodynamic instabilities, the production of runaway electron distributions, and other problems which can all be addressed without achieving high levels of fusion power.

ITER began at the Geneva Summit in November 1985, when the idea of a collaborative international project to develop fusion energy for peaceful purposes was proposed by Gorbachev of the former Soviet Union and President Ronald Reagan. The European Union, Japan, the Soviet Union and the USA jointly pursued the design for this large international fusion facility. Conceptual design work began in 1988, followed by increasingly detailed engineering design phases until the final design was approved in 2001. The People's Republic of China and the Republic of Korea joined the Project in 2003, followed by India in 2005. The ITER installation is under construction near Aix-en-Provence in southern France. First plasma is expected in December 2025.

As progress in confinement and plasma heating continues to improve, it is possible to consider more exotic scenarios such as the reaction

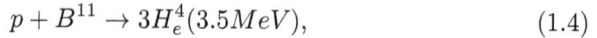

$$p + B^{11} \to 3H_e^4(3.5MeV), \tag{1.4}$$

with the enormous advantage that no neutrons are emitted, the fusion products all consisting of charged alpha particles, thus reducing wall damagage and making shielding much simpler. In addition the fuel is abundantly available, as opposed to tritium. The cross section of the pB^{11} nuclear reaction exhibits a very narrow resonance peak at incident energy E 148 keV, and a broader resonance peak at E 586 keV [34], thus requiring a much hotter plasma than for the DT reaction. Nevertheless this scenario is presently under development at TriAlpha in California.

This book is restricted to the physics of toroidal confinement. The presentation reflects the order of concerns in designing a toroidal plasma device. Chapter 1 concerns the nature of a toroidal magnetic field configuration and its destruction by perturbations. Chapter 2 discusses the form of, and the requirements for, plasma equilibrium, followed in Chapter 3 by a brief introduction to single particle orbits in such equilibria using guiding center theory. Chapter 4 is an introductory analysis of mode particle interaction. Chapter 5 discusses particle resonances. For a perturbation to have an effect on the particle population there must exist a class of particles that can periodically experience the same force. For this to happen the particles must regularly return to the same location in the machine, with the period of the particle motion matching the period of the perturbation.

The questions of linear stability of equilibria to ideal modes and to resistive modes are discussed in Chapters 6 and 7. Chapter 8 treats the subject of the interaction of these modes with the particle distribution, including the effect of a high energy particle population on magnetohydrodynamic stability and the subsequent modification of the particle distribution by the modes. Chapter 9 departs from guiding center theory to treat the full particle cyclotron motion. The nonlinear evolution of ideal and resistive modes and the consequences for maintaining plasma discharges are treated in Chapter 10. Chapter 11 treats mode chirping, a common event often observed in tokamaks, leading to local distribution modification and often also preceeding particle avalanche.

Chapter 12 gives an introduction to the subject of transport. In Chapter 13 the theory of the effect of particle collisions on mode-particle resonances is discussed. Chapter 14 analyzes avalanches which occurred in the Princeton tokamak NSTX. Chapter 15 introduces the concept of using flowing liquid metal on the surfaces in contact with the plasma for the purpose of reducing particle recycling from the walls, a process that absorbs a significant amount of heating power, and Chapter 16 an introduction to the method of phase integrals, a very useful tool for the investigation of plasma instabilities.

The equations defining resistive magnetohydrodynamics (MHD) are Ampere's law,

$$\nabla \times \vec{B} = \vec{j}, \tag{1.5}$$

Faraday's law,

$$\nabla \times \vec{E} = -\partial_t \vec{B}, \tag{1.6}$$

and Ohm's law,

$$\vec{E} + \vec{v} \times \vec{B} = \eta \vec{j}. \tag{1.7}$$

with $\vec{B}$ the magnetic field, $\vec{E}$ the electric field, $\vec{j}$ plasma current, $\vec{v}$ plasma velocity, and η the resistivity.

In addition, for the description of the plasma, which is approximated as a fluid, we need the equation of continuity for the density ρ,

$$\partial_t \rho + \nabla \cdot (\rho \vec{v}) = 0, \tag{1.8}$$

an equation of state to describe the evolution of the scalar plasma pressure p,

$$(\partial_t + \vec{v} \cdot \nabla) \left(\frac{p}{\rho^\gamma} \right) = 0, \tag{1.9}$$

and the equation of motion for the charged fluid,

$$\rho(\partial_t + \vec{v} \cdot \nabla)\vec{v} = -\nabla p + \vec{j} \times \vec{B}. \tag{1.10}$$

Some problems are treated in the approximation of neglecting the plasma resistivity (a high temperature plasma has roughly the same conductivity as metallic copper), in which case the equations describe ideal MHD.

For the purposes of thermonuclear fusion, the plasma must remain approximately in equilibrium for times long compared to either ideal or resistive motion. It is Eq. 1.10 which limits the form such equilibria can take, and for a time independent flow-free state it can be conveniently written in tensor form

$$\partial_{x_i} T_{ik} = 0 \tag{1.11}$$

with

$$T_{ik} = p_\perp \left(\delta_{ik} - \frac{B_i B_k}{B^2} \right) + p_\parallel \frac{B_i B_k}{B^2} \tag{1.12}$$

where $p_\perp = p + B^2/2$ and $p_\parallel = p - B^2/2$, as can be verified by substitution. This formulation of the equilibrium condition makes it apparent that the magnetic field pressure, $B^2/2$, acts as though it exerts an expansion force in the transverse direction and a tension in the parallel direction. The form such equilibria can take will be examined in Chap. 2.

It is beyond the scope of this book to discuss the domain of validity of resistive MHD as defined by the above six equations. They will simply be accepted as a starting point. Roughly speaking, the results of the theory are valid provided the system size and the wavelengths of the modes discovered are large compared to particle gyroradii, the modes have growth rates which are larger than diamagnetic drift frequencies, and the particle collisionality is large enough so that the pressure is isotropic, but sufficiently collisionless so that the resistivity is small. Some sections of the book, regarding, for example, the modification of modes to include a real frequency due to Larmor radius effects or to coupling of the plasma with energetic particles, are in fact outside the domain of resistive magnetohydrodynamics, but only modestly so.

Derivations of ideal MHD from the Boltzman equation and from single particle orbit theory along with further references, can be found in the review by Freidberg (1982). It is known that the predictions of MHD are valid in a domain of collisionality much larger than one has a right to expect. Some of the generalizations of ideal MHD are discussed briefly in Chap. 4, the main result being that stability conditions derived from MHD are not significantly changed by the extensions.

Rationalized units with c = 1 are used throughout. In addition, by expressing distances in terms of the characteristic scale length of the problem considered and magnetic field in terms of the on-axis toroidal field,

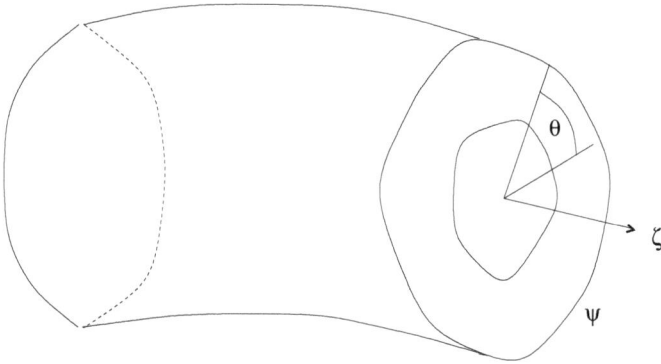

Fig. 1.2   General toroidal coordinates.

equations quickly reduce to expressions involving the fundamental Alfvén time for the system under consideration.

1.2   GENERAL COORDINATES

To describe a toroidal magnetic configuration it is convenient to use coordinates defined by the field itself. In addition to simplifying the description of the field, and providing a general theory for the description of all toroidal fusion devices (tokamak, spheromak, toroidal pinch, stellarator, heliac, etc.) the magnetic coordinates are closely related to canonical variables for the Hamiltonian description of the first order guiding center particle motion in the field.

The advantages gained by using a coordinate system which is defined by, and is natural for, the description of the magnetic field outweigh those gained by using a standard orthonormal system. The fields and equilibria must ultimately be described in a convenient laboratory coordinate system, so the general coordinates are taken to be functions of the euclidean coordinates. Because of the necessarily toroidal topology of the magnetic field surfaces, the coordinates are taken to be toroidal in form.

Introduce general coordinates ψ, θ, ζ, as shown in Fig. 1.2. Surfaces of constant ψ are taken to consist topologically of nested tori, which necessarily possess an axis which will usually be designated by $\psi = 0$. Surfaces of constant θ define a general poloidal angle and ζ defines the "toroidal" direction. In some cases ζ will be simply equal to the geometrical toroidal angle ϕ, but it is convenient to keep it as a general toroidal coordinate. The surface label ψ is taken to be increasing outward, and thus the system ψ, θ, ζ defines a right-handed coordinate system.

1.3   BASIS VECTORS, METRIC TENSOR

Begin with functions defining the transformation from general coordinates to orthonormal euclidean coordinates, $\vec{r}(\psi, \theta, \zeta)$ with $\vec{r} = (x, y, z)$. Define a covariant basis

$$\vec{e}_\psi = \partial_\psi \vec{r} \quad \vec{e}_\theta = \partial_\theta \vec{r} \quad \vec{e}_\zeta = \partial_\zeta \vec{r} \tag{1.13}$$

with Jacobian

$$\mathcal{J} = \vec{e}_\psi \cdot (\vec{e}_\theta \times \vec{e}_\zeta). \tag{1.14}$$

Define also the contravariant basis

$$\vec{e}^\psi = \nabla\psi, \quad \vec{e}^\theta = \nabla\theta, \quad \vec{e}^\zeta = \nabla\zeta. \tag{1.15}$$

The matrices of the differential coordinate transformation

$$\begin{pmatrix} \frac{\partial x}{\partial \psi} & \frac{\partial x}{\partial \theta} & \frac{\partial x}{\partial \zeta} \\ \frac{\partial y}{\partial \psi} & \frac{\partial y}{\partial \theta} & \frac{\partial y}{\partial \zeta} \\ \frac{\partial z}{\partial \psi} & \frac{\partial z}{\partial \theta} & \frac{\partial z}{\partial \zeta} \end{pmatrix} \quad \begin{pmatrix} \frac{\partial \psi}{\partial x} & \frac{\partial \psi}{\partial y} & \frac{\partial \psi}{\partial z} \\ \frac{\partial \theta}{\partial x} & \frac{\partial \theta}{\partial y} & \frac{\partial \theta}{\partial z} \\ \frac{\partial \zeta}{\partial x} & \frac{\partial \zeta}{\partial y} & \frac{\partial \zeta}{\partial z} \end{pmatrix} \tag{1.16}$$

are inverses of one another, $\vec{e}^\alpha \cdot \vec{e}_\beta = \delta^\alpha_\beta$ and $\vec{e}^\alpha \vec{e}_\alpha = I$ where I is the unit matrix, $I^k_j = \delta^k_j$ and α, β refer to ψ, θ, ζ, the indices j, k refer to x, y, z, and the summation convention is used for repeated indices. In addition we find

$$\vec{e}_\alpha \times \vec{e}_\beta = \mathcal{J}\epsilon_{\alpha\beta\gamma}\vec{e}^\gamma, \quad \vec{e}^\alpha \times \vec{e}^\beta = \frac{1}{\mathcal{J}}\epsilon^{\alpha\beta\gamma}\vec{e}_\gamma \tag{1.17}$$

with $\epsilon_{\alpha\beta\gamma}$ antisymmetric under odd permutations of indices and $\epsilon_{\psi\theta\zeta} = 1$. These tensors satisfy the relation $\epsilon_{\alpha\beta\gamma}\epsilon^{\alpha\rho\sigma} = \delta^\rho_\beta\delta^\sigma_\gamma - \delta^\sigma_\beta\delta^\rho_\gamma$. Note that the tensor δ^α_β is invariant under all coordinate transformations, but $\epsilon_{\alpha\beta\gamma}$ is invariant only under a restricted set of transformations.

Any vector can be written $\vec{v} = v^\alpha \vec{e}_\alpha = v_\alpha \vec{e}^\alpha$ and v_α, v^α are called the covariant and contravariant components, respectively, given by $v_\alpha = \vec{e}_\alpha \cdot \vec{v}$ and $v^\alpha = \vec{e}^\alpha \cdot \vec{v}$. The notation is consistent with the transformation properties of the vectors under coordinate transformation. Quantities with raised indices are called contravariant. The differential distance produced by displacements in ψ, θ, ζ is given by the chain rule, $d\vec{r} = \vec{e}_\alpha d\alpha$ so the differential distance is given by $ds^2 = d\vec{r} \cdot d\vec{r} = \vec{e}_\alpha \cdot \vec{e}_\beta d\alpha d\beta$ and thus the metric tensor is given by

$$g_{\alpha\beta} = \vec{e}_\alpha \cdot \vec{e}_\beta, \quad with \quad g^{\alpha\beta} = \vec{e}^\alpha \cdot \vec{e}^\beta. \tag{1.18}$$

It follows immediately that $g_{\alpha\beta}g^{\beta\kappa} = \delta_\alpha^\kappa$ and that $\vec{e}_\alpha = g_{\alpha\beta}\vec{e}^\beta$. Also for any vector $\vec{v} = v^\alpha \vec{e}_\alpha = v_\alpha \vec{e}^\alpha$ we have $v_\alpha = g_{\alpha\beta}v^\beta$, *i.e.* the metric tensor can be used to raise and lower indices of basis vectors and components. Note that $\vec{v} = v^\alpha \vec{e}_\alpha = v_\alpha \vec{e}^\alpha$ is coordinate system invariant, with the basis vectors transforming covariantly (contravariantly) while the components transform contravariantly (covariantly).

Note that

$$\frac{1}{\mathcal{J}} = \nabla\psi \cdot (\nabla\theta \times \nabla\zeta). \tag{1.19}$$

The relation of $\mathcal{J}$ to the volume element follows easily by calculating the volume of a small parallelepiped constructed with $d\vec{r}_1 = \vec{e}_\psi d\psi$, $d\vec{r}_2 = \vec{e}_\theta d\theta$, $d\vec{r}_3 = \vec{e}_\zeta d\zeta$. Then $dV = d\vec{r}_1 \cdot (d\vec{r}_2 \times d\vec{r}_3) = \mathcal{J} d\psi d\theta d\zeta$.

The Jacobian is related to the metric tensor through $\mathcal{J}^{-2} = det(g^{\alpha\beta})$. To see this consider the matrix

$$M_l^\alpha = e_l^\alpha \tag{1.20}$$

with $\alpha = \psi, \theta, \zeta$ and $l = x, y, z$. Then we have

$$det M = \frac{1}{\mathcal{J}}. \tag{1.21}$$

Further note that

$$g^{\alpha\beta} = (MM^T)^{\alpha\beta} \tag{1.22}$$

from which it follows that

$$det(g^{\alpha\beta}) = (det M)^2 = \frac{1}{\mathcal{J}^2} \tag{1.23}$$

and

$$det(g_{\alpha\beta}) = \mathcal{J}^2. \tag{1.24}$$

For most purposes in plasma physics second-rank tensors can be written as dyads, *i.e.* the product of two vectors. For this reason the technique of using either a covariant or a contravariant representation for the vector depending on the nature of the calculation to be done is sufficient to simplify the result. Here we introduce the Christoffel symbols, which describe the differential spatial dependence of the basis vectors, and provide an alternate means of calculation. The basis vectors are functions of space, so we define $\partial_\alpha \vec{e}_\beta = \{_{\alpha\beta} \,^\tau\}\vec{e}_\tau$, where $\{_{\alpha\beta} \,^\tau\}$ is the Christoffel symbol of the second kind. Take the dot product with $\vec{e}^\kappa$, giving $\{_{\alpha\beta} \,^\kappa\} = \vec{e}^\kappa \cdot \partial_\alpha \vec{e}_\beta$, but $\vec{e}_\beta = \partial_\beta \vec{r}$ and so $\{_{\alpha\beta} \,^\kappa\} = (1/2)\vec{e}^\kappa \cdot [\partial_\alpha \vec{e}_\beta + \partial_\beta \vec{e}_\alpha]$.

Use $g_{\kappa s}$ to lower the index, giving $g_{\kappa s}\{_{\alpha\beta} \,^s\} = [\alpha\beta, \kappa]$, which is the Christoffel symbol of the first kind. Thus

$$[\alpha\beta, \kappa] = \frac{1}{2}\vec{e}_\kappa \cdot [\partial_\alpha \vec{e}_\beta + \partial_\beta \vec{e}_\alpha] \tag{1.25}$$

or

$$[\alpha\beta, \kappa] = \frac{1}{2}[\partial_\alpha g_{\kappa\beta} - \vec{e}_\beta \cdot \partial_\alpha \vec{e}_\kappa + \partial_\beta g_{\alpha\kappa} - \vec{e}_\alpha \cdot \partial_\beta \vec{e}_\kappa] =$$
$$\frac{1}{2}[\partial_\alpha g_{\kappa\beta} + \partial_\beta g_{\alpha\kappa} - \partial_\kappa(\partial_\beta \vec{r} \cdot \partial_\alpha \vec{r})] \tag{1.26}$$

giving an expression for the Christoffel symbol in terms of the metric tensor,

$$[\alpha\beta, \kappa] = \frac{1}{2}[\partial_\alpha g_{\kappa\beta} + \partial_\beta g_{\alpha\kappa} - \partial_\kappa g_{\alpha\beta}]. \tag{1.27}$$

1.4   VECTOR OPERATORS

Vector calculus in general coordinates is facilitated by using a representation for the vectors involved which is appropriate for the calculation to be done. Simplification arises because of the properties of the gradient operator, namely $\nabla \times \nabla f = 0$ and $\nabla \cdot (\nabla A \times \nabla B) = 0$. Thus to calculate the divergence of a vector write it in terms of a covariant basis.

The gradient of a scalar is given by $\nabla f = \vec{e}^\alpha \partial_\alpha f$. The divergence of a vector is simply expressed in terms of its components. Write

$$\vec{v} = J[(\vec{v} \cdot \nabla\psi)\nabla\theta \times \nabla\zeta + (\vec{v} \cdot \nabla\theta)\nabla\zeta \times \nabla\psi + (\vec{v} \cdot \nabla\zeta)\nabla\psi \times \nabla\theta.] \tag{1.28}$$

Taking the divergence of this expression we find

$$\nabla \cdot \vec{v} = \frac{1}{J}\partial_\alpha(J\vec{v} \cdot \nabla\alpha). \tag{1.29}$$

The curl of a vector is easily calculated if written in terms of a contravariant basis. Write

$$\vec{v} = v_\psi \nabla \psi + v_\theta \nabla \theta + v_\zeta \nabla \zeta. \tag{1.30}$$

Taking the curl of this expression and then taking the dot product with the three contravariant basis vectors we find

$$(\nabla \times \vec{v}) \cdot \nabla \zeta = \frac{1}{\mathcal{J}} (\partial_\psi v_\theta - \partial_\theta v_\psi) \tag{1.31}$$

and cyclic permutations of ψ, θ, ζ.

For the Laplacian of f simply take the divergence of the vector $\vec{v} = \nabla f$. We have from above $\nabla \cdot \vec{v} = \frac{1}{\mathcal{J}} \partial_\alpha (\mathcal{J} v^\alpha)$. Since $v_\beta = \partial_\beta f$ we have $v^\alpha = g^{\alpha\beta} \partial_\beta f$ giving

$$\nabla^2 f = \frac{1}{\mathcal{J}} \partial_\alpha (\mathcal{J} g^{\alpha\beta} \partial_\beta f). \tag{1.32}$$

1.5 MAGNETIC FIELD REPRESENTATION

There are many different ways to represent a magnetic field. We derive here a representation of a general magnetic field which is useful for two reasons. First, it displays manifestly the Hamiltonian character of the magnetic field line trajectories. Secondly, it is the form which we will find most useful for discussing MHD equilibrium and particle trajectories.

Write $\vec{B} = \nabla \times \vec{A}$, since $\nabla \cdot \vec{B} = 0$ (Poincaré, 1892). Write the vector potential in terms of the coordinates ρ, θ, ζ with $\vec{r}(\rho, \theta, \zeta)$

$$\vec{A} = A_\rho \nabla \rho + A_\theta \nabla \theta + A_\zeta \nabla \zeta. \tag{1.33}$$

Then, define $\partial_\rho G = A_\rho$. Since $\nabla G = \partial_\rho G \nabla \rho + \partial_\theta G \nabla \theta + \partial_\zeta G \nabla \zeta$ we have

$$\vec{A} = \nabla G + (A_\theta - \partial_\theta G) \nabla \theta + (A_\zeta - \partial_\zeta G) \nabla \zeta \tag{1.34}$$

or

$$\vec{A} \equiv \nabla G + \psi \nabla \theta - \psi_p \nabla \zeta \tag{1.35}$$

and thus

$$\vec{B} = \nabla \psi \times \nabla \theta - \nabla \psi_p \times \nabla \zeta. \tag{1.36}$$

Using this form we can easily show that the field line trajectories are Hamiltonian in nature. The magnetic field lines are defined by $d\psi/d\zeta = \vec{B} \cdot \nabla\psi / \vec{B} \cdot \nabla\zeta$, $d\theta/d\zeta = \vec{B} \cdot \nabla\theta / \vec{B} \cdot \nabla\zeta$ or

$$\frac{d\psi}{d\zeta} = -\frac{\nabla\psi \cdot (\nabla\psi_p \times \nabla\zeta)}{(\nabla\psi \times \nabla\theta) \cdot \nabla\zeta} \qquad \frac{d\theta}{d\zeta} = -\frac{\nabla\theta \cdot (\nabla\psi_p \times \nabla\zeta)}{(\nabla\psi \times \nabla\theta) \cdot \nabla\zeta}. \qquad (1.37)$$

But the gradient of ψ_p is given by

$$\nabla\psi_p = \partial_\psi\psi_p\nabla\psi + \partial_\theta\psi_p\nabla\theta + \partial_\zeta\psi_p\nabla\zeta. \qquad (1.38)$$

Thus

$$\frac{d\psi}{d\zeta} = -\partial_\theta\psi_p, \qquad \frac{d\theta}{d\zeta} = \partial_\psi\psi_p. \qquad (1.39)$$

This is of Hamiltonian form with $\psi_p(\psi, \theta, \zeta)$ the Hamiltonian, ψ the momentum, θ the coordinate, and ζ the time.

1.6   MAGNETIC SURFACES

A two dimensional surface defined by a function $f(\vec{r}) = constant$ is said to be a magnetic surface if at any point the magnetic field lies within the surface, *i.e.* $\vec{B} \cdot \nabla f = 0$. The existence of such surfaces in magnetic confinement devices, or at least the existence of approximate magnetic surfaces over a large fraction of the plasma volume, is an essential requirement for long-term confinement. The existence of such surfaces has been shown only under fairly restrictive conditions (Morozov and Solov'ev, 1966). They are known to exist everywhere or in all but a small part of the plasma volume only when there exists a symmetry or approximate symmetry. This is easily demonstrated in cylindrical geometry for translational, axial, and helical symmetry. Writing the field in terms of the vector potential, $\vec{B} = \nabla \times \vec{A}$, the surfaces are defined by $A_z(r, \theta) = constant$ in case $\vec{A}$ is translationally invariant in z, $rA_\theta(r, z) = constant$ in case $\vec{A}$ is independent of θ, and $A_z(r, \theta - \alpha z) + \alpha r A_\theta(r, \theta - \alpha z) = constant$ in case $\vec{A}$ is helically invariant; *i.e.* depends only on the variables $r, \theta - \alpha z$. It is readily verified that these equations define surfaces to which $\vec{B}$ is tangent.

According to a theorem by Kolmogorov (1957), small perturbation of a symmetric case leaves well defined magnetic surfaces existing everywhere except in a small volume proportional to the square root of the perturbation, where the field topology changes to include magnetic islands and

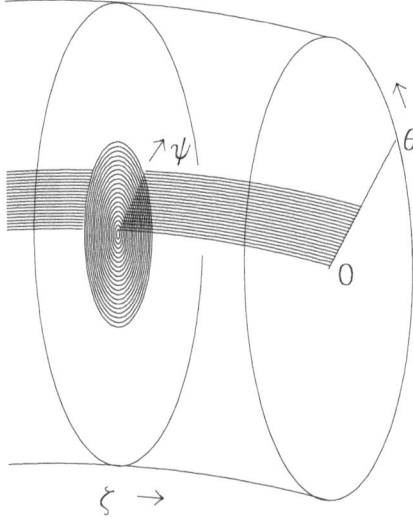

Fig. 1.3   Toroidal and poloidal surfaces defining flux ψ and ψ_p.

can also assume a stochastic character. The onset of stochasticity will be further discussed in Sec. 1.7.

Consider the representation of $\vec{B}$ given by Eq. 1.36. Forming $\vec{B} \cdot \nabla \psi$ and $\vec{B} \cdot \nabla \psi_p$ it is easily seen that ψ_p is a magnetic surface only if ψ_p is independent of ζ, *i.e.* $\psi_p = \psi_p(\psi, \theta)$, and ψ is a magnetic surface only if ψ_p is independent of θ, *i.e.* $\psi_p = \psi_p(\psi, \zeta)$. Thus both ψ and ψ_p are magnetic surfaces if $\psi_p = \psi_p(\psi)$.

The simplest way to guarantee the existence of approximate surfaces is to take the confinement device to be a small perturbation of a system in which $\psi_p = \psi_p(\psi)$. Consider such a system, with ζ the toroidal angle, and ψ_p a function of ψ alone, so that $\vec{B} \cdot \nabla \psi = \vec{B} \cdot \nabla \psi_p = 0$. Let $d\psi_p/d\psi \equiv 1/q(\psi)$. From Eqs. 1.19, 1.36 the toroidal flux is then $(1/2\pi) \int (\vec{B} \cdot \nabla \zeta) \mathcal{J} d\zeta d\theta d\psi = 2\pi\psi$, and the poloidal flux is $(1/2\pi) \int (\vec{B} \cdot \nabla \theta) \mathcal{J} d\zeta d\theta d\psi = 2\pi\psi_p$, where the integration extends from the magnetic axis to the flux surface ψ. Shown in Fig. 1.3 is a poloidal surface (fixed θ), and a toroidal surface (fixed ζ), each bounded by the magnetic axis and a flux surface defined by ψ. The fluxes ψ and ψ_p are defined by the total magnetic flux passing through these surfaces. This differs from the definition used by some authors in that we have taken both ψ and ψ_p to be zero at the magnetic axis, so the surfaces are limited by the magnetic axis and ψ. Note that the assumption of nested magnetic surfaces does not imply axisymmetry, and also note that generally θ and ζ are not the usual geometrically defined angular variables, it is only required that the position in space be periodic with period 2π in each of them and that the volume defined by them be topologically toroidal.

1.7   MAGNETIC SURFACE DESTRUCTION

To illustrate the way in which nested toroidal magnetic surfaces are destroyed, introduce a small perturbation of the poloidal flux function, Fourier decompose it in the variables θ, ζ, and consider a single harmonic

$$\psi_p = \int \frac{d\psi}{q} + V\cos(n\zeta - m\theta), \tag{1.40}$$

and the equations for the field lines become, from Eq. 1.39

$$\frac{d\psi}{d\zeta} = -Vm\sin(n\zeta - m\theta), \qquad \frac{d\theta}{d\zeta} = \frac{1}{q(\psi)}. \tag{1.41}$$

The function $q(\psi)$, referred to as the safety factor in tokamak theory, thus defines the field helicity in the θ, ζ variables on the surface ψ . If $q' = dq/d\psi$ is not zero the field possesses shear. Note from Eq. 1.36 that $\vec{B}\cdot\nabla\psi = mV\sin(n\zeta - m\theta)/\mathcal{J}$, i.e. the perturbation V introduces a component of $\vec{B}$ directed across the original flux surfaces. If $q(\psi)$ is not close to the rational m/n with $V \neq 0$, ψ is oscillatory and $q(\psi)$ can be approximated as constant, giving $\theta = \zeta/q + \theta_0$ and $\psi - \psi_0 = V\cos(n\zeta - m\theta)/(n - m/q)$. The flux surfaces are distorted but they remain topologically nested surfaces. However, if $mq(\psi) - n\zeta \simeq 0$ the resonant denominator creates large excursions in ψ and this solution is not valid. In this case expand $q(\psi)$ about ψ_0, with $q(\psi_0) = m/n$, and introduce the variable $Q = n\zeta - m\theta$. We then find

$$dQ = \left(\frac{mq'}{q^2}\right)(\psi - \psi_0)d\zeta \tag{1.42}$$

$$d\psi = -mV\sin(Q)d\zeta$$

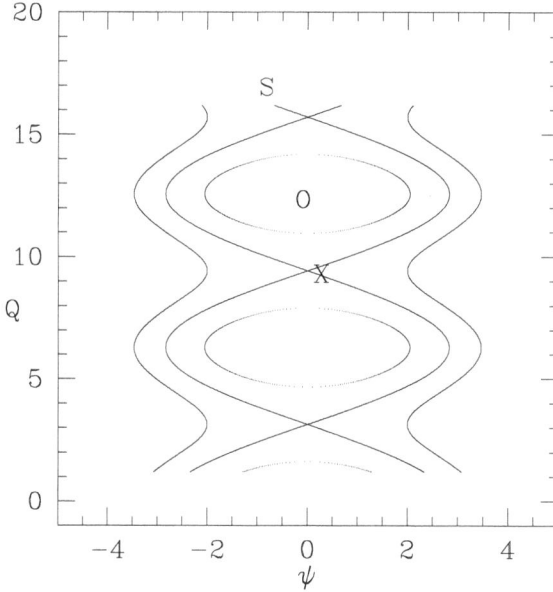

Fig. 1.4   Broken flux surfaces due to a single island chain showing separatrix S, X-point and O-point.

where $q' = dq/d\psi$, evaluated at $\psi = \psi_0$. Integrating these equations gives

$$(\psi - \psi_0)^2 = \frac{2q^2 V}{q'}[cos(Q) + k] \tag{1.43}$$

where the constant k is determined by the initial position. (Note that the flux surfaces cannot be found by setting the Hamiltonian ψ_p constant, since this is a "time" dependent Hamiltonian.)

The points in the ψ, θ plane at $\zeta = 0$ formed by successive transits of a field line are called a Poincaré plot. If q is rational the Poincaré plot will consist only of a finite set of the points on the surface given by Eq. 1.43, but if q is irrational the whole surface will be covered. The points given by $cos(Q) = 1$, $k = -1$, and those with $cos(Q) = -1$, $k = 1$ are periodic points of period m. Expanding the cosine function near $cos(Q) = +1$ gives

$$(\psi - \psi_0)^2 = \frac{2q^2 V}{q'}(1 - Q^2/2 + k), \tag{1.44}$$

i.e. for $Vq' > 0$ the points lie on elliptic manifolds about the periodic points. They form either a continuous curve or discrete points depending on the value of q. Similarly, near the points $cos(Q) = -1$ the points lie on hyperbolic manifolds. Thus there is a chain of elliptic and hyperbolic points at $\psi = \psi_0$. This island chain is separated from the topologically toroidal surfaces by a separatrix which passes through the X-points. The separatrix is described by choosing the integration constant $k = 1$, giving islands with width (twice the maximum value of $(\psi - \psi_0)$)

$$\delta\psi = 4\left(\frac{Vq^2}{q'}\right)^{1/2}. \tag{1.45}$$

In Fig. 1.4 are shown the resulting flux surfaces for a single island chain.

An equivalent pendulum for this structure is given by the canonical transformation generated by $F_3(p, Q, t) = -npt/m + pQ/m$ with $p, q, t = \psi - \psi_r, \theta, \zeta$ and $q = -\partial_p F_3$, $P = -\partial_Q F_3$ giving $P = -p/m$, $Q = n\zeta - m\theta$ (see Goldstein, 1953 p. 242). The new Hamiltonian is $K = H + \partial_t F_3$. Hamilton's equations are also left invariant with the transformation $t \to at$, $P \to aP$. Taking $a = m\sqrt{q'/q^2}$ gives

$$K = -\frac{P^2}{2} + V\cos Q \tag{1.46}$$

with $P^2 = mq'(\psi - \psi_r)^2/q^2$. The equations of motion are then $dP/dt = -\partial K/\partial Q$, $dQ/dt = \partial K/\partial P = P$ and the phase plot is given by surfaces of constant $K = E$. The period for an orbit trapped in the island is then

$$T = \frac{2}{\sqrt{2V}} \int_{-Q_b}^{Q_b} \frac{dQ}{\sqrt{\cos Q - \cos Q_b}} \tag{1.47}$$

with $2V\cos Q_b = -E$, and Q_b defining the maximum excursion in Q. Letting $\cos Q = 1 - \sin^2(Q/2)$ with $k = \sin(Q_b/2)$ and $k\sin\phi = \sin(Q/2)$ we find that

$$T = \frac{1}{\sqrt{V}} \int_0^{\pi/2} \frac{d\phi}{1 - k^2\sin^2\phi} = \frac{1}{\sqrt{V}} K(k) \tag{1.48}$$

with $K(k)$ an elliptic integral. In the deeply trapped limit we find the period to be $T \simeq 2\pi/\sqrt{V}$. Using the relation between the time and ζ we find the internal q_I for rotation about the island O-point,

$$q_I = \frac{d\zeta}{d\theta_I} = \frac{q}{m\sqrt{q'V}}, \tag{1.49}$$

with θ_I the angle around the island O-point. For an orbit near the separatrix $k \simeq 1$ and we find

$$t = \frac{1}{\sqrt{2V}} \int_0^Q \frac{dQ}{\sqrt{cosQ + 1}} = \frac{1}{\sqrt{V}} ln[tan(\pi/4 + Q/4)], \qquad (1.50)$$

or equivalently $Q = 4atan(e^{\omega_0 t}) - \pi$ with $\omega_0 = \sqrt{V}$. The period for a complete oscillation near the X-point is approximately $T = 4ln(4/k')/\sqrt{V}$ with $k' = \sqrt{1 - k^2} \ll 1$.

The Hamiltonian describing an exact pendulum, Eq. 1.46, gives rise to an exactly defined separatrix between the island interiors and the topologically unperturbed exterior. However, if the system is additionally perturbed the separatrix is no longer exact, and becomes a band of chaotic trajectories of nonzero width. The separatrix width can be estimated by the addition of a small perturbation to the pendulum equation, and finding the modification of an orbit near the separatrix. Thus consider the Hamiltonian

$$H = \frac{p^2}{2} + V cosQ - \epsilon cos(aQ - t - t_0). \qquad (1.51)$$

Use an unperturbed orbit near the separatrix to find changes due to ϵ,

$$\Delta H = \int_{-\infty}^t dt \partial_t H. \qquad (1.52)$$

The unperturbed orbit is $Q = 4atan[e^{\omega_0 t}] - \pi$ with $\omega_0 = \sqrt{V} \ll 1$, giving

$$\partial_t H = -\epsilon[sin(aQ - t)cos(t_0) - sin(t_0)cos(aQ - t)] \qquad (1.53)$$

and the first term contributes nothing in the limit of large t since $Q(t)$ is odd. Let $s = \omega_0 t$, giving

$$\Delta H = \frac{\epsilon sin(t_0)}{\omega_0} \int_{-\infty}^s cos[aQ(s) - s/\omega_0]ds \qquad (1.54)$$

and $Q(s) = 4atan(e^s) - \pi$. This integral is called a Melnikov–Arnold integral (Melnikov, 1963; Arnold, 1964), and it has both oscillatory and secular parts. As an example, consider the integral

$$\mathcal{A} = \int_{-\infty}^s cos[Q(s)/2 - Q_0 s]ds. \qquad (1.55)$$

Let $x = 2atan(e^s) = Q/2 + \pi/2$, giving $e^s = (e^{ix} - 1)/(ie^{ix} + i)$, and $e^{iQ/2} = -ie^{ix} = (1 + ie^s)/(i + e^s)$.

$$\mathcal{A} = Re \int_{-\infty}^{s} e^{-iQ_0 s} \frac{1 + ie^s}{i + e^s} ds. \tag{1.56}$$

This integral has simple poles at $e^s = -i$, $s = -i\pi/2 - 2i\pi n$. Closing the contour in the lower half plane, $\mathcal{A}$ is the integral over the contour minus contributions from vertical pieces. The contributions from the vertical parts of the contour are, letting $s = S + iy$ with $S \gg 1$,

$$\int_{0}^{-\infty} idy e^{-iQ_0 S} e^{Q_0 y} i \sim \frac{1}{Q_0} e^{-iQ_0 S} \tag{1.57}$$

giving an oscillatory contribution. The contribution from the dominant pole at $s = -i\pi/2$ is $\mathcal{A} \simeq -4\pi e^{-\pi Q_0/2}$, and $Q_0 = 1/\omega_0$, giving

$$\Delta H \sim \frac{\epsilon}{\sqrt{V}} sin(t_0) e^{-\frac{\pi}{2\sqrt{V}}} + oscillatory\ terms. \tag{1.58}$$

Orbits near the separatrix thus suffer a shift in position given by ΔH of effectively random sign because of the phase t_0, and the fact that the period is becoming logarithmically infinite as one approaches the separatrix. This leads to a small stochastic band about the original separatrix, with a width which is nonanalytic in V. As the magnitude of V increases the separatrix itself broadens, eliminating good Kolmogorov, Arnold, Moser (KAM) surfaces, i.e. those magnetic surfaces which retain the original topology of the unperturbed field, in its vicinity (see Kolmogorov, 1957).

In the presence of many Fourier harmonics the magnetic flux surfaces break up into island chains at each rational surface where $\delta_{nm} \neq 0$ with the island widths given above. In between the island chains the flux surfaces remain topologically toroidal although distorted. Actually nonlinear interaction of two island chains with $q = m/n$ and m'/n' produces a smaller island structure at the surface $q = (m + m')/(n + n')$ and so on. Note that Eq. 1.41 is normally highly nonlinear, because the perturbation amplitude V is a function of ψ. Thus for arbitrarily small perturbations all rational surfaces in this Fibonacci sequence produced by the mode numbers of the original perturbations break up into island chains. However, as long as the perturbations are small, the total volume of the island structure is small, proportional to $\sqrt{\delta}$, and magnetic surfaces which are topologically toroidal exist almost everywhere. This was shown rigorously by Kolmogorov, Arnold and Moser, and the surviving toroidal surfaces are called KAM surfaces. To see this in a nonrigorous manner, note that the island width is proportional to the square root of the perturbation, and the new islands produced at order k are given by the fractions $[lm + (k - l)m']/[ln + (k - l)n']$ for $0 < l < k$, so there are $k - 1$ of them. The total island width is then proportional to $\sum_k (k - 1)\alpha^{k/2}$, which converges for $\alpha < 1$, giving a finite width occupied by islands. As long as this width is small compared to the whole domain some KAM surfaces still exist. The KAM surfaces are important because no field lines can cross them. Their existence prevents the magnetic field from wandering in ψ. As the perturbation strength increases, island widths grow until neighboring island chains overlap. Chirikov (1979) gave this overlap as a criterion for the loss of the last KAM surface and the stochastic wandering of the magnetic field. In fact the last KAM surface vanishes significantly earlier due to the nonlinearly produced secondary islands, but the overlap criterion provides a rough first estimate for the onset of stochasticity.

The separation between two neighboring rational surfaces is $\delta\psi \simeq (\Delta q)/q'$. Equating this to the island width, Eq. 1.45, we find the overlap condition

$$V = \frac{1}{16}\left(\frac{\Delta q}{q}\right)^2 \frac{1}{q'}. \tag{1.59}$$

Thus for fixed mode spectrum Δq, island overlap, and thus the onset of stochasticity, occurs at smaller perturbation amplitude for large shear, q' large.

Above stochastic threshold the field trajectories, although deterministic, are apparently chaotic, and the behavior is described in terms of statistical quantities. Two such quantities are the magnetic diffusion, D, giving the rate at which field trajectories diffuse across the original flux surfaces ψ, and the Kolmogorov entropy h, describing the rate at which nearby trajectories on average exponentially diverge. The fact that the Kolmogorov entropy is positive above stochastic threshold guarantees that trajectories are exponentially sensitive to initial conditions, introducing apparent chaos in a deterministic system.

In Fig. 1.5 is shown an example of magnetic surface destruction due to perturbations δ_{nm} for $m/n = 8/4, 9/4$. The higher order islands are due to nonlinear interaction of these two modes.

An island chain at the rational surface $q = m/n$ has associated with it a periodic orbit of length m, made up of the elliptic points of the islands

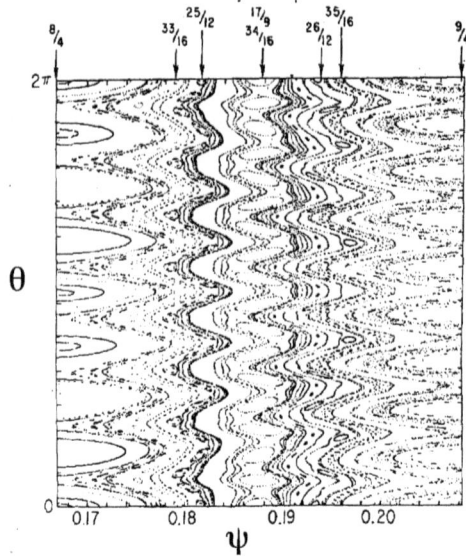

Fig. 1.5   Magnetic field structure due to perturbations.

in the chain, that is, m toroidal circuits are required before returning to the initial elliptic point. Consider the discrete map of points in the ψ, θ plane consisting of the transformation produced by one toroidal circuit. Orbits in the neighborhood of the elliptic points can be computed in the linear differential approximation. The domain of this approximation is called the tangent space. A toroidal transit is then represented by a matrix M acting on the initial vector $(\delta\psi, \delta\theta)$. Since $\nabla \cdot \vec{B} = 0$, this map is area preserving and hence $\det M = 1$. Equivalently, from the Hamiltonian nature of the trajectories, the Liouville theorem guarantees that phase space area is conserved. The eigenvalues of M depend only on its trace. Greene (1979) has investigated the criterion for the vanishing of the last KAM surface for some discrete maps. He has defined the residue

$$R = \frac{2 - trace M}{4}.$$ (1.60)

Clearly in the limit of $\delta\vec{B} = 0$ the residue is zero. The eigenvalues of M

are given in terms of the residue through

$$\lambda = 1 - 2R \pm 2[R(R-1)]^{1/2}. \tag{1.61}$$

As $\delta\vec{B}$ increases from zero, the residue increases monotonically from zero. When R is between zero and one, λ is complex with magnitude one, and the tangent-space orbits may take many toroidal circuits to rotate about the elliptic periodic points. Writing $\lambda = e^{i2\pi/q_I}$, q_I is the ratio of toroidal transits divided by the number of revolutions about the periodic points. As the residue increases monotonically, q_I decreases monotonically from infinity. As R approaches one, the ellipses forming the nearby orbits flatten until at $R = 1$ the opposite sides touch, and the original elliptic point becomes an X-point, the island thus bifurcating into two elliptic points separated by the new X-point. At this point the two complex conjugate eigenvalues coalesce at $\lambda = -1$ and become real, and $q_I = 2$. This process is referred to as period doubling (Lichtenberg and Lieberman, 1992) since the two new elliptic points have a periodicity twice that of the original elliptic point, and generally after the doubling some orbits exist with the new periodicity. This means that in the power spectrum of the system (the Fourier transform in time of the amplitude squared) there appears a new frequency at half the existing one.

As the strength of the peturbation is increased, elliptic points with larger values of m and n are destabilized, becoming X-points, and at each such destabilization stable orbits of double the periodicity of the destroyed elliptic point appear. A KAM surface at an irrational q is destabilized when all island chains m/n are destabilized in the limit $m/n \to q$. Examination of the destabilization of island chains with very large values of m and n then gives an accurate estimate of stochastic threshold (Greene, 1979). This behavior is to a large degree independent of the details of the map M.

In the case of analytically given discrete maps, the residue associated with an island chain can be calculated analytically, and the bifurcation of various order periodic orbits studied as a function of perturbation amplitude. This permits the calculation of the extent of stochastic domains. In the case of maps defined by toroidal magnetic field configurations the matrix M must be obtained by integration of the differential equations, and even the location of the periodic elliptic points is generally beyond analytic representation. However, the different domains – toroidal, island, and stochastic– can be differentiated numerically. Consider two nearby points

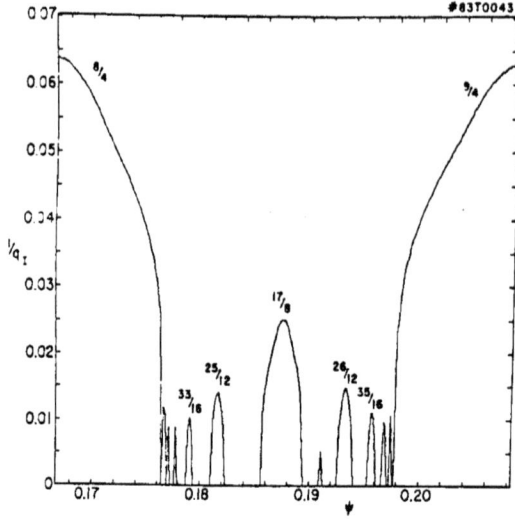

Fig. 1.6   Local q_I due to magnetic islands.

in the ψ, θ plane with separation

$$\vec{\delta} = \vec{r}_1 - \vec{r}_2 = \delta_0(cos\theta_I, sin\theta_I).$$ (1.62)

Advance the points together toroidally. If they are within an island, they rotate about each other on the average with $\Delta\theta_I = \Delta\zeta/q_I$. If they are within a stochastic domain, on the average they separate exponentially with $\delta = \delta_0 e^{hR\Delta\zeta}$, with h the Kolmogorov entropy. This exponential separation on average is a result of the fact that phase space has become densely filled with X-points. Finally, if good magnetic surfaces exist on the scale of the separation of the points, they will not rotate, and will separate only linearly due to the difference in shear at the two surfaces.

The Kolmogorov entropy is a statistical quantity and must be averaged over various initial conditions. The island rotational transform q_I, is, on the other hand, well defined locally. In Fig. 1.6 is shown the resulting numerical determination of the island internal q_I for the field of Fig. 1.5.

Since the rotation about the O-point is non uniform, an accurate determination of q_I is made by advancing in ζ until the trajectory rotates completely around the periodic point. The largest value of q_I that can be

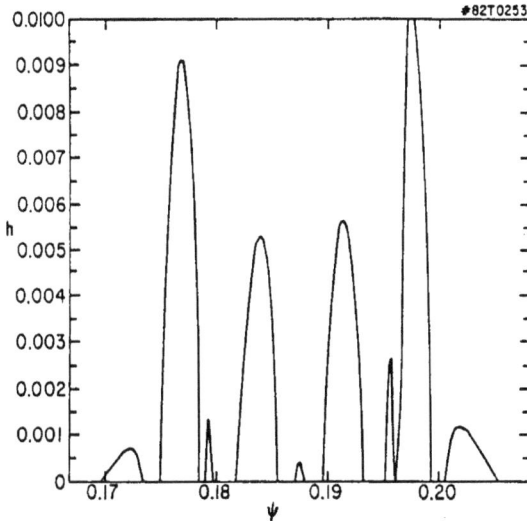

Fig. 1.7 Kolmogorov entropy in stochastic bands near separatrices.

observed is restricted by the number of toroidal circuits made. The method suffices to pick out the first four orders in the Fibonacci series. In Fig. 1.7 is shown the determination of the Kolmogorov entropy using the same procedure. In this example it is found that significant stochastic domains exist in the vicinity of the separatrices of the largest islands, that is, the 8/4 and 9/4 chains, as well as the 17/8. Owing to the finite number of toroidal circuits used to determine h, there is a minimum value, given by nonstochastic orbit excursions, below which h is not significant.

Although toroidal flux surfaces can be destroyed by an arbitrarily small perturbation, for small perturbations the island and stochastic domains are confined between surviving toroidal flux surfaces. This means that an appropriately defined concept of approximate closed flux surfaces is stable with respect to small perturbations. It is in this sense that the equilibrium fields of Chap. 2 must be understood.

1.8   THE STANDARD MAP

It has been found that the approach to chaos, or stochasticity, of deterministic Hamiltonian systems exhibits quite universal behavior, independent of
the form of the Hamiltonian. Thus both the behavior of the magnetic field
and of the particle orbits can be better understood by the study of simple
Hamiltonian systems. The simplest such system, the standard map, introduced independently by Chirikov (1979) and Taylor (unpublished), can be
obtained by simplifying the field stepping equations of Sec. 1.7 for a single
harmonic perturbation by considering steps in ζ of 2π and taking constant
shear. Suitably normalized variables then gives

$$x' = x + v$$
$$v' = v + \epsilon sin(x') \tag{1.63}$$

where v is the cross field (ψ) direction and x the poloidal angle. The prime
on the x in the second equation is necessary to ensure that the map has
determinant one. The safety factor is $q = 2\pi/v mod(2\pi)$ and ranges from
∞ to 1 in the interval $v = 0, 2\pi$.

Islands appear at $v = 2k\pi$, k integer, of width $\Delta v = 4\sqrt{\epsilon}$. The Chirikov
island overlap criterion then gives for the vanishing of the last KAM surface
$\epsilon = \pi^2/4 \simeq 2.47$. The nonlinear effects lower this threshold considerably,
and toroidal flux surfaces exist only for $\epsilon < 0.9716...$ (Greene, 1979). The
most stable KAM surface is at $q = (\sqrt{5} - 1)/2$, which is that irrational
which is "farthest away" from the rationals.

For large ϵ it is possible to analytically calculate both the diffusion in
v and the Kolmogorov entropy. The diffusion rate is simply that obtained
by making a random phase approximation for the displacement in v, $D =
(\Delta v)^2/2t$ with t the number of steps, and using $\overline{sin^2 x} = 1/2$ we find

$$D_{QL} = \frac{\epsilon^2}{4} \tag{1.64}$$

where QL stands for quasilinear.

The period doubling point can be found analytically. From Eq. 1.63 we
find the tangent map to be given by the matrix

$$M = \begin{pmatrix} 1 & 1 \\ k & 1+k \end{pmatrix} \tag{1.65}$$

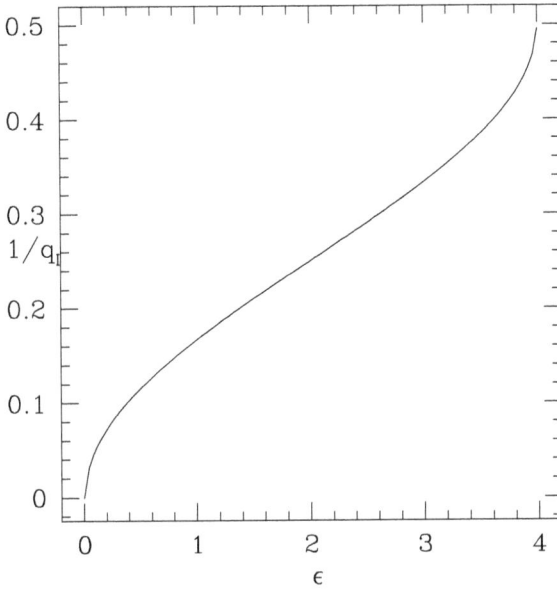

Fig. 1.8   Internal q_I for the standard map vs ϵ.

with $k = \epsilon cosx$. The elliptic fixed point is at $x = \pi$ and the eigenvalues of M are $\lambda = 1 + k/2 \pm \sqrt{k^2/4 + k}$. Period doubling occurs at $\epsilon = 4$ where $q_I = 2$. We find

$$\frac{2\pi}{q_I} = atan\left(\frac{\sqrt{\epsilon - \epsilon^2/4}}{1 - \epsilon/2}\right). \qquad (1.66)$$

In Fig. 1.8 is shown the behavior of the internal winding number q_I for the standard map up to the period doubling point $\epsilon = 4$.

For large k we have $\lambda_\pm \to -k, -1/k$. Thus the maximum stretching of distance between points is $\lambda_+ = -k$. Then since repeated stretching gives exponential growth

$$h \simeq \frac{< ln\lambda^2 >}{2} \qquad (1.67)$$

with the average taken over the trajectory. The stretching is primarily in the x direction, giving

$$h \simeq \frac{1}{\pi} \int dw ln[(\epsilon cosw)^2] = ln(\epsilon/2). \tag{1.68}$$

Long time correlations modify D significantly from the quasilinear value, and the result can be obtained analytically (Rechester and White, 1980). Write the Chirikov–Taylor map as $x_j = x_{j-1} + v_{j-1}, v_j = v_{j-1} + \epsilon sin(x_j)$. The distribution of particles (field lines) at position x_j, v_j at time t_j is given by

$$P(x_j, v_j, t_j) = \int dv_{j-1} dx_{j-1} \delta(v_j - v_{j-1} - \epsilon sin(x_j)) \times$$
$$\delta(x_j - x_{j-1} - v_{j-1}) P(x_{j-1}, v_{j-1}, t_{j-1}). \tag{1.69}$$

Now iterate this expression, obtaining

$$P(x_T, v_T, t_T) =$$
$$\int dv_{T-1} dx_{T-1} \delta(v_T - v_{T-1} - \epsilon sin(x_T)) \delta(x_T - x_{T-1} - v_{T-1})$$
$$\cdots dv_0 dx_0 \delta(v_1 - v_0 - \epsilon sin(x_1)) \delta(x_1 - x_0 - v_0) P(x_0, v_0, t_0). \tag{1.70}$$

Now take the initial distribution to be uniform in x but located at $v = 0$, $P(x_0, v_0, t_0) = \delta(v_0)$. We wish to calculate the diffusion in v. Use the delta functions to do the integrals in v_k, giving

$$P(x_T, v_T, T) = \int dx_{T-1} \cdots dx_0 \delta(v_T - S_T) \prod_{k=1}^{T} \delta(x_k - x_{k-1} - S_{k-1}) \tag{1.71}$$

with $S_k = \epsilon \sum_1^k sin(x_p)$. The diffusion constant D is given by the limit of $T \to \infty$ of

$$D = \frac{1}{2T} \int dv_T \int \frac{dx_T}{2\pi} v_T^2 P(x_T, v_T, T) \tag{1.72}$$

or

$$D = \frac{1}{2T} \int \frac{dx_0 \cdots dx_T}{(2\pi)} S_T^2 \prod_{k=1}^{T} \delta(x_k - x_{k-1} - S_{k-1}). \tag{1.73}$$

Substitute for the δ functions $\delta(z) = \sum_{-\infty}^{\infty} e^{imz}/(2\pi)$, giving

$$D = \frac{1}{2T} \int \frac{dx_0 \cdots dx_T}{(2\pi)^{T+1}} S_T^2 \sum_{m_1} \cdots \sum_{m_T} e^{i \sum_1^T m_k(x_k - x_{k-1} - S_{k-1})}. \tag{1.74}$$

But

$$\sum_{1}^{T} m_k(x_k - x_{k-1} - S_{k-1}) \quad = \quad \sum_{k=1}^{T} m_k(x_k - x_{k-1})$$

$$-\epsilon[(m_2 + \cdots m_T)sinx_1 + (m_3 + \cdots m_T)sinx_2 + \cdots m_T sinx_{T-1}]. \quad (1.75)$$

Now use $e^{\pm izsinx} = \sum_{-\infty}^{\infty} J_n(z)e^{\pm inx}$, and we obtain

$$D = \frac{1}{2T} \int \frac{dx_0 \cdots dx_T}{(2\pi)^{T+1}} S_T^2 \sum_{m_1} \cdots \sum_{m_T} e^{i\sum_1^T m_k(x_k - x_{k-1})}$$

$$\sum_{n_1} \cdots \sum_{n_{T-1}} J_{n_1}(\epsilon(m_2 + m_3 + \cdots + m_T))e^{-in_1 x_1} \cdots$$

$$J_{n_p}(\epsilon(m_{p+1} + \cdots + m_T))e^{-in_p x_p} \cdots$$

$$J_{n_{T-1}}(\epsilon m_T)e^{-in_{T-1} x_{T-1}}. \quad (1.76)$$

Now for large ϵ use the fact that $J_n(\epsilon) \sim 1/\sqrt{\epsilon}$. Thus dominant contributions are terms in the sum with few Bessel functions different from $J_0(0)$. There is one term with all $m_k = 0$ and all $n_k = 0$ giving a contribution to D of

$$D_0 = \frac{1}{2T} \int \frac{dx_0 \cdots dx_T}{(2\pi)^{T+1}} S_T^2 = \epsilon^2/4, \quad (1.77)$$

which we recognize as the quasilinear value. The only terms which contain a single Bessel function with nonzero argument are those arising from

$$\vec{m} = (0, 0, \cdots, 0, -1, 1, 0, \cdots, 0) \quad (1.78)$$

with the -1 appearing in the lth place, and there are $T - 1$ such terms, plus an additional $T - 1$ similar terms with -1,1 replaced by 1,-1, giving

$$D_1 = \frac{T-1}{T} \int \frac{dx_0 \cdots dx_T}{(2\pi)^{T+1}} S_T^2 e^{i(-2x_l + x_{l-1} + x_{l+1})} \sum_{n_l} J_{n_l}(\epsilon)e^{in_l x_l}. \quad (1.79)$$

Only the term with $n_l = 2$ survives. Expanding S_T and the exponential we find for large T

$$D_1 = -\epsilon^2 \frac{J_2(\epsilon)}{2}. \quad (1.80)$$

The terms containing two Bessel functions can also be found. They are given by $\vec{m} = (0, 0, \cdots, 0, -1, 0, 1, 0, \cdots, 0), \vec{m} = (0, 0, \cdots, 0, 1, -2, 1, 0, \cdots, 0)$,

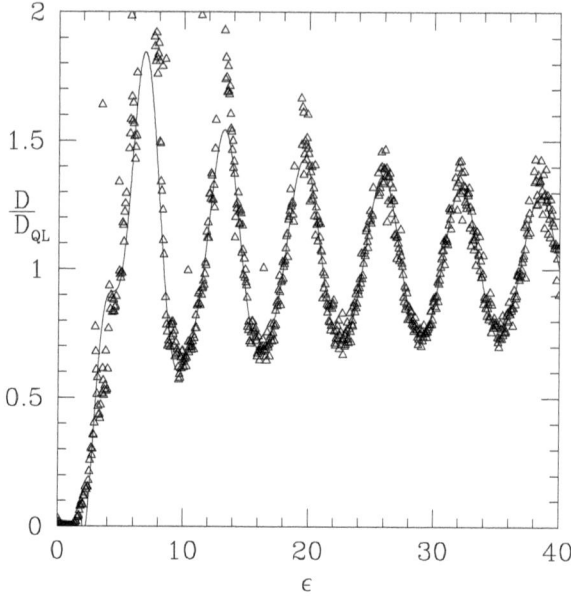

Fig. 1.9   Diffusion in the standard map.

and $\vec{m} = (0, 0, \cdots, 0, -1, 1, 1, -1, 0, \cdots, 0)$. The final result to order J^2 is

$$\frac{D}{D_{QL}} = 1 - 2J_2(\epsilon) - 2J_1^2(\epsilon) + 2J_3^2(\epsilon) + 2J_2^2(\epsilon). \tag{1.81}$$

In Fig 1.9 is shown a numerical evaluation of the diffusion constant versus ϵ, and a plot of the analytic result given above. The Chirikov–Taylor map, because of its symmetries, possesses stronger correlations than many other maps, but the phenomenon of ringing above stochastic threshold is observed also in other maps.

For some values of ϵ very close to multiples of π there exist, even for large values of ϵ, small islands in phase space which give rise to nondiffusive motion for initial conditions within the islands, and no matter how small the islands are these trajectories dominate any evaluation of transport for large time, resulting in nondiffusive motion. See Karney *et al.* (1982).

Also of interest is the behavior near threshold, which exhibits a universal character, valid for a large class of Hamiltonian systems. It was found, both numerically (Rechester *et al.* 1981) and analytically (Mackay *et al.* 1984) that near threshold the diffusion has the form

$$D \simeq a(\epsilon - \epsilon_c)^p \qquad (1.82)$$

with $p \simeq 3$, and ϵ_c the threshold value. Furthermore, the critical exponent $p \simeq 3$ appears to be independent of the particular map.

Near threshold, trajectories undergo long time sticking in complex structures consisting of nested chains of islands with narrow stochastic bands between them. This phenomenon leads to nondiffusive transport $x^2 \sim t^p$, with both hyperdiffusion, $p > 1$, and subdiffusion, $p < 1$. For a description of this phenomenon see White *et al.* (1998).

1.9   Problems

1. Find the set of coordinate transformations $x^k \rightarrow x'^k$ that leave $\epsilon_{i,j,k}$ invariant.

2. In a reversed field pinch the field relaxes to a Taylor state, with $\vec{j} = \mu\vec{B}$, μ a constant. Assume cylindrical symmetry, so that $\psi_p = \psi_p(r)$ and $\nabla\phi = \hat{\phi}/R$ with $R \gg r$. Find explicitly the covariant and contravariant representations of $\vec{B}$.

3. For any map $x' = f(x,y)$, $y' = g(x,y)$, the matrix describing the map of the differentials dx, dy is called the tangent space map. From the tangent space map derived from the standard map calculate the internal $q_I(\epsilon)$ associated with the O-point, and the eigenvalues at the X-point. Find the value of ϵ at which the O-point bifurcates. Find the values of q_I at Chirikov overlap and at stochastic threshold, $\epsilon \simeq 1$.

4. Consider the Fermi model for the acceleration of cosmic rays, consisting of a particle bouncing elastically between two walls oscillating with velocity $v_W = v_0 \sin\omega t$. Let $\psi = \omega t$ and describe the motion by a map giving successive values of the particle velocity u and collision phase ϕ, the value of ωt when the particle hits a wall. Show that linearization in u results in the standard map.

5. Consider a straight stellarator with three pairs of helical windings. The vacuum field is $\vec{B} = B_0 \hat{z} + \lambda B_0 \nabla[r^3 cos(3\theta - kz)]$.

   Assume $kr \ll 1$. *i.e.* work close to the axis. Find the magnetic surfaces $\psi = $ constant, with $\vec{B} \cdot \nabla\psi = 0$. Find a condition on λ such that the flux surfaces are topologically circles (no X-point) inside radius r. Use this bound on λ. Find the rotational transform

$$\left\langle \frac{\Delta\theta}{k\Delta z} \right\rangle$$

where <> means an average over z, since the transform is not a constant in z. Note that the equations for the field lines depend only on r and $\zeta = 3\theta - kz$. (Rosenbluth lecture notes)

6. A dipole magnetic field can be written as $\vec{B} = \nabla(\vec{M} \cdot \vec{r}/r^3)$, with $\vec{M}$ the dipole moment. Find the poloidal flux function $\psi_p(r, \theta)$ with $\vec{M} \cdot \vec{r} = Mrcos\theta$ such that

$$\vec{B} = \nabla\phi \times \nabla\psi_p.$$

Find the equations for the field lines.

1.10   References

General

- Goldstein, H., Classical Mechanics, (Harvard University, Cambridge, Mass, 1953).
- Jassby, D., D. Meade, private communication (2000).
- Lawson, J.D., Proc. Phys. Soc. B 70, 6 (1957).
- Morse, P. M., and H. Feshbach, Methods of Theoretical Physics (McGraw-Hill, New York 1953).
- Poincaré, H., Les Nouvelles Methodes de la Mecanique Celeste (1892) (Engl. Transl. NASA, Washington, 1967).

Magnetohydrodynamics

- Bateman, G., MHD Instabilities (MIT, Cambridge, Mass, 1978).
- Boyd, T. J. M., and J. J. Sanderson, Plasma Dynamics (Barnes and Noble, New York, 1969).
- Braginski, S. I., in Reviews of Plasma Physics, vol. 1, edited by M. A. Leontovitch (Consultants Bureau, New York, 1965).
- Freidberg, J. P., Rev. Mod. Phys. 54, 801 (1982).
- Freidberg, J. P., Ideal Magnetohydrodynamics (Plenum Press, New York, 1987).
- Krall, N. A., and A. W. Trivelpiece, Principles of Plasma Physics (McGraw-Hill, New York, 1973).
- Montgomery, D.C., and D. A. Tidman, Plasma Kinetic Theory, (McGraw-Hill, New York, 1964).
- Wesson, J., Tokamaks (Clarendon Press, Oxford, 1987).

Magnetic Surfaces

- Kolmogorov, A. N., in Proc. of the International Congress of Mathematicians, Amsterdam (North Holland, Amsterdam, 1957) Vol. 1, p. 315.
- Morozov, A. I., and L. S. Solov'ev in Reviews of Plasma Physics, (Consultants Bureau, New York, Vol. 2) (Transl. from Russian: Voprosy Teorii Plazmy (Atomizdat, Moscow, 1963).

Stochasticity

- Arnold, V. I., Sov. Math. Dokl. 5, 581 (1964).
- Chirikov, B. V., Phys. Rep. 52, 263 (1979).
- Escande, D. F., Phys. Rep. 121, 166 (1985).
- Greene, J. M., J. Math. Phys. 20, 1183 (1979).
- Lichtenberg, A. J., and M. A. Leiberman, Regular and Stochastic Motion (Springer-Verlag, New York, 1992).
- Mackay, R. S., J. D. Meiss, and I. C. Percival, Phys. Rev. Lett. 52, 697 (1984).
- Melnikov, V. K., Dokl. Akad. Nauk 148, 1257 (1963).
- Rechester, A. B. and R. B. White, Phys. Rev. Lett. 44, 1586 (1980).
- Karney, C. F. F., A. B. Rechester, and R. B. White, Phys. Rev. 23, 2664 (1981).
- Rechester, A. B., M. N. Rosenbluth, and R. B. White, Physica. [4D], 425 (1982).
- White, R. B., S. Benkadda, S. Kassibrakis, and G. M. Zaslavsky, Chaos 8, 757 (1998).

Chapter 2

Equilibrium

2.1 INTRODUCTION

The magnetic configuration chosen for confinement must be in equilibrium with the plasma it contains. The plasma, thrusting against the magnetic field but remaining attached to it, distorts it and the combined system is in equilibrium when the condition $\nabla p = \vec{j} \times \vec{B}$ holds everywhere. Since this equation is linear in the pressure but quadratic in the field strength, it is the ratio $\beta = 2p/B^2$, the plasma beta, which characterizes the equilibrium. As seen, a general toroidal field with nested flux surfaces is characterized by the poloidal and toroidal flux functions, related by the safety factor q. The ratio of toroidal and poloidal field strength turns out to be restricted by considerations of equilibrium, stability, and particle confinement. The limit $q \to \infty$, toroidal field only, does not possess an equilibrium, as we will find. In addition, in a fusion device using deuterium and tritium for fuel a minimum poloidal field is required to confine the 3.5 MeV fusion product alpha particles produced. For economic reasons it is desirable to design a device with the minimum possible magnetic field strength, and given this constraint on the poloidal field it is advantageous to use very low values of q. We will find in Chap. 4 that in a tokamak values of q below one are prohibited by stability considerations. The tokamak generates its poloidal field entirely through plasma current, and operates within the restrictions given by these considerations, with $q \sim 1$. Other configurations such as the stellarator, reversed field configuration, or spheromak have either externally generated poloidal fields or self-generated fields producing a very different range of q values than a tokamak, and different stability requirements.

Many of the essential features of toroidal equilibrium can be obtained by expanding the equations in the inverse aspect ratio of the torus. For very low β the plasma pressure deforms the vacuum field only slightly, but typically for β of a few percent the plasma pushes strongly outward, compressing the field on the outside of the torus. If the equilibrium is not axisymmetric but more complicated, as in a stellarator, the outward shift of the magnetic surfaces can produce magnetic islands and stochastic domains, destroying the confinement properties of the equilibrium. Thus there can exist limits on the possible values of β obtainable in some configurations which arise even before the stability of the equilibrium is discussed.

The problem of the existence of equilibria in non axisymmetric configurations is not solved in general. See Grad (1967) for a review. Berk *et al.* (1986) showed that sharp boundary equilibria exist for fully three-dimensional configurations if β is sufficiently small. We will not examine the question of existence of equilibria, since we begin by assuming the existence of nested toroidal magnetic flux surfaces, and use this fact to derive the form of the magnetic field.

2.2 THE VIRIAL THEOREM

The equilibrium condition allows one to equate a positive definite integral involving plasma and magnetic pressure to an integral over the surrounding surface. The simplest derivation (Shafranov, 1966) follows from the tensor expression of the equilibrium condition,

$$\partial_{x_i} T_{ik} = 0 \tag{2.1}$$

with

$$T_{ik} = p_\perp \left(\delta_{ik} - \frac{B_i B_k}{B^2} \right) + p_\parallel \frac{B_i B_k}{B^2} \tag{2.2}$$

where $p_\perp = p + B^2/2$, $p_\parallel = p - B^2/2$.

The virial theorem follows from the identity

$$\partial_{x_i}(x_k T_{ik}) = T_{kk} + x_k \partial_{x_i} T_{ik}. \tag{2.3}$$

The second term on the right hand side is zero at equilibrium. Integrating this equation over a volume V bounded by surface S we find

$$\int_V T_{kk} dV = \oint_S T_{ik} x_k dS_i. \tag{2.4}$$

Substituting the expressions for T_{ik} and for $p_\perp$, $p_\parallel$, we then find

$$\int (3p + B^2/2)dV = \oint \left[(p + B^2/2)\vec{r} \cdot d\vec{S} - (\vec{B} \cdot \vec{r})(\vec{B} \cdot d\vec{S}) \right] . \quad (2.5)$$

Now consider the volume to entirely contain the plasma, so that $p = 0$ outside it. If there are no rigid current-carrying conductors the surface integral vanishes if the surface is removed to infinity, since the magnetic field falls off at least as $1/r^3$. Thus the relation cannot be satisfied, and any configuration in equilibrium must be supported by rigid external magnets.

2.3 FIELD LINE CURVATURE

The curvature of a field line is given by

$$\vec{\kappa} = (\hat{b} \cdot \nabla)\hat{b} \qquad (2.6)$$

with $\hat{b} = \vec{B}/B$. Expanding we find

$$B^2\vec{\kappa} = (\vec{B} \cdot \nabla)\vec{B} - \frac{\vec{B}}{2B^2}(\vec{B} \cdot \nabla)B^2. \qquad (2.7)$$

Now from the identity

$$\vec{B} \times (\nabla \times \vec{B}) = \frac{\nabla B^2}{2} - (\vec{B} \cdot \nabla)\vec{B} \qquad (2.8)$$

and using the equilibrium condition $\vec{j} \times \vec{B} = \nabla p$ we find

$$B^2\vec{\kappa} = \frac{\nabla B^2}{2} + \nabla p - \frac{\vec{B}}{2B^2}(\vec{B} \cdot \nabla)B^2. \qquad (2.9)$$

Further noting that $\vec{B} \cdot \nabla p = 0$ this can be written as

$$B^2\vec{\kappa} = \nabla \left[\frac{B^2}{2} + p \right] - \frac{\vec{B}}{B^2}(\vec{B} \cdot \nabla) \left[\frac{B^2}{2} + p \right]. \qquad (2.10)$$

Thus the curvature is the component of $\nabla[B^2/2 + p]$ orthogonal to $\vec{B}$. This leads to a simple physical interpretation (Greene, unpublished). That part of the gradient of the total energy density $B^2/2 + p$, which is orthogonal to the surface, is supported by the field line tension through the curvature $\vec{\kappa}$. If the component of $\vec{\kappa}$ normal to the magnetic surface, $\vec{\kappa} \cdot \vec{n}$, is negative (directed toward the plasma interior) then a local displacement of the plasma and field lines outwards increases the local curvature, putting more stress on the field line tension. This is referred to as unfavorable curvature, and $\vec{\kappa} \cdot \vec{n} > 0$ is referred to as favorable curvature.

2.4   GENERAL 3-D EQUILIBRIA

Consider a field configuration of nested flux surfaces labeled by ψ, topologically toroidal although they may be significantly deformed in space. Instead of using the cylindrical angle ϕ for our coordinate variable we will use general poloidal and toroidal coordinates θ, ζ. Since $\vec{B}$ is orthogonal to $\nabla\psi$ use the Clebsch representation*

$$\vec{B} = \nabla\psi \times \nabla V \qquad (2.11)$$

and choose $V = \theta - \zeta/q(\psi) + \lambda(\psi, \theta, \zeta)$.

Choose $2\pi\psi$ to be the toroidal flux inside the surface ψ. The element of surface area in a toroidal section is $d\vec{S}_t = \mathcal{J}\nabla\zeta d\theta d\psi$, with the Jacobian given by $\mathcal{J}^{-1} = \nabla\psi \cdot \nabla\theta \times \nabla\zeta$, giving

$$2\pi\psi = \int \vec{B} \cdot \nabla\zeta \mathcal{J} d\psi d\theta = \int d\psi d\theta + \int_{\theta_0}^{\theta_0+2\pi} \frac{\partial\lambda}{\partial\theta} d\psi d\theta \qquad (2.12)$$

for any θ_0, and thus λ is periodic in θ.

Now choose $2\pi\psi_p = 2\pi \int d\psi/q(\psi)$ to be the poloidal flux inside the magnetic surface. The element of surface area in a poloidal section is $d\vec{S}_p = \mathcal{J}\nabla\theta d\zeta d\psi$, giving

$$2\pi\psi_p = \int \vec{B} \cdot \nabla\theta \mathcal{J} d\psi d\zeta = \int d\psi_p d\zeta + \int_{\zeta_0}^{\zeta_0+2\pi} \frac{\partial\lambda}{\partial\zeta} d\psi_p d\zeta \qquad (2.13)$$

for any ζ_0, and thus λ is periodic in ζ. The surface labels ψ and ψ_p can be used interchangeably, according to which is more convenient for the equilibrium under study, with $d\psi/d\psi_p = q(\psi)$.

---

*Divergence $\vec{B} = 0$, so write $\vec{B} = \nabla \times \vec{F}$. Then also require that $\vec{B} \cdot \nabla\psi = 0$. Write $\vec{F} = f_\psi \nabla\psi + f_\theta \nabla\theta + f_\zeta \nabla\zeta$, giving $\partial_\zeta f_\theta - \partial_\theta f_\zeta = 0$. Let $f_\theta = \partial_\theta G$, then $f_\zeta = \partial_\zeta G$ and $\vec{F} = \nabla G + f_\psi \nabla\psi - \partial_\psi G \nabla\psi \equiv \nabla G - V\nabla\psi$ and $\vec{B} = \nabla\psi \times \nabla V$.

We then have the contravariant representation for $\vec{B}$,

$$\vec{B} = (1 + \lambda'_\theta)\nabla\psi \times \nabla\theta + (1/q - \lambda'_\zeta)\nabla\zeta \times \nabla\psi. \tag{2.14}$$

Straight field line coordinates, meaning that $\vec{B} \cdot \nabla\zeta / \vec{B} \cdot \nabla\theta = q(\psi_p)$ is a flux function, independent of θ, ζ, are often used for equilibrium and stability studies. A straight field line system can be obtained either with a modification of the toroidal or the poloidal coordinate. The usual choice is to replace $\zeta - q\lambda$ with ζ in V, equivalent to setting λ equal to zero in all subsequent equations. This is possible because of the periodicity of the function λ. The variable θ is still free to be modified in a way to select the form of the Jacobian. In these "straight field line variables", $d\zeta/d\theta = \vec{B} \cdot \nabla\zeta / \vec{B} \cdot \nabla\theta = q(\psi_p)$ along a field line, and the contravariant representation for $\vec{B}$ is somewhat simplified,

$$\vec{B} = \nabla\psi \times \nabla\theta + \nabla\zeta \times \nabla\psi_p = \frac{\vec{e}_\zeta}{\mathcal{J}} + \frac{\vec{e}_\theta}{q\mathcal{J}} \tag{2.15}$$

with $d\psi/d\psi_p = q(\psi_p)$. We also find

$$\vec{B} \cdot \nabla = \frac{1}{q\mathcal{J}}(\partial_\theta + q\partial_\zeta). \tag{2.16}$$

Now we make use of the equilibrium property of $\vec{B}$. We assume a scalar pressure equilibrium, the generalization to a tensor pressure being straightforward. Parallel electron heat conduction is generally very rapid compared to other processes we wish to consider, so that temperature and pressure are uniform following a magnetic field line. At every surface where q is irrational a field line covers the surface ergodically, and thus the pressure is uniform over the surface, $p = p(\psi)$. At surfaces where q is rational, however, one cannot conclude this, and the pressure could be different on each field line. At first glance this appears to be a very unphysical situation, and one to dismiss offhand, but it is a little more subtle than that. Recall that arbitrarily small perturbations produce chains of islands at rational surfaces. In the presence of the islands a plasma pressure which has a different value at the elliptic points than at the hyperbolic points is quite physical. Thus this question is related to the physical process involved in the perturbations and the nature of the cross field transport in the vicinity of the rational surfaces with low values of m, n. For fixed perturbation amplitude the island size decreases with increasing m, n until finally the island width is comparable to the particle gyro radius or some other scale defining the breakdown of ideal MHD. Ideal MHD must be regarded as a theory applying in an average sense for scale sizes larger than the islands produced by the fluctuations. The effect of the fluctuations existing as perturbations of the MHD equilibrium solutions is to make the transport anomalous from the point of view of ideal MHD. For the present we thus consider an equilibrium with perfect flux surfaces with $p = p(\psi)$.

From the equilibrium condition

$$\vec{j} \times \vec{B} = \nabla p \tag{2.17}$$

with $p = p(\psi)$ it follows that $\vec{j} \cdot \nabla \psi = 0$. Write $\vec{j} = \nabla \psi \times \nabla W$, and choose $W = \bar{I}'(\psi)\theta + \bar{g}'(\psi)\zeta + \alpha(\psi, \theta, \zeta)$. Then

$$\vec{j} = (\bar{I}'(\psi) + \alpha'_\theta)\nabla\psi \times \nabla\theta + (\bar{g}'(\psi) + \alpha'_\zeta)\nabla\psi \times \nabla\zeta \tag{2.18}$$

with primes indicating derivatives with respect to the subscripted variable. Choose $2\pi\bar{I}$ to be the toroidal current inside ψ, and $2\pi\bar{g}$ to be the poloidal current outside ψ. Then as is the case with λ, α is periodic in θ, ζ. Then $\nabla \times \vec{B} = \vec{j}$ gives the covariant representation for $\vec{B}$,

$$\vec{B} = g\nabla\zeta + I\nabla\theta + \delta\nabla\psi = g\vec{e}^\zeta + I\vec{e}^\theta + \delta\vec{e}^\psi \tag{2.19}$$

with $< g >= \bar{g}$, $< I >= \bar{I}$ and brackets indicate averaging over θ and ζ,

$$I(\psi,\theta,\zeta) = \bar{I}(\psi) + \sigma'_\theta, \; g(\psi,\theta,\zeta) = \bar{g}(\psi) + \sigma'_\zeta, \; \delta(\psi,\theta,\zeta) = \sigma'_\psi - \alpha \tag{2.20}$$

where σ and α are arbitrary functions of ψ, θ, ζ periodic in θ and ζ. Here f'_α denotes $\partial_\alpha f$.

Using $\vec{e}^\alpha \times \vec{e}^\beta = \frac{1}{\mathcal{J}}\epsilon^{\alpha\beta\gamma}\vec{e}_\gamma$,

$$\vec{j} = -(\bar{I}'(\psi) + \alpha'_\theta)\frac{\vec{e}_\zeta}{\mathcal{J}} + (\bar{g}'(\psi) + \alpha'_\zeta)\frac{\vec{e}_\theta}{\mathcal{J}} \tag{2.21}$$

We then easily find

$$\vec{j} \times \vec{B} = [\bar{I}'(\psi) + \alpha'_\theta + (\bar{g}'(\psi) + \alpha'_\zeta)q]\vec{e}^\psi/(q\mathcal{J}). \tag{2.22}$$

Equating this to $\nabla p = p'\vec{e}^\psi$ gives

$$\bar{I}'(\psi) + \alpha'_\theta + (\bar{g}'(\psi) + \alpha'_\zeta)q = p'q\mathcal{J}. \tag{2.23}$$

Averaging in ζ and θ we find that $\bar{I}$, $\bar{g}$, and p' are related through

$$\bar{I}' + \bar{g}'q = \mathcal{J}_{00}p'q, \quad f_{00} = \int_0^{2\pi} \int_0^{2\pi} f d\zeta d\theta/(4\pi^2). \tag{2.24}$$

Note that for a general equilibrium we have

$$\partial_\zeta I = \partial_\theta g. \tag{2.25}$$

From $\vec{B} \cdot \nabla\psi = 0$ we find

$$\delta(\psi, \theta, \zeta) = \frac{-(I\nabla\theta \cdot \nabla\psi + g\nabla\zeta \cdot \nabla\psi)}{|\nabla\psi|^2}. \tag{2.26}$$

Thus δ is related to the nonorthogonality of the coordinate system. Note that we can also write the covariant representation in the form

$$\vec{B} = \bar{g}(\psi)\nabla\zeta + \bar{I}(\psi)\nabla\theta + \nabla\sigma - \alpha\nabla\psi. \tag{2.27}$$

The equilibrium is determined by the flux functions $q(\psi)$, $\bar{g}(\psi)$, $\bar{I}(\psi)$, and the two functions of three variables $\alpha(\psi, \theta, \zeta)$ and $\sigma(\psi, \theta, \zeta)$. However, these functions are not all independent. Dot the covariant and contravariant expressions for $\vec{B}$, giving $B^2 \mathcal{J}q = gq + I$. Averaging in ζ and θ we find

$$q(\mathcal{J}B^2)_{00} = \bar{g}q + \bar{I}. \tag{2.28}$$

We also obtain the magnetic differential equation for the 3-D function σ,

$$q\mathcal{J}B^2 - q(\mathcal{J}B^2)_{00} = q\sigma'_\zeta + \sigma'_\theta. \tag{2.29}$$

Further we find the magnetic differential equation for the 3-D function α,

$$\alpha'_\theta + q\alpha'_\zeta = \tilde{\mathcal{J}}p'q \tag{2.30}$$

with $\tilde{\mathcal{J}} = \mathcal{J} - \mathcal{J}_{00}$. Primes here indicate derivatives with respect to ψ. Changing the flux variable to ψ_p introduces factors of q in the derivative terms and in the change from $\mathcal{J}$ to $\mathcal{J}_p$. Equilibria are typically determined by choosing two of the scalar functions $q(\psi)$, $p(\psi)$, $\bar{g}(\psi)$, and the two functions of three variables $\alpha(\psi, \theta, \zeta)$ and $\sigma(\psi, \theta, \zeta)$.

Taking Fourier transforms of α, σ using $e^{i(m\theta - n\zeta)}$ we find

$$(m - nq)\sigma_{mn} = (\mathcal{J}B^2)_{mn}q \tag{2.31}$$

$$(m - nq)\alpha_{mn} = \mathcal{J}_{mn}p'q. \tag{2.32}$$

Now evaluate the parallel component of the current. Dotting the co-variant expression for $\vec{B}$ into $\vec{j}$ we find

$$\frac{j_\parallel}{B} = \frac{q(g\bar{I}' - I\bar{g}')}{gq + I}$$

$$-\sum_{mn}' \frac{(nI + mg)}{I + gq} cos(m\theta - n\zeta) \left[\frac{p'q\mathcal{J}_{mn}}{m - nq} + q\delta(m - nq)h_{mn} \right], \quad (2.33)$$

with h_{mn} unknown. The first term is the force free part of $j_\parallel$ and the term proportional to p' is the Pfirsch–Schlüter current.

For $\mathcal{J}_{mn} \neq 0$, a finite pressure gradient drives an infinite parallel current at the rational surface. Elimination of this singularity would require the vanishing of the pressure gradient at rational surfaces. Magnetic perturbations would produce small magnetic islands and stochastic layers at rational surfaces, decreasing in size as m, n increase, giving a local flattening of the pressure profile in the vicinity of such surfaces as a correct physical solution. However, in the limit of perfectly nested surfaces, there result sheet currents at the rational surfaces, or equivalently, discontinuities in the gradient of ψ. The function h_{mn} is nonzero only if ψ possesses step function singularities, *i.e.* if there exist current sheets in the plasma at rational values of q.

These covariant and contravariant representations are used in the Hamiltonian formulation of the guiding center drift motion in Chap. 3. There are several different straight field line coordinate systems used in the literature, distinguished by the choice of the form of the Jacobian. For numerical calculations this choice can be quite relevant, as it determines the distribution of grid points in the discrete coordinate mesh. The Hamada coordinates have a Jacobian which is a function of ψ alone. In Boozer coordinates, $B^2 \mathcal{J}$ is a function of ψ. One simple choice is to take $\mathcal{J} = 1$, so that the volume element is given simply by $d\theta d\zeta d\psi_p$. A Jacobian $\mathcal{J} \sim X^2$ is used in the Princeton Equilibrium and Stability (PEST) code (Grimm *et al.*, 1976), one of the numerical codes developed for the calculation of tokamak equilibrium and stability, and ζ is chosen equal to the toroidal angle ϕ.

2.4.1   *Hamada coordinates*

Hamada coordinates (Hamada, 1962) are a particular form of straight field line coordinates. Angular coordinates can be chosen so that in addition to

$\vec{B}$ also the current is straight in the coordinates, *i.e.* $\vec{j} \cdot \nabla\zeta / \vec{j} \cdot \nabla\theta$ is a flux function. Begin with Eq. 2.18,

$$\vec{j} = \nabla\psi \times \nabla\alpha + \nabla\bar{I} \times \nabla\theta + \nabla\bar{g} \times \nabla\zeta, \qquad (2.34)$$

and make the substitution $\theta \to \theta + \beta$. Then

$$\frac{\vec{j} \cdot \nabla\zeta}{\vec{j} \cdot \nabla\theta} = -\frac{\bar{I}' + \alpha'_\theta + \bar{I}'\beta'_\theta}{\bar{g}' + \alpha'_\zeta + \bar{I}'\beta'_\zeta} \qquad (2.35)$$

and thus choosing $\beta = -\alpha/\bar{I}'$ we find

$$\frac{\vec{j} \cdot \nabla\zeta}{\vec{j} \cdot \nabla\theta} = -\frac{\bar{I}'}{\bar{g}'}, \qquad (2.36)$$

giving both $\vec{B}$ and $\vec{j}$ straight in these coordinates. The current then takes the form

$$\vec{j} = \bar{I}'\nabla\psi \times \nabla\theta + \bar{g}'\nabla\psi \times \nabla\zeta. \qquad (2.37)$$

Now use pressure balance $\vec{j} \times \vec{B} = p'\nabla\psi$ and dot with $\nabla\theta \times \nabla\zeta$ giving

$$(\vec{j} \cdot \nabla\theta)(\vec{B} \cdot \nabla\zeta) - (\vec{j} \cdot \nabla\zeta)(\vec{B} \cdot \nabla\theta) = p'/\mathcal{J} \qquad (2.38)$$

or

$$\bar{g}' - \bar{I}'/q = p'\mathcal{J} \qquad (2.39)$$

from which we conclude that the Jacobian $\mathcal{J}$ is a function of the flux surface ψ only.

2.4.2   *Boozer coordinates*

Boozer coordinates are a particular form of straight field line coordinates. Angular coordinates can be chosen so that $g = \bar{g}(\psi)$ and $I = \bar{I}(\psi)$. Change variables to $\zeta' = \zeta + q(\psi)w$, $\theta' = \theta + w$. Write $\vec{B} = \bar{g}\nabla\zeta + \bar{I}\nabla\theta + \nabla\sigma - \alpha\nabla\psi$. Then we have $\vec{B} = \bar{g}(\nabla\zeta' - q\nabla w) + \bar{I}(\nabla\theta' - \nabla w) + \nabla\sigma + \alpha'\nabla\psi$. Combining terms and setting $(\bar{g}q + \bar{I})w = \sigma$ we find in this coordinate system (Boozer, 1981)

$$\vec{B} = \bar{g}(\psi)\nabla\zeta' + \bar{I}(\psi)\nabla\theta' + \delta'\nabla\psi \qquad (2.40)$$

and the Jacobian satisfies $B^2\mathcal{J} = \bar{g}q + \bar{I}$, which is a function of ψ only. Note also that the contravariant representation of $\vec{B}$ is the same in all straight field line coordinates, $\vec{B} = \nabla\psi \times \nabla\theta' + \nabla\zeta' \times \nabla\psi_p$.

2.4.3 *Cylindrical coordinates*

It is not always convenient to choose straight field line coordinates. Cylindrical coordinates use the toroidal angle ϕ for ζ, so that in Eq. 2.11 $V = q(\psi_p)\theta - \phi + \lambda(\psi_p, \theta, \zeta)$. Requiring $2\pi\psi_p$ to be the poloidal flux, and $2\pi\psi$ to be the toroidal flux, still makes λ periodic in θ and ϕ and we still have $d\psi/d\psi_p = q$, but the local helicity is no longer given by q but by $\vec{B} \cdot \nabla\phi / \vec{B} \cdot \nabla\theta = (q + \lambda'_\theta)/(1 - \lambda'_\phi)$. The vector potential is

$$\vec{A} = \psi\nabla\theta + \psi_p\nabla\phi - \lambda\nabla\psi_p, \qquad (2.41)$$

and the covariant representation is

$$\vec{B} = g\nabla\phi + I\nabla\theta + \delta\nabla\psi. \qquad (2.42)$$

Simplification results from the fact that $\nabla\phi \cdot \nabla\psi = \nabla\phi \cdot \nabla\theta = 0$.

2.4.4 *Zakharov coordinates*

Zakharov coordinates are not straight field line coordinates, but they put the magnetic field in a particularly simple form. Write $\vec{B} = \bar{g}\nabla\zeta + \bar{I}\nabla\theta + \nabla\sigma - \alpha\nabla\psi$, and substitute $\zeta = \zeta' + w$, $\theta = \theta' + v$. Setting $\nabla\sigma - \alpha\nabla\psi + \bar{g}\nabla w + \bar{I}\nabla v = 0$, a solution is given by

$$w = \frac{\bar{I}(f' - \alpha) - \bar{I}'(f - \sigma)}{\bar{I}\bar{g}' - \bar{g}\bar{I}'}, \qquad v = \frac{-\bar{g}(f' - \alpha) + \bar{g}'(f - \sigma)}{\bar{I}\bar{g}' - \bar{g}\bar{I}'}, \qquad (2.43)$$

where $f = f(\psi)$ is arbitrary, leaving the simple form

$$\vec{B} = \bar{g}(\psi)\nabla\zeta' + \bar{I}(\psi)\nabla\theta'. \qquad (2.44)$$

Notice that the denominator in Eq. 2.43 is proportional to the force-free part of the parallel current, Eq. 2.33.

2.5 STELLARATORS

A quasi-symmetric configuration (Nuhrenberg and Zille, 1988) is one in which which the field magnitude is independent of ζ, *i.e.* $\partial_\zeta B = 0$. However, it has been shown by Garren and Boozer (1991) that a general equilibrium cannot be quasi-symmetric except locally on a single flux surface, and although quasi-symmetry can be approximate, it necessarily fails by terms of third order in the aspect ratio on other surfaces.

Equilibria are found numerically using the Variational Moments Equilibrium Code (VMEC) (Hirschman, 1986) by fixing the bounding surface for ψ (the edge of the plasma) and its value on this surface, and prescribing the functions $q(\psi)$, $p(\psi)$.

Consider the cylindrical coordinate system X, Z, ϕ, right handed in that order, with Z the vertical axis and ϕ the toroidal angle. VMEC uses the magnetic coordinates ψ, θ, ϕ, which are not straight field line coordinates. The metric is given by

$$ds^2 = (dX)^2 + (dZ)^2 + X^2 d\phi^2. \tag{2.45}$$

Writing $dX = \partial_\psi X d\psi + \partial_\theta X d\theta$, $dZ = \partial_\psi Z d\psi + \partial_\theta Z d\theta$, we have

$$ds^2 = g_{\alpha\beta} d\alpha d\beta \tag{2.46}$$

with

$$g_{\alpha\beta} = \partial_\alpha X \partial_\beta X + \partial_\alpha Z \partial_\beta Z + \delta_{\alpha\phi} \delta_{\beta\phi} X^2 \tag{2.47}$$

and $\mathcal{J} = \sqrt{det[g_{\alpha\beta}]}$. The minimization is then carried out using Fourier expansions

$$X = \sum X_{mn}(\psi) cos(m\theta - n\phi), \tag{2.48}$$

$$Z = \sum Z_{mn}(\psi) sin(m\theta - n\phi), \tag{2.49}$$

$$\lambda = \sum \lambda_{mn}(\psi) sin(m\theta - n\phi), \tag{2.50}$$

where the equilibrium is assumed to have symmetry about a horizontal axis in at least the two toroidal planes $\phi = 0, \pi$. This is called standard stellarator symmetry, is not a significant restriction, and can easily be relaxed

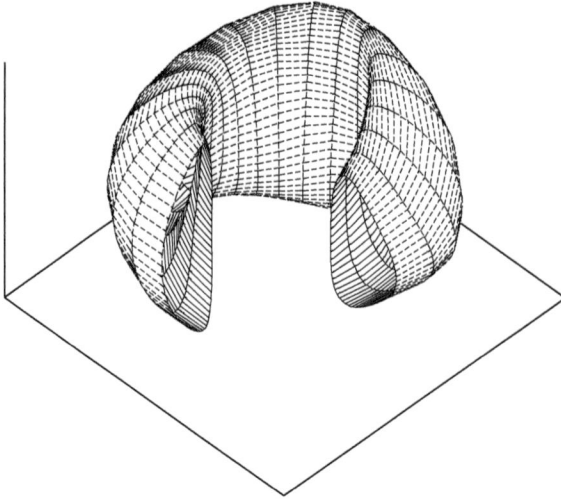

Fig. 2.1   A stellarator equilibrium with weak external windings.

if desired. Now we find

$$B^\phi = \vec{B} \cdot \nabla\phi = \frac{1 + \partial_\theta \lambda}{\mathcal{J}_p},$$
(2.51)

$$B^\theta = \vec{B} \cdot \nabla\theta = \frac{q - \partial_\zeta \lambda}{\mathcal{J}_p},$$
(2.52)

where $\mathcal{J}_p = \mathcal{J}/q$ and

$$B^2 = (B^\theta)^2 g_{\theta\theta} + 2B^\theta B^\zeta g_{\zeta\theta} + (B^\zeta)^2 g_{\zeta\zeta}.$$
(2.53)

The functions X, Z, λ are then iterated upon and adjusted to minimize the energy $\int (B^2/2 + p)d^3x$, subject to the constraint that the plasma edge is fixed in space, *i.e.* for $\psi = \psi_E$ the surface described by $X(\psi_E, \theta, \phi)$, $Z(\psi_E, \theta, \phi)$ is fixed. Minimization of the energy with respect to variation of the flux surface ψ guarantees that the obtained configuration satisfies $\vec{j} \times \vec{B} = \nabla p$. If the pressure has a nonzero gradient at a rational surface $q = m/n$ then ψ will have discontinuous derivatives, related to current sheets developing to guarantee that no islands occur at this surface.

In Fig. 2.1 is shown a stellarator equilibrium obtained using VMEC. This is a stellarator that is similar to the Aries tokamak (Najmabadi, 1997)

W7X

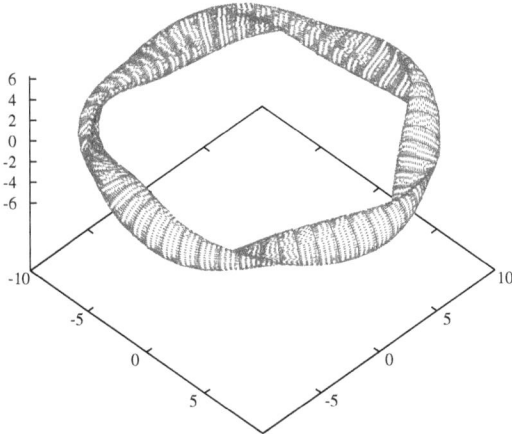

Fig. 2.2   The W7X stellarator equilibrium.

with about 20 percent of the transform coming from external windings. Most of the field helicity comes from the plasma current.

In Fig. 2.2 is shown the W7X stellarator equilibrium operating in Germany (Beidler, 1990). See also section 5.3. It is characterized by being nearly quasi-symmetric, the condition that the magnitude of the field B possess a ignorable coordinate, and thus depends only on ψ, θ, although the configuration itself, as is evident from the figure, does not possess a symmetry.

One important omission of representations assuming nested flux surfaces is the impossibility of representing divertors. In order to direct the flow of escaping particles from a toroidal device the magnetic field is often configured to possess a separatrix, with one or two X-points bounding the plasma. Since the representation of the magnetic field used assumes nested magnetic surfaces the treatment of the domain outside the separatrix is excluded from our analysis. The numerical extension of equilibrium codes and guiding center codes to handle fields not possessing nested surface topology is non trivial. One such equilibrium code is PIES (Reiman *et al.*, 1999).

2.6 AXISYMMETRIC EQUILIBRIA

Now assume axisymmetry, all scalar functions independent of ζ. Also use straight field line variables, equivalent to setting λ equal to zero. In these variables, $d\zeta/d\theta = \vec{B} \cdot \nabla\zeta/\vec{B} \cdot \nabla\theta = q(\psi)$ along a field line, and the contravariant representation for $\vec{B}$, Eq. 2.14, is somewhat simplified

$$\vec{B} = \nabla\psi \times \nabla\theta + \nabla\zeta \times \nabla\psi_p = \frac{[\vec{e}_\zeta + \vec{e}_\theta/q]}{\mathcal{J}}. \tag{2.54}$$

It is also useful to note the vector potential associated with $\vec{B}$, with $\vec{B} = \nabla \times \vec{A}$,

$$\vec{A} = \psi\nabla\theta + \psi_p\nabla\zeta. \tag{2.55}$$

Using Eq. 2.25 we find that axisymmetry implies that g is also independent of θ,

$$g = g(\psi_p). \tag{2.56}$$

The function σ must be independent of ζ, so $I = I(\psi, \theta)$ in general, but particular coordinates can be chosen to make I independent of θ, equilalent to setting σ to zero. Taking α independent of ζ, Eqs. 2.20, 2.22, and 2.23 show that σ and α can be absorbed in the generalization of $I(\psi)$ to $I(\psi, \theta)$ so they will be ignored.

Consider the cylindrical coordinate system X, Z, ϕ, right-handed in that order, with Z the vertical axis and ϕ the toroidal angle, the equilibrium symmetric in the variable ϕ. Toroidal symmetry implies that $\nabla\psi$ and $\nabla\theta$ are orthogonal to $\nabla\phi = \hat{\phi}/X$.

The ignorable coordinate ζ is related to ϕ through $\zeta = \phi - \nu(\psi, \theta)$, and the straight field line condition $q = \vec{B} \cdot \nabla\zeta/\vec{B} \cdot \nabla\theta$ gives, using $\vec{B} \cdot \nabla\zeta = \vec{B} \cdot \nabla\phi - \partial_\theta\nu\vec{B} \cdot \nabla\theta$, and $\vec{B} \cdot \nabla\phi = g/X^2$, and $\vec{B} \cdot \nabla\theta = 1/(\mathcal{J}q)$,

$$\frac{\partial\nu}{\partial\theta} = \frac{gq\mathcal{J}}{X^2} - q. \tag{2.57}$$

The solution $\nu(\psi, \theta)$ involves an arbitrary function of ψ, reflecting the fact that the field lines are constrained to surfaces of $\psi = $ constant. Note that the Jacobian in the ϕ, ψ, θ system is equal to that in the ζ, ψ, θ system, since $\mathcal{J}^{-1} = \nabla\zeta \cdot (\nabla\psi \times \nabla\theta) = \nabla\phi \cdot (\nabla\psi \times \nabla\theta)$.

Consider the force balance equation in the ϕ, ψ, θ system, giving

$$\mathcal{J}(j^\theta B^\phi - j^\phi B^\theta) = p'. \tag{2.58}$$

Using Eq. 2.19 we find the contravariant component

$$B^\phi = \vec{B} \cdot \nabla\phi = \frac{g}{X^2} \tag{2.59}$$

and from Eq. 2.54

$$B^\theta = \vec{B} \cdot \nabla\theta = \frac{1}{q\mathcal{J}}. \tag{2.60}$$

Again using the contravariant representation, Eq. 2.54 gives

$$\vec{j} \cdot \nabla\phi = \nabla \cdot (\vec{B} \times \nabla\phi) = \nabla \cdot [(\nabla\zeta \times \nabla\psi_p) \times \nabla\phi], \tag{2.61}$$

and using $(\vec{A} \times \vec{B}) \times \vec{C} = (\vec{A} \cdot \vec{C})\vec{B} - (\vec{C} \cdot \vec{B})\vec{A}$,

$$\vec{j} \cdot \nabla\phi = \nabla \cdot [(\nabla\phi \cdot \nabla\zeta)\nabla\psi_p] \tag{2.62}$$

which gives simply

$$\vec{j} \cdot \nabla\phi = \nabla \cdot \frac{\nabla\psi_p}{X^2}, \tag{2.63}$$

and using the covariant representation, Eq. 2.19 we find

$$\vec{j} \cdot \nabla\theta = \nabla \cdot (\vec{B} \times \nabla\theta) = \nabla g \cdot (\nabla\zeta \times \nabla\theta) = -\frac{g'}{\mathcal{J}}. \tag{2.64}$$

Force balance becomes, upon substituting these expressions into Eq. 2.58 and changing to the variable ψ_p,

$$\nabla \cdot \frac{\nabla\psi_\mathbf{p}}{X^2} + \mathbf{p'} + \frac{\mathbf{gg'}}{X^2} = 0 \tag{2.65}$$

where now primes indicate derivatives with respect to ψ_p. This is the Grad–Shafranov equation in magnetic coordinates. Note that it is an invariant expression, it can be evaluated using any coordinate system. Using, for example, the variables X, Z the first term in Eq. 2.65 takes the form $(1/X)d/dX(1/X)d\psi_p/dX + (1/X^2)d^2\psi_p/dZ^2$. Freedom in the form of the equilibrium consists in choosing two of the three functions $g(\psi)$, $p(\psi)$, and $q(\psi)$, related through Eq. 2.24 and the boundary conditions.

The Grad–Shafranov equation requires elliptic boundary conditions to fully determine the solution. A common procedure is to give ψ on some bounding surface. This is equivalent to prescribing the magnetic field normal to this surface as well as determining the arbitrary constant in the definition of ψ, if it is not taken to be zero at the magnetic axis. Different equilibria then result by changing the surface value of ψ and the shape of the bounding surface.

The determination of ψ on this surface is equivalent to the determination of an external field, fixed by some external, rigid, conductors. An alternate means of specifying the boundary conditions is to determine the boundary value of ψ through these external circuits.

Note that through the functions $p(\psi_p)$ and $g(\psi_p)$ this equation is nonlinear, and thus quite nonstandard. However, as we will see in the next sections, for large aspect ratio and low plasma pressure the nonlinearities are small corrections.

2.7 TOKAMAK ORDERING

The tokamak ordering is defined through the small parameter $\epsilon = a/R$, the inverse aspect ratio. The local helicity q is taken to be of order unity and thus B_θ/B_ϕ is of order ϵ. The plasma beta, $\beta = 2p/B^2$, is taken to be of order ϵ^2 (low beta) or of order ϵ (high beta). As a small parameter, the inverse aspect ratio ϵ is not an impressive quantity in which to make an asymptotic expansion of a theory, typically having a value of $1/4$. Nevertheless, the development of MHD in this parameter has proved extremely useful for the understanding of toroidal confinement devices. For one thing, qualitatively different physical phenomena enter at each order in the expansion, and only on rare occasions have higher order terms reversed the qualitative understanding gained by lower order analysis. In addition, the ϵ expansion made possible the numerical analysis of the time evolution of large scale modes which would not otherwise have been feasible. The very rapid magnetosonic waves appear only at higher order, and truncation of the expansion allows numerical computations to proceed in much larger time steps. The equations of resistive MHD have been developed to 5th order in ϵ, and it has been shown that this development is maximal, *i.e.* there are no qualitatively different physical effects to be gained by going any further (Izzo *et al.*, 1985). Present codes such as MH3D do not expand in aspect ratio, see, for example, Park and Monticello (1990).

The plasma pressure does not play an important role in equilibria and instabilities if β is of order ϵ^2, but it does for β of order ϵ. Thus the two cases are considered separately, beginning with the low β phenomena.

2.8 THE SHAFRANOV EQUILIBRIUM

The Grad–Shafranov equation for ψ is

$$\nabla \cdot \frac{\nabla \psi}{qX^2} + q\frac{dp}{d\psi} + \frac{gqdg/d\psi}{X^2} = 0 \qquad (2.66)$$

with $d\psi/d\psi_p = q(\psi_p)$ and $2\pi\psi$ is the toroidal flux. Let us consider the ordering of terms in the Grad–Shafranov equation in a low β equilibrium, *i.e.* with $\beta \sim \epsilon^2$ with $\epsilon = a/R$ and a the minor radius, R the major radius of the torus. The toroidal flux ψ is $\sim B_\phi a^2 \sim \epsilon^2$. Thus $\nabla(1/X^2)g\nabla\psi \sim O(1)$, since ψ goes from 0 to its maximum in $\Delta X = a$, and $qdp/d\psi \sim O(1)$, since p goes from its maximum to zero in $\Delta\psi = \epsilon^2$. Thus $g = 1 + g_2(\psi)$ and $g_2(\psi) \simeq O(\epsilon^2)$.

Now attempt solution of the Grad–Shafranov equation order by order in ϵ with the boundary condition given by a circular conducting wall. To lowest order the equilibrium constant ψ surfaces are concentric circles. To second order the equilibrium surfaces consist of shifted circles (Shafranov, 1966), and to third order they have elliptical and triangular distortion (Green *et al.*, 1971). We will solve the equilibrium only to second order in ϵ. We assume that the flux surfaces consist of shifted circles and verify that a solution of the Grad–Shafranov equation is obtained.

Take the major radius as the unit of distance, and normalize the magnetic field to unity on the major axis. Thus write

$$X = 1 + r\cos\theta - \Delta(r), \qquad (2.67)$$

$$Z = r\sin\theta \qquad (2.68)$$

with $\Delta(0) = 0$, *i.e.* $X = 1$ is the magnetic axis location with magnetic surfaces as shown in Fig. 2.3. We choose this description for the convenience of defining $X = 1$ to be the magnetic axis location, with the location of the outermost flux surface given by Δ. Physically, note that if the outermost flux surface is fixed by a conducting metal wall it is the magnetic axis which shifts its position in space according to the value of Δ, which as we will see

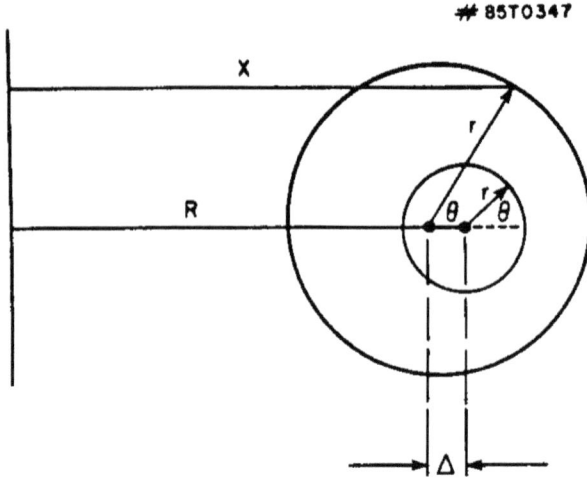

Fig. 2.3   Flux surfaces for a second order equilibrium.

depends on the plasma pressure. Since $\vec{B} \cdot \hat{\phi} = 1 + g_2$ the toroidal flux is $\psi = (r^2/2)[1 + O(\epsilon^2)]$, the order ϵ^2 corrections contribute only to ellipticity and triangularity, and thus $\nabla\psi = r\nabla r$ within our approximation. We must evaluate $\nabla \cdot [\nabla\psi/(X^2 q)]$, and will do the evaluation in the r, θ, ϕ system using Eq. 1.29.

The Jacobian follows from the partial derivatives

$$\frac{\partial X}{\partial r} = cos\theta - \Delta', \qquad \frac{\partial X}{\partial \theta} = -rsin\theta, \tag{2.69}$$

$$\frac{\partial Z}{\partial r} = sin\theta \qquad \frac{\partial Z}{\partial \theta} = rcos\theta \tag{2.70}$$

where $\Delta' = d\Delta/dr \sim \epsilon$. Thus

$$\begin{pmatrix} \frac{\partial r}{\partial X} & \frac{\partial r}{\partial Z} \\ \frac{\partial \theta}{\partial X} & \frac{\partial \theta}{\partial Z} \end{pmatrix} = \frac{1}{r(1 - \Delta' cos\theta)} \begin{pmatrix} rcos\theta & rsin\theta \\ -sin\theta & cos\theta - \Delta' \end{pmatrix}. \tag{2.71}$$

From this expression we can calculate the necessary expressions to substitute into the Grad–Shafranov equation.

$$\nabla r = \frac{cos\theta}{1 - \Delta' cos\theta} \hat{X} + \frac{sin\theta}{1 - \Delta' cos\theta} \hat{Z} \tag{2.72}$$

$$\nabla \theta = -\frac{sin\theta}{r(1 - \Delta' cos\theta)} \hat{X} + \frac{cos\theta - \Delta'}{r(1 - \Delta' cos\theta)} \hat{Z} \tag{2.73}$$

$$\mathcal{J}^{-1} = \nabla\phi \cdot (\nabla r \times \nabla\theta) = \frac{1}{rX(1 - \Delta' cos\theta)} \tag{2.74}$$

$$\frac{1}{\mathcal{J}} \frac{\partial}{\partial\theta} \left(\frac{\mathcal{J}\nabla\theta \cdot \nabla\psi}{X^2 q} \right) = -\frac{\Delta' cos\theta}{q} + O(\epsilon^2) \tag{2.75}$$

$$\frac{1}{\mathcal{J}} \frac{\partial}{\partial r} \left(\frac{\mathcal{J}\nabla r \cdot \nabla\psi}{X^2 q} \right) = \frac{1}{rX(1 - \Delta' cos\theta)} \frac{\partial}{\partial r} \left[\frac{r^2}{q} (1 + \Delta' cos\theta - rcos\theta) \right] \tag{2.76}$$

$$\nabla \cdot \left(\frac{1}{X^2 q} \nabla\psi \right) = -\frac{\Delta' cos\theta}{q} + \frac{1}{r} \left(\frac{r^2}{q} \right)' (1 + 2\Delta' cos\theta - 2rcos\theta)$$
$$+ \frac{r}{q} (\Delta'' cos\theta - cos\theta). \tag{2.77}$$

Substituting into the Grad–Shafranov equation and examining those terms which are independent of θ, we find

$$p' + gg' + \frac{f}{r}(rf)' = 0 \tag{2.78}$$

where $f = r/q$, and primes denote differentiation by r. This equation states that of the three functions p, q, g, only two can be chosen freely, defining the equilibrium.

The terms in the Grad–Shafranov equation which are independent of θ reflect radial pressure balance, keeping the plasma isolated from the surrounding walls, and have to do only with cylindrical geometry. Note that solutions exist both in the limit $q \to \infty$, which means toroidal field only, or a θ pinch configuration, with pressure balance given by $p' + gg' = 0$, which is $(d/dr)(p + B_\phi^2/2) = 0$, and in the limit $q \to 0$, which means poloidal field only, or a Z pinch configuration. In this case pressure balance is given by $p' + (f/r)(rf)' = 0$ or $(d/dr)(p + B_\theta^2/2) + B_\theta^2/r = 0$.

Terms proportional to $cos\theta$ reflect toroidal pressure balance, and we will find that in toroidal geometry the θ-pinch equilibrium does not exist. Terms in $cos\theta$ give

$$\Delta'' + \left[\frac{2(rf)'}{rf} - \frac{1}{r}\right]\Delta' - \frac{2(rf)'}{f} - 1 - \frac{2rgg'}{f^2} = 0 \qquad (2.79)$$

where again primes indicate differentiation with respect to r. This equation determines Δ, and thus the physical location of the flux surfaces in terms of the two functions defining the equilibrium. Eliminate gg' in this equation and multiply by rf^2, then note

$$(rf^2\Delta')' = \left[\frac{(rf)^2\Delta'}{r}\right]' = rf^2\left[\Delta'' + (\frac{2(rf)'}{rf} - \frac{1}{r})\Delta'\right]. \qquad (2.80)$$

Thus

$$\Delta' = \frac{1}{rf^2}\int_0^r (f^2 - 2rp')rdr. \qquad (2.81)$$

Note that normally $p' < 0$ so $\Delta' > 0$. The constant of integration is zero, otherwise Δ' is unbounded as $r \to 0$.

These terms have a simple interpretation. Integrating the pressure term by parts we find

$$\Delta' = \left(\frac{l}{2} + \beta_\theta\right)r \qquad (2.82)$$

where $l = [2/(r^2B_\theta^2)]\int_0^r B_\theta^2 rdr = 2W_{mag}/I^2$ is the internal inductance per unit length, I the current, and $\beta_\theta = (2/B_\theta^2)(\bar{p} - p)$, with $\bar{p} = (2/r^2)\int_0^r prdr$ the average pressure. Note that at the plasma edge, where $p = 0$, β_θ is the poloidal beta, β_p. Thus the internal inductance and the plasma beta both increase the rate of change of the outward shift with respect to the minor radius.

In the same manner, the expression for Δ'' can be simplified. Differentiating Δ' we find

$$\Delta'' = 1 - (3 - 2s)(\beta_\theta + l/2) - \frac{2p'q^2}{r} \qquad (2.83)$$

where $s = rq'/q$ is the magnetic shear.

From Eq. 2.81 we find an expression for the Shafranov shift

$$\Delta = \int_0^r \frac{dr}{rf^2} \int_0^r (f^2 - 2rp')r dr \qquad (2.84)$$

and again the constant of integration is zero since we have taken $\Delta(0) = 0$. The shift is positive, and results in a compression of magnetic flux on the outside edge of the torus, increasing the magnitude of the poloidal magnetic field to oppose the outward thrust of the plasma pressure.

Note that there is a shift even in the limit $p = 0$. This is due to toroidal bending and the effect is known as the hoop force, caused by the compression of the poloidal field due to toroidal bending. For nonzero pressure $\Delta \to \infty$ if $f = (r/q) \to 0$. The condition $q \to \infty$ means no poloidal field, *i.e.* in a tokamak no plasma current, or a θ-pinch configuration, and there is no equilibrium in this case. Thus a toroidal device must either possess helical windings (stellarator) or have a toroidal current (tokamak) to produce the poloidal field to be in equilibrium.

Now find all functions in the representations of $\vec{B}$. The Jacobian is given in Eq. 2.74. Then $\zeta = \phi - \nu$ with $\partial_\theta \nu = (g\mathcal{J}/X^2) - q$, giving the expression

$$\partial_\theta \nu = \frac{gq}{X}(1 - \Delta'cos\theta) - q \qquad (2.85)$$

and the expression for the magnetic field simplifes to

$$\vec{B} = \nabla\phi \times \nabla\psi_p + (q + \partial_\theta\nu)\nabla\psi_p \times \nabla\theta, \qquad (2.86)$$

and the first term is the poloidal component and the second term the toroidal component. Further $\nabla\phi = \hat{\phi}/X$ is orthogonal to $\nabla\psi_p$ so the poloidal field is $B_p = r/[Xq(1 - \Delta'cos\theta)]$, or approximately

$$B_p = \frac{r}{q(1 + rcos\theta - \Delta'cos\theta)}. \qquad (2.87)$$

The $rcos\theta$ term is due to toroidal bending and weakens B_p on the outside, and the $\Delta'cos\theta$ is due to the outward shift and strengthens B_p on the outside. The outward shift thus produces a vertical field, by strengthening B_p outside and weakening it inside due to the compression of flux surfaces caused by the shift. In the absence of a conducting shell this vertical field must be supplied by external coils. Alternatively, external coils can supply increasing vertical field as the plasma pressure is increased, so that the

Shafranov shift is avoided and the plasma center remains in its original position. From the expression for B_p we find the vertical field magnitude

$$B_v \simeq \frac{r}{q}\Delta' = I\left(\frac{l}{2} + \beta_\theta\right). \tag{2.88}$$

Note from the expression for B_p that if $\Delta' > 1$, or $\beta_\theta > R/a$, the vertical field exceeds the poloidal field, and we might expect a separatrix to invade the plasma at the inside edge of the torus (Yoshikawa, 1977). For some time this was believed to set an equilibrium limit to β, of $\beta \sim a/(Rq^2)$. As we will see this does not set a real limit provided the plasma is heated on a time scale faster than the resistive skin time, so that the separatrix cannot invade the plasma.

From the cross product

$$\nabla r \times \nabla\theta = \frac{1}{r(1 - \Delta'cos\theta)}\hat{\phi}, \tag{2.89}$$

we calculate the magnitude of the toroidal field $B_\phi = (q + \partial_\theta\nu)/(q(1 - \Delta'cos\theta))$, or

$$B_\phi = \frac{(1 - rcos\theta - \Delta'cos\theta)}{1 - \Delta'cos\theta} \simeq 1 - rcos\theta \simeq \frac{1}{X}. \tag{2.90}$$

Thus $B_\phi = 1/X + O(\epsilon^2)$. Thus to first order in ϵ, $\vec{B} = \nabla\phi + \nabla\phi \times \nabla\psi_p$, Δ is negligible, and $g = 1$, $\delta = 0$, and $I = r^2/q$, with the field magnitude $B = 1 - rcos\theta$. This is the cylindrical equilibrium.

Keeping the next order in the expression for $\vec{B}$ we find

$$\vec{B} = (1 + g_2)\nabla\phi + \frac{r^2\nabla\theta}{qX(1 - \Delta'cos\theta)} + \frac{\Delta'sin\theta\nabla\psi_p}{X(1 - \Delta'cos\theta)}, \tag{2.91}$$

and g_2 is the solution to $p' + g' + (1/q)(r^2/q)' = 0$ with $g_2 = 0$ at the tokamak wall, *i.e.*

$$g_2 = -p(r) - \int_a^r \frac{1}{q}\left(\frac{r^2}{q}\right)' dr. \tag{2.92}$$

It is useful to evaluate these expression for a model equilibrium, with a parabolic pressure profile, $p = p_o(1 - r^2/a^2)$, and a constant current profile, $q = constant$. Substituting, we find

$$\bar{p} = \frac{2}{r^2}\int_0^r prdr = p_0\left(1 - \frac{r^2}{2a^2}\right), \tag{2.93}$$

$$\beta_\theta = \frac{q^2 p_0}{a^2}, \tag{2.94}$$

$$g_2 = \frac{a^2}{qR^2}\left(1 - \frac{r^2}{a^2}\right)(1 - \beta_\theta). \tag{2.95}$$

Thus there are two modifications to the strength of the toroidal field caused by the plasma. One is due to the plasma current and appears also for $\beta_\theta = 0$. It is caused by the fact that the current follows field lines, and thus for $q \neq \infty$ there is a poloidal current, which acts to increase the toroidal field. The decrease of the toroidal field due to β_θ is caused by the diamagnetic current, which will be derived using a kinetic theory analysis. For $\beta_\theta > 1$ the diamagnetic effects dominate, g_2 becomes negative, and the plasma decreases the toroidal field strength below the large aspect ratio limit value.

The Shafranov shift in this case is given by $\Delta(r)$ with

$$\Delta' = \left(\frac{1}{4} + \beta_\theta\right)r. \tag{2.96}$$

2.9 CYLINDRICAL TOKAMAK EQUILIBRIA

For very large aspect ratio, with a circular conducting wall, and at low β, toroidal effects are negligible and the equilibrium is approximately cylindrical. All expressions can be obtained from the Shafranov equilibrium neglecting second order terms. The Jacobian becomes $\mathcal{J} = r$. The magnetic field, normalized to its value on axis, has the covariant form

$$\vec{B} = \nabla\phi + \frac{r^2}{q(r)}\nabla\theta \tag{2.97}$$

with the major radius $R = 1$, and $r \ll 1$. $\nabla\phi = \hat{\phi}$, $\nabla\theta = \hat{\theta}/r$. Terms neglected in $\vec{B}$ are of order r^2. The contravariant form is

$$\vec{B} = \frac{r}{q(r)}\nabla\phi \times \nabla r + r\nabla r \times \nabla\theta = \frac{\vec{e}_\theta}{q(r)} + \vec{e}_\phi. \tag{2.98}$$

The toroidal flux function is

$$\psi(r) = \int \vec{B} \cdot \nabla\phi r dr = \frac{r^2}{2} \tag{2.99}$$

and the poloidal flux function is

$$\psi_p(r) = \int \vec{B} \cdot \nabla\theta r dr = \int \frac{rdr}{q}. \tag{2.100}$$

A general equilibrium is given by two arbitrary functions, which can be chosen to be the toroidal field profile and the q profile. In the large aspect ratio limit the toroidal field is constant so the choice of $q(r)$ determines the equilibrium. In cylindrical geometry the value of q at the plasma edge is simply related to the total current through $q = 2\pi a^2 B_\phi / RI$. (With B in Gauss, a and R in centimeters, and I in amps, this becomes $q = 5a^2 B_\phi / RI$.)

The equilibrium current $\nabla \times \vec{B}$ is toroidal, given by

$$\vec{j} = \frac{1}{r}\frac{d}{dr}\left(\frac{r^2}{q}\right)\nabla\phi. \tag{2.101}$$

Note that in large aspect ratio the pressure is of order ϵ^2, $p \sim r^2/R^2$. A convenient class of equilibria are given by

$$j_\phi = \frac{j_0}{[1 + (r/r_0)^{2\nu}]^{1+1/\nu}}, \tag{2.102}$$

$$q = q_0[1 + (r/r_0)^{2\nu}]^{1/\nu}. \tag{2.103}$$

Note from Eq. 2.101 that $j_0 = 2/q_0$. The cases $\nu = 1, 2, 4$ are referred to as peaked, rounded, and flat profiles, respectively. For large r the vanishing of the current density implies that $q \sim r^2$.

Any field can be written

$$\vec{B} = \nabla\psi \times \nabla\theta + \nabla\phi \times \nabla\psi_p, \tag{2.104}$$

which is also

$$\vec{B} = \nabla\phi + \nabla\phi \times \nabla_\perp\psi_p \tag{2.105}$$

where $\nabla_\perp = \nabla r \partial_r + \nabla\theta\partial_\theta$. As shown previously, if ψ_p is a function of r alone it describes nested magnetic surfaces. Perturbations about the equilibrium by taking $\psi_p = \psi_p(r, \theta, \phi)$ produce distortions of the flux surfaces and magnetic islands at resonant values of q. From Eq. 2.105 we find in general for this representation

$$\vec{j} = \nabla_\perp^2 \psi_p \nabla\phi - \partial_\phi \nabla_\perp\psi_p. \tag{2.106}$$

Because the equilibrium is symmetric in both θ and ϕ it is useful to express field perturbations in Fourier harmonics of the form

$$f(r)e^{i(m\theta - n\phi)}. \tag{2.107}$$

The wave vector is thus $\vec{k} = \nabla(m\theta - n\phi)$. The helicity, m/n, is preserved in nonlinear products of terms with the same helicity, and is thus useful also in nonlinear analysis. Introduce the helicity vector

$$\vec{h} = \nabla\phi + \frac{nr^2}{m}\nabla\theta \tag{2.108}$$

and the helicity variable $\tau = \theta - (n/m)\phi$ with $\vec{h} \cdot \nabla\tau = \vec{h} \cdot \vec{k} = 0$. It is convenient to introduce the helical flux

$$\psi_h = \int rdr\vec{B} \cdot \nabla\tau = \psi_p - \left(\frac{n}{m}\right)\frac{r^2}{2}. \tag{2.109}$$

The magnetic field is given by

$$\vec{B} = \vec{h} + \nabla\phi \times \nabla\psi_h \tag{2.110}$$

and ψ_h is a magnetic surface function, $\vec{B} \cdot \nabla\psi_h = 0$ provided that $\psi_h = \psi_h(r, \tau)$. From Eqs. 2.106, 2.109 we find

$$\vec{j} = \left(\nabla_\perp^2\psi_h + \frac{2n}{m}\right)\nabla\phi - \partial_\phi\nabla_\perp\psi_h. \tag{2.111}$$

Write the helical flux in terms of an equilibrium and a single harmonic

$$\psi_h = \psi_0(r) + \psi_{m,n}(r)sin(m\theta - n\phi). \tag{2.112}$$

Note from Eq. 2.109 that $\psi_0' = r(1/q - n/m) = (r/m)\vec{k} \cdot \vec{B}_0$.

We find a magnetic island at $r = r_s$ with $q(r_s) = m/n$ with width

$$\Delta r = 4\left(\frac{-\psi_{m,n}}{\psi_0''}\right)^{1/2} \tag{2.113}$$

with all quantities evaluated at $r = r_s$. Stability of an equilibrium to ideal perturbations, which are restricted to have $\psi_{m,n}(r_s) = 0$, so that flux surface perturbation occurs but there is minimal island formation at the rational surface, will be studied in Chap. 4, and stability to resistive, large island producing perturbations in Chap. 5.

2.10 HIGH BETA EQUILIBRIUM

For low plasma pressure, and in particular for $\beta \ll 1$, the equilibrium field is essentially determined by the geometry and the plasma current. As the plasma pressure is increased, it begins to play a role in the shaping of the equilibrium. Because of the nonzero ion gyro radius, a toroidal shell of ions produces two equal poloidal currents, negative just outside and positive just inside the shell. In the presence of a density gradient the contributions from neighboring shells do not cancel and there is a net local poloidal diamagnetic current, acting to decrease B_ϕ. The plasma pressure begins to dig a magnetic well in the magnetic field and the effect becomes important when the poloidal beta exceeds unity. This effect can be seen by using the lowest order expressions for the field and substituting them into the equilibrium equation, which includes the plasma pressure.

Substituting the expressions for $f = r/q = B_\theta/B_\phi$ and $q = rB_\phi/(RB_\theta)$, the Grad–Shafranov equation becomes

$$p' + \left(\frac{B_\phi^2}{2}\right)' + \left(\frac{B_\theta^2}{2}\right)' + \frac{B_\theta^2}{r} = 0 \tag{2.114}$$

neglecting terms of order ϵ^4. Multiply by r, integrate over volume, giving

$$\int r^2 \frac{d}{dr}\left(p + \frac{B_\phi^2}{2} + \frac{B_\theta^2}{2}\right)dr + \int drrB_\theta^2 = 0, \tag{2.115}$$

or upon integrating by parts

$$\frac{a^2}{2}(B_\phi^2 + B_\theta^2)|_a - \int\left(p + \frac{B_\phi^2}{2}\right)2rdr = 0. \tag{2.116}$$

Dividing by $B_\theta^2(a)$ we find

$$\frac{a^2}{2} + \frac{a^2 B_\phi^2(a)}{2B_\theta^2(a)} - \frac{a^2 \beta_p}{2} - \frac{1}{B_\theta^2(a)}\int_0^a B_\phi^2 rdr = 0. \tag{2.117}$$

Now use $B_\phi^2(a) - B_\phi^2(r) \simeq 2B_\phi(a)(B_\phi(a) - B_\phi(r))$ to find

$$\beta_p = 1 + \frac{4B_\phi}{B_\theta^2 a^2}\int[B_\phi(a) - B_\phi(r)]rdr. \tag{2.118}$$

The change in toroidal flux due to the plasma pressure is given by $\delta\Phi_\phi = \pi\int r[B_\phi - B_\phi(a)]dr$ and thus $\beta_p = 1 - B_\phi 2\delta\Phi_\phi/(B_\theta^2 a^2\pi)$. Again

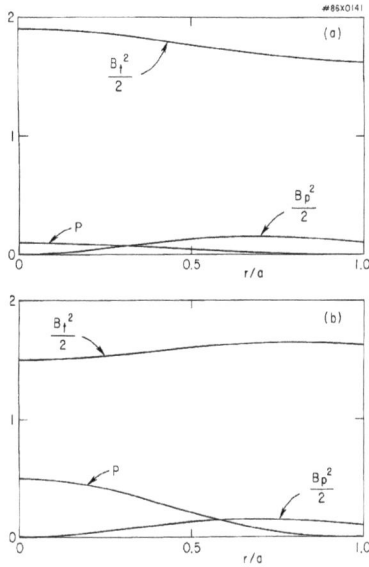

Fig. 2.4   Magnetic energy and pressure in (a) low β and (b) high β tokamak equilibria.

using $I = 2\pi a B_\theta$ we find

$$\beta_p = 1 - \frac{8\pi B_\phi}{I^2}\delta\Phi_\phi. \qquad (2.119)$$

Thus the result found in Sec. 2.8 for the particular equilibrium $q = constant$ and a parabolic pressure profile is more general; for $\beta_p < 1$ the plasma is paramagnetic, the toroidal flux is increased by the plasma pressure, and for $\beta_p > 1$ the plasma is diamagnetic, the toroidal flux is decreased by the plasma pressure. In Fig. 2.4 are shown typical toroidal and poloidal magnetic energy density and plasma pressure for a) $\beta_p < 1$ and b) $\beta_p > 1$.

A useful class of analytic high β equilibria has been obtained by Greene (1988) and Kleibergen and Goedbloed (1988). They consider large aspect ratio, with a circular conducting wall, and retain only leading order terms in the Grad–Shafranov equation. Let $X = R_0 + r\cos\theta$ where R_0 is the geometric center of the plasma. Terms of order $1/r$ in Eq. 2.65 relate the diamagnetic current and the pressure. Incorporating them we find

$$\nabla^2 \psi_p + 2R_0 r p'(\psi_p)\cos\theta + g_0 g_2' = 0 \qquad (2.120)$$

with

$$g(\psi) = g_0 \left[1 - \frac{R_0^2 p}{g_0^2} - \frac{1}{2}\left(\frac{R_0^2 p}{g_0^2}\right)^2 + \frac{g_2}{g_0} + ...\right], \qquad (2.121)$$

the terms included in $g(\psi_p)$ accounting for the diamagnetism. The magnetic axis is at $r = \Delta$, $\theta = 0$, and Δ is not assumed small. To obtain analytic solutions we choose special profiles which linearize Eq. 2.120. Namely p' is taken constant, and $g_0 g_2'$ linear in ψ_p. We then find

$$\frac{1}{r}\partial_r r \partial_r \psi_p + \frac{1}{r^2}\partial_\theta^2 \psi_p + A(1 - \Gamma_1)\psi_p = A(1 + Br\cos\theta) \qquad (2.122)$$

with boundary conditions $\psi_p = 0$, $\partial_r \psi_p = 0$ at $r = \Delta$, and A and B are eigenvalues to be determined. Eq. 2.122 can be solved directly, giving

$$\psi_p = \frac{1}{1 + cJ_0(\lambda)}[1 + cJ_0(\lambda r) + d(J_1(\lambda r) - rJ_1(\lambda))\cos\theta] \qquad (2.123)$$

with $\lambda = [A(1-\Gamma_1)]^{1/2}$, $c = -[J_0(\sigma) - J_1(\sigma)/\sigma - J_1(\lambda)/\lambda]/e$, $d = -J_1(\sigma)/e$, where $e = J_1^2(\sigma) - J_0(\sigma)J_1(\sigma)/\sigma + J_1^2(\sigma) - [J_0(\sigma) + \sigma J_1(\sigma)]J_1(\lambda)/\lambda$, and $\sigma = \lambda\Delta$. These equilibria are a generalization of a one-paramenter set ($\lambda = 0$) first found by Haas (1972). They exhibit the equilibrium limit caused by the occurrence of a separatrix for $\beta_\theta > 1/\epsilon$, which was observed with the second order equilibrium. For high β they have a current profile which does not vanish at $r = a$, having current reversal on the high field side and a large positive current on the low field side.

2.11 FLUX CONSERVING EQUILIBRIA

As seen in Sec. 2.8, the Grad–Shafranov equation can be solved by choosing any two of the functions $p(\psi_p)$, $q(\psi_p)$, $g(\psi_p)$, these functions being related by force balance. However, a solution does not exist for any arbitrary choice of any pair of these functions. If $p(\psi_p)$ is chosen to be very large for a given $g(\psi_p)$, the vertical field needed to contain the plasma can become large enough to cancel the poloidal field at the inside edge of the torus. This happens when $\beta_p > 1/\epsilon$. Note that the expression for the poloidal field derived in Sec. 2.8, $B_\theta = (1 - r\cos\theta + \Delta'\cos\theta)$, does vanish for $\beta_\theta \sim 1/\epsilon$

at $\theta = \pi$, although the expression is of course not valid for Δ' this large. The resulting separatrix shrinks the domain of topologically circular flux surfaces and hence the area of the confinement region.

The vanishing of B_θ implies $q = \infty$ and this suggests that equilibria with higher pressure could be constructed by specifying $q(\psi_p)$ and $p(\psi_p)$ and solving for the necessary toroidal field strength $g(\psi_p)$. Since $q(\psi_p)$ is conserved during plasma heating in an infinitely conducting plasma, *i.e.* on a time scale short with respect to the resistive diffusion time, one can imagine a sequence of equilibria with increasing pressure, all with the same q profile. This concept is called the flux conserving tokamak and was developed analytically by Clarke and Sigmar (1977) and numerically by Dory and Peng (1977). As the pressure is increased by changing temperature or density the q profile remains fixed, but the Shafranov shift causes more and more compression of the flux at the outward edge of the equilibrium. Since $\nabla p = \vec{j} \times \vec{B}$ for equilibrium, the increased pressure gradient is balanced by the spatial compression of the poloidal flux between the shifted magnetic axis and the conducting wall. There appears to be no equilibrium limit in β using this method. However, for very high β the equilibria have very strongly localized current sheets flowing close to the outside edge of the plasma, and cannot be regarded as very physical.

2.12 EQUILIBRIUM SCALING

The properties of the Grad–Shafranov equation can be used to produce a family of equilibria from a single given equilibrium. The simplest such scaling is to multiply all components of the magnetic field by a scale factor s, and the pressure by s^2 to keep β constant. Thus $g \to sg$, $dp/d\psi_p \to sdp/d\psi_p$, $\psi_p \to s\psi_p$ and the Grad–Shafranov equation is unchanged. A less trivial modification is to change the toroidal field without changing the poloidal field or the pressure, thus changing the total β, but not the poloidal β, which, as we saw in Sec. 2.8, determines the Shafranov shift. Accomplish this by choosing a new function $g_N(\psi_p)$ with $g_N^2 = g^2 + c$. Choosing c such that $g_n(\psi_{pw}) = sg(\psi_{pw})$ we find $g_N^2 = g^2 + (s^2 - 1)g(\psi_{pw})$ with ψ_{pw} the outside (wall) poloidal flux surface. Leave the pressure unchanged. Since gg' is unchanged, the Grad–Shafranov equation, and hence ψ_p itself, are unchanged, but the toroidal field, toroidal flux, and hence q and the toroidal β are modified.

2.13   EQUILIBRIUM TYPES

All toroidal configurations for confinement with reactor potential must approximate one in which a symmetry guarantees the existence of toroidal magnetic flux surfaces and hence gives a possibility of good confinement. Some toroidal devices presently being studied do not in fact satisfy this requirement, namely stellarators which are far from quasi-symmetry, such as LHD in Japan, and reversed field pinches (RFP), such as the RFX in Italy. The RFP are inherently unstable and have a large level of magnetic turbulence. Even so, an RFP is not too far from a symmetric state, with the addition of a turbulent magnetic fluctuation spectrum. Thus the general equilibrium formulation presented above is valid for most toroidal equilibria as a first approximation, to which it is possible to add perturbations. The various types of equilibria are distinguished only by differences in the functions $I(\psi)$, $g(\psi)$, $q(\psi)$, the Jacobian $\mathcal{J}$, and by their physical form, given by $X(\psi,\theta,\zeta)$, $Z(\psi,\theta,\zeta)$. This means that much of instability analysis, the derivation of guiding center equations for the particle motion, and neoclassical diffusion, which do not make use of the specific form of these functions, are valid for all types of equilibria. However, the means of forming them, the nature of the fixed field coils necessary to maintain them, the poloidal and toroidal field strength profiles, and the topology of the field lines outside of the last closed flux surface, are very different, so much so that they have their own names and research efforts.

The spheromak is an axisymmetric equilibrium with a simply connected bounding surface and a vanishing toroidal field ($q = 0$) at the bounding surface. No external coils link the spheromak and it generates its own internal toroidal field. The chamber wall is spheroidal. For a detailed description see Bellan (2000).

The tokamak, as described in detail above, is ideally an axisymmetric system with a radially increasing q profile of order 1 on axis. The toroidal field coils link the plasma so the chamber is toroidal. Tokamaks with very small aspect ratio are called spherical tokamaks, and since transition from low beta to high beta occurs for $\beta \sim \epsilon$ and ϵ can be of order $1/3$, they are capable of very high beta, and devices have exceeded beta of 50 percent.

The stellarator has very small toroidal current, essentially only that developed by the bootstrap current. Both the toroidal and poloidal fields are produced by field coils, so the chamber is toroidal. The q profile is typically much less than 1 and decreasing outward. In general, a stellarator does not possess a symmetry, but the only real candidates for reactors are the quasi symmetric configurations, which are not spatially symmetric but in which the magnitude of the magnetic field does approximately possess an ignorable coordinate.

The reversed field pinch (Bodin and Newton, 1980; Baker and Quinn, 1981; P. Martin *et al*, 2009) has a q profile which decreases radially, becoming negative near the outer wall, this reversal of the toroidal field giving it its name. The chamber wall is toroidal.

The field reversed configuration, or theta pinch, has no toroidal field, only coils which produce the poloidal field. The chamber wall is spheroidal.

2.14 STEPPED PRESSSURE EQUILIBRIA

The assumption of nested flux surfaces in a toroidal equilibrium is of course unrealistic. Always to a small degree, and often to a much larger extent, the magnetic surfaces are broken up into island and stochastic domains by magnetic perturbations as described in Chap. 1, making the assumption of nested unbroken flux surfaces a crude first approximation.

The global character of a magnetic field line is characterized by the local field helicity q. If $q = m/n$, a rational number, then the corresponding surface is foliated by periodic orbits which close on themselves after m toroidal transits, having undergone n poloidal transits. If q is an irrational number, then the field line defines a flux surface, covered ergodically by a single quasi-periodic orbit, a KAM surface.

The periodic orbits are fragile. Resonant magnetic fields associated with geometric deformation destroy almost all of the periodic orbits; magnetic islands form, and regions of chaotic magnetic field lines emerge. As the magnitude of the geometric deformation increases, the size of the magnetic islands increases, the volume of the chaotic seas increases, and each given KAM surface becomes more geometrically deformed until a critical point is reached, at which point the surface is continuous but no longer smooth. These critical tori form fractal boundaries between the chaotic seas associated with different island chains.

For the purpose of constructing a robust and efficient numerical solution of well defined, 3-D MHD equilibria with nonintegrable fields, it is best to work with smooth functions, and to employ an algorithm that does not depend on resolving the infinitely complicated structure of phase space. So the class of functions for p and $\vec{B}$ is extended beyond globally continuous functions, since non-trivial, continuous functions that satisfy force balance are necessarily fractal. Solving $\nabla p = \vec{j} \times \vec{B}$ for chaotic fields gives a non differentiable pressure p. Instead, restrict the solutions to include functions that are continuous and smooth, except for a finite set of discontinuities, which can be easily managed numerically.

Some progress has been made in the construction of stepped-pressure equilibria as extrema of a multi-region, relaxed MHD energy functional that combines elements of ideal MHD and Taylor relaxation. The model allows magnetic islands and stochastic domains and is compatible with Hamiltonian chaos theory. A specific equilibrium state is characterized by the pressure, *i.e.* p is considered to be a supplied, input function. The computational challenge is to then determine the magnetic field that is consistent with the given pressure and boundary. To admit numerically tractable solutions for $\vec{B}$, it is necessary to restrict the class of admissible functions for p; and to guarantee that $\vec{B}$ is consistent with a given p, topological constraints on $\vec{B}$ must be enforced.

A discrete invariant partition of phase space is constructed. Choose a set of smooth, KAM tori, $\mathcal{I}_l$, where $l = 1, 2, \ldots N_V$, that partitions phase space into N_V invariant toroidal or annular subvolumes, $\mathcal{V}_l$. Each $\mathcal{V}_l$ is an invariant set under the field line map, but not necessarily an ergodic invariant set because the field may not be totally chaotic inside it. This describes a simplified, discrete partition which greatly simplifies the equilibrium problem and leads naturally to stepped-pressure equilibria, where the plasma is modeled as a set of nested volumes.

To constrain the relaxation of the magnetic field in each $\mathcal{V}_l$, Taylor relaxation is employed, *i.e.* the most conserved invariant for a weakly resistive plasma is taken to be the helicity,

$$K_l \equiv \int_{\mathcal{V}_l} \vec{A} \cdot \vec{B} \, dv, \qquad (2.124)$$

where $\vec{A}$ is a vector potential, $\vec{B} = \nabla \times \vec{A}$, is considerd to be differentiable and a single valued function of position. The helicity is related to the Gauss linking number: it reflects how knotted, or twisted, the magnetic field lines are. In each region $\mathcal{V}_l$ in the field thus satisfies the Beltrami equation, $\nabla \times \vec{B} = \mu \vec{B}$, giving $\vec{j} \times \vec{B} = \nabla p = 0$.

Thus in each region, $\mathcal{V}_l$, use a single constant p_l. Each $\mathcal{V}_l$ is simply or doubly connected with a smooth boundary and it is a simple computational task to solve $\nabla \times \vec{B}_l = \mu_l \vec{B}_l$ in each $\mathcal{V}_l$. The constraint that $\vec{n} \cdot \vec{B} = 0$ holds on the $\mathcal{I}_l$, but otherwise the topology of the field in each $\mathcal{V}_l$ is unconstrained. The pressure is thus piecewise continuous with finite pressure jumps on the finite set $\mathcal{I}_l$, and the total pressure must be continuous across the $\mathcal{I}_l$, *i.e.* $[[p + B^2/2]] = 0$.

The multi-regional relaxed energy functional to be extremized is then

$$F = \sum_l [W_l - \mu K_l/2] \tag{2.125}$$

where the energy local to each volume is

$$W_l \equiv \int_{\mathcal{V}_l} \left(\frac{p}{\gamma - 1} + \frac{B^2}{2} \right) dv. \tag{2.126}$$

In each $\mathcal{V}_l$, variations in the pressure and the field, and the geometry of the interfaces, are allowed in order to extremize the MHD energy functional. These variations are arbitrary, except for (i) the mass-entropy constraint, $p_l V_l^\gamma = const$; (ii) helicity conservation in each $\mathcal{V}_l$; (iii) the interfaces must remain tangential to the magnetic field; and (iv) the magnetic fluxes are conserved.

The enclosed toroidal fluxes in each volume and the poloidal fluxes in each annular region constrain the magnetic field from being trivial. Gauge freedom is used to specify the loop integrals of $\oint \vec{A} \cdot d\vec{l}$ on each interface.

The pressure and tangential field are discontinuous across the interfaces, but these comprise a finite set of measure zero and so p and B^2 are both integrable functions, and the discontinuities are easily accomodated in the numerical discretization. In order to retain some control over the equilibria, the $\mathcal{I}_l$ are considered to be preserved as ideal barriers that restrict both pressure transport and field transport. Rather than continuously constraining the topology, the topology is discretely constrained.

In each $\mathcal{V}_l$, the mass and entropy constraints usually used in ideal MHD do not apply to individual fluid elements, but apply instead to the entire volume, giving the isentropic, ideal-gas constraint,

$$p_l V_l^\gamma = a_l, \tag{2.127}$$

where V_l is the volume of $\mathcal{V}_l$ and a_l is a constant. The internal energy in $\mathcal{V}_l$ is $\int_{\mathcal{V}} p_l/(\gamma - 1) \, dv = a_l V_l^{(1-\gamma)}/(\gamma - 1)$, and the first variation of this due to a deformation, $\vec{\xi}$, of the boundary is $-p \int_{\partial \mathcal{V}} \vec{\xi} \cdot d\vec{s}$.

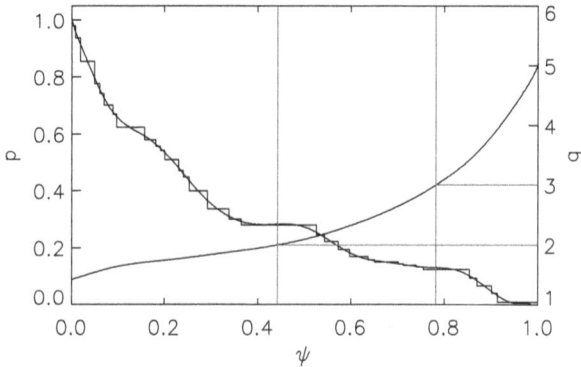

Fig. 2.5 Pressure profile (smooth) from a DIIID reconstruction using STELLOPT and stepped-pressure approximation. Also shown is the q profile.

To avoid a problem with small denominators, enforce the condition that the interfaces have irrational transform. The problematic Pfirsch–Schlüter currents are eliminated because the pressure gradient is identically zero across the resonances. The δ-function currents are also not present because the topology of the field is unrestricted at the rational surfaces, *i.e.* magnetic islands are allowed to form.

Solutions obtained by minimizing the energy functional are multi-volume, sharp-boundary states, called stepped-pressure equilibria. Stepped-pressure profiles are sufficiently general to represent observed profiles to within experimental error. Localized regions of flat pressure indicate the presence of magnetic islands or stochastic domains.

For the numerical implementation of the solution, Fourier representation is employed for all doubly-periodic, scalar functions. An initial guess for the geometry of a set of N_V nested, toroidal surfaces, $\mathcal{I}_l$, is assumed given. The $\mathcal{I}_l$ may be described by $R(\theta, \zeta)\,\hat{R} + Z(\theta, \zeta)\,\hat{k}$, where $\hat{R} \equiv \cos\theta\,\hat{i} + \sin\theta\,\hat{j}$, and $\hat{i}$, $\hat{j}$, and $\hat{k}$ are the Cartesian unit vectors.

To construct global equilibria, the $R_{l,j}$ and $Z_{l,j}$ describing the interface geometry are varied to extremize the energy functional and/or satisfy force balance. Tangential geometric variations merely change the angular parametrization of the interfaces and do not change the interface geometry, and so do not affect the energy functional, but *do* alter the $\{R_{l,j}, Z_{l,j}\}$. This freedom may be exploited to obtain a preferred angle parametrization using spectral width (Hirschman, 1986).

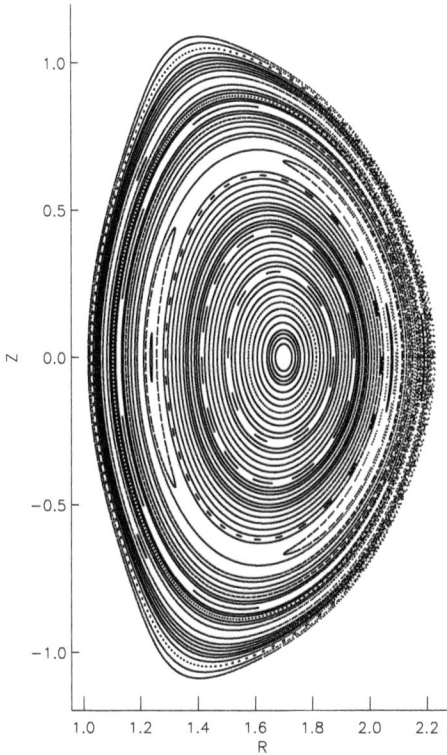

Fig. 2.6 Poincaré plot of a DIIID equilibrium with perturbed boundary, calculated using SPEC, including magnetic islands and stochastic domains.

As an illustration, consider a stepped-pressure equilibrium consistent with the boundary and profiles obtained via a 3-D STELLOPT reconstruction (Lazerson, 2012) of an up-down symmetric DIII-D experimental shot with applied resonant magnetic perturbation (RMP) fields. The reconstruction process seeks to infer the experimental configuration by adjusting the MHD equilibrium by varying the plasma boundary and the pressure and current profiles, so that derived quantities (such as Thomson scattering, motional Stark effect polarimetry, and magnetic diagnostics) match the experimental measurements. Because of the applied error fields, and the plasma response to these error fields, the reconstructed boundary is slightly, but significantly, perturbed from axisymmetry. The pressure and q-profiles derived from the reconstruction are shown in Fig. 2.5. It is interesting to observe that the reconstructed pressure profile appears quite flat across the lowest order rational surfaces. Furthermore, the locations of locally maximum pressure gradient appear to coincide with strongly-irrational surfaces.

The stepped-pressure equilibrium is computed using the reconstructed boundary, a stepped, $N_V = 32$ approximation (Fig. 2.5) to the reconstructed pressure profile, and the reconstructed q-profile. The rotational transforms of the interfaces are chosen by selecting the most noble irrationals that are within range. The Fourier resolution is $M = 10$ and $N = 6$, and the total radial sub-grid resolution is 279. A Poincaré plot is shown in Fig. 2.6; most visible is a $q = 2$ island where VMEC has the $q = 2$ rational surface. In this calculation, the boundary is constrained to remain a fixed, good flux surface.

2.15   Problems

1. Expand the poloidal flux in Z and $X^2 - R^2$ to third order. Approximate $p' = a + a'\psi_p$ and $gg'/R^2 = b + b'\psi_p$ and determine ψ_p as a series solution of the Grad–Shafranov equation. Show that if the constants are chosen such that the coefficient of $(X^2 - R^2)^3$ vanishes the solution is exact. Solov'ev (1968).

2. The second order Shafranov tokamak equilibrium with circular conducting walls is given by $\vec{B} = \nabla\zeta \times \nabla\psi_p + q\nabla\psi_p \times \nabla\theta$ with

$$\nabla\psi_p = \frac{r}{q}\frac{(\cos\theta\hat{X} + \sin\theta\hat{Z})}{1 - \Delta'\cos\theta}$$

$$\nabla\theta = \frac{1}{r}\frac{(-\sin\theta\hat{X} + (\cos\theta - \Delta')\hat{Z})}{1 - \Delta'\cos\theta}$$

and $\zeta = \phi - \nu$ with $\partial_\theta\nu = (gq/X)(1 - \Delta'\cos\theta) - q$, $g(r) = 1 + g_2$, and $p' + g' + (1/q)(r^2/q)' = 0$. To first order in ϵ the covariant representation of $\vec{B}$ is given by $g = 1$, $I = r^2/q$, $\delta = 0$, and the field magnitude is $B = 1 - r\cos\theta$.

a. Find the covariant representation to order ϵ^2.

b. Note that to this order $I = I(r, \theta)$. Find a new poloidal angle $\theta_c = \theta + \epsilon K(r, \theta)$ such that $I = I(r)$. Find the Jacobian in the ψ_p, θ_c, ζ system.

3. Calculate the curvature to second order in ϵ for the second order equilibrium field of Sec. 2.8.

4. Show that a magnetic field configuration which minimizes the magnetic field energy within the volume v under the condition $\int_v \vec{A}\cdot\vec{B}d\tau = constant$, where $\vec{B} = \nabla \times \vec{A}$, is force free within v.

2.16 References

Magnetic coordinates

- Baker, D. A., and W. E. Quinn, in Fusion, edited by E. Teller (Academic, New York, 1981), Vol I, Part A, Chap. 7
- Beidler, C., Fusion Technology 17, 148 (1990).
- Bodin, H. A. B., and A.A. Newton, Nuclear Fusion 20, 1255 (1980).
- Boozer, A. H., Phys. Fluids 24, 1999 (1981).
- Boozer, A. H., Phys. Fluids 26, 1288 (1983).
- Greene, J. M., Commun. Pure Appl. Math 36, 537 (1983).
- Greene, J. M., J. L. Johnson, M. D. Kruskal and L. Wilets, Phys. Fluids 5, 1063 (1962).
- Hamada, S., Nucl. Fusion 2, 23 (1962).
- White, R. B., and M. S. Chance, Phys. Fluids 27, 2455 (1984).

Plasma equilibria

- Bellan, P. M., Spheromaks (Imperial College Press, London, 2000).
- Berk, H. L., J. P. Freidberg, X. Llobet, P. J. Morrison, and J. A. Tataronis, Phys. Fluids 29, 3281 (1986).
- Boozer, A. H., Phys. Fluids 24, 1999 (1981).
- Dory, R. A., and Y.-K.M. Peng, Nucl. Fusion 17, 21 (1977).
- Garren, D. A. and A. H. Boozer, Phys. Fluids 10, 2822 (1991).
- Grad, H., Proc. Symp. Appl. Math. 18, 162 (1967).
- Grad, H., Phys. Fluids 10, 137 (1967).
- Greene, J. M., Plasma Physics and Controlled Fusion 30, 327 (1988).
- Greene, J. M., J. L. Johnson, K. E. Weimer, Phys. Fluids 14, 67 (1971).
- Greene, J. M., and J. L. Johnson, Phys. Fluids 5, 510 (1962).
- Greene, J. M., and J. L. Johnson, Advances in Theoretical Physics (Academic Press, NY, 1965) Vol. 1 p. 195.
- Grimm, R. C., J. M. Greene, and J. L. Johnson in Methods in Computational Physics (Academic, New York, 1976) Vol. 16, p. 253.
- Haas, F. A., Phys. Fluids 15, 141 (1972).
- Hirshman, S. P., W. I. Vanrij, Comput. Phys. Commun. 43, 143 (1986).

- Kleinberger, R., and J. P. Goedbloed, Plasma Physics and Controlled Fusion 30, 339 (1988).
- Martin, P., L. Apolloni, M. E. Puiatti, J. Adamek *et al.* Nuclear Fusion 49, 104019 (2009).
- Najmabadi, F., and the Aries Team, Fusion Engineering and Design, 38, 3 (1997).
- Nuhrenberg, J. and R. Zille, Phys. Lett. A 129, 113 (1988).
- Reiman, A., and H. S. Greenside, J. Comput. Phys. 75, 423 (1988).
- Shafranov, V. D., Reviews of Plasma Physics, (Consultants Bureau, New York, 1966) Vol. 2, p. 103.
- Solov'ev, L. S., Zh Ehksp. Teor. Fiz. 53, 626 (1967), Sov. Phys. JETP 26, 400 (1968).
- Yoshikawa, S., Phys. Fluids 20, 706 (1977).

MHD simulation

- Izzo, R. D., A. Monticello, J. DeLucia, W. Park, *et al.*, Phys. Fluids 28, 903 (1985).
- Park, W., and D. Monticello, Nucl. Fusion 30, 2413 (1990).
- Rosenbluth, M. N., D. A. Monticello, H. R. Strauss, and R. B. White, Phys. Fluids 19, 1987 (1976).
- Strauss, H. R., Phys. Fluids 20, 1354 (1977).

Plasma rotation

- Bhattacharjee, A., in Theory of Fusion Plasmas, Varenna, Italy, Soc. Italiana di Fisica (1988).
- Clemente R. A., and R. Farengo, Phys. Fluids 27, 776 (1984).
- Mashke, E. K., and H. Perrin, Plasma Phys. 22, 579 (1980).

Stepped pressure equilibria

- Berk, H. L., J. P. Freidberg, X. Llobet, P. J. Morrison *et al.*, Phys. Fluids 29, 3281 (1986)
- Cappello S. and D. Biskamp, Nucl. Fus. 36, 571 (1996).
- Dewar, R. L., M. J. Hole, M. McGann, R. Mills *et al.*, Entropy 10, 621 (2008).
- Finn, J. M. and T. M. Antonsen, Comments Plasma Phys. Control. Fusion, 9, 111 (1985).

- Finn, J. M. and L. Chacon, Phys. Plasmas 12, 054503 (2005).
- Hirshman S. P. and H. K. Meier, Phys. Fluids 28, 1387 (1985).
- Hudson S. R., Phys. Plasmas 14, 052505 (2007).
- Hudson S. R. and J. Breslau, Phys. Rev. Lett.100, 095001 (2008).
- Hudson S. R., Phys. Plasmas 17, 114501 (2010).
- Lazerson, S.A., Submitted, Phys. Plasmas (2012).
- Lichtenberg A.J. and M.A. Liberman, Regular and Chaotic Dynamics, 2nd ed. (Springer-Verlag, New York, 1992).
- Mackay, R. S., J. D. Meiss, I. C. Percival, Phys. Rev. Lett. 52, 697 (1984).
- MacKay R. S., Physica D. 13, 55 (1984).
- McGann, M., S. R. Hudson, R. L. Dewar, G. von Nessi, Phys. Lett. A (2010).
- Meiss, J. D., Rev. Mod. Phys. 64, 795 (1992).
- Meiss J. D., Phys. Rev. A., 34, 2375 (1986).
- Taylor J. B., Phys. Rev. Lett. 33, 1139 (1974).
- Taylor J. B., Rev. Mod. Phys. 58, 741 (1986).

Chapter 3

Guiding Center Motion

3.1 INTRODUCTION

The motion of charged particles in electromagnetic fields is one of the oldest problems of plasma physics. The evaluation of the confinement of particles in fusion devices and the determination of particle diffusion requires following particle orbits for very long times. In order to integrate particle motion for the required length of time, an expansion of the equations of motion in the gyro radius must be made, and the rapid gyro phase motion averaged over, leaving equations for the guiding center motion. The Larmor radius of charged particles in a reactor plasma is small compared to the plasma dimensions and the radius of curvature of the magnetic field. Even fusion products with energies of a few MeV have Larmor radii 1/20 of the minor radius of a typical reactor design. Thus charged particle motion is well described using the drift approximation. The guiding center motion is determined by the mass, charge and energy of the particle and the field strength and configuration. These first four, however, enter the equations in one combination only: the gyro radius, given by $\rho \simeq 10^2 \sqrt{mE}/(BZ)$cm where the energy E is in electron Volts, Z is the charge, B is in Gauss, and mass in units of the proton mass. Thus there is actually only a single parameter characterizing a particle insofar as its guiding center motion is concerned, and it is convenient to use units in which this is apparent. Introducing units of time given by ω_0^{-1}, where $\omega_0 = eB/(mc)$ is the on axis gyro frequency, and units of distance given by the major radius R, the basic unit of energy becomes $m\omega_0^2 R^2$, which can also be written as $(mv^2/2)(2R^2/\rho^2)$, the gyro radius is $\rho = v/B \ll 1$, and the magnetic moment $\mu = v_\perp^2/(2B)$ is

75

of order ρ^2. Particle motion both along and across the field lines is of order ρ, but to leading order the cross field motion is the gyro motion, and cross field drift is of order ρ^2.

The Lagrangian for guiding center motion was derived by Littlejohn (1983) using Lie algebra, capable of producing the correct expression to all orders in the gyro radius, depending only on one's stamina. The method consists of an order by order expansion in gyro radius, at each step adding exact time derivatives to the Lagrangian in order to produce simplification at that order. Exact time derivatives do not change the Lagrangian, and this method can be carried out to any order. The Lagrangian for the motion of a charged particle in an electromagnetic field is

$$L = [\vec{A}(x,t) + \vec{v}] \cdot \dot{\vec{x}} - H(\vec{v}, \vec{x}) \tag{3.1}$$

with $\vec{B} = \nabla \times \vec{A}$, and the Hamiltonian $H = \vec{v}^2/2 + \Phi(\vec{x}, t)$. Putting the Lagrangian in the form $\sum p_i \dot{q}_i - H$ we recognize the canonical momentum for a charged particle in a electromagnetic field to be $\vec{p} = \vec{v} + \vec{A}$. For purposes of deriving equations of motion, the momentum variable $\vec{v}$ is considered independent of the velocity $\dot{\vec{x}}$, even though the equations of motion in the end give $\vec{v} = \dot{\vec{x}}$. Now explicitly separate the motion along the field from the cross field motion through

$$\vec{v} = v_\parallel \hat{b} + w\hat{c} \tag{3.2}$$

with $\hat{b} = \vec{B}/B$, and $\hat{c} = -sin\xi\hat{e}_1 - cos\xi\hat{e}_2$, with $\hat{e}_1$ and $\hat{e}_2$ unit vectors orthogonal to $\vec{B}$ and to each other chosen so that $\hat{e}_1 \times \hat{e}_2 = \hat{b}$, and ξ the gyro phase, and w the magnitude of the perpendicular velocity $\vec{v}_\perp$.*

Define the particle gyro center through

$$\vec{x} = \vec{X} + \frac{w\hat{a}}{B} \tag{3.3}$$

with $\hat{a} = cos\xi\hat{e}_1 - sin\xi\hat{e}_2$, so that $\hat{c} \times \hat{a} = \hat{b}$, $\hat{a} \times \hat{b} = \hat{c}$, i.e. $\hat{a}, \hat{b}, \hat{c}$ form a right handed orthonormal basis, and all quantities on the right are evaluated at

---

*In a toroidal device with everywhere nonzero B_ϕ, such as a tokamak, a globally defined continuous set $\hat{e}_j$ can be defined by taking $\hat{e}_1 \sim \vec{X} \times \hat{b}$ with $\vec{X}$ the cylindrical radial coordinate. In general, such a set of vectors exists for any toroidal system for which $\vec{B}$ does not vanish at a point. See Burby and Qin, (2012).

the guiding center $\vec{X}$. The Lagrangian then becomes

$$L = [\vec{A}(\vec{x},t) + v_\parallel \hat{b} + w\hat{c}] \cdot \left[\dot{\vec{X}} + \frac{d}{dt}\left(\frac{w\hat{a}}{B}\right)\right] - H. \tag{3.4}$$

Now make an expansion in the small parameter w. We will assume that all spatial scales are large enough compared to w so that the time variations of the field viewed by the particle, as well as all explicit time variations of the field, are slow compared to the gyration frequency $d\xi/dt$. Write

$$\vec{A}(x,t) \simeq \vec{A}(\vec{X},t) + \frac{w\hat{a}}{B} \cdot \nabla \vec{A}(\vec{X},t) \tag{3.5}$$

and substitute into the Lagrangian, giving

$$L = \left[\vec{A} + v_\parallel \hat{b} + w\hat{c}\right] \cdot \dot{\vec{X}} + v_\parallel \hat{b} \cdot \frac{d}{dt}\left(\frac{w\hat{a}}{B}\right) + w\hat{c} \cdot \frac{d}{dt}\left(\frac{w\hat{a}}{B}\right) +$$
$$\left(\frac{w\hat{a}}{B} \cdot \nabla\right)\vec{A} \cdot \frac{d}{dt}\left(\frac{w\hat{a}}{B}\right) + \vec{A} \cdot \frac{d}{dt}\left(\frac{w\hat{a}}{B}\right) + \left(\frac{w\hat{a}}{B} \cdot \nabla\right)\vec{A} \cdot \dot{\vec{X}} - H \tag{3.6}$$

where all $\vec{A}$ are now evaluated at $\vec{X}, t$. Now add to the Lagrangian the perfect derivative dS/dt with $S = -(w/B)\hat{a} \cdot \vec{A}(\vec{X},t)$, and use the fact that $d\vec{A}/dt = \dot{\vec{X}} \cdot \nabla \vec{A} + \partial \vec{A}/\partial t$. Further note in taking time derivatives that $d\hat{a}/dt = \dot{\xi}\hat{c}$ plus terms small in the ordering of the time dependence of $\vec{B}$ compared to $\dot{\xi}$. The Lagrangian then reduces to

$$L = [\vec{A} + v_\parallel \hat{b}] \cdot \dot{\vec{X}} + \frac{w^2 \dot{\xi}}{B} - \frac{w}{B}\hat{a} \cdot \frac{\partial \vec{A}}{\partial t} +$$
$$\frac{w\dot{w}}{B^2}(\hat{a} \cdot \nabla)\vec{A} \cdot \hat{a} + \frac{w^2 \dot{\xi}}{B^2}(\hat{a} \cdot \nabla)\vec{A} \cdot \hat{c} - H. \tag{3.7}$$

Now again add a perfect derivative to L, of the form dS/dt with $S = -(w^2/2B^2)(\hat{a} \cdot \nabla)\vec{A} \cdot \hat{a}$, and note that

$$(\hat{a} \cdot \nabla)\vec{A} \cdot \hat{c} - (\hat{c} \cdot \nabla)\vec{A} \cdot \hat{a} = -B. \tag{3.8}$$

To see this, choose a local coordinate system $(\hat{c}, \hat{a}, \hat{b}) = (\hat{x}, \hat{y}, \hat{z})$. Then this equation is simply $\partial_y A_x - \partial_x A_y = -B_z$. We then have $L = [\vec{A} + v_\parallel \hat{b}] \cdot \dot{\vec{X}} + w^2 \dot{\xi}/2B - w\hat{a} \cdot \partial_t \vec{A}/B - H$, plus terms of second order in w not involving $\dot{\xi}$, which are thus second order in gyro radius and are dropped. Now average over the fast gyro motion time scale. The $\hat{a} \cdot \partial_t \vec{A}$ term vanishes, leaving to second order the Lagrangian

$$L = \left[\vec{A} + v_\parallel \hat{b}\right] \cdot \dot{\vec{X}} + \frac{w^2 \dot{\xi}}{2B} - H(v_\parallel, w, \vec{X}, t), \tag{3.9}$$

from which we conclude that $w^2/2B \equiv \mu$ is a constant of the motion, and that $\dot{\xi}$ is constant, and the Hamiltonian $H = v_\parallel^2/2 + \mu B(\vec{X}) + \Phi(\vec{X}, t)$ is also evaluated at the guiding center. No assumptions were made about the form of the magnetic field, this expression is valid independent of the existence of magnetic surfaces, equilibrium, or any other conditions on $\vec{B}$, except that it be slowly varying in space and time compared to the scales given by ρ and $\dot{\xi}$.

Historically, Morozov and Solov'ev (1963) showed that the guiding center velocity could be put in the convenient form

$$\vec{v} = \rho_\parallel (\vec{B} + \nabla \times \rho_\parallel \vec{B}) \tag{3.10}$$

where $\rho_\parallel = v_\parallel/B$ is the "parallel gyro radius", and in this equation must be regarded as a function of position, *i.e.*

$$\rho_\parallel = [2W - 2\mu B(\vec{r}) - 2\Phi(\vec{r})]^{1/2}/B(\vec{r}) \tag{3.11}$$

where W is the particle energy and $\Phi(\vec{r}, t)$ the electric potential. It is readily verified that this expression includes the $\vec{E} \times \vec{B}$, ∇B, and curvature drifts. See for example Northrop (1963) and Schmidt (1966). The electric field is assumed small, and polarization drift terms, proportional to both $\vec{E}$ and the gyro radius, are neglected.

However, this expression suffers the serious fault of not being Hamiltonian, and giving rise to motion which does not obey Liouville's theorem, and in some cases not even conserving the energy. The equations resulting from the Lagrangian Eq. 3.9 partially correct this by the addition of a denominator $1 + \rho_\parallel \hat{b} \cdot \nabla \times \hat{b}$, with $\hat{b} = \vec{B}/B$. This correction was first found by Bañõs (1967). This factor is a scalar and thus does not modify the particle trajectories, only the time required to traverse a given portion of the trajectory, and thus for most problems is not a relevant modification. However, for investigation of long time particle diffusion, it is essential that the motion satisfy the Liouville theorem.

3.2   LAGRANGIAN AND HAMILTONIAN FORMULATION

Begin with the guiding center Lagrangian, Eq. 3.9, which we put in the form

$$L = (\vec{A} + \rho_\| \vec{B}) \cdot \vec{v} + \mu \dot{\xi} - H \tag{3.12}$$

with $\vec{v}$ the guiding center velocity, $\vec{B} = \nabla \times \vec{A}$, $\rho_\| = v_\|/B$ the normalized parallel velocity, μ the magnetic moment, ξ the gyrophase,

$$H = \rho_\|^2 B^2/2 + \mu B + \Phi \tag{3.13}$$

the Hamiltonian and Φ the electric potential. The field magnitude B and the potential may be functions of ψ_p, θ and ζ.

The units are defined by the on-axis gyro frequency (time) and the major radius (distance). For a stellarator or other non axisymmetric equilibrium this may be chosen to be the major radius at a particular toroidal angle.

Rewrite Eq. 2.14 in the form $\vec{B} = \nabla \times (\psi \nabla \theta - \psi_p \nabla \zeta) = \nabla \times \vec{A}$ and substitute $\vec{A}$ and Eq. 2.19 for $\vec{B}$ into the Lagrangian $L(\rho_\|, \psi_p, \theta, \zeta, \dot{\rho}_\|, \dot{\psi}_p, \dot{\theta}, \dot{\zeta})$, giving

$$L = (\psi + \rho_\| I)\dot{\theta} + (\rho_\| g - \psi_p)\dot{\zeta} + \mu \dot{\xi} + \delta q \rho_\| \dot{\psi}_p - H. \tag{3.14}$$

Lagrange's equations are

$$\frac{d}{dt} \frac{\partial L}{\partial \dot{q}} = \frac{\partial L}{\partial q}. \tag{3.15}$$

We then find

$$\begin{vmatrix} 0 & \delta q & I & g \\ -\delta q & 0 & F & C \\ -I & -F & 0 & 0 \\ -g & -C & 0 & 0 \end{vmatrix} \begin{vmatrix} \dot{\rho}_\| \\ \dot{\psi}_p \\ \dot{\theta} \\ \dot{\zeta} \end{vmatrix} = \begin{vmatrix} \partial_{\rho_\|} H \\ \partial_{\psi_p} H \\ \partial_\theta H \\ \partial_\zeta H \end{vmatrix} \tag{3.16}$$

where $C = -1 + \rho_\|(g'_{\psi_p} - q\delta'_\zeta)$, $F = q + \rho_\|(I'_{\psi_p} - q\delta'_\theta)$, and we have used Eq. 2.25, $g'_\theta = I'_\zeta$. (Here we use the notation f'_α to denote $\partial_\alpha f$.)

Inverting this equation gives

$$
\begin{vmatrix} \dot\rho_\| \\ \dot\psi_p \\ \dot\theta \\ \dot\zeta \end{vmatrix} = \frac{1}{D} \begin{vmatrix} 0 & 0 & C & -F \\ 0 & 0 & -g & I \\ -C & g & 0 & -q\delta \\ F & -I & q\delta & 0 \end{vmatrix} \begin{vmatrix} \partial_{\rho_\|} H \\ \partial_{\psi_p} H \\ \partial_\theta H \\ \partial_\zeta H \end{vmatrix}
\tag{3.17}
$$

with denominator

$$
D = gq + I + \rho_\|(gI'_{\psi_p} - Ig'_{\psi_p} - gq\delta'_\theta + Iq\delta'_\zeta).
\tag{3.18}
$$

Note that there are no derivatives of I or g with respect to ζ or θ.

In simplest form, in an axisymmetric configuration, ($\partial_\zeta f = 0$ for all equilibrium functions f), in straight field line coordinates, and without field perturbation or magnetic field ripple, the equations become

$$
\dot\theta = \frac{\rho_\| B^2}{D}(1 - \rho_\| g') + \frac{g}{D}\left[(\mu + \rho_\|^2 B)\frac{\partial B}{\partial \psi_p} + \frac{\partial \Phi}{\partial \psi_p}\right],
\tag{3.19}
$$

$$
\dot\psi_p = -\frac{g}{D}\left[(\mu + \rho_\|^2 B)\frac{\partial B}{\partial \theta} + \frac{\partial \Phi}{\partial \theta}\right],
\tag{3.20}
$$

$$
\dot\rho_\| = -\frac{(1 - \rho_\| g')}{D}\left[(\mu + \rho_\|^2 B)\frac{\partial B}{\partial \theta} + \frac{\partial \Phi}{\partial \theta}\right],
\tag{3.21}
$$

$$
\dot\zeta = \frac{\rho_\| B^2}{D}(q + \rho_\|[I'_{\psi_p} - q\delta'_\theta]) - \frac{I}{D}\left[(\mu + \rho_\|^2 B)\frac{\partial B}{\partial \psi_p} + \frac{\partial \Phi}{\partial \psi_p}\right]
$$
$$
+ \frac{q\delta}{D}\left[(\mu + \rho_\|^2 B)\frac{\partial B}{\partial \theta} + \frac{\partial \Phi}{\partial \theta}\right].
\tag{3.22}
$$

Notice that the function δ modifies ψ_p, θ, and $\rho_\|$ only through a renormalization of the time interval, through the denominator D. Thus the projection of the orbit into the poloidal plane is independent of this function. We will show that also the toroidal precession is independent of δ, so its omission produces only a nonsecular redefinition of the guiding center. These equations are realized with a fourth order Runge–Kutta stepping algorithm in the numerical code ORBIT.

To calculate the second order drift of a trapped particle, integrate ζ over one transit of the orbit in θ. Note that for this calculation it does not matter whether the orbit is closed in the ψ_p, θ plane,

$$\Delta\zeta = \int \dot{\zeta}\,dt = \oint \frac{\dot{\zeta}}{\dot{\theta}}\,d\theta. \tag{3.23}$$

Then from Eqs. 3.19, 3.22, to first order in $\rho_{\parallel}$

$$\Delta\zeta = \oint d\theta(q + \rho_{\parallel}[I'_{\psi_p} - q\delta'_\theta + g'q])$$

$$- \oint d\theta \frac{(\mu + \rho_{\parallel}^2 B)}{\rho_{\parallel} B^2} \left[(I + gq)\frac{\partial B}{\partial \psi_p} - q\delta \frac{\partial B}{\partial \theta} \right]. \tag{3.24}$$

Using energy conservation with $E = \rho_{\parallel}^2 B^2/2 + \mu B$ to find $(\mu + \rho_{\parallel}^2 B)\partial_\theta B = -\rho_{\parallel} B^2 \partial_\theta \rho_{\parallel}$, the terms in δ can be written as

$$- \oint d\theta \partial_\theta [q\delta\rho_{\parallel}] = 0. \tag{3.25}$$

Thus neglecting δ produces only a nonsecular change in the guiding center motion. Furthermore, this periodic motion does not involve the trajectory in the poloidal plane, only the speed. Dropping the term in δ, $C = -1 + \rho_{\parallel} g'_{\psi_p}$, $F = q + \rho_{\parallel} I'_{\psi_p}$ and $D = gq + I + \rho_{\parallel}(gI'_{\psi_p} - Ig'_{\psi_p})$, and the Lagrangian takes the form

$$L = (\psi + \rho_{\parallel} I)\dot{\theta} + (\rho_{\parallel} g - \psi_p)\dot{\zeta} + \mu\dot{\xi} - H, \tag{3.26}$$

and the canonical coordinates and momenta are immediately identified to be θ, ζ and P_θ, P_ζ, with $P_\zeta = \rho_{\parallel} g - \psi_p$, $P_\theta = \psi + \rho_{\parallel} I$. These expressions are immediately recognizable as the covariant components of the sum of particle momentum and vector potential, just as in classical electrodynamics, Eq. 3.1. Thus in all the following we will simply drop terms in δ.

In simplest form, in an axisymmetric configuration, in straight field line coordinates, and without field perturbation or magnetic field ripple, the equations then become

$$\dot{\theta} = \frac{\rho_{\parallel} B^2}{D}(1 - \rho_{\parallel} g') + \frac{g}{D}\left[(\mu + \rho_{\parallel}^2 B)\frac{\partial B}{\partial \psi_p} + \frac{\partial \Phi}{\partial \psi_p} \right], \tag{3.27}$$

$$\dot{\psi}_p = -\frac{g}{D}\left[(\mu + \rho_{\parallel}^2 B)\frac{\partial B}{\partial \theta} + \frac{\partial \Phi}{\partial \theta} \right], \tag{3.28}$$

$$\dot{\rho}_{\parallel} = -\frac{(1 - \rho_{\parallel}g')}{D}\left[(\mu + \rho_{\parallel}^2 B)\frac{\partial B}{\partial \theta} + \frac{\partial \Phi}{\partial \theta}\right], \tag{3.29}$$

$$\dot{\zeta} = \frac{\rho_{\parallel}B^2}{D}(q + \rho_{\parallel}I'_{\psi_p}) - \frac{I}{D}\left[(\mu + \rho_{\parallel}^2 B)\frac{\partial B}{\partial \psi_p} + \frac{\partial \Phi}{\partial \psi_p}\right]. \tag{3.30}$$

The terms in $\partial\Phi/\partial\psi_p$, $\partial\Phi/\partial\theta$, $\partial\Phi/\partial\zeta$ are easily recognized as describing $\vec{E} \times \vec{B}$ drift.

The equations of motion are also given by the Hamiltonian form

$$\dot{\theta} = \frac{\partial H}{\partial P_\theta} \qquad \dot{P}_\theta = -\frac{\partial H}{\partial \theta}$$
$$\dot{\zeta} = \frac{\partial H}{\partial P_\zeta} \qquad \dot{P}_\zeta = -\frac{\partial H}{\partial \zeta}. \tag{3.31}$$

Several different atttempts at constructing a simple Hamiltonian formulation of the guiding center motion were made over the years. White *et al.* (1982) used Boozer coordinates (Boozer, 1981) to find simple equations for the guiding center motion. Canonical variables for toroidal geometry were found by White and Chance (1984) using the Solov'ev expression for the drift velocity to find the time derivatives of variables, e.g. $\dot{\psi} = \vec{v} \cdot \nabla\psi$, and then writing the obtained expressions for $\dot{\psi}$, $\dot{\theta}$, $\dot{\zeta}$, $\dot{\rho}_{\parallel}$ in terms of partial derivatives of the Hamiltonian, and finding the Darboux transformation to put the equations in canonical form. Littlejohn (1985) and White (1990) subsequently pointed out that the Lagrangian formalism made this process much less arduous, significantly more transparent, and also allows for a simplification of the canonical variables through a redefinition of the guiding center. Meiss and Hazeltine (1990) found a means of producing canonical variables using small changes in the angular coordinates, but this method is impractical for numerical implementation. White and Zakharov (2003) found a simple transformation of the toroidal angle that produces exact canonical variables and is easily implemented numerically. As seen above, simply discarding the function δ produces a nonsecular redefinition of the guiding center, and puts the equations in Hamiltonian form.

Now let us compare these equations with the motion resulting from the Morozov–Solov'ev expression, and show that they agree to second order in ρ. For simplicity we neglect the potential Φ, its inclusion is left as an exercise. Calculate $\vec{v} \cdot \nabla\theta$ using the Banõs-corrected Morozov–Solov'ev expression,

$$\vec{v} = \frac{\rho_\|(\vec{B} + \nabla \times \rho_\|\vec{B})}{1 + \rho_\|\hat{b} \cdot \nabla \times \hat{b}} \tag{3.32}$$

where now $\rho_\| = \rho_\|(\psi_p, \theta, \zeta)$ through $E = \rho_\|^2 B^2/2 + \mu B + \Phi$. First evaluate $\hat{b} \cdot \nabla \times \hat{b}$. Use the covariant expression for $\vec{B}$ to find

$$\frac{\vec{B} \cdot \nabla \times \vec{B}}{B^2} = \frac{-g'I + I'g - g\partial_\theta\delta + I\partial_\zeta\delta}{gq + I}. \tag{3.33}$$

Similarly use the curl $\nabla \times \rho_\|\vec{B} = \nabla\rho_\| \times \vec{B} + \rho_\|\nabla \times \vec{B}$ and the expression $\partial_\alpha\rho_\| = -(\mu + \rho_\|^2 B)\partial_\alpha B/(\rho_\| B^2)$, for $\alpha = \theta, \zeta, \psi_p$, giving

$$\nabla\theta \cdot (\vec{B} + \nabla \times \rho_\|\vec{B}) = \frac{B^2}{gq + I} + \frac{g(\mu + \rho_\|^2 B)}{\rho_\|(gq + I)}\partial_{\psi_p} B$$

$$-\frac{\mu + \rho_\|^2 B}{\rho_\|(gq + I)}\delta\partial_\zeta B - \frac{\rho_\| B^2 g'}{gq + I} + \frac{\rho_\| B^2 B^2}{gq + I}\partial_\zeta\delta. \tag{3.34}$$

Combining these expressions and dropping all derivatives with respect to ζ, we find to order ρ^2

$$\vec{v} \cdot \nabla\theta = \dot{\theta}_L \tag{3.35}$$

with $\dot{\theta}_L$ the Lagrangian expression, given by Eq. 3.19. The equivalence of the Hamiltonian and the conventional guiding center drift equations is shown for all other variables in a similar manner.

The formalism was also extended to include the treatment of fields not possessing good magnetic surfaces. This is conveniently done in perturbation theory since a small perturbation of a system in which good surfaces exist can give rise to quite stochastic fields.

Note that the equations of motion, Eqs. 3.27–3.29, exactly conserve the Hamiltonian given by Eq. 3.13, and in using these equations all orders in ρ are to be kept. Higher order terms in ρ are dropped only for the comparison of the Hamiltonian results with the lowest order drift motion.

To relate the terms in these equations to the conventional expressions for grad B drift, curvature drift, and $\vec{E} \times \vec{B}$ drift, simply find the form of these expressions in our coordinate system. Grad B drift is given by

$$\vec{v}_G = \mu\frac{\vec{B} \times \nabla B}{B^2}. \tag{3.36}$$

Calculate the contribution to $\dot{\theta}$, given by $\dot{\theta} = \vec{v}_G \cdot \nabla \theta$, giving

$$\dot{\theta} = \mu \frac{[g\nabla\zeta \times \partial_{\psi_p} B\nabla\psi_p]}{B^2} \cdot \nabla\theta = \frac{\mu g \partial_{\psi_p} B}{gq + I} \tag{3.37}$$

equal to the term in Eq. 3.19 within terms of third order in ρ.

Curvature drift is given by

$$\vec{v}_c = \rho_\parallel^2 B \vec{B} \times \vec{\kappa}. \tag{3.38}$$

Using Eq. 2.9 for the curvature we find the contribution to $\dot{\theta}$ to be given by $B\dot{\theta} = \rho_\parallel^2 \vec{B} \times [\nabla B^2/2 + \nabla p] \cdot \nabla\theta$ or

$$\dot{\theta} = \frac{\rho_\parallel^2 B^2 g}{gq + I}[B\partial_{\psi_p} B + p']. \tag{3.39}$$

But we also have $\mathcal{J}p' = g'q + I'$. Expand $D = (gq + I)[1 + \rho_\parallel(gI' - Ig')/(gq + I)]$ in $\rho_\parallel$. Terms in $\rho_\parallel^2$ in Eq. 3.19 then become

$$\frac{\rho_\parallel^2 B^2 g}{D}\left[B\partial_{\psi_p} B - \frac{g'q + I'}{gq + I}\right]. \tag{3.40}$$

The terms in g' and I' disagree with those in the Hamiltonian equations, but they are also higher order in inverse aspect ratio. The Hamiltonian equations give expressions which guarantee the Liouville theorem. The approximate expressions do not satisfy this.

3.3   ORBIT TYPES IN AXISYMMETRIC EQUILIBRIA

In the absence of an electric field the particle energy takes the form

$$W = \frac{[P_\zeta + \psi_p(P_\zeta, P_\theta)]^2 B^2}{2g^2} + \mu B. \tag{3.41}$$

From the equations of motion derived in the last section we see that P_ζ is constant if in the system ζ is an ignorable coordinate. Note this does not imply symmetry of the configuration, but only the symmetry of the field magnitude B, often refered to as quasi-symmetry. This integral of the motion considerably simplifies the particle orbits. In this case, W is a function of P_θ and θ. But $\rho_\parallel = (2W - 2\mu B)^{1/2}/B$ is also a function of ψ, θ so $W = constant$ defines closed curves in the ψ, θ plane. Thus in the case of quasi-symmetry the particle motion is restricted to a two dimensional

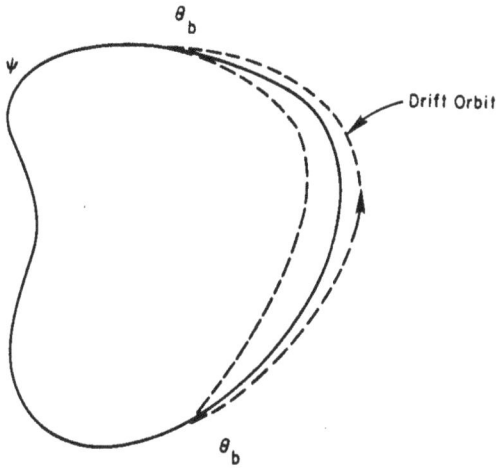

Fig. 3.1 A trapped particle drift orbit.

drift surface characterized by the values of W, μ, regardless of the shape of the flux surface cross section. Note that this is true also in the presence of electric and magnetic perturbations, as long as they are independent of the ignorable coordinated ζ, but we will treat only the case with these perturbations absent.

The motion across the flux surfaces reduces to

$$\dot{\psi}_p = -\frac{g}{D}(\mu + \rho_\parallel^2 B)\frac{\partial B}{\partial \theta}, \tag{3.42}$$

which is second order in ρ and usually odd in θ, since B increases in the direction of decreasing X. To first order in ρ the particle follows the field lines and to next order it drifts across the surfaces, with $\dot{\psi}_p < 0$ for $0 < \theta < \pi$ and $\dot{\psi}_p > 0$ for $-\pi < \theta < 0$, *i.e.* inward drift in the upper half of the torus and outward drift in the lower half. As it moves around θ towards decreasing X it will normally encounter increasing field magnitude, and may encounter a mirror point, where $\rho_\parallel = 0$ because $B(\psi_p, \theta_b) = W/\mu$, θ_b the bounce angle. If such a point exists the particle is called trapped, otherwise it is called passing. Because of drift, this point is not equivalent to an extremum of θ. Since drifts across ψ are proportional to ρ^2, the orbit of a trapped particle has the form of a banana with width $\sim \rho$, as shown in Fig. 3.1. To calculate the drift use conservation of P_ζ and neglect variation of g along the orbit, which gives a higher order correction in ρ, to find $\Delta\psi_p = g\Delta\rho_{\|}$. For a trapped particle $\rho_{\|}$ is zero near the banana tip. The banana width along the orbit is then by conservation of P_ζ given by $\Delta\psi_p = g\rho_{\|}$ where $\rho_{\|} = (2W - 2\mu B)^{1/2}/B$. For a passing particle $\rho_{\|}$ does not vanish, so the excursion from the flux surface has the form $\Delta\psi_p = g[\rho_{\|} - \rho_{\|}(0)]$, and is normally much smaller than for a trapped particle.

The time for the orbit to traverse a range of θ is in general given by

$$T = \int dt = \oint \frac{d\theta}{\dot\theta} \simeq \oint \frac{d\theta D}{\rho_{\|} B^2}, \tag{3.43}$$

where terms higher order in ρ/R have been dropped. The bounce or transit frequency is $\omega_b = 2\pi/T$ with the range of integration being $(-\theta_b, \theta_b)$ for a trapped particle and $(0, 2\pi)$ for a passing particle. For a general axisymmetric equilibrium and general particle energy it is necessary to use the constants of the motion, E, μ, P_ζ, to perform this integration around the drift orbit.

Note that the direction of the drift across the flux surfaces is dependent on the sign of $\rho_{\|}$. Thus particles with opposite velocities, initially at the same point at $\theta = 0$, describe different banana orbits. An ion moving toroidally in the direction of the plasma current (co-moving) begins its trajectory at the outermost point of the banana orbit, whereas a counter-moving ion begins its trajectory at the innermost point. The opposite is true for electrons.

Now classify orbits according to whether they are trapped or passing, and confined or not. We have the two constants of the motion

$$E = \rho_{\|}^2 B^2/2 + \mu B, \tag{3.44}$$

$$P_\zeta = g\rho_{\|} - \psi_p. \tag{3.45}$$

The poloidal flux range is $0 < \psi_p < \psi_w$. Assume the magnetic field to be decreasing outward, so that B_{max} occurs at ψ_w, $\theta = \pi$ and B_{min} occurs at ψ_w, $\theta = 0$, with B_0, the value on axis, between these values. Because of the drift direction, the loss boundary for co-moving particles occurs for orbits touching the right hand edge at the midplane,

$$\frac{(P_\zeta + \psi_w)^2 B_{min}^2}{2g^2(\psi_w)} + \mu B_{min} - E = 0. \tag{3.46}$$

This is a parabola in P_ζ, μ with intercepts at $\mu = 0$ at $P_\zeta = -\psi_w \pm \sqrt{2E}g(\psi_w)/B_{min}$ and the maximum at $\mu = E/B_{min}$. Note that for fixed E, μ, the two signs of the square root correspond to positive and negative parallel velocity.

The loss boundary for counter-moving particles occurs for orbits touching the left hand edge at the midplane,

$$\frac{(P_\zeta + \psi_w)^2 B_{max}^2}{2g^2(\psi_w)} + \mu B_{max} - E = 0. \tag{3.47}$$

This is a parabola in P_ζ, μ, with intercepts at $\mu = 0$ at $P_\zeta = -\psi_w \pm \sqrt{2E}g(\psi_w)/B_{max}$ and the maximum at $\mu = E/B_{max}$. Orbits which pass through the magnetic axis satisfy

$$\frac{P_\zeta^2 B_0^2}{2g^2(0)} + \mu B_0 - E = 0. \tag{3.48}$$

This is a parabola in P_ζ, μ, with intercepts at $\mu = 0$ at $P_\zeta = \pm\sqrt{2E}g(0)/B_0$ and the maximum at $\mu = E/B_0$. Again for fixed E, μ, the two signs of the square root correspond to positive and negative parallel velocity.

Orbits in the ψ_p, θ plane are defined by $f(\psi_p, \theta) = (P_\zeta + \psi_p)^2 B^2/(2g^2) + \mu B - E = 0$. For a separatrix to exist $\nabla f = 0$, and thus they can only exist at the midplane, $\partial_\theta B = 0$. The separatrix orbit and a banana orbit near the separatrix are shown in Fig. 3.2. Note that $\dot\theta = 0$ (the banana tip) and $v_\parallel = 0$ (velocity reversal) do not occur at the same point. The parallel velocity changes sign at the point of minimum B, which is not necessarily at the extreme value of θ, but this correction is higher order in gyro radius, *i.e.* for zero banana width these two points coincide.

Now add the trapped-passing boundary. To leading order in ρ the separatrix occurs at the point where the parallel velocity vanishes. The condition $\rho_\parallel = 0$ gives $P_\zeta = -\psi_p$, and $E = \mu B(\psi_p, \theta)$. Using the fact that the separatrix must occur at the midplane, $\partial_\theta B = 0$, which we take to be at $\theta = 0, \pi$, we find the two curves,

$$\mu B(\psi_p, 0) = E, \qquad P_\zeta = -\psi_p \tag{3.49}$$

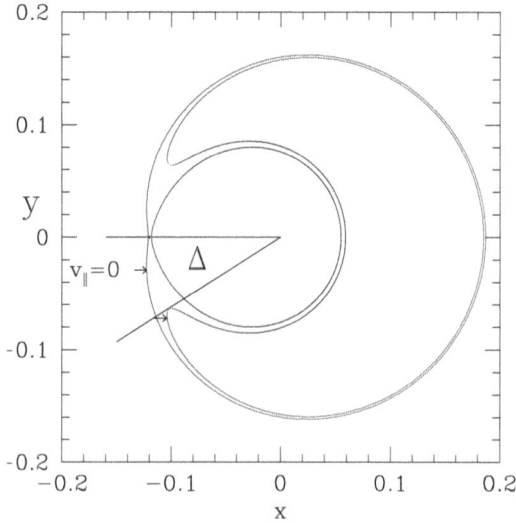

Fig. 3.2   Separatrix and near separatrix orbits.

and

$$\mu B(\psi_p, \pi) = E, \qquad P_\zeta = -\psi_p. \qquad (3.50)$$

Near the magnetic axis $\psi_p \simeq 0$ we can use the large aspect ratio expansion to find $dB/d\psi_p \to \infty$ and thus these two curves meet the axis vertically. Along the vertical line $P_\zeta = -\psi_w$, the parallel velocity $\rho_\parallel$ vanishing implies approximate contact of a banana tip with the wall, giving the left boundary of the trapped particle domain. (Since the banana tip is not exactly the point $\rho_\parallel = 0$ the actual boundary is very closely nearby, given by the contact of that part of the orbit with the largest value of ψ_p with the wall.) But trapped particles are lost when the outer leg of the orbit, with positive pitch, makes contact with the right edge. The vertical line $P_\zeta = -\psi_w$ is the left boundary of lost banana orbits.

These curves are shown in Fig. 3.3. Intersection with the right wall is shown with a solid line. Intersection with the left wall is shown with a long dash. The trapped-passing boundary is shown with a short dash, and intersection with the magnetic axis is shown with alternating short and long dashes. In addition the domains are labelled, with T-L indicating trapped-lost, T-C indicating trapped-confined, $P_\pm - L$ indicating co- and counter

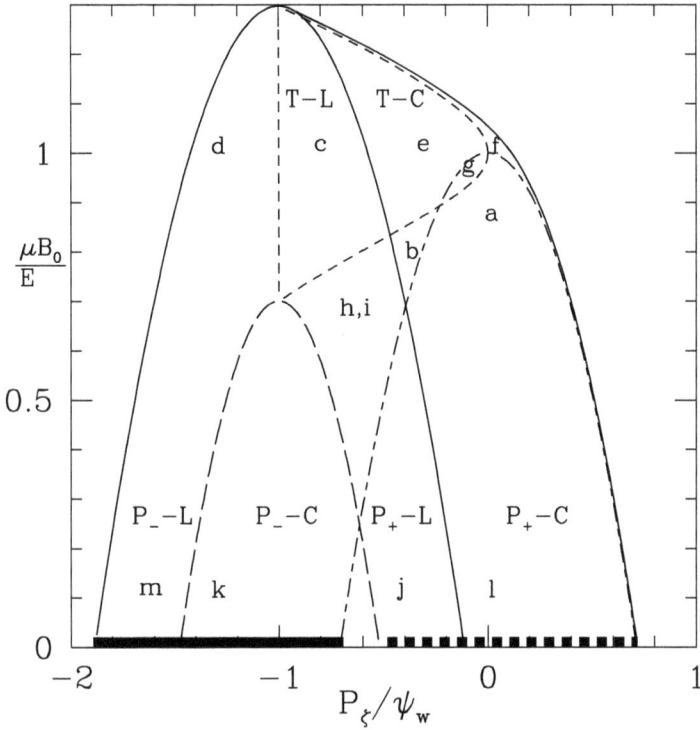

Fig. 3.3   Orbit types in the plane of P_ζ, μ, for fixed high energy.

passing-lost, and $P_\pm - C$ indicating co- and counter passing-confined. This figure is for a JET equilibrium (Wesson, 1997), using high energy particles to show clearly domains which only exist for large banana width. The gyro radius for this figure is 9 cm and the major radius is 299 cm.

To characterize the orbits note that for $\mu = 0$, $i.e.$ passing particles, the range of P_ζ is given by $-\psi_w + \sqrt{2Eg}/B_{max} < P_\zeta < \sqrt{2Eg}/B_0$ for $\rho_\| > 0$, co-passing particles, indicated by dashed shading along the bottom of Fig. 3.3 and bounded on the left by the intersection with the left wall and on the right by the magnetic axis. For counter-passing particles $\rho_\| < 0$ the domain is $-\psi_w - \sqrt{2Eg}/B_{min} < P_\zeta < -\sqrt{2Eg}/B_0$ and is indicated by solid shading, bounded on the left by the intersection with the right wall and on the right by the magnetic axis. Co-moving passing orbits are

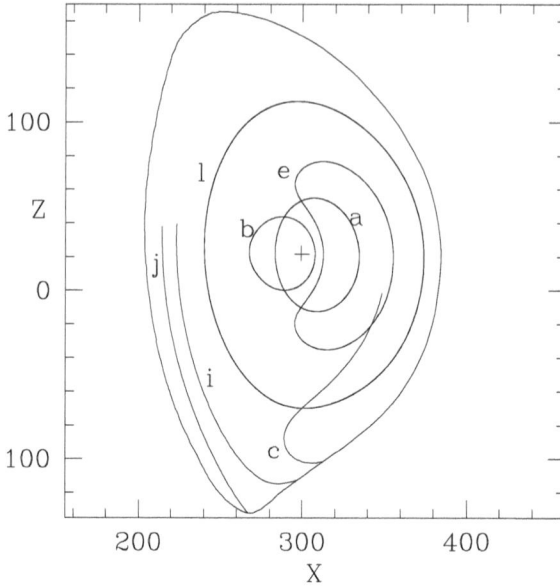

Fig. 3.4   Trajectories of different orbit types, high energy.

shifted by drift to the right, the largest excursion in ψ occurs at the right
wall, so co-moving orbits to the left of the right wall intersection curve
(larger ψ_p) in Fig. 3.3 are lost. Similarly, counter-moving particles are
shifted left by drift, and counter-moving orbits to the left of the left wall
intersection curve (larger ψ_p) are lost. Beginning with a co- or counter-
passing orbit with $\mu = 0$ on the bottom edge of the boundary we can move
to other domains, identifying the orbit type by whether we have passed a
line indicating contact with the wall, passing through the magnetic axis, or
transition from passing to trapped.

The upper trapped-passing boundary does not describe the maximum
value of P_ζ for a fixed value of μ, clearly given by orbits with positive $\rho_\parallel$.
To find this maximum write

$$\frac{dP_\zeta}{d\psi_p} = -1 - \frac{g[\rho_\parallel^2 B + \mu]}{\rho_\parallel B^2}\partial_{\psi_p}B + g'\rho_\parallel = 0. \tag{3.51}$$

Along with $E = \rho_\parallel^2 B^2/2 + \mu B$ this gives for any ψ_p the value of $\rho_\parallel$, and hence
both μ and P_ζ. We can find two asymptotic limits of this curve. First take

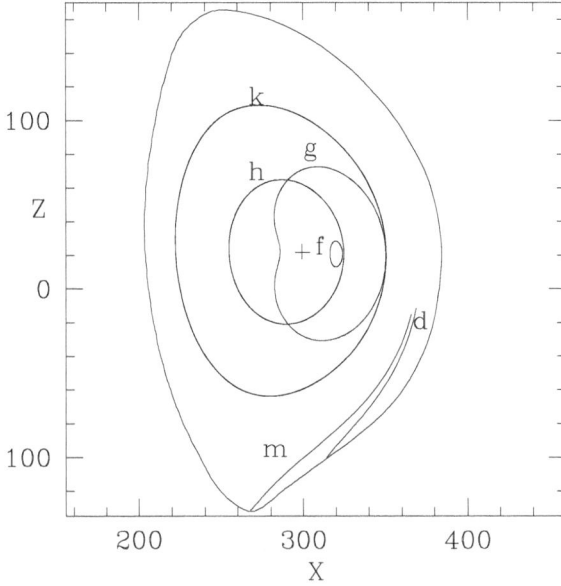

Fig. 3.5   Trajectories of different orbit types, high energy.

$\mu \to 0$, $\rho_\parallel \to \rho$. Then the only possible balance is given by $\partial_{\psi_p} B \sim 1/\rho$, which can happen only near axis. Near axis we can use the large aspect ratio expansion, $\partial_{\psi_p} B = -q/r$, giving that $r \sim \rho$. Thus $\psi_p \sim \rho^2$ and we find $P_\zeta \to g\rho$ for $\mu \to 0$. Thus this curve asymptotes to the right side of the parabola given by Eq. 3.48. Now consider the limit $\rho_\parallel^2 \ll \mu$. Then find that balance requires $\partial_{\psi_p} B \ll 1/\rho$. Consistent with this take $\psi_p \to \psi_w$ giving $\mu = E/B_{min}$ and $P_\zeta = -\psi_w$, and thus the curve asymptotes to the right side of the curve given by Eq. 3.46 and this maximum adds only a thin domain to the right upper edge of Fig. 3.3, as shown.

Various orbit trajectories and their location in the P_ζ, μ plane are shown in Figs. 3.3, 3.4, and 3.5 obtained with the numerical code ORBIT, which numerically integrates the guiding center equations, is written in Fortran 90, and is capable of compilation on many different platforms (White, 2000). Orbits a and l are co-moving confined passing, and j and i are co-moving lost passing orbits. Orbit e is trapped confined and orbit c is trapped lost.

Orbits k and h are counter-moving confined passing orbits and d and m are counter-moving lost passing orbits. From the trapped orbits, such

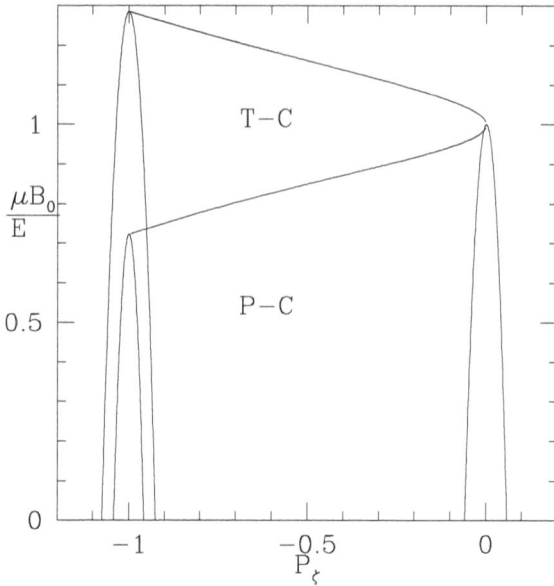

Fig. 3.6   Orbit types in the plane of P_ζ, μ for fixed low energy.

as orbit e, moving to the right crosses the trapped-passing boundary, but
not the magnetic axis. Thus orbit f always has positive parallel velocity,
but it does not circle the magnetic axis. These are refered to as stagnation
orbits, but of course they precess toroidally. There are also counter-moving
stagnation orbits, located on the high field side. Similarly, from orbit e
moving downward it is possible to cross the magnetic axis boundary without
crossing the trapped-passing boundary, leading to orbit g, which is trapped
(parallel velocity changes sign) but still circles the magnetic axis due to
drift. These orbits are called potato orbits, resembling trapped particles,
but with large, almost circular paths in the ψ_p, θ plane. Orbit b is across
the trapped-passing boundary from orbit e, and is counter-moving and
confined. Orbits h and i have practically the same values of P_ζ, μ but i is
co-moving lost and h is counter-moving and confined. Inside the domain
occupied by h and i all co-moving orbits are confined and all counter-
moving orbits are lost. Of course orbits such as f, g exist only due to
drift motion, and these domains are small or entirely missing for lower
energy.

In Fig. 3.6 is shown a similar plot for low energy particles. There are large domains of passing-confined (P-C) and trapped-confined (T-C) orbits. The special stagnation and potato orbit domains are too small to see.

From the equations of motion, Eq. 3.19–3.22, we see that $\dot{\rho}_{\parallel}$ and $\dot{\psi}_p$ vanish at the midplane where $\partial_\theta B = 0$. Setting the equation for $\dot{\theta}$ to zero then gives fixed points in the poloidal plane, orbits that do not move in this plane. However, some of these points are stable and some are unstable points. Orbits in small domains in the vicinity of the stable fixed points define the stagnation orbits. In Fig. 3.7 is an example equilibrium where are shown some of the fixed points corresponding to stable stagnation points. These orbits ocupy two dimensional manifolds in the P_ζ, E, μ volume and hence have measure zero, but are located inside domains of stagnation orbits with nonzero measure.

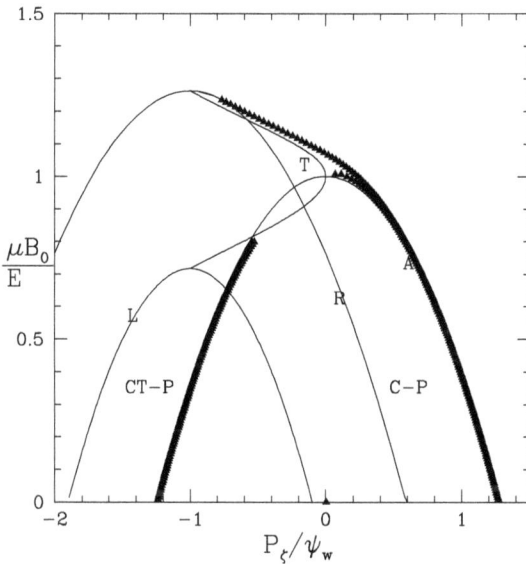

Fig. 3.7   Fixed point stagnation orbits shown with dark triangles. The left branch consists of counter-passing orbits not circling the magnetic axis, situated in the inner midplane, very near the axis. The right branch consists of co-passing orbits which do not circle the axis, situated in the outer midplane.

In the figure the magnetic axis (A), the right edge (R), and the left edge (L) are labeled. Co-passing (C-P) orbits occupy the domain between the right edge and the axis, counter-passing orbits (CT-P) the domain between the axis and the left edge. Also the domain of trapped orbits (T) is labeled. Those orbits on the right edge of the domain correspond to stable fixed points on the outer midplane, $\theta = 0$, and are co-moving. Starting in the co-passing domain, they are obtained by moving through the right parabolic boundary labeled A, which defines orbits passing through the axis with positive pitch. These co-passing orbits, initially circling the axis, move through it to become co-passing orbits not circling the axis, but stagnant in the outer midplane. The points in the lower left correspond to stable fixed points on the inner midplane, $\theta = \pi$, and are counter-moving. Starting in the domain of confined counter-passing orbits and moving to the right we again cross the parabola defining the axis, but on the branch with negative pitch, moving into the narrow domain neighboring it. This narrow domain of stagnation orbits is slightly outside of the original counter-passing domain because these orbits do not circle the axis.

In very small aspect ratio devices, where the poloidal magnetic field can be as strong as the toroidal field, there can exist other unusual orbits, so a new classification must be made using these methods for any equilibria strongly deviating from large aspect ratio devices. See for example Mikkelsen *et al.* 1997.

Since modes with frequency much smaller than the cyclotron frequency do not change μ, the modification of the distribution occurs in the P_ζ, E plane, for a mode with toroidal mode number n and frequency ω along lines $E - P_\zeta \omega/n = c$. Thus it is also of interest to map out the domains of orbit types for a fixed value of μ.

An example of this plane is shown in Fig. 3.8. Shape and size of the various domains changes with the equilibrium parameters, but the general topology is always similar to that shown. The plane of P_ζ, E is shown for $\mu B_0 = 50$ keV with B_0 the magnetic field on axis. The apex of the parabolas are at $E = \mu B_{min}$ (label c) for the high field side (left edge, parabola with label L), $E = \mu B_0$ (label b) for the magnetic axis (parabola with label A), and $E = \mu B_{min}$ (label c) for the low field side (right edge, parabola with label R). All confined particle domains are shaded. The confined counter-passing and co-passing orbits share a common triangular region, in which they have the same values of P_ζ and E but opposite signs of pitch. The small eye-shaped region near point b consists of potato orbits, particles for which $v_\parallel$ vanishes along the orbit, but which circle the magnetic

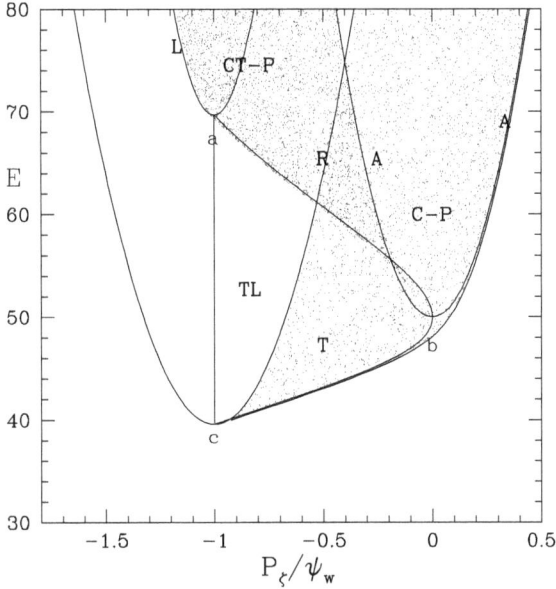

Fig. 3.8   Plane of P_ζ, E with $\mu B_0 = 50$ keV, showing domains of confined particles. Shown are co-passing (C-P), counter-passing (CT-P), trapped (T) and trapped-loss (T-L) domains. The apexes of the parabolas are at $E = \mu B_{min}$ (c), $E = \mu B_0$ (b), $E = \mu B_{max}$ (a).

axis due to drift. The trapped particle domain borders both the co-passing and the counter-passing domains along its upper boundary. The co-passing stagnation orbit domain is seen along the lower right boundary, largest near point b.

3.4   INTEGRAL INVARIANTS

The guiding center equations give the particle motion to first order in gyro radius. In analytical dynamics exact constants of the motion are usually obtained by finding canonical variables. The Noether theorem states that invariance of the Hamiltonian with respect to translations in a variable q means that the conjugate variable p, which is the generator for infinitesimal translations in q, is conserved. A systematic method of finding canonical variables, and thus adiabatic invariants to any order in ρ, has been devel-

oped using the Lie algebra of the generators (Littlejohn, 1982). Even with
these powerful methods, however, proceeding beyond the guiding center
approximation is quite involved. Within the guiding center approximation,
constants of the motion are readily identified through the Hamiltonian formalism. A charged particle has three degrees of freedom, and this is the
maximum number of adiabatic invariants the motion can possess. In deriving the guiding center equations it is assumed that any time dependence of
the field is slow compared to the gyro motion, and that spatial variation is
small on a scale of the gyro radius. This is reflected by the appearance of
the magnetic moment μ as a conserved quantity in the Lagrangian, and the
fact that H is independent of gyro phase. An earlier proof of the invariance
of μ under these conditions can be found in Northrop (1963).

Each adiabatic invariant is associated with a periodicity. In a symmetric
equilibrium, besides the transverse invariant, or magnetic moment μ associated with the gyro period, the other two invariants are quickly identified
using the canonical variables. Because of symmetry the integral $\oint P_\theta d\theta$
around the bounce orbit is a constant of the motion, since the orbit is a
closed curve in the P_θ, θ plane. Using the first order expression for $\dot{\theta}$ and
$\rho_\| B dt = dl$ to give $\psi = gq[\rho_\| - \rho_\|(0)] + \psi_0 + O(\rho^2)$ we find

$$\oint P_\theta d\theta = \oint d\theta[\psi_0 - gq\rho_\|(0)] + \oint v_\| dl. \qquad (3.52)$$

Within terms of second order in the gyro radius, the first term is zero for a
trapped particle, and a constant, approximately equal to the toroidal flux
contained within the orbit for a passing particle, and the second term is
the familiar longitudinal invariant J, associated with the periodic bounce
motion. Finally, P_ζ itself is invariant in systems which are quasi-symmetric.

The adiabatic invariants are useful in that they are approximately conserved even if the symmetry in question is broken. What is essential is that
the symmetry breaking terms produce small deviation of the inital closed
orbit in one period, *i.e.* that the gyro orbit, or the banana orbit, or the
toroidal drift surface, almost closes. The requirement that the orbit be
almost periodic is clearly related to the nature of the trajectories in the
space of the canonical variables. If the particle trajectories are stochastic
the orbit is far from periodic and the invariants are not conserved. The loss
of the adiabatic invariants is related to the Chirikov overlap of resonances
in the space of trajectories.

Although in an idealized symmetric equilibrium all three invariants are
conserved, perturbations of the field that break symmetry can destroy the

invariance of both P_ζ and J. Examples of this will be seen in Chap. 9, where it is shown how the loss of these invariants can greatly increase particle loss rates.

The magnetic moment μ requires very small scale or high frequency perturbations for its destruction, and it is much better preserved.

The guiding center equations result from an asymptotic expansion of the equations of motion in the gyro radius, meaning that truncation of the expansion at order ρ^n gives the particle position correctly within terms of order ρ^{n+1}. In uniform magnetic and electric fields the expansion for the transverse invariant, or magnetic moment μ, truncates exactly at $\mu = v_\perp^2/2B$, since all higher order terms involve gradients of B. In addition, the magnetic moment μ is constant to all orders, *i.e.* the expansion for μ in powers of ρ is a true asymptotic series. This means that if a particle, initially in a region of uniform field, where μ is exactly $v_\perp^2/2B$, traverses a region of time and space dependent fields and returns, the change in μ vanishes faster than any power in ρ, being for example of the form $e^{-a/\rho}$.

To illustrate the approximate conservation of an invariant in the presence of a magnetic perturbation consider the full guiding center equations including a perturbation of the form given by Eq. 3.87. The calculation for a general perturbation is more complicated, but not essentially different. The adiabatic invariant J is associated with $I = \int P_\theta d\theta$. Differentiate with respect to time, giving

$$\frac{dI}{dt} = \oint \left[\dot{P}_\theta \frac{\partial H}{\partial P_\theta} + P_\theta \frac{d}{dt} \frac{\partial H}{\partial P_\theta} \right] dt. \tag{3.53}$$

Integrate the second term by parts, giving

$$\frac{dI}{dt} = P_\theta(t) \frac{\partial H}{\partial P_\theta}(t) - P_\theta(t-\tau) \frac{\partial H}{\partial P_\theta}(t-\tau) \tag{3.54}$$

where τ is the bounce period. If $\alpha = 0$ the orbit is closed and I is conserved. Due to the perturbation α the orbit does not close, and the gap can be calculated directly from the equations of motion, giving to lowest order in ρ and α

$$\Delta\theta = \oint d\theta g \frac{\partial\alpha}{\partial\psi_p}$$

$$\Delta\psi_p = -\oint d\theta \left[I \frac{\partial\alpha}{\partial\zeta} + g \frac{\partial\alpha}{\partial\theta} \right]. \tag{3.55}$$

Here and in the following we use the unperturbed closed orbit to evaluate the integrals. Taking account of the perturbed orbit would only add terms of order α^2. Substituting expressions for P_θ, and $\partial H/\partial P_\theta$ into Eq. 3.54 and expanding to lowest order in $\Delta\theta$, $\Delta\psi_p$ we find

$$\frac{dI}{dt} = \rho_\| G(\psi_p, \theta, \alpha) + O(\rho^2) \qquad (3.56)$$

where G is a linear functional of α. From Eq. 3.56 we have the important result that to first order in ρ, dI/dt is periodic with zero average value, and

$$\frac{1}{T} \oint \frac{dI}{dt} dt = O(\rho^2) \qquad (3.57)$$

and thus I is conserved to order ρ^2, except for periodic oscillation at the bounce frequency, also in a nonsymmetric field. This property is demonstrated by Northrop (1979) by writing $\oint dt\, dI/dt = \oint \oint dt\, dt'\, K(t, t')$ and showing that to leading order in ρ K is antisymmetric in t, t'. A proof by Littlejohn (1982) makes use of Lagrange and Poisson brackets.

3.5   TOROIDAL PRECESSION

The average toroidal precession of an orbit is given by the time average of $\dot{\phi} = \dot{\zeta} + \dot{\nu}$ over one orbital transit period, $T = \oint dt$. Converting to integrals over θ and ψ_p, the term in ν integrates to zero because it is single valued on the torus and the orbit is closed in this plane. Thus the average toroidal precession is given by $\bar{\dot{\zeta}} = (1/T) \oint dt\dot{\zeta}$. Substitute the value of $\dot{\zeta}$ given by Eq. 3.22 and convert to an integral over θ, giving

$$\bar{\dot{\zeta}} = \frac{1}{T} \oint q d\theta - \frac{(gq + I)}{T} \oint \frac{\Delta}{\rho_\|} d\theta + \frac{(I' + g'q)}{T} \oint d\theta \rho_\| \qquad (3.58)$$

where

$$T = \oint d\theta \frac{D}{\rho_\| B^2}, \qquad \Delta = \frac{(\rho_\|^2 B + \mu)}{B^2} \partial_{\psi_p} B,$$

where terms of order ρ^3 have been neglected in taking the functions I, g outside the integral. The term $\oint q d\theta$ is due to toroidal motion following the field line and for trapped particles can be estimated through the shear as $gq' \oint \rho_\| d\theta$ with $q' = dq/d\psi_p$. The term in $I' + g'q$ is generally very small.

For trapped particles the toroidal drift reduces to

$$\bar{\dot{\zeta}} = \frac{2gq'\sqrt{2W}}{T} \int_{-\theta_b}^{\theta_b} d\theta \frac{\sqrt{1 - B(\theta)/B(\theta_b)}}{B}$$
$$- \frac{gq + I}{2T} \sqrt{2W} \int_{-\theta_b}^{\theta_b} d\theta \frac{2 - B(\theta)/B(\theta_b)}{[1 - B(\theta)/B(\theta_b)]^{1/2}} \frac{1}{B^2} \frac{\partial B}{\partial \psi_p}. \tag{3.59}$$

Normally $\partial B/\partial \psi_p$ is negative for deeply trapped particles (θ_b small) and thus the toroidal precession is positive. Either a very large bounce angle, so that the particle spends much time near $\theta = \pi$, where $\partial B/\partial \psi_p$ is positive, or sufficient plasma pressure to create a magnetic well and make $\partial B/\partial \psi_p$ positive also near $\theta = 0$, can reverse the sign of the toroidal drift.

The toroidal drift is a result of the cross field drift described in Sec. 3.3. The component across the flux surfaces is odd in θ, resulting in the finite banana width, but due to the helicity of $\vec{B}$ there is a toroidal component and it is even in θ, resulting in a net toroidal precession of the orbit.

The toroidal drift is directly related to the first adiabatic invariant. From $\rho_\parallel^2 = (2W - 2\mu B)/B^2$ find $\partial \rho_\parallel/\partial W|_{\psi,\mu} = 1/B^2 \rho_\parallel$ and $\partial \rho_\parallel/\partial \psi_p|_{W,\mu} = -\Delta/g\rho_\parallel$, giving

$$\bar{\dot{\zeta}} = \frac{\partial J/\partial \psi_p|_{W,\mu}}{\partial J/\partial W|_{\psi,\mu}} \tag{3.60}$$

where $J = \oint D\rho_\parallel d\theta$ is the longitudinal invariant. The denominator expression $\partial J/\partial W|_{\psi,\mu}$ is simply the bounce time T. Thus if the unit of time is taken to be T, corresponding to bounce averaging, the toroidal drift equation

$$\bar{\dot{\zeta}} = \frac{\partial J}{\partial \psi_p} \tag{3.61}$$

becomes of Hamiltonian form, with J the Hamiltonian. Thus we have a sequence of contractions, beginning with the Hamiltonian for charged particle motion, with three conjugate variables, P_x, P_y, P_z and x, y, z. The first contraction, through gyro-phase averaging, over time large compared to the gyro period, results in the Hamiltonian guiding center motion, with two conjugate variables, P_ζ, P_θ and ζ, θ. For both of these systems the Hamiltonian is the particle energy. The final contraction, by averaging over the bounce time, leads to the single conjugate variables, ψ_p, ζ, and the Hamiltonian is the longitudinal invariant J.

3.6 LARGE ASPECT RATIO

In a lowest order tokamak equilibrium as given in Sec. 2.9, where the magnetic field has the form $B = 1 - r cos\theta$, the parallel and perpendicular velocities have the form $v_\perp^2 = 2\mu(1 - r cos\theta)$, $v_\parallel^2 = 2W - 2\mu + 2\mu r cos\theta$. For narrow banana width the passing particles are those with $v_\parallel^2/v_\perp^2 > 2r/R$ at $\theta = 0$. This defines forward and backward cones. For an isotropic distribution the fraction of trapped particles is then approximately $(2r/R)^{1/2}$. The particle drifts in a lowest order tokamak equilibrium have a very simple form. Substituting the equilibrium expressions into the equations of motion in Sec. 3.3 we find for the second order drifts

$$r\dot\theta = -(\mu + \rho_\parallel^2 B)cos\theta$$
$$\dot r = -(\mu + \rho_\parallel^2 B)sin\theta, \qquad (3.62)$$

from which it follows that the horizontal drift is zero, and the vertical drift is $\dot z = -(\mu + \rho_\parallel^2 B)$, *i.e.* the drift motion is vertically downward throughout the orbit. Here "downward" refers to the direction of magnetic field at the toroidal axis $R = 0$ produced by the toroidal current flowing in the plasma. The sign is of course reversed for electrons.

Notice that this drift is independent of q. If there is no poloidal field, $(q = \infty)$, the drift leads quickly to a charge separation, giving a vertical electric field, producing an $\vec E \times \vec B$ drift outwards of both species. On the level of the particle orbits this is the reason for the nonexistence of the equilibrium in this case, as found in Sec. 2.8. A poloidal field carries the orbit around in θ so that the vertical drift moves it radially inward for half the orbit and radially outward for the other half, and the orbits are closed in the r, θ plane.

The banana width is readily found from the analysis in Sec. 3.3 for a banana with $\theta_b = \pi/2$ to be $\Delta r = 2(R/r)^{1/2}q\rho$. For a passing particle the maximum drift from the flux surface is $\Delta r = q(R/r)\Delta\rho_\parallel \simeq 2q\rho$.

The time for the orbit to traverse a range of θ is given by Eq. 3.43, with the bounce or transit frequency $\omega_b = 2\pi/T$ and the range of integration being one full bounce for a trapped particle and $(0, 2\pi)$ for a passing particle.

It is worth evaluating the bounce and transit frequencies in a lowest order tokamak equilibrium to find the scaling and dependence. Ignoring drift motion from the surface r (narrow banana width or low energy approximation) and substituting the lowest order expression $B = B_0(1 - r cos\theta)$

and for consistency using the lowest order expressions $g = 1 + O(r^2/R^2)$ and $I = r^2/q$, we find the period $T = q \int d\theta / \sqrt{2E - 2\mu B_0(1 - r\cos\theta)}$ or

$$T = 2q \int_{-\theta_b}^{\theta_b} d\theta \frac{1}{\sqrt{2\mu B_0 r(\cos\theta - \cos\theta_b)}} \tag{3.63}$$

with $\cos\theta_b = (\mu B_0 - E)/\mu B_0 r$ and the restriction that this number have magnitude less than 1, *i.e.* this expression is valid for trapped particles only. Let $\kappa = \sin(\theta_b/2)$ and substitute $\kappa\sin\phi = \sin(\theta/2)$, whereupon we find the elliptic integral

$$\int_0^{\pi/2} \frac{d\phi}{\sqrt{1 - \kappa^2\sin^2\phi}} = K(\kappa). \tag{3.64}$$

The period then reduces to $T = 4qK(\kappa)/\sqrt{\mu B_0 r}$ with κ restricted to $[0,1]$. The bounce frequency is

$$\omega_b = \frac{\pi\sqrt{\mu B_0 r}}{2qK(\kappa)}. \tag{3.65}$$

In the deeply trapped limit $\kappa = 0$, $K = \pi/2$ and restoring time in units of ω_0, distance in units of R, and energy in units of $m\omega_0^2 R^2$, we have for the bounce frequency

$$\omega_b(0) = \sqrt{\frac{\mu B_0}{mc^2}} \sqrt{\frac{r}{R}} \frac{c}{Rq}, \tag{3.66}$$

which is easily evaluated using $mc^2 = 931$ MeV per proton. Also $\omega_0 = 9.58 \times 10^3 ZB/m$ with Z the charge in proton units and B in Gauss.

For passing particles write $B = B_0[1-r+r(1-\cos\theta)]$ and define B_e to be the field strength at $\theta = 0$, $B_e = B_0(1-r)$, giving $B = B_e + 2B_0 r\sin^2(\theta/2)$. We then discover that $(E - \mu B_e)/(2\mu B_0 r) = \kappa^2$ giving

$$T = \frac{q}{\kappa\sqrt{\mu B_0 r}} \int_0^\pi \frac{d\theta}{(1 - \sin^2(\theta/2)/\kappa^2)} = \frac{2qK(1/\kappa)}{\kappa\sqrt{\mu B_0 r}}, \tag{3.67}$$

and the transit frequency is

$$\omega_t = \frac{\pi\kappa\sqrt{\mu B_0 r}}{qK(1/\kappa)}. \tag{3.68}$$

In Fig. 3.9 are shown the bounce and transit frequencies as a function of the parameter κ, normalized to the deeply trapped value. Particles on the boundary between trapped and passing ($\kappa = 1$) spend infinite time

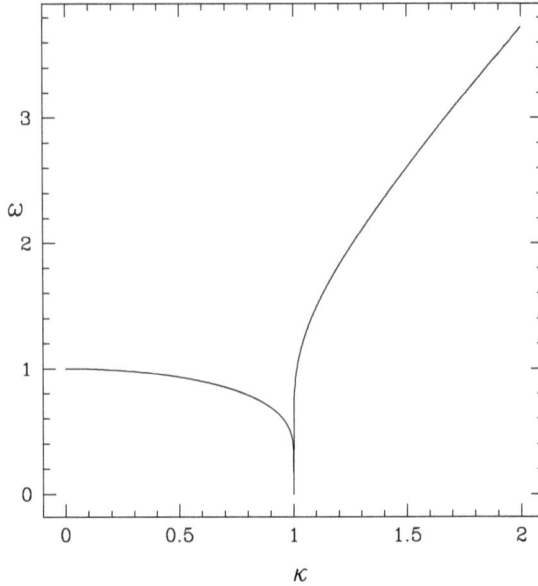

Fig. 3.9   Bounce ($\kappa < 1$) and transit ($\kappa > 1$) frequency.

at the reflection point $\theta = \pi$, and thus $\omega_b \to 0$. The approach to this point is very steep, because for $\kappa \simeq 1$, or $\kappa' = \sqrt{1 - \kappa^2} \ll 1$, we have $K \simeq ln(4/\kappa') + (\kappa'^2/4)ln(4/\kappa')$, giving a lograthmic approach to the point $\kappa = 1$. Note that in the limit of purely passing particles ($\kappa \to \infty$, $\mu \to 0$) one must use $\kappa\sqrt{2\mu B_0 r} \to \sqrt{E}$ giving $\omega_t \to v_\parallel/qR$. For a derivation giving also the action associated with the bounce motion see Brizard (2011).

The toroidal precession rate ω_d is, from Eq. 3.59,

$$\omega_d = \frac{4q^2 s\rho}{r^{3/2}T}\int_0^{\theta_b} d\theta(cos\theta - cos\theta_b)^{1/2} + \frac{4\mu q^2}{r^{3/2}\rho T}\int_0^{\theta_b} d\theta \frac{cos\theta}{(cos\theta - cos\theta_b)^{1/2}} \quad (3.69)$$

with $s = r(dq/dr)/q$ the shear, which upon similar substitution of variables becomes

$$\omega_d = \frac{2qs\rho^2}{r}\frac{[E(\kappa) + (\kappa^2 - 1)K(\kappa)]}{K(\kappa)} + \frac{\mu q}{r}\frac{[2E(\kappa) - K(\kappa)]}{K(\kappa)} \quad (3.70)$$

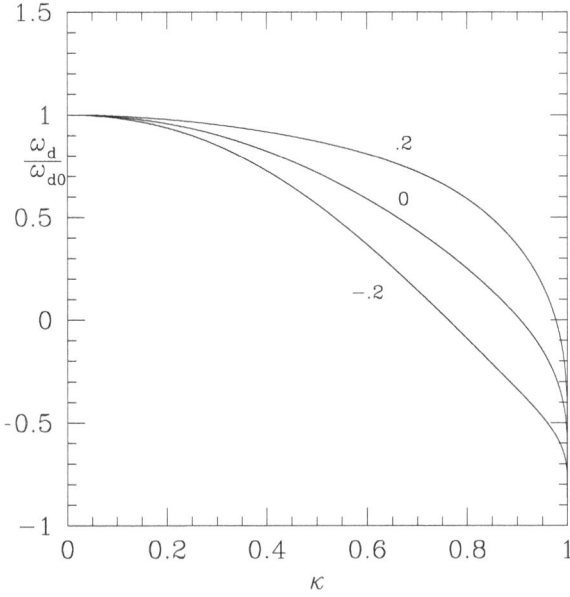

Fig. 3.10   Toroidal precession rate vs κ, with shear $s = -.2, 0, .2$.

with the elliptic integral $E(\kappa) = \int_0^{\pi/2} d\phi \sqrt{1 - \kappa^2 sin^2\phi}$. The toroidal motion in half a bounce simplifies to

$$d\zeta \simeq 2\sqrt{2}(\rho/R)q^2(R/r)^{3/2}[2E(\kappa) - K(\kappa)]. \tag{3.71}$$

The deeply trapped limit is $\kappa \ll 1$, and in this limit $K \simeq \pi/2 + \pi\kappa^2/8$, $E \simeq \pi/2 - \pi\kappa^2/8$ giving

$$\omega_d \simeq \frac{2Eqs\kappa^2}{r} + \frac{Eq(1 - \kappa^2)}{r} \tag{3.72}$$

and the value at $\kappa = 0$ is $\omega_{d0} = Eq/r$, which upon restoring the units for energy $(m\omega_0^2 R^2)$, distance (R), and time (w_0^{-1}) becomes $\omega_{d0} \simeq Eq/(mRr\omega_0)$. A convenient way to express the $\kappa = 0$ limit is $\omega_{d0} = (E/mc^2)qc^2/(rR\omega_0)$, recalling that $mc^2 = 931$ MeV per proton and that $\omega_0 = 9.58 \times 10^3 ZB/m$ with Z the charge in proton units, B in Gauss and m in proton masses. In this expression, if ω_0 is regarded as a signed quantity, the correct direction of precession is obtained also for electrons.

The barely trapped limit is given by $\kappa \simeq 1$, or $\kappa' = \sqrt{1 - \kappa^2} \ll 1$, in which case $K \simeq ln(4/\kappa') + (\kappa'^2/4)ln(4/\kappa')$, and $E \simeq 1 + (\kappa'^2/2)ln(4/\kappa')$, giving

$$\omega_d = \frac{2qs\rho^2}{rln(4/\kappa')} - \frac{\mu q}{r}\left(1 - \frac{2}{ln(4/\kappa')} - \frac{\kappa'^2}{4}\right). \tag{3.73}$$

For large bounce angle $\kappa' \to 0$ the precession reverses. In Fig. 3.10 is shown the typical toroidal precession as a function of κ, for values of shear equal to $s = -.2, 0, .2$. Note that for any value of shear the precession in the limit of barely trapped is exactly the negative of the deeply trapped result, although the log dependence makes the approach to this value very slow.

In the presence of a radial electric field the dominant effect of the field on a trapped particle is to modify the parallel velocity on the inside and outside of the banana orbit. Use $\rho_\| = \sqrt{2E - 2\mu B - 2\Phi}/B$ to find approximately $\partial_{\psi_p}\rho_\| = -\partial_{\psi_p}\Phi/(B^2\rho_\|)$, giving

$$\langle\dot{\zeta}\rangle = \frac{1}{T}\oint\frac{\rho_\| B^2 q}{D}dt = -\frac{B^2 q}{DT}\oint\frac{\partial_{\psi_p}\Phi\Delta\psi_p}{\rho_\|}dt. \tag{3.74}$$

Then use $\Delta\psi_p = g\rho_\|$ for the banana orbit to find

$$\langle\dot{\zeta}\rangle = -\frac{gq\partial_{\psi_p}\Phi}{D}\frac{1}{T}\oint dt. \tag{3.75}$$

Thus to leading order in large aspect ratio precession of a trapped particle due to a radial electric field is given by

$$\frac{\omega_\phi}{\omega_0} = -\frac{q}{r}\frac{\rho^2}{2E}\partial_r\Phi \tag{3.76}$$

with Φ the electric potential, ω_ϕ the toroidal rotation rate, ω_0 the gyro frequency, ρ the gyro radius, and E the particle energy.

3.7   DIAMAGNETIC CURRENT

As seen in Sec. 2.10, a plasma develops a poloidal current, acting to decrease the toroidal field and thus leading to a lower energy state. In this section we consider this diamagnetic current from a kinetic viewpoint. For simplicity consider a uniform temperature plasma with density $n(r)$, equal to zero outside some radius. To calculate the poloidal current we consider

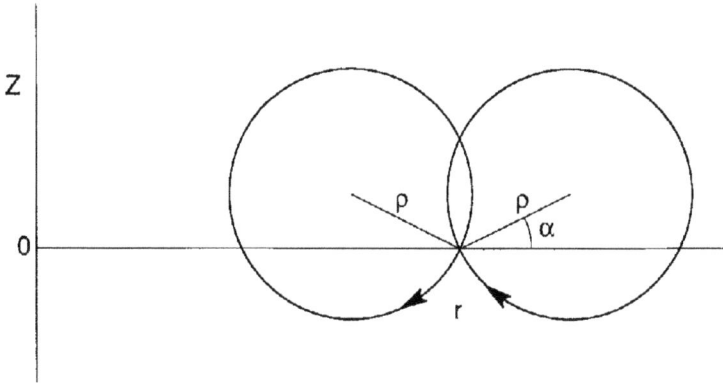

Fig. 3.11   Particles of gyro radius ρ contributing to poloidal current.

all particles passing through the plane $\theta = 0$ between $r, r + dr$. The current density produced by such particles is given by

$$j_\theta(r) = \frac{e\omega_0}{2\pi} \int_0^\pi \rho\cos\alpha d\alpha [n(r + \rho\cos\alpha) - n(r - \rho\cos\alpha)] \qquad (3.77)$$

with ρ the gyro radius and ω_0 the gyro frequency. See Fig. 3.11 for the geometry. To leading order in the gyro radius this is

$$j_\theta(r) = \frac{1}{B} \frac{dp}{dr} \qquad (3.78)$$

where $p = (1/2)nm\rho^2\omega_0^2$ is the plasma pressure. This agrees with the diamagnetic current found from force balance in Sec. 2.8.

Now consider the total diamagnetic current flowing in the plasma. By integrating Eq. 3.77 over r we find

$$I = \frac{e\omega_0}{2\pi} n(0)\pi\rho^2, \qquad (3.79)$$

which is of course the current due to those particles which encircle the magnetic axis, these being the only ones which carry a net poloidal current.

3.8   CONFINEMENT OF FUSION ALPHA PARTICLES

The confinement of high energy particles places a restriction on the mini-mum poloidal field with which a toroidal device can operate. This follows from requiring that the width of the banana orbit be smaller than the plasma radius. From Sec. 3.3 using conservation of P_ζ and energy we find the banana width in poloidal flux to be

$$\Delta\psi_p \approx g\rho\sqrt{\frac{B_{max}}{B_{min}} - 1} \qquad (3.80)$$

where B_{max} is the field strength at the bounce point, B_{min} the value at the outer midplane, and ρ the gyro radius. The device must have more poloidal flux than this. For a large aspect ratio tokamak this reduces to

$$2(R/r)^{1/2}q\rho < r. \qquad (3.81)$$

This gives a condition on the poloidal field, and thus in a large aspect ratio tokamak a minimum plasma current, of

$$I > 2 \times 10^6 \sqrt{r/R} \ amps \qquad (3.82)$$

for confinement of 3.5 MeV α particles. This condition is minimal, it is also necessary that there are no other loss channels capable of producing rapid loss. Such loss channels can be introduced by symmetry breaking due to either the presence of harmonic components in the equilibrium field, discussed in Sec. 12.7, or by the presence of instabilities producing waves capable of resonating with the alpha population, discussed in Sec. 8.1 and in Sec. 8.18.

3.9   MAGNETIC PERTURBATIONS

3.9.1   *Toroidal field ripple*

At this point, although the equilibrium is taken axisymmetric, we may allow for a ζ dependence of B, to account for toroidal field ripple, and also a toroidal dependence of the electric potential Φ. The equations of motion are then

$$\dot{\rho}_\parallel = \frac{C}{D}\left[(\mu + \rho_\parallel^2 B)\frac{\partial B}{\partial\theta} + \frac{\partial\Phi}{\partial\theta}\right] - \frac{F}{D}\left[(\mu + \rho_\parallel^2 B)\frac{\partial B}{\partial\zeta} + \frac{\partial\Phi}{\partial\zeta}\right], \quad (3.83)$$

$$\dot{\psi}_p = -\frac{g}{D}\left[(\mu + \rho_\parallel^2 B)\frac{\partial B}{\partial \theta} + \frac{\partial \Phi}{\partial \theta}\right] + \frac{I}{D}\left[(\mu + \rho_\parallel^2 B)\frac{\partial B}{\partial \zeta} + \frac{\partial \Phi}{\partial \zeta}\right], \quad (3.84)$$

$$\dot{\theta} = \frac{-C\rho_\parallel B^2}{D} + \frac{g}{D}\left[(\mu + \rho_\parallel^2 B)\frac{\partial B}{\partial \psi_p} + \frac{\partial \Phi}{\partial \psi_p}\right], \quad (3.85)$$

$$\dot{\zeta} = \frac{F\rho_\parallel B^2}{D} - \frac{I}{D}\left[(\mu + \rho_\parallel^2 B)\frac{\partial B}{\partial \psi_p} + \frac{\partial \Phi}{\partial \psi_p}\right]. \quad (3.86)$$

Although the ripple can be made small by increasing the number of toroidal field coils in a device, even small ripple can lead to stochastic loss of high energy trapped particles. This effect will be discussed in Sec. 12.7.

3.9.2 *Flute modes*

As shown in Sec. 1.7, very small resonant perturbations of an equilibrium field can produce large islands and stochastic domains in the magnetic field. It is useful to extend the guiding center formalism to include field perturbations so that particle motion can be studied when magnetic surfaces are not well defined. To this end we consider field perturbations of a restricted nature. The generalization is straightforward but tedious, and arbitrary perturbations are unnecessary for most problems. In a general stellarator the magnitude of B depends on all three variables. In axisymmetric approximation the magnitude of B in a tokamak depends only on ψ, θ, but simple modification of the scalar magnitude of B, depending on ζ, describes magnetic ripple produced by discrete toroidal field coils and introduces magnetic mirroring. All tearing and shear Alfvén perturbations are perturbations of $\vec{B}$ primarily orthogonal to the original $\vec{B}$, a type of perturbation known as a flute mode. These latter can be described through

$$\delta\vec{B} = \nabla \times \alpha\vec{B} \quad (3.87)$$

with α an arbitrary function of ψ_p, θ, ζ, t. This form represents exactly the $\nabla\psi$ component of any perturbation, which is the component responsible for perturbations of the flux surfaces. From the analysis in Sec. 1.7, if α has the form $\alpha = \alpha_{mn}sin(n\zeta - m\theta)$ it produces a magnetic island at ψ_p

with $q(\psi_p) = m/n$, of width

$$\delta\psi_p = 4\left(\frac{(mg + nI)\alpha_{mn}}{ns}\right)^{1/2} \qquad (3.88)$$

where $s = q'/q$ is the local shear and primes denote derivatives with respect to the poloidal flux ψ_p.

The equations of motion are changed only by the substitution of $\vec{A}$ by $\vec{A} + \alpha\vec{B}$, into the Lagrangian, Eq. 3.26, giving

$$L = (\psi + \rho_\| I + \alpha I)\dot\theta + (\rho_\| g + \alpha g - \psi_p)\dot\zeta + \mu\dot\xi - H. \qquad (3.89)$$

Note that this is still in canonical Hamiltonian form $L = \sum p_k\dot{q}_k - H$, only the canonical momenta have been modified by α. Lagrange's equations then give

$$\begin{vmatrix} 0 & 0 & I & g \\ 0 & 0 & F & C \\ -I & -F & 0 & K \\ -g & -C & -K & 0 \end{vmatrix}\begin{vmatrix} \dot\rho_\| \\ \dot\psi_p \\ \dot\theta \\ \dot\zeta \end{vmatrix} = \begin{vmatrix} \partial_{\rho_\|} H \\ \partial_{\psi_p} H \\ \partial_\theta H + I\partial_t\alpha \\ \partial_\zeta H + g\partial_t\alpha \end{vmatrix} \qquad (3.90)$$

where $C = -1 + (\rho_\| + \alpha)g'_{\psi_p} + g\alpha'_{\psi_p}$, $K = g\alpha'_\theta - I\alpha'_\zeta$, $F = q + (\rho_\| + \alpha)I'_{\psi_p} + I\alpha'_{\psi_p}$, and we have used Eq. 2.25, $g'_\theta = I'_\zeta$. (Here we use the notation f'_α to denote $\partial_\alpha f$.)

Inverting this equation gives

$$\begin{vmatrix} \dot\rho_\| \\ \dot\psi_p \\ \dot\theta \\ \dot\zeta \end{vmatrix} = \frac{1}{D}\begin{vmatrix} 0 & -K & C & -F \\ K & 0 & -g & I \\ -C & g & 0 & 0 \\ F & -I & 0 & 0 \end{vmatrix}\begin{vmatrix} \partial_{\rho_\|} H \\ \partial_{\psi_p} H \\ \partial_\theta H + I\partial_t\alpha \\ \partial_\zeta H + g\partial_t\alpha \end{vmatrix} \qquad (3.91)$$

with denominator $D = gF - IC$,

$$D = gq + I + (\rho_\| + \alpha)(gI'_{\psi_p} - Ig'_{\psi_p}). \qquad (3.92)$$

The equations become

$$\dot\rho_\| = \frac{C}{D}\left[(\mu + \rho_\|^2 B)\frac{\partial B}{\partial\theta} + \frac{\partial\Phi}{\partial\theta}\right] - \frac{K}{D}\left[(\mu + \rho_\|^2 B)\frac{\partial B}{\partial\psi_p} + \frac{\partial\Phi}{\partial\psi_p}\right]$$
$$- \frac{F}{D}\left[(\mu + \rho_\|^2 B)\frac{\partial B}{\partial\zeta} + \frac{\partial\Phi}{\partial\zeta}\right] - \frac{\partial\alpha}{\partial t}, \quad (3.93)$$

$$\dot{\psi}_p = \frac{K\rho_\| B^2}{D} - \frac{g}{D}\left[(\mu + \rho_\|^2 B)\frac{\partial B}{\partial \theta} + \frac{\partial \Phi}{\partial \theta}\right] + \frac{I}{D}\left[(\mu + \rho_\|^2 B)\frac{\partial B}{\partial \zeta} + \frac{\partial \Phi}{\partial \zeta}\right] \quad (3.94)$$

$$\dot{\theta} = \frac{-C\rho_\| B^2}{D} + \frac{g}{D}\left[(\mu + \rho_\|^2 B)\frac{\partial B}{\partial \psi_p} + \frac{\partial \Phi}{\partial \psi_p}\right], \quad (3.95)$$

$$\dot{\zeta} = \frac{F\rho_\| B^2}{D} - \frac{I}{D}\left[(\mu + \rho_\|^2 B)\frac{\partial B}{\partial \psi_p} + \frac{\partial \Phi}{\partial \psi_p}\right], \quad (3.96)$$

and the $\partial_t \alpha$ term in Eq. 3.93 takes into account any explicit time dependence of the perturbation.

This formalism is applicable for modes with frequency small compared to the cyclotron frequency. If the mode is an ideal MHD mode the induced electric field parallel to $\vec{B}$ is shorted out by the rapid response of the electrons, and with the field perturbation given by $\delta\vec{B} = \nabla \times \alpha\vec{B}$ with $\alpha = \sum_{mn} \alpha_{m,n} e^{i(n\zeta - m\theta - \omega t)}$ it is necessary to add an electric potential Φ to cancel the parallel electric field induced by $d\vec{B}/dt$, with

$$\sum_{m,n} \omega B\alpha_{m,n} e^{i(n\zeta - m\theta - \omega t)} - \vec{B} \cdot \nabla\Phi/B = 0, \quad (3.97)$$

where we have neglected terms of order α^2. In Boozer coordinates, using the Fourier decomposition $\Phi = \sum_{m,n} \Phi_{m,n}(\psi_p) e^{i(n\zeta - m\theta - \omega t)}$, the solution is

$$(gq + I)\omega\alpha_{m,n} = (nq - m)\Phi_{m,n}, \quad (3.98)$$

but in general coordinates where $I = I(\psi, \theta)$, the solution is complicated by the coupling of different poloidal harmonics.

In introducing perturbations, it is necessary to keep in mind that Fourier decomposition is not coordinate independent. In Fig. 3.12 are shown Boozer and equal arc coordinates in a shaped equilibrium. In Fig. 3.13 are shown the Poincaré plots of a single $m = 2, n = 1$ perturbation, *i.e.* $\alpha = \alpha_{21}(\psi_p)cos(2\theta - \zeta)$ in Boozer and equal arc coordinates. Note that in the ψ_p stepping equation there is a term of the form $I\rho_\| B^2\alpha'_\zeta$. In Boozer coordinates I is independent of θ, but in equal arc coordinates I contains $m/n = 1/0$ as well as the $m/n = 0/0$ harmonic. Thus in equal arc coordinates the 1/0 couples to the 2/1 to produce a $m/n = 3/1$ mode, visible at

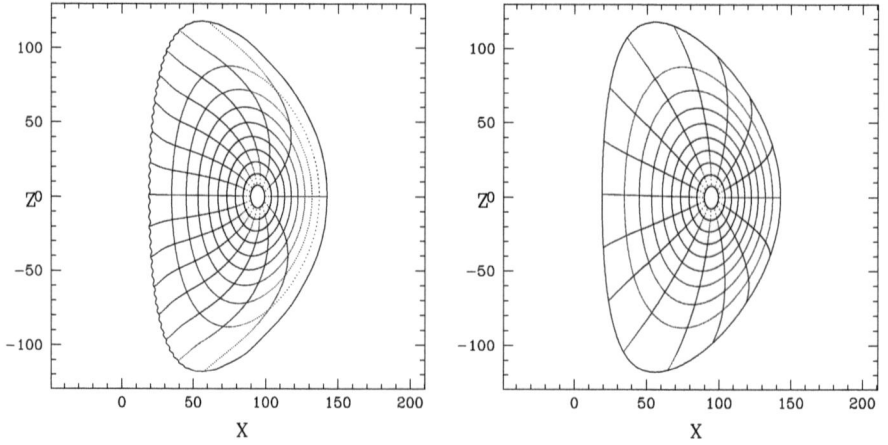

Fig. 3.12   Boozer (left) and equal arc (right) coordinates.

the $q = 3$ surface at $\psi_p = .175$. The 2/1 and 3/1 harmonics couple to produce 1/1 which couples with the 2/1 to produce a higher order $m/n = 3/2$ mode at $q = 3/2$, $\psi_p = .12$.

In the presence of a single mode, with α and the Hamiltonian functions of $n\zeta - \omega t$ we have

$$\omega \dot{P}_\zeta = n\dot{H}. \tag{3.99}$$

This condition restricts the motion of particles in the P_ζ, E plane due to the action of a mode, defining the possible diffusion in this plane for a single mode, and numerical error in this relation produces incorrect particle diffusion.

It is worth confirming this relation. We have

$$H = \frac{\rho_\parallel^2 B^2}{2} + \mu B + \Phi, \quad \dot{H} = \partial_{\psi_p} H \dot{\psi}_p + \partial_\theta H \dot{\theta} + \partial_{\rho_\parallel} H \dot{\rho}_\parallel + \partial_t H, \tag{3.100}$$

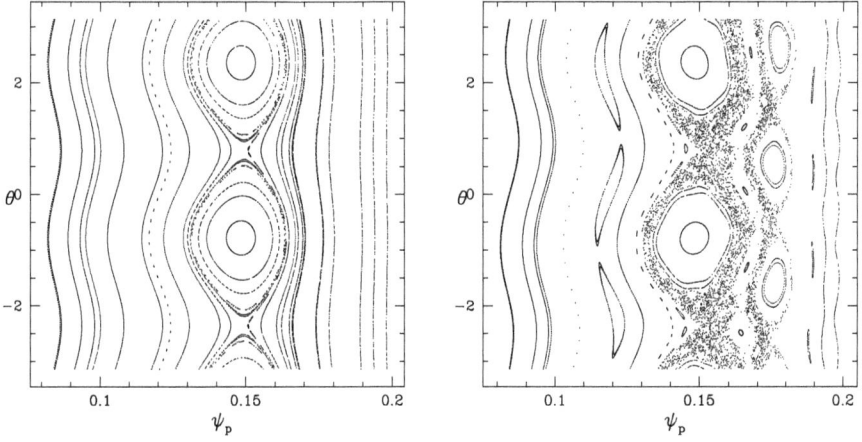

Fig. 3.13   A single harmonic $m = 2, n = 1$ perturbation in Boozer (left) and equal arc (right) coordinates.

and from the Lagrangian we find

$$P_\zeta = g\rho_\| - \psi_p + g\sum_{mn} \alpha_{mn}(\psi_p)sinQ_{mn},$$

$$\dot{P}_\zeta = (g'\rho_\| - 1)\dot{\psi}_p + g\dot{\rho}_\| + \sum_{mn} \alpha'_{mn}sinQ_{mn}\dot{\psi}_p$$

$$+ \sum_{mn}[n\alpha_{mn}\dot{\zeta} - m\alpha_{mn}\dot{\theta} - \omega\alpha_{mn}]cosQ_{mn}, \qquad (3.101)$$

with $\alpha = \sum_{mn} \alpha_{mn}(\psi_p)sinQ_{mn}$ and $Q_{mn} = n\zeta - m\theta - \omega t$.

We find after some algebra

$$\dot{H} = -\partial_{\rho_\|}H\partial_t\alpha + \partial_t H, \qquad\qquad \dot{P}_\zeta = \partial_{\rho_\|}H\alpha'_\zeta - \partial_\zeta H \quad (3.102)$$

and we then confirm Eq. 3.99 for the case of a single n value provided both α and the Hamiltonian are functions of the combination $n\zeta - \omega t$. Note that the changes in energy and in P_ζ are proportional to $\rho_\| B^2\alpha$, whereas the equations for stepping the variables of Eq. 3.93-3.96, in particular that for $\rho_\|$, include terms of order $\rho_\| B^2$, typically four orders of magnitude larger than this. Thus using these equations the accuracy of the time evolution of E and P_ζ is compromised by the necessary cancellation of large terms in the Runge–Kutta process and we find that Eq. 3.99 is satisfied typically only within about one percent. A method of advancing E and P_ζ

as primary variables, hence greatly increasing the accuracy of the guiding center equations has been developed by White *et al.* (2013).

We have assumed positive charge, so these equations give the trajectories of ions. The trajectories of electrons are slightly different. Aside from the different gyro radius and thus energy normalization, the normalization of time to the gyro frequency makes it necessary to change the sign of $\rho_\parallel$, and the overall sign of the time derivatives to obtain electron trajectories. Also the sign of the electric potential Φ must be changed. This is equivalent to changing the sign of the order ρ^2 drift terms, giving the well-known result that the grad B and curvature drift are charge dependent, but the $\vec{E} \times \vec{B}$ drift is not.

3.9.3 *Ideal MHD modes*

Introduce an ideal perturbation with fluid displacement $\vec{\xi}$ giving a magnetic field perturbation linear in ξ of the form $\delta\vec{B} = \nabla \times (\vec{\xi} \times \vec{B})$. Now expand $\vec{\xi}$ in the covariant basis $\vec{\xi} = \xi^\alpha \vec{e}_\alpha$, and make a Fourier decomposition of $\vec{\xi}$ through $\xi^\psi = \sum_{mn} \xi^\psi_{mn} sin(Q)$, $\xi^\theta = \sum_{mn} \xi^\theta_{mn} cos(Q)$, and $\xi^\zeta = \sum_{mn} \xi^\zeta_{mn} cos(Q)$ with $Q = n\zeta - m\theta - \omega t$ and $\nabla \cdot \vec{\xi} = \frac{1}{\mathcal{J}_p} \partial_\alpha (\mathcal{J}_p \xi^\alpha)$.

Write the guiding center Lagrangian for a charged particle in a magnetic field

$$L = (\vec{A} + \rho_\parallel \vec{B}) \cdot \vec{v} - H \tag{3.103}$$

with $\rho_\parallel = v_\parallel / B$, $\vec{A}$ the vector potential, $\vec{v}$ the particle velocity, and H the Hamiltonian

$$H = \frac{\rho_\parallel^2 B^2}{2} + \mu B + \Phi \tag{3.104}$$

with μ the magnetic moment and Φ the electric potential. The equilibrium vector potential is $\vec{A} = \psi \nabla\theta - \psi_p \nabla\zeta$ and the perturbation $\delta\vec{A} = \vec{\xi} \times \vec{B} = \xi^\alpha B^\beta \epsilon_{\alpha\beta\gamma} \mathcal{J}_p \vec{e}^\gamma$. The Lagrangian becomes

$$L = (\psi \vec{e}^\theta - \psi_p \vec{e}^\zeta + \xi^\alpha B^\beta \epsilon_{\alpha\beta\gamma} \mathcal{J}_p \vec{e}^\gamma + \rho_\parallel B_\alpha \vec{e}^\alpha) \cdot \vec{v} - H. \tag{3.105}$$

Using $B^{\psi_p} = 0$, $B^\theta = 1/\mathcal{J}_p$, $B^\zeta = q/\mathcal{J}_p$ this simplifies to

$$L = (\psi + \rho_\parallel I - q\xi^{\psi_p})\dot{\theta} + (\rho_\parallel g - \psi_p + \xi^{\psi_p})\dot{\zeta} + (q\xi^\theta - \xi^\zeta)\dot{\psi}_p - H \tag{3.106}$$

where we have dropped δ, not modifying the particle trajectory in the poloidal plane and giving rise only to periodic oscillations in the toroidal precession.

Largange's equations are

$$\frac{d}{dt}\frac{\partial L}{\partial \dot{q}} = \frac{\partial L}{\partial q} \tag{3.107}$$

giving

$$\begin{pmatrix} 0 & -A & -C & 0 \\ A & 0 & -F & I \\ C & F & 0 & g \\ 0 & -I & -g & 0 \end{pmatrix} \begin{pmatrix} \dot{\psi}_p \\ \dot{\theta} \\ \dot{\zeta} \\ \dot{\rho}_\| \end{pmatrix} = \begin{pmatrix} -\partial_{\psi_p} H - q\partial_t \xi^\theta + \partial_t \xi^\zeta \\ -\partial_\theta H + q\partial_t \xi^{\psi_p} \\ -\partial_\zeta H - \partial_t \xi^{\psi_p} \\ -\partial_{\rho_\|} H \end{pmatrix} \tag{3.108}$$

with

$$A = q + \rho_\| I' - \partial_{\psi_p}(q\xi^{\psi_p}) - q\partial_\theta \xi^\theta + \partial_\theta \xi^\zeta,$$
$$C = \rho_\| g' - 1 + \partial_{\psi_p}\xi^{\psi_p} - q\partial_\zeta \xi^\theta + \partial_\zeta \xi^\zeta,$$
$$F = \partial_\theta \xi^{\psi_p} + q\partial_\zeta \xi^{\psi_p}, \tag{3.109}$$

which we invert to find

$$\begin{pmatrix} \dot{\psi}_p \\ \dot{\theta} \\ \dot{\zeta} \\ \dot{\rho}_\| \end{pmatrix} = \frac{1}{D} \begin{pmatrix} 0 & g & -I & -F \\ -g & 0 & 0 & C \\ I & 0 & 0 & -A \\ F & -C & A & 0 \end{pmatrix} \begin{pmatrix} -\partial_{\psi_p} H - q\partial_t \xi^\theta + \partial_t \xi^\zeta \\ -\partial_\theta H + q\partial_t \xi^{\psi_p} \\ -\partial_\zeta H - \partial_t \xi^{\psi_p} \\ -\partial_{\rho_\|} H \end{pmatrix}, \tag{3.110}$$

with $D = Ag - IC$.

In Chap. 6 it is shown that representing an ideal mode through $\vec{\xi}$ or through $\nabla \times \alpha \vec{B}$ are equivalent for the purposes of investigating mode-particle resonances.

We find after some algebra

$$D\dot{H} = \omega\partial_{\rho_\|} H(g'q + I')\xi^\psi cos(Q) + \omega\partial_\theta Hg(q\xi^\theta - \xi^\zeta)sin(Q)$$
$$-\omega\partial_{\psi_p} H(gq + I)\xi^\psi cos(Q) + D\partial_t H$$
$$D\dot{P}_\zeta = n\partial_{\rho_\|} H(g'q + I')\xi^\psi cos(Q) + n\partial_\theta Hg(q\xi^\theta - \xi^\zeta)sin(Q)$$
$$-n\partial_{\psi_p} H(gq + I)\xi^\psi cos(Q) - D\partial_\zeta H \tag{3.111}$$

and we then confirm Eq. 3.99 for the case of a single n value provided both $\vec{\xi}$ and the Hamiltonian are functions of the combination $n\zeta - \omega t$.

3.9.4   *Kinetic Poincaré plots*

We are interested in the case of the interaction of particles of arbitrary pitch with modes of nonzero frequency. It is fairly easy to assess the effect of a particular mode on a particle distribution by examining a Poincaré plot for a particular choice of either co-moving and trapped or counter-moving particles, which we refer to as a kinetic Poincaré plot to distinguish it from a plot of the magnetic field. Points are plotted in the poloidal cross section whenever $n\zeta - \omega_n t = 2\pi k$ with k integer.

The toroidal motion then gives successive Poincaré points in the poloidal cross section ψ_p, θ, or better, since P_ζ and E are constant in the absence of perturbations, the P_ζ, θ plane or the E, θ plane. Individual modes produce islands in the phase space of the particle orbits, which through phase mixing produce local flattening of the particle distribution. In addition, overlap of these islands, the Chirikov criterion, leads to stochastic transport of particles. Such a plot shows the canonical division of orbits into those following good KAM surfaces, isolated islands bounded by separatrices, and stochastic domains. In an ideal situation with a single perturbation the separatrix is a well defined boundary, but in an actual equilibrium it is broadened into a thin stochastic layer by toroidal coupling or nonlinear coupling to other perturbations.

Points are plotted in the poloidal cross section whenever $n\phi - \omega t = 2\pi k$ with k integer. Neglecting the effect of the drift in modifying the toroidal motion then gives successive Poincaré points, with $\Delta\phi = \omega_t \Delta t$ satisfying

$$n\omega_t - \omega = 2\pi/\Delta t. \tag{3.112}$$

For there to be a periodic fixed point in θ with period m' we also require $\Delta\theta = 2\pi l/m'$ with l integer. But we also have $\Delta\theta = \omega_t \Delta t/q$, giving

$$[n - m'/ql]\omega_t = \omega, \qquad q = \frac{m'/l}{n - \omega/\omega_t}, \tag{3.113}$$

this last equation determining the location of the resonance. Note that the poloidal mode number m does not appear in this expression. A resonance appears whenever there exist integers m', l such that this relation can be satisfied. Thus for $q > m'/ln$ resonance occurs with a co-moving passing particle and for $q < m'/ln$ it occurs for a counter-moving passing particle. Since the Alfvén frequency is generally large, only rapidly moving particles are capable of resonating with these modes, and important interaction occurs only for high energy heating particles or for fusion products such as alpha particles. Note that for co-moving passing particles ($\omega_t > 0$), increasing the mode frequency ω increases the q value of the resonance, and increasing energy or pitch (and thus ω_t) decreases the q value of the resonance. These islands exist in real space and in the energy variable. However Eqs. 3.113 were calculated by ignoring the precession and drift motion, and are for a large aspect ratio circular equilibrium. In order to examine the effect of resonance for arbitrary energy and pitch, as well as in a general equilibrium, we use a general numerical method of displaying these resonances.

Kinetic energy $E = v^2/2$ is not conserved since the mode is time dependent. For a single value of n the Hamiltonian depends only on the combination $n\zeta - \omega t$ and thus $\dot{P}_\zeta = n\dot{E}/\omega$, so the distribution must be initiated with a fixed value of μ and

$$\omega P_\zeta - nE = c, \tag{3.114}$$

where $P_\zeta = g(\psi_p)v_\parallel/B - \psi_p$ is the canonical toroidal momentum, ζ a straight field line toroidal coordinate, and the covariant expression for the magnetic field is $\vec{B} = g(\psi_p)\nabla\zeta + I\nabla\theta + \delta\nabla\psi_p$. The expression $E - \omega P_\zeta/n$ is simply the particle energy in the frame rotating with the mode. A Poincaré plot with particles of fixed μ and energy E does not give a coherent plot, it contains intersecting surfaces, since it is really an overlaying of plots with different values of c. All particles must be initiated with the same values of μ and c, which means that the energy depends on position.

Choosing the energy to be E_0 at the magnetic axis where $\psi_p = 0$, the pitch on axis is $\lambda_0 = \sqrt{1 - \mu B_0/E_0}$, and $c = g(0)\omega\lambda_0 v_0/B_0 - nE_0$ and for any surface the pitch is $\lambda = \pm\sqrt{1 - \mu B(\psi_p, \theta)/E}$, and finally the velocity on surface ψ_p is the solution to

$$\omega(gv_\parallel/B - \psi_p) - n(v_\parallel^2/2 + \mu B) = c. \tag{3.115}$$

Given ψ_p and θ this is simply a quadratic equation for $v_\parallel$. Particle initiation can be done using $\theta = 0$ for all co-passing and trapped orbits and $\theta = \pi$ for counter-passing orbits. Choosing θ does not restrict the mode-particle phase since the toroidal angle ζ can be taken to be random.

A kinetic Poincaré plot shows islands indicating resonance of the particles with the perturbation, and it includes all nonlinear couplings, deviation of the orbits from a single drift surface, and particle precession rates. It is applicable for the case of a general numerical equilibrium.

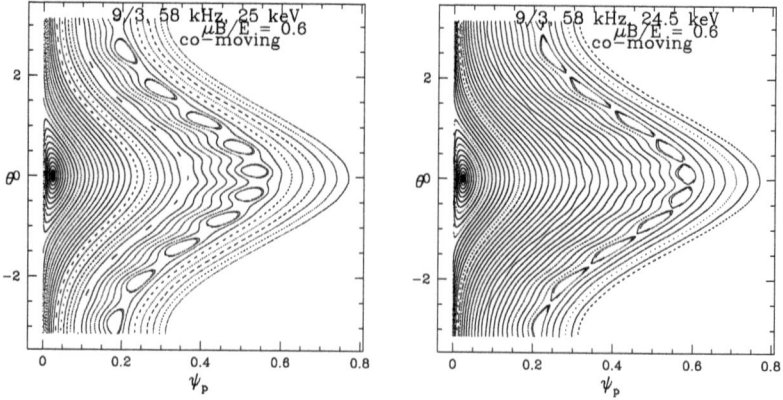

Fig. 3.14   Kinetic Poincaré plots for mode $m/n = 9/3$, showing energy dependence of the $m' = 10$ resonances.

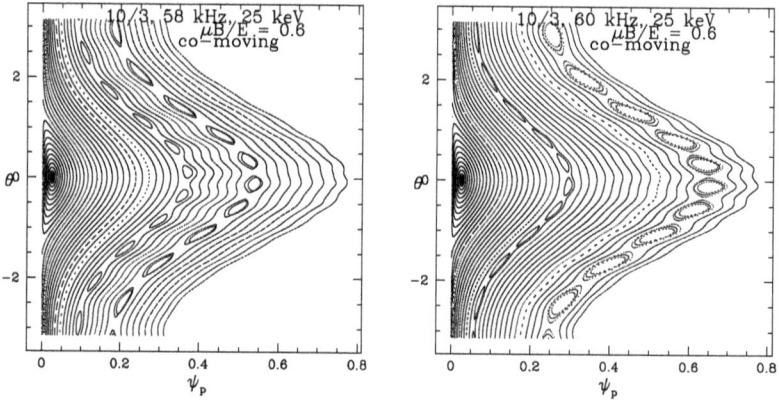

Fig. 3.15   Kinetic Poincaré plots for mode $m/n = 10/3$, showing frequency dependence of $m' = 10$ resonances. Note resonant surface unchanged from the $m = 9$ mode for 58 kHz.

In Figs. 3.14 and 3.15 are shown kinetic Poincaré plots for different modes in a reversed q profile equilibrium, illustrating islands produced by modes with $\delta B_r/B$ of order 10^{-4}. Energies and pitch are chosen to reflect a particular beam distribution. Shown is the dependence on energy, frequency, and m values. Note that the number of islands in the chain is not simply given by the mode number m. We will denote the number of islands in the chain with m'. Also note that the surfaces mapped out by the trajectories are drift surfaces, the magnetic axis is not at the center of these orbits.

All these plots have values of $\mu B/E$ approximately equal to 0.6, the dominant value in the beam distribution. Fig. 3.14 demonstrates the resonance position shift of an $m' = 10$ island chain due to a change in energy, decreasing the energy increases the q value, and for this radius the surface moves outward. This resonance was produced by a single harmonic with $m = 9$, and $n = 3$. The first plot of Fig. 3.14 and the first of Fig. 3.15, produced by a single harmonic with $m = 10$ and $n = 3$, show that the resonance position and the value of m' is independent of the m value. The two plots of Fig. 3.15 show the frequency dependence of the resonance surface, to a larger q value for increasing frequency, inward motion for the inside resonance, and outward motion for the outside resonance, away from the minimum q value.

Note that for compressional modes the toroidal canonical momentum is $P_\zeta = \partial_\zeta L = \rho_\parallel g - \psi_p - a$ and Eqs. 3.114 and 3.115 must be modified accordingly.

3.10   SCATTERING AND ENERGY DIFFUSION

For simulations of any length it is necessary to include the effects of scattering and energy diffusion of the test particles by the background species, in general consisting of the ion species, the electrons, and other additional ion species, the impurities. For many applications, where the background species is of nearly constant temperature and density, constant values for pitch angle scattering rate and energy diffusion rate are sufficient. A Monte Carlo algorithm given by Boozer and Kuo-Petravic and applied each time step δt produces a Gaussian diffusion in pitch

$$\lambda' = \lambda(1 - \nu\delta t) \pm \sqrt{(1 - \lambda^2)\nu\delta t} \qquad (3.116)$$

with ν the local energy-dependent collision frequency of the test particles on the background, and $\lambda = v_\parallel/v$ the value of the pitch. The plus and minus sign is chosen randomly for each collision. This expression must always have $\nu\delta t \ll 1$ to obtain a proper Gaussian diffusion. If ν is large one performs N steps of this operator using $\delta t/N$.

For collisions with a background species of varying temperature and density one must use the calculation of the four relaxation processes for a stream of test particles (α), interacting with a background Maxwellian (β). The processes are:
slowing down

$$dv_\alpha/dt = -\nu_s v_\alpha, \tag{3.117}$$

transverse diffusion

$$d(v_\alpha - v_\alpha)^2_\perp/dt = \nu_\perp v^2_\alpha, \tag{3.118}$$

parallel diffusion

$$d(v_\alpha - v_\alpha)^2_\parallel/dt = -\nu_\parallel v^2_\alpha, \tag{3.119}$$

and energy loss

$$dv^2_\alpha/dt = -\nu_E v^2_\alpha, \tag{3.120}$$

with

$$\nu_s = (1 + m_\alpha/m_\beta)\psi(x)\nu_0, \tag{3.121}$$

$$\nu_\perp = 2[(1 - 1/2x)\psi(x) + \psi'(x)]\nu_0, \tag{3.122}$$

$$\nu_\parallel = [\psi(x)/x]\nu_0, \tag{3.123}$$

$$\nu_E = 2[(m_\alpha/m_\beta)\psi(x) - \psi'(x)]\nu_0, \tag{3.124}$$

where $\nu_0 = 4\pi e^2_\alpha e^2_\beta \lambda n_\beta/m^2_\alpha v^3_\alpha$, $x = m_\beta v^2_\alpha/(2kT_\beta)$, and $\lambda = ln(\Lambda)$ is the Coulomb logarithm.

The Rosenbluth function

$$\psi(x) = \frac{2}{\sqrt{\pi}} \int_0^x dt\sqrt{t}e^{-t} \tag{3.125}$$

has an approximation asymptotically valid in the limits of both small and large x and with a relative error of $d\psi/\psi < 10^{-2}$ over the whole range of x

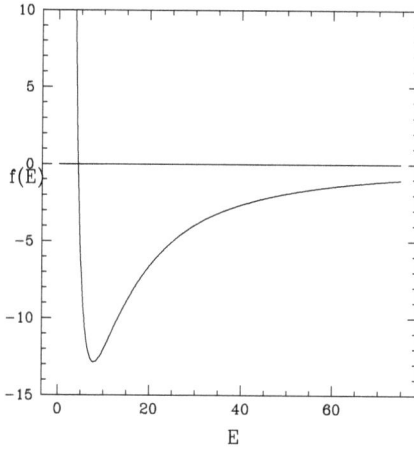

Fig. 3.16   Energy drag function for Monte Carlo simulations.

given by[†]

$$\psi(x) = 1 - 1/(1 + p),$$ (3.126)

with $p = x^{1.5}[a_0 + a_1 x + a_2 x^2 + a_3 x^3 + a_4 x^4]$, and $a_0 = .75225$, $a_1 = -.238$, $a_2 = .311$, $a_3 = -.0956$, $a_4 = .0156$.

A Monte Carlo operator for the energy diffusion given by Boozer and Kuo-Petravic is

$$dE = f(E)E dt \pm 2\sqrt{ET\nu dt}$$ (3.127)

with the function $f(E) = -\nu_0[g(x) + g'(x)]$. Here ν_0 is the Coulomb log modified collision frequency and $g(x) = 2(m_\alpha/m_\beta)\psi(x) - 2\psi'(x)$. The plus and minus signs must be taken randomly each application of this operator, this term providing energy diffusion. $f(E)$ is shown in Fig. 3.16 as a function of energy.

Even for energies equal to the background Maxwellian this function must exert some drag to keep the diffusion term from causing the energy to move above that of the background species. For small energy the function

[†]R. B. White, 2000, unpublished.

becomes large and positive, since at very low energy a particle will be heated by any collision and dE/E diverges at zero.

These Monte Carlo routines are present in the numerical guiding center code ORBIT including dependence on density and temperature profiles, but it is often possible to use the simple forms with constant collision frequencies.

3.11   Problems

1. Verify the Hamiltonian guiding center drift in a general toroidal field for the variable $\rho_\parallel$.

In a large aspect ratio tokamak:

2. Find the effect of the Shafranov shift on the toroidal precession rate of trapped particles.

3. Assume a quadratic safety factor profile $q(r) = q_0 + cr^2/a^2$ and quadratic density and temperature profiles, $i.e.$ $p = p_0[l - (r^2/a^2)]^2$. Find the central plasma β necessary to produce precession reversal of deeply trapped particles, and the plasma radius where it first occurs.

4. A trapped particle is heated with no input of angular momentum. Calculate the time averaged outward shift of the banana orbit, and compare it to the toroidal precession rate. Keep only lowest order in inverse aspect ratio.

5. Use the representation of the dipole field found in Problem 6, Chap. 1, to find canonical variables for guiding center drift. Find the drift equations. Find expressions for the bounce period and the bounce averaged drift frequency, and approximate them for very small bounce angle.

6. An ion is subject to the static magnetic and electric fields $\vec{B} = \hat{\theta}2I/r$, $\vec{E} = \hat{z}E$, where r, θ, and z are the cylindrical coordinates, and I and E are constants. At time t=0, the ion velocity is $(v_r, v_\theta, v_z) = (W, U, 0)$.
   a. Find canonical variables for guiding center drift.
   b. Find the guiding-center velocity of the ion.
   c. Examine whether your solution conserves energy.

7. Find the minimum gyro radius which permits a stagnation orbit, $\dot{\theta} = \dot{\psi}_p = 0$, and find the location of the orbit. Use a simple large aspect ratio tokamak equilibrium.

8. From the equations of motion verify the expressions given by Eq. 3.102.

3.12   References

- Alfvén, H., Ark. Mat. Astron. Fys. 27h, 1 (1940).
- Alfvén, H., Cosmical Electrodynamics (Clarendon Press, Oxford, 1950).
- Banõs, A., J. Plasma Phys. 1, 305 (1967).
- Boozer, A. H., Phys. Fluids 24, 1999 (1981).
- Boozer, A. H., Phys. Fluids 27, 2441 (1984).
- Boozer A. H. and G. Kuo-Petravic, Phys. Fluids 24, 851 (1981).
- Brizard, A. J., Phys. Plasmas 18, 022508 (2011).
- Burby, J. and H. Qin , Phys. Plasmas 19, 052106 (2012).
- Garren D. A. and A. H. Boozer, Phys. Fluids B 3(10), 2822 (1991).
- Goldstein, H., Classical Mechanics (Harvard University, Cambridge, Mass., 1953).
- Hsu, C. T. and D. J. Sigmar, Physics of Fluids 4, 1492 (1992).
- Kruskal, M., J. Math. Phys. 3, 80 (1962).
- Littlejohn, R. G., Phys. Fluids 24, 1730 (1981).
- Littlejohn, R. G., J. Plasma Phys. 29, 111 (1983).
- Littlejohn, R. G., Physica Scipta T2/1, 119 (1982).
- Meiss, J. D. and R. D. Hazeltine, Phys Fluids B 2, 2563 (1990)
- Mikkelsen, D. R, R. B. White, R. J. Akers, S. M. Kaye *et al.*, Phys. Plasmas 4, 3667 (1997).
- Morozov, A. I., and L.S. Solov'ev, in Reviews of Plasma Physics (Consultants Bureau, New York 1968), Vol. 2. Translated from Russian; Voprosy Teorii Plazmy (Atomizdat, Moscow, 1963).
- Northrop, T. G., The Adiabatic-Motion of Charged Particles, edited by E. Marshak (Interscience, New York, 1963).
- Rabinovitch, M.S. in Handbook of Plasma Physics, edited by M.N. Rosenbluth and R. Z. Sagdeev (North Holland, Amsterdam, 1983) Vol. 1, p. 3.
- Schmidt, G., Physics of High Temperature Plasmas (Academic Press, New York, 1966).
- Wesson, J., Tokamaks (Clarendon, Oxford, 1997) pp. 581–603.
- White, R. B., A. H. Boozer, and R. Hay, Phys. Fluids 25, 575 (1982).
- White, R. B., and M. S. Chance, Phys. Fluids 27, 2455 (1984).
- White, R. B., Phys. Fluids B[2] 4, 845 (1990).
- White, R. B. and L. E. Zakharov, Phys. Plasmas 10, 573 (2003).

- White, R. B., G. Spizzo and M. Gobbin, Plas. Phys. Control. Fusion xx, xx, (2013).
- White, R. B., (2000) The code ORBIT is available by anonymous ftp. Type "ftp ftp.pppl.gov" and for user type "anonymous", for password give your e-mail address. Then change directory through "cd /pub/white/Orbit" after which "get *" will retrieve all files.

Chapter 4

Particle Response to Modes

4.1 INTRODUCTION

In a symmetric plasma confinement device a perturbation such as an Alfvén mode produces a resonance island in the particle orbits which flattens the local energy and density gradient due to rotation about the elliptic points of the resonance. The location of the resonance is mode frequency and particle energy dependent, and particles away from the resonance are not affected. A projection onto the Poincaré subspace allows the evaluation of resonance widths and internal rotation rates. In a stellarator without symmetry, because of the toroidal dependence of the equilibrium magnetic field a mode instead produces chaos in many orbits passing where the mode amplitude is large. The mode growth rate and saturation are affected by the local chaos.

4.2 PERTURBED HAMILTONIAN

High energy ions are a necessary ingredient in fusion-grade plasmas. Energetic particles (EPs) result from auxiliary heating mechanisms, such as neutral beam injection and ion cyclotron resonant heating, as well as from fusion-born alpha particles. In order to sustain a burning plasma, the EPs should be confined at least during the characteristic time they need to transfer a substantial fraction of their energy to the thermal reacting species via collisions. The free energy associated with the EP pressure gradient, however, can resonantly destabilize magnetohydrodynamic modes, jeopardizing the macroscopic plasma confinement and potentially leading to energy losses with accompanying EP loss, thereby halting fusion reactions and harming plasma facing components. In this chapter we analyze in general terms how the magnetic structure and associated particle orbits and resonances of fast ions in tokamaks and stellarators can lead to radial transport.

4.2.1 *Guiding Center Equations*

We investigate the effect of Alfvén modes using a guiding center formalism described in chapter 3. These are among the most important magnetic perturbations commonly occuring in plasma confinement devices, with perturbations of $\vec{B}$ primarily orthogonal to the equilibrium $\vec{B}$. They can be described through $\delta\vec{B} = \nabla \times \alpha\vec{B}$ with α an arbitrary function of ψ, θ, ζ, t. Using magnetic coordinates the equilibrium field becomes $\vec{B} = \nabla \times (\psi\nabla\theta - \psi_p\nabla\zeta) = g(\psi_p)\nabla\zeta + I(\psi_p)\nabla\theta + \delta\nabla\psi_p$, where θ and ζ are poloidal and toroidal angles, respectively, ψ is the toroidal and ψ_p the poloidal flux, related through the field line helicity $q(\psi_p)$ by $d\psi/d\psi_p = q(\psi_p)$.

Introducing units of time given by ω_0^{-1}, where $\omega_0 = eB/(mc)$ is the on axis gyro frequency, and units of distance given by the major radius R, the basic unit of energy becomes $m\omega_0^2 R^2$, which can also be written as $(mv^2/2)(2R^2/\rho^2)$, the gyro radius is $\rho = v/B \ll 1$, and the magnetic moment $\mu = v_\perp^2/(2B)$ is of order ρ^2.

The evolution of an orbit in time is given by

$$\begin{vmatrix} \dot{\rho}_\| \\ \dot{\psi}_p \\ \dot{\theta} \\ \dot{\zeta} \end{vmatrix} = \frac{1}{D} \begin{vmatrix} 0 & -K & C & -F \\ K & 0 & -g & I \\ -C & g & 0 & 0 \\ F & -I & 0 & 0 \end{vmatrix} \begin{vmatrix} \partial_{\rho_\|} H \\ \partial_{\psi_p} H \\ \partial_\theta H + I\partial_t\alpha \\ \partial_\zeta H + g\partial_t\alpha \end{vmatrix} \tag{4.1}$$

where $\rho_\| = v_\|/B$, the parallel velocity normalized to the magnetic field, $F = q + (\rho_\| + \alpha)I' + I\alpha'_{\psi_p}$, $C = -1 + (\rho_\| + \alpha)g' + g\alpha'_{\psi_p}$, $K = g\alpha'_\theta - I\alpha'_\zeta$, and $D = Fg - CI = gq + I + (\rho_\| + \alpha)(gI' - Ig')$. We use the shorthand notation $\partial f/\partial x = f'_x$.

4.2.2 *Mode Structure*

For simplicity we consider a single mode harmonic with $\alpha = \alpha_{mn}(\psi_p)sin(n\zeta - m\theta - \omega t)$. An ideal mode must have no induced electric field parallel to $\vec{B}$. This determines the potential Φ associated with the perturbation α, giving

$\Phi = \phi_{mn}\cos(n\zeta - m\theta - \omega t)$ with $\phi_{mn}(\psi_p) = (gq + I)\omega\alpha_{mn}(\psi_p)/(nq - m)$. Also $\alpha_{mn}(\psi_p) = (m/q - n)\xi_{mn}^\psi(\psi_p)/(mg + nI)$ where $\vec{\xi}$ is the ideal displacement. The magnitude of the perturbed field is approximately given by $m\xi$. The canonical toroidal momentum is $P_\zeta = -\psi_p + g(\psi_p)(\rho_\| + \alpha)$, the poloidal momentum is $P_\theta = \psi + I(\psi_p)(\rho_\| + \alpha)$. The Hamiltonian is $H = (P_\zeta + \psi_p)^2 B^2/(2g^2) + \mu B + \Phi$, Helical momentum is readily constructed using the relevant combination of P_ζ and P_θ.

4.2.3 *Hamiltonian Expansion*

We expand the Hamiltonian in α and drop terms of second order in α, giving an unperturbed part plus an interaction term, $H = H_0 + H_I$ with $H_0 = \rho_\|^2 B^2/2 + \mu B$ and

$$H_I = \left[\frac{\rho_\| B^2(nq - m)}{gq + I} - \omega\right]\frac{\phi_{mn}(\psi_p)}{\omega}e^{i(n\zeta - m\theta - \omega t)}. \tag{4.2}$$

We have from Eqs. 4.1

$$H_I = [n\dot\zeta - n\dot\zeta_d - m\dot\theta + m\dot\theta_d - \omega]\frac{\phi_{mn}(\psi_p)}{\omega}e^{i(n\zeta - m\theta - \omega t)}, \tag{4.3}$$

with $\dot\zeta_d$ and $\dot\theta_d$ the drift terms, second order in $\rho_\|$. Also

$$\frac{d}{dt}\phi_{mn}(\psi_p)e^{i(n\zeta - m\theta - \omega t)}$$
$$= [-i\phi'_{nm}(\psi_p)\dot\psi_p + (n\dot\zeta - m\dot\theta - \omega)\phi_{nm}(\psi_p)]e^{i(n\zeta - m\theta - \omega t)}. \tag{4.4}$$

But $E_0(t) = \frac{d}{\omega dt}\phi_{mn}(\psi_p)e^{i(n\zeta - m\theta - \omega t)}$ is a periodic adiabatic energy, due to the particle oscillating in the mode, and cannot contribute to mode particle energy exchange. Thus we drop this contribution, and

$$H_I = [-n\dot\zeta_d + m\dot\theta_d]\frac{\phi_{mn}}{\omega}\cos(n\zeta - m\theta - \omega t)$$
$$- \sum_{m,n}\frac{\phi'_{mn}\dot\psi_p}{\omega}\sin(n\zeta - m\theta - \omega t) \tag{4.5}$$

is entirely due to the drift motion. The transverse Alfvén wave can resonantly exchange energy with a particle only through the cross field drift motion.

4.3 SYMMETRIC EQUILIBRIA

To examine resonance phenomona we construct a Poincaré plot, found by plotting points of a trajectory in the wave frame, where $n\zeta - \omega t = 2\pi k$ with k integer. From Eqs. 4.1 a time step requires the expressions for the four time derivatives for $\rho_\parallel$, ψ_p, θ, and ζ, or equivalently, $\rho_\parallel$, P_ζ, θ, and ζ. In the wave frame $\dot{\zeta}$ is given by ω, and since in a tokamak $H = H(n\zeta - \omega t)$ also $n\dot{E} = \omega \dot{P}_\zeta$. In an equilibrium with some other symmetry there is a different associated canonical momentum. From this equation and the condition that $nE - \omega P_\zeta = C$ for all particles in this frame we find $\dot{\rho}_\parallel$ in terms of $\dot{\psi}_p$ and $\dot{\theta}$. Thus the space of the Poincaré section for each value of C is two dimensional, the space of P_ζ, θ. Different values of C define different ranges of E and P_ζ. Poincaré sections with coherent KAM surfaces exist over the whole space of orbits.

Note that both terms in H_I are dominantly even in θ. Drift motion in ζ and θ is proportional to $\partial_\psi B$ and the magnetic field has primarily a $cos(\theta)$ dependance, giving terms even in θ with primary harmonics of $m\pm 1$. Instead $\dot{\psi}_p$ is given by $\partial_\theta B$, and is thus odd in θ, so also the ϕ'_{mn} term is even in θ, and the Poincaré Hamiltonian takes the form

$$\mathcal{H} = \frac{\rho_\parallel^2 B^2}{2} + \mu B + \sum_p H_p(\psi_p) cos(p\theta). \qquad (4.6)$$

The multiple harmonics are produced by the product of the drift motion with the original perturbation due to α.

We project to the two dimensional Poincaré space of P_ζ, θ with $nE - \omega P_\zeta = C$, using the Routhian (Goldstein, 1980)

$$R = L - P_\zeta \dot{\zeta} = (\psi + \rho_\parallel I)\dot{\theta} - \mathcal{H}. \qquad (4.7)$$

To illustrate the resonance islands and the induced mixing, we drop terms of higher order in gyro radius to system size, so the Routhian becomes

$$R = L - P_\zeta \dot{\zeta} = \psi \dot{\theta} + \frac{\omega}{n} P_\zeta - H_I \qquad (4.8)$$

and $d\psi = -q(\psi_p)dP_\zeta$.

The Lagrange equations of motion in the projected Poincaré section become

$$q\dot{P}_\zeta = \sum_p pH_p sin(p\theta) \tag{4.9}$$

$$q\dot{\theta} = \omega/n - \sum_p \frac{\partial H_p}{\partial P_\zeta} cos(p\theta). \tag{4.10}$$

Note that the Poincaré Hamiltonian is not the particle energy. Surfaces of constant $\mathcal{H}$ give the KAM particle orbits in the Poincaré plane, but they are not surfaces of constant particle energy. Both P_ζ and E vary over these orbits with $nE - \omega P_\zeta = constant$.

The fixed hyperbolic points of this map are given by $\dot{P}_\zeta = \dot{\theta} = 0$ in the 2D Poincaré section. The first equation determines the values of θ. The elliptic points are at $p\theta = 2\pi k$ at low energy, and displaced from these values as energy increases.

Then $\partial_{P_\zeta}\mathcal{H} = 0$ determines the value of P_ζ at the resonance, different for each value of p. A perturbation with a given value of p has an effect on the particle orbits only in the vicinity of the resonance, at values of P_ζ far from the resonance the effect of the perturbation is negligible. Thus one can examine the resonances one at a time, ignoring all other values of p, unless they are so close that they overlap.

Using the full toroidal equations of motion Eqs. 4.1 we find

$$\partial_{P_\zeta}\mathcal{H} = -\dot{\zeta} + q\dot{\theta}. + \mathcal{H}'_p \tag{4.11}$$

In this equation $\dot{\zeta}$ and $\dot{\theta}$ refer to the full toroidal motion, not to motion in the Poincaré section. Both $\dot{\zeta}$ and $\dot{\theta}$ are closely given by ω_ζ and ω_θ, the mean values of the transit frequencies, simply equal to the inverse of the transit times, and $n\omega_\zeta = \omega$. Thus

$$\partial_{P_\zeta}\mathcal{H} = -\omega_\zeta + q(\psi_p)\omega_\theta + \mathcal{H}'_p, \tag{4.12}$$

so for small perturbation amplitude and low energy the resonance is located at the value of P_ζ such that $q(\psi_{pr}) = \omega_\zeta/\omega_\theta$ and to leading order in $\rho_\parallel$, $dP_\zeta = -d\psi_p$ so $\mathcal{H}'' \simeq -q'(\psi_p)\omega_\theta$.

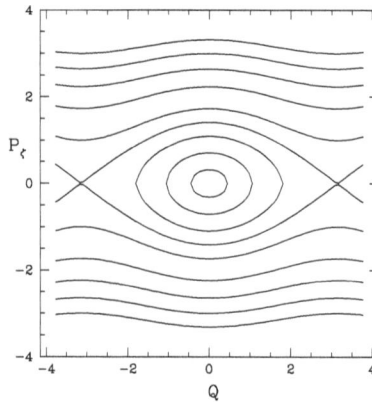

Fig. 4.1 Poincaré KAM surfaces of a resonance. $Q = p\theta$. The elliptic point at the center is a fixed point. The separatrix passes through the X-points, and particles inside it are trapped. Far from the resonance the orbits are practically unaffected. P_0 is taken to be 0.

4.3.1   The Poincaré Hamiltonian

Thus an expansion of the Poincaré Hamiltonian about the resonance surface P_0 associated with $\mathcal{H}_p$ has the form

$$\mathcal{H} = \mathcal{H}(P_0) + \frac{\mathcal{H}''}{2}(P_\zeta - P_0)^2 + \mathcal{H}_p cos(p\theta). \qquad (4.13)$$

Surfaces of constant $\mathcal{H}$ are given by $P_\zeta - P_0 = \sqrt{2\mathcal{H}_p/\mathcal{H}''}$ $\sqrt{cos(m\theta) - cos(m\theta_0)}$, where $cos(m\theta_0)$ defines the surface chosen, and are shown in Fig. 4.1. Particles rotate about the elliptic point following these surfaces. The maximum resonance width is thus $\Delta P_\zeta = 2\sqrt{2\mathcal{H}_p/\mathcal{H}''}$.

The above equations give the form of a resonance should it exist in the plasma. But for the resonance to actually be present, the flux surface ψ_{pr} with $q(\psi_{pr}) = \omega_\zeta/\omega_\theta$ must be in the plasma. Since ω_ζ and ω_θ are energy dependent, resonances can move in and out of the plasma as a function of particle energy.

4.3.2 Internal Rotation

To find the rate of rotation about the elliptic point, differentiate Eq. 4.13 with respect to time, giving

$$(P_\zeta - P_0)\dot{P}_\zeta = V sin(Q)\dot{Q} \qquad (4.14)$$

with $V = \mathcal{H}_p/\mathcal{H}''$ and $Q = p\theta$. Note that this equation shows that the effect of the perturbation is negligible away from the resonance surface P_0, with $\dot{P}_\zeta$ quickly tending to zero.

Examining $P_\zeta > P_0$ and $Q > 0$ the rotation about the elliptic point is clockwise if $V > 0$ and counterclockwise if $V < 0$. But $\dot{P}_\zeta = \partial_\theta \mathcal{H}/D = -p\mathcal{H}_p sin(Q)/D$ giving

$$\dot{Q} = \sqrt{2p\mathcal{H}_p\mathcal{H}''/D}\sqrt{cos(Q) - cos(Q_0)}. \qquad (4.15)$$

The time to complete an orbit around the elliptic point is $T = 4\int_0^{Q_0} dQ/\dot{Q}$ giving

$$T = 4\sqrt{\frac{D}{2p\mathcal{H}_p\mathcal{H}''}}\int_0^{Q_0}\frac{dQ}{\sqrt{cos(Q) - cos(Q_0)}} \qquad (4.16)$$

For small Q_0 this is $T = 4\pi/\sqrt{D/2p\mathcal{H}_p\mathcal{H}''}$ and for $Q_0 = \pi$ the integral diverges. The frequency about the elliptic point is proportional to the island width, or the square root of the perturbation amplitude, and it goes to zero as the separatrix is approached. The elliptic "O-point" and hyperbolic "X-point" are the exact solutions of the resonance condition. Particles are trapped within the separatrix that passes through the X-point. If any additional perturbations are present the separatrix is not a single surface, but has an exponentially small width consisting of chaotic orbits, see chapter 1 section 1.7.

4.3.3 A large Aspect Ratio Tokamak

For simplicity, consider these results for a large aspect ratio tokamak, with $B = 1 - rcos(\theta)$, $\psi = r^2/2$, $dr/d\psi_p = q(r)/r$, $g(\psi_p) = 1$, $I = r^2/q(r)$, and $P_\zeta = \rho_\parallel - \psi_p$. The drift terms $\dot{\theta}_d$ and $\dot{\zeta}_d$ are of the form $(\rho_\parallel^2 B + \mu)\partial_\theta B$. To leading order in inverse aspect ratio this produces terms of first order in $cos(\theta)$, but using energy conservation we see that $\rho_\parallel$ contributes terms of all orders in $cos(\theta)$. To leading order

$$\dot{\theta}_d = -\frac{2E - \mu}{r}cos(\theta), \qquad \dot{\zeta}_d = \frac{r(2E - \mu)}{q}cos(\theta), \qquad (4.17)$$

and to leading order $\phi_{mn} \simeq -\omega\xi^\psi_{mn}/m$, and the leading contribution to H_I is

$$H_I = \frac{2E - \mu}{2rm}\xi^\psi_{mn}[cos(n\zeta - (m + 1)\theta - \omega t) + cos(n\zeta - (m - 1)\theta - \omega t)] \quad (4.18)$$

The Poincaré time independent Hamiltonian becomes

$$\mathcal{H} = \frac{\rho_\parallel^2 B^2}{2} + \mu B + \frac{2E - \mu}{2rm}\xi^\psi_{mn}(\psi_p)[cos((m + 1)\theta) + cos((m - 1)\theta)] \quad (4.19)$$

producing resonances at $nq(r) = m \pm 1$. Resonances at other nearby values of q have smaller widths, but the coupling to other values of q increases with particle energy. Then at resonance $m + 1$

$$\partial_{P_\zeta}\mathcal{H} = -\omega\zeta + q(\psi_p)\omega_\theta + d\mathcal{H}_{m+1}/d\psi_p. \quad (4.20)$$

Assume that the mode eigenfunction is very broad, so that the derivative of ξ^ψ_{mn} can be neglected, then

$$\frac{d\mathcal{H}_{m+1}}{d\psi_p} \simeq \frac{q(2E - \mu)\xi^\psi_{mn}}{2mr^3}. \quad (4.21)$$

Thus using $\omega_\theta \simeq \sqrt{2E}/(2\pi r)$ for a deeply passing orbit and $d^2\mathcal{H}/d^2P_\zeta \simeq q'(\psi_p)\rho/r$ and E is of order ρ^2 so $d\mathcal{H}_{m+1}/d\psi_p \simeq \xi^\psi_{mn}\rho^2/(mr^3)$, typically ignorable, and the island width is

$$\Delta P_\zeta = \sqrt{\frac{\mathcal{H}_{m+1}}{\mathcal{H}''}} \simeq 2\sqrt{\frac{q\rho^3\xi^\psi_{mn}}{q'(\psi_p)mr}}. \quad (4.22)$$

Now consider resonances as a function of energy. For resonance with a mode a particle orbit must periodically return to experience the same perturbation, so that the mode can reinforce its action on successive passages. This requires, since the mode is a function of θ and $n\zeta - \omega t$

$$(n\omega_\zeta - \omega)T = 2\pi p, \qquad \omega_\theta T = 2\pi l \quad (4.23)$$

where ω_θ is the mean particle poloidal frequency, ω_ζ is the mean particle toroidal frequency, ω the mode frequency, n the toroidal mode number, T

some time interval, and p, l integers. This gives

$$p = \frac{(n\omega_\zeta - \omega)l}{\omega_\theta}. \tag{4.24}$$

This equation must be solved with integers p, l. Both ω_ζ and ω_θ scale as the square root of the energy so p tends to minus infinity as $E \to 0$. At low energy $\omega_\zeta/\omega_\theta = q(\psi_p)$ and as energy increases the ratio becomes larger than this for co-moving orbits, and smaller for counter moving. Both ω_ζ and ω_θ increase approximately linearly in particle velocity, so as energy increases $\omega_\zeta/\omega_\theta$ must decrease, meaning the resonance location must move to a smaller value of q.

The motion of particles that are trapped inside the separatrix of a wave-particle resonance becomes ergodic. Eventually, on average, all particles trapped by the wave adopt the constants-of-motion P_ζ, E of the exact resonance. Initially there is a gradient in energy or density versus ψ responsible for destabilizing the mode. Because of the variation of the flow rate as a function of distance from the elliptic point, regions of high energy or density are put in close vicinity of regions of low energy or density. In a few rotation periods ergodicity is achieved, and the domain inside the separatrix is uniform. See Chapter 13, section 13.6 for the dynamics of mode saturation due to the flattening of the local gradients in the presence of collisions.

4.4 STELLARATORS

4.4.1 *Introduction*

In a symmetric equilibrium an Alfvén mode produces a regular island structure in the orbits in the plane of P_ζ, θ in the coordinate system moving with the mode, where P_ζ is the toroidal canonical momentum in the case of toroidal symmetry, or the appropriate canonical momentum P_c in case of another symmetry. (One can also use ψ or E instead of P_ζ and ζ instead of θ). This can be seen by looking at the time stepping equations for an orbit. By moving to the frame $n\zeta - \omega t = 2\pi k$ the equations of motion depend only on the values of P_c and θ. The magnetic moment μ is fixed and $nE - \omega P_c$ is constant. Thus the time step and future history of any orbit passing through this point is unique, the trajectory in this plane is determined, trajectories cannot cross one another. An expansion of the Poincaré Hamiltonian in the vicinity of a resonance Eq. 4.13 gives the structure of an elliptic point and a well defined island structure, with simple unperturbed lines away from the resonance.

4.4.2 Chaotic Poincaré sections

However, in a stellarator without a symmetry the plot is not coherent. In the presence of an Alfvén perturbation, because the time step also depends on the value of ζ through the equilibrium $B(\psi, \theta, \zeta)$ and ζ is given at each time by $n\zeta - \omega t = 2\pi k$, a projection onto a Poincaré plane moving with the mode yields only chaos, different orbits passing through a point P_c, θ have different future histories because of the toroidal variation of B, typically of order $1/10$, much larger than δB from the mode. There is no frame in which the orbits are not chaotic. Particles affected by the mode are much more likely to be lost than in a system with a symmetry.

Consider motion in ψ and $\rho_\parallel$. Drop I, giving the toroidal current, small in a stellarator, giving

$$\dot{\psi} = mgB(\psi, \theta, \zeta)v_\parallel \alpha(\psi)cos(\Omega), \qquad \dot{\rho}_\parallel = v_\parallel^2 \partial_\psi B(\psi, \theta, \zeta)\xi(\psi)sin(\Omega) \quad (4.25)$$

with $\Omega = n\zeta - m\theta - \omega t$. A Poincaré section is chaotic. In a rest frame $\zeta = \zeta_0$, $\dot{\psi}$ depends on the time a particle passes this point, and in the mode frame $n\zeta - \omega t = 2k\pi$ the result depends on the value of B at this ζ. Thus there is in any frame a random part of the step in ψ and in $\rho_\parallel$; given by

$$\dot{\psi} = m\delta(B)v_\parallel \alpha(\psi), \qquad \dot{\rho}_\parallel = v_\parallel^2 \delta(\partial_\psi B)\xi(\psi) \quad (4.26)$$

with δX equal to X minus the mean value of X over ζ. The chaotic part of the step depends on the product of the equilibrium modulation of B and the perturbation α, zero if either is zero. Since the step is random in direction, it leads to diffusion

$$< \psi^2 > = Dt/2 \quad (4.27)$$

with $D = [mg\delta(B)v_\parallel \alpha(\psi)/2q]^2 dt$ for ψ and $[\delta(\partial_\psi B)v_\parallel^2 \xi(\psi)/2q]^2 dt$ for $\rho_\parallel$ with dt the magnitude of a time step. If B is independent of ζ a coherent Poincaré island is produced by α.

Note that this chaos is not the product of an overlap of two islands a la Chirikov. There is no threshold in the value of α, the step is random for any value of α not zero. There is thus a particularly strong effect of a small Alfvén mode due to the fact that the particle motion is determined by the combined effect of the eigenmode and the equilibrium B modulation, which is normally many orders of magnitude larger.

In a symmetric device the effect of an Alfvén mode can be investigated in detail because the Hamitonian guiding center formalism allows a projection onto the two dimensional space of a Poincaré plot, giving explicit expressions for resonance location, width, and rotation about an elliptic point. An Alfvén mode of the form $sin(n\zeta - m\theta - \omega t)$ produces primary resonances located with low particle energy at $q(\psi_p) = (m \pm 1)/n$, but higher order terms couple also to produce islands at other flux values. At low mode amplitude these islands remain small, with other surfaces unchanged, leading to a small modification of the unperturbed KAM surfaces. Instead, in a nonsymmetric equilibrium, because of the toroidal dependence of B, chaos in high energy particle orbits is produced wherever the mode amplitude is significant, acting on particles of all energy and canonical momentum. An unstable Alfvén mode can have a much stronger effect on plasma confinement than in a symmetric equilibrium because the effect of the mode also includes the equilibrium modulation of B. The effect of collisions, or any form of orbital chaos, on mode growth and saturation is discussed in chapter 13.

4.5 References

- H. Goldstein, Classical Mechanics (Addison Wesley, 1980)
- R. B. White, A. Bierwage, and S. Ethier, Phys. Plasmas 29, 052511 (2022)
- M. Landreman and E. Paul, Phys. Rev. Lett. 128 (2022)
- K. Toi, F. Watanabe, T. Tokuzawa, K. Ida, S. Morita, T. Ido, A. Shimizu, M. Isobe, K. Ogawa, and et al, Phys. Rev. Lett. 105, 145003 (2010)
- V. N. Duarte, N. N. Gorelenkov, R. B. White, and H. L. Berk, Phys. Plasmas 26, 120701 (2019)
- X. Wang, S. Briguglio, P. Lauber, V. Fusco, and F. Zonca, Phys. Plasmas 23, 012514 (2016)
- X. Wang and S. Briguglio, New Journal of Physics 18, 085009 (2016)
- C. Slaby, A. Koenies, R. Kleiber, and J. M. GarcaRegana, Nuclear Fusion 58, 082018 (2018)
- H. L. Berk, B. N. Breizman, and M. Pekker, Phys. Rev. Lett. 76, 1256 (1996)
- V. N. Duarte and N. N. Gorelenkov, Nucl. Fusion 59, 044003 (2019)
- J. B. Lestz and V. N. Duarte, Phys. Plasmas 28, 062102 (2021)
- N. N. Gorelenkov and R. B. White, Plas Phys. Controlled Fusion 55, 015007 (2013)

Chapter 5

Resonance

5.1 INTRODUCTION

Particle resonances in magnetic fusion confinement devices produce local flattening of the particle distribution, and if they become sufficiently large, or if nearby resonances overlap, can lower the high energy distribution to the point of reducing the fusion rate and even producing particle loss to the walls. In some cases it is desirable to modify a particle distribution in a restricted range of energy, for example to remove cool alpha particle ash from a fusion reaction while not affecting higher energy particles. We examine the energy dependence and radial location of resonances in a number of equilibria. In section 5.2 we consider tokamak resonances. In section 5.3 we examine stellarator resonances.

5.2 TOKAMAK RESONANCES

The existence and location of resonances is only the first step in the analysis of their possible deleterious or beneficial effects. At a given resonance, an instability can develop only if it is driven by local gradients which can provide the energy for instability growth, or if an external drive is provided by antennae. Here we examine only the first necessary condition, how to determine whether there exists a resonance which can cause problems, or be utilized. In section 5.2.1 we discuss the guiding center formalism used, in section 5.2.2 is an analysis of resonance conditions along with examples of resonances in different fusion devices, and in section 11.3 are the conclusions.

5.2.1   *Guiding Center Formalism*

The equilibrium magnetic field is given by

$$\vec{B} = g\nabla\zeta + I\nabla\theta + \delta\nabla\psi_p, \tag{5.1}$$

where θ and ζ are poloidal and toroidal coordinates and ψ_p is the poloidal flux, and in an axisymmetric equilibrium using Boozer coordinates g and I are functions of ψ_p only.

An incompressible electromagnetic flute mode perturbation in a torus can be put in the form $\delta\vec{B} = \nabla \times \alpha\vec{B}$. For ideal modes there must also be an electric potential Φ chosen to cancel the parallel electric field induced by $d\vec{B}/dt$. The perturbation quantities have Fourier expansions

$$\alpha = \sum_{m,n} A_n \alpha_{m,n}(\psi_p) sin(\Omega_{mn}), \quad \Phi = \sum_{m,n} A_n \Phi_{m,n}(\psi_p) sin(\Omega_{mn}), \tag{5.2}$$

with n the toroidal mode number, m the poloidal mode number and $\Omega_{mn} = n\zeta - m\theta - \omega_n t - \phi_n$, ω_n the mode frequency, and ϕ_n the phase. Vanishing of the parallel electric field requires

$$\sum_{m,n} \omega_n B\alpha_{m,n} cos(\Omega_{mn}) - \vec{B}\cdot\nabla\Phi/B = 0,$$

giving in Boozer coordinates

$$(gq + I)\omega_n\alpha_{mn} = (nq - m)\Phi_{mn}.$$

The perturbation α is related to the ideal displacement $\vec{\xi}$, through

$$\alpha_{mn} - \frac{(m/q - n)}{(mg + nI)}\xi_{mn}^{\psi}.$$

The functions $\alpha_{m,n}(\psi_p)$ are the eigenfunctions of ideal magnetohydrodynamic (MHD) modes, and the amplitude A_n gives the maximum ideal displacement of the mode, normalized to the major radius. The modes can be either produced by external antennae, or naturally occuring unstable modes in the plasma.

The guiding center Hamiltonian is $H = (\rho_\| - \alpha)^2 B^2/2 + \mu B + \Phi$, with μ the magnetic moment, $\rho_\| = v_\|/B$ the normalized parallel velocity, and Φ the electric potential associated with the perturbation.

In simplest form, in an axisymmetric configuration, in straight field line coordinates, and without field perturbation or magnetic field ripple, the equations of particle motion become

$$\dot{\theta} = \frac{\rho_\| B^2}{D}(1 - \rho_\| g') + \frac{g}{D}(\mu + \rho_\|^2 B)\frac{\partial B}{\partial \psi_p}, \tag{5.3}$$

$$\dot{\psi}_p = -\frac{g}{D}(\mu + \rho_\|^2 B)\frac{\partial B}{\partial \theta}, \tag{5.4}$$

$$\dot{\rho}_\| = -\frac{(1 - \rho_\| g')}{D}(\mu + \rho_\|^2 B)\frac{\partial B}{\partial \theta}, \tag{5.5}$$

$$\dot{\zeta} = \frac{\rho_\| B^2}{D}(q + \rho_\| I'_{\psi_p}) - \frac{I}{D}(\mu + \rho_\|^2 B)\frac{\partial B}{\partial \psi_p}. \tag{5.6}$$

where $D = gq + I + \rho_\|(gI'_{\psi_p} - Ig'_{\psi_p})$. Note that the denominator D does not modify the ratio of the toroidal and poloidal velocities. These equations, derived in Chapter 3, are used by means of a third order Runge Kutta formalism in the code ORBIT.

5.2.2   *Orbit resonance*

For resonance with a mode a particle orbit must periodically return to experience the same perturbation, so that the mode can reinforce its action on successive passages. This requires, since the mode is a function of θ and $n\zeta - \omega t$

$$(n\omega_\zeta - \omega)T = 2\pi p, \qquad \omega_\theta T = 2\pi l \tag{5.7}$$

where ω_θ is the mean particle poloidal frequency, ω_ζ is the mean particle toroidal frequency, ω the mode frequency, n the toroidal mode number, T a time interval giving l poloidal transits, and p, l integers. This gives

$$p = \frac{(n\omega_\zeta - \omega)l}{\omega_\theta}. \tag{5.8}$$

This equation must be solved with integers p, l.

Many resonances of interest occur in the space of co-passing particles, and in fact all of the resonances we will consider in the examples in the next section are in this domain. The particle orbit helicity h is

$$h = \frac{\omega_\zeta}{\omega_\theta}. \tag{5.9}$$

From Eqs 5.3, 5.6 we see that except for drift terms proportional to $\partial B / \partial \psi_p$ which changes sign according to whether the orbit is to the left or to the right of the magnetic axis and thus tend to cancel when averaged over the orbit, and terms in g' and I'_{ψ_p}, both small, we have at low energy $h \simeq q(\psi_p)$. At higher energy we note that the drift term for θ is larger and has the opposite sign of that for ζ. High energy co-moving particles spend more time to the right of the axis, where $\partial B / \partial \psi_p < 0$, so $\dot\theta$ decreases and $\dot\zeta$ increases with energy, causing h to increase. The opposite is the case for counter moving ions. In the determination of p and l the time T gives a full poloidal period. Since the equilibrium is axisymmetric this give the correct mean value for both frequencies. To determine $\dot\theta$ and $\dot\zeta$ orbits are followed for many poloidal transits.

Both ω_ζ and ω_θ are approximately linear in particle parallel velocity, since except for higher order drift terms Eqs. 5.3 and 5.6 are proportional to $v_{\parallel}$. Both approach zero when the energy approaches μB, so at low energy and positive mode frequency p is $-\infty$, increasing with energy until asymptotically for large velocity we find $p \simeq nlh$.

From Eq. 5.8 we can see that the integer p is the number of poloidal elliptic points of the resonance. Consider a time interval T such that $nl\omega_\zeta T = 2\pi c$ with c integer, Then we find $\omega_\theta T = 2\pi c/p - \omega l T/p$. Thus $\theta = \omega_\theta T$ for $c = 0, 1, 2, ...p-1$ can take on p values in the interval $[0, 2\pi]$ about $-\omega l T/p$. These are all resonance points, corresponding to the different elliptic points in θ. We also find the toroidal angle values to be $\omega_\zeta T = 2\pi c/nl$, giving nl elliptic points toroidally with $c = 0, 1, 2, ...nl-1$. Values of $l > 1$ correspond to resonances with higher toroidal mode periodicity than n. Because of particle drift motion, the resonance periodicity in θ need not be given by one of the poloidal mode harmonics m, except at very low energy. The drifts can cause coupling to higher or lower poloidicity. For high energy particles the poloidal drift motion is primarily $cos(\theta)$, leading principally to coupling to harmonics with $m \pm 1$. Equilibrium shape can also influence this spectrum.

For resonance p must be integer, but it is also necessary that there exist a poloidal harmonic capable of coupling to the particle motion, the orbit averaged mode amplitude giving the coupling strength should be non-zero. The perturbation eigenfunction can be caused by an external antenna or by local destabilization by a particle gradient at some particular energy. In either case the eigenfunction acts on particles of all energies, and the

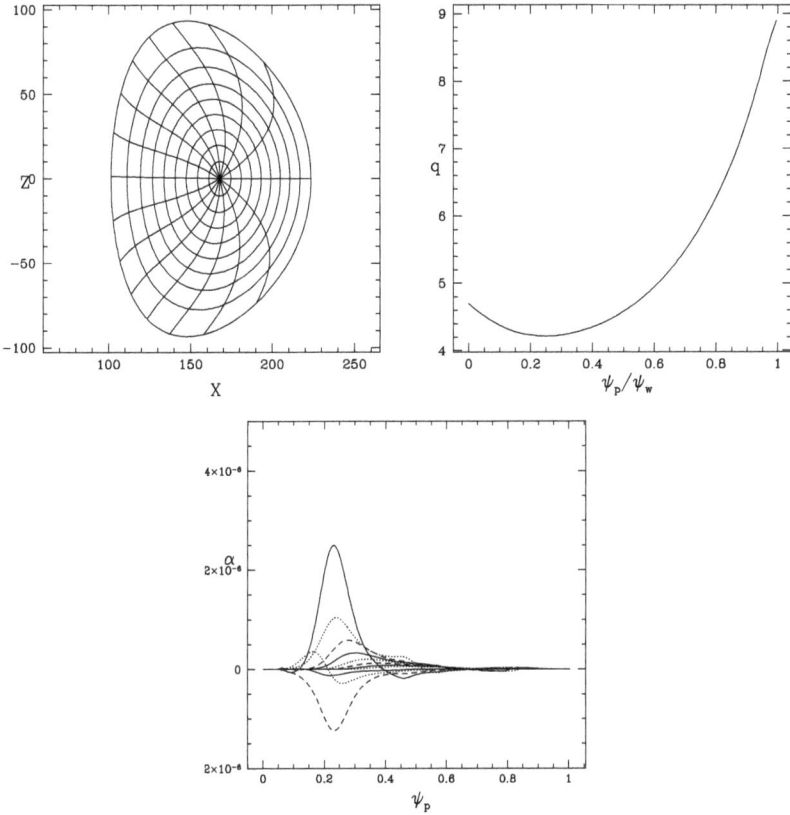

Fig. 5.1 Equilibrium, q profile, and harmonics for the $n = 3$, TAE mode with $10 \leq m \leq$ 23, observed in DIII-D in shot 122117.

resonance condition must be considered to determine the effect on the distribution.

5.2.3   DIII-D Resonances

There is an interesting set of resonances found in a DIII-D equilibrium which illustrates some of the complex behavior possible. The poloidal harmonics of the perturbation and the q profile are shown in Fig. 5.1, an 81 kHz Toroidal Alfvén Eigenmode (TAE) with $n = 3$ and $10 < m < 23$. (See R. White and A. Bierwage, 2021) To find resonances for this mode we use

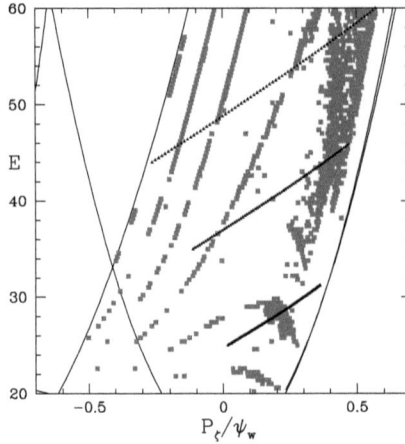

Fig. 5.2   The plane of P_ζ, E with $\mu B_0 = 14$ keV, with paths for kinetic Poincaré plots (black) and resonances due to an Alfvén mode (red) as a function of P_ζ and energy in KeV.

the method of phase vector rotation to determine domains in which KAM surfaces are broken. The method consists of following the vector between two particles in the space of P_ζ, θ. If good KAM surfaces exist, this vector cannot rotate by more than π, so rotation beyond this indicates either a resonance island or a chaotic domain. A full co-moving particle distribution in minor radius is used, with a fixed value of magnetic moment μB_0, with B_0 the on-axis magnetic field. See R. B. White, Plasma Phys. Control. Fusion 53, (2011), and R. B. White, Communications in Nonlinear Science and Numerical Simulation 17, 2200 (2012).

In Fig. 5.2 is a plot of the resonances due to this mode, shown in red. The domain shown is that of co-passing particles, bounded on the right by the magnetic axis and on the left by the last closed flux surface, with $\mu B_0 = 14$ keV. A particle orbit in an axisymmetric tokamak is completely determined by the values of the energy E, the magnetic moment μ, and the toroidal canonical momentum P_ζ, ie by one point in this plot. In addition, kinetic Poincaré plots showing resonances for a mode with toroidal mode number n and frequenty ω can be made only along lines given by $nE - \omega P_\zeta = constant$. Plots are shown for the lines in black in Figs. 5.4, 5.5. A Poincaré plot is useful to verify the values of p and l. They can also be obtained by choosing a point (orbit) determined to be a resonance

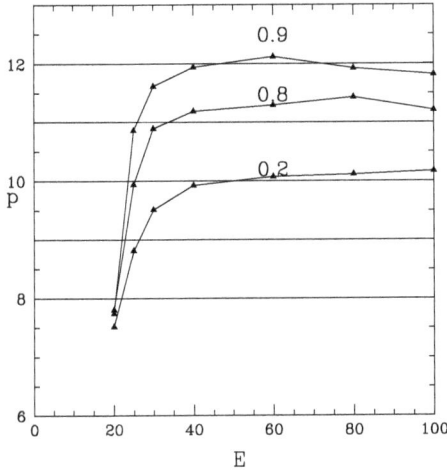

Fig. 5.3   Number of elliptic points p versus particle energy E at different plasma radii, $\psi_p/\psi_w = 0.2, 0.8, 0.9$. Asymptotic values are reached over the whole plasma by $E = 40$ keV. Values of $p = 8, 9$ are only possible at energies below 30 keV.

by the method of phase vector rotation (red points), and following this orbit to directly find ω_θ and ω_ζ and attempting to fit Eq. 5.8. Naturally numerical accuracy can enter into this determination, whereas a Poincaré plot showing the number of elliptic points in the resonance is irrefutable. For a complete analysis of Poincaré structure see chapter 4, section 4.3.

Resonances can be found for any value of μ, and there is small variation in the plot of resonances as a function of μ. This will be demonstrated in detail in subsection 5.2.6.

Occasional high order resonances produce isolated broken KAM surfaces, so one must examine the red dots looking for connected domains, ignoring the isolated red herrrings. The lowest significant resonance domain on the right, near the magnetic axis with $P_\zeta/\psi_w = 0.1$ and $\psi_p/\psi_w \simeq 0.2$, at 21 keV, has 8 elliptic points, $p = 8$, as determined by a Poincaré plot. Moving up in energy, there is a gap without any resonance, and then a resonance around 27 keV with $p = 9$. This resonance begins at about 25 keV at $P_\zeta/\psi_w = -.2$, and is visible by a partly broken line of red points. As the energy increases, this sequence moves inward in minor radius (to the right in P_ζ). This is because $p = n\omega_\zeta/\omega_\theta - \omega/\omega_\theta$ must remain at $p = 9$. As the energy increases, ω_θ increases, making the second term smaller. To compensate this the first term, proportional to q, must move to smaller q values. It does this until it reaches the minimum value of q at $P_\zeta/\psi_w = .2$. At this point it is joined by a second resonance which has emerged from the magnetic axis, two resonances with the same value of p existing because of the non monotone q profile. The two resonances have opposite internal rotations, and they mutually annihilate at about $E = 30$, since at higher energy p becomes larger than 9. This process will be examined in more detail in subsection 5.2.5.

To illustrate the energy dependence of the resonances of Fig. 5.2, in Fig. 5.3 is shown $p = (n\omega_\zeta - \omega)/\omega_\theta$ as a function of particle energy for three different flux surfaces, with possible resonances at integer values. The values of ω_θ and ω_ζ were directly determined by launching particles and determining the slope of the plots of $\theta(t)$ and $\zeta(t)$. Many poloidal transits were followed to give high accuracy. As the particle energy increases, both ω_ζ and ω_θ increase approximately linearly in parallel velocity.

The mode transitions from one resonance to another, with the value of p increasing. This explains the fact that in Fig. 5.2 the resonances with $p = 8$ and $p = 9$ have a limited range in energy. Above the $p = 9$ resonance there is a gap in energy up to about 35 keV and then extending without limit above that a strong resonance with $p = 10$. These resonances are all at small values of minor radius where the mode perturbation is large. The radial position of a resonance, and thus the local helicity can change due to small changes in the relative values of ω_ζ and ω_θ due to drift terms. The curves in Fig. 5.3 are at fixed values of ψ_p. Asymptotically for large E we find $p = n\omega_\zeta/\omega_\theta$, proportional to the local value of the particle helicity h. At asymptotic energies small changes in position can easily change p by a small amount to make it integer. The asymptotic value is reached at different energies at different flux surfaces. Once this occurs the resonance continues to exist for arbitrarily high energy, the drift terms in $\dot\theta$ and $\dot\zeta$ causing small changes in radial location.

There are also visible red dots indicating resonances from about 25 keV upward on the left of Fig. 5.2 with $p = 12$ and 11. They exist near the plasma edge where the q value is higher. At higher energies they both move slightly inward, to lower q values, but by 40 keV p has saturated and they continue to higher energy at a fixed flux value. There is also a resonance with $p = 10$, starting at $P_\zeta\psi_w = -0.3$ at 25 keV. At higher energy it moves gradually inward, and joins the other $p = 10$ resonance at about 70 keV.

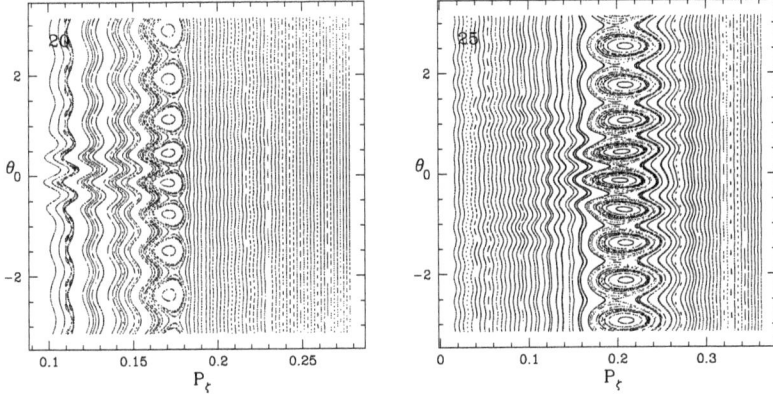

Fig. 5.4 Poincaré plots showing resonance at small minor radius at $E = 21$ keV ($p = 8$) and $E = 25$ keV ($p = 9$) in Fig. 5.2. Here and in the following P_ζ is normalized to ψ_w.

Particle motion with a mode with a single n value and fixed constant frequency ω conserves the quantity $\omega P_\zeta - nE$. Poincaré plots must be made for particles with a particular value of this quantity, as well as the magnetic moment μ. In Figs. 5.4, 5.5 are four kinetic Poincaré plots showing the nature of the resonances along lines with $\omega P_\zeta - nE = constant$, with energies at the left edge of the Poincaré plot of 20 and 25 keV and 35 and 44 keV. The phase vector rotation plot is seen to accurately indicate the location and size of resonances. The plots of Fig. 5.4 are restricted in scope, and show only the resonance located at small minor radius.

In the first plot of Fig. 5.5, beside the strong resonance at $P_\zeta = .45$ with $p = 10$ one can also see a weak resonance crossed by the line $\omega P_\zeta - nE = constant$ in Fig. 5.2 at $P_\zeta = -.02$ also with $p = 10$. In the second plot of Fig. 5.5 one can also see the three weak resonances crossed by the line $\omega P_\zeta - nE = constant$ in Fig. 5.2, one at $P_\zeta = -.2$ with $p = 12$, one at $P_\zeta = .02$ with $p = 11$, and one at $P_\zeta = .22$ with $p = 10$, in addition to the large resonance at $P_\zeta = .45$, also with $p = 10$. These additional resonances are very weak because, in spite of conditions for the existence of a resonance due to the values of ω_θ and ω_ζ, the perturbation amplitude is very small at large minor radius.

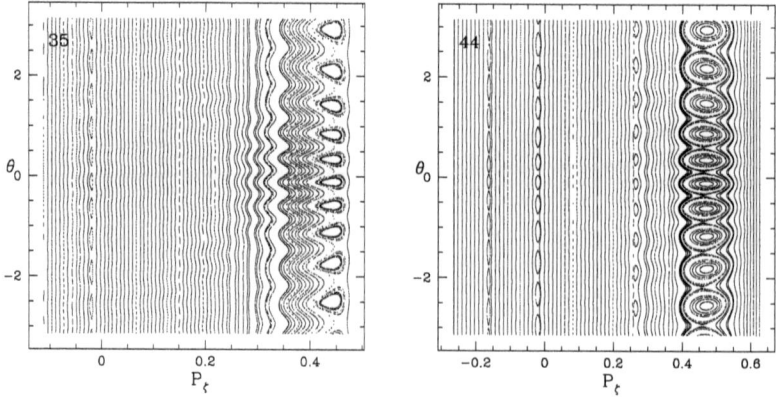

Fig. 5.5 Poincaré plots associated with lines originating in Fig. 5.2 at $E = 35$ keV showing two $p = 10$ resonances, one at $P_\zeta = -.02$ and one at $P_\zeta = .45$, and a plot associated with the line originating at $E = 44$ keV showing three resonances, one at $P_\zeta = -.2$ with $p = 12$, one at $P_\zeta = .02$ with $p = 11$, one at $P_\zeta = .22$ with $p = 10$, and the large resonance at $P_\zeta = .45$, also with $p = 10$.

5.2.4   *Low Frequency modes in ITER*

We consider an equilibrium based on the ITER baseline scenario at Q=10, $q_{95} = 3.0$, type-I ELMy H-mode (inductive plasma). The equilibrium and profile for q for ITER38530D46 at 6 *sec*, are shown in Fig. 5.6. The field on axis is 52.583 kG.

The modes considered in this case were for an investigation of channeling, with low frequency, in an attempt to provide resonances for low energy particles without disturbing the high energy population. We examined 6 kHz modes with $n = 5$ and m values of 4, 5, and 6. Unfortunately, an examination of the equation for resonance and the calculation of ω_θ and ω_ζ, Eq. 5.8 shows that the mode frequency of 6 kHz is negligible even at particle energies of 20 keV, being a factor of five smaller than $n\omega_\zeta$. In Fig. 5.7 is shown one of the resonances at low energy and at an energy of one MeV. Aside from the shape change due to increased particle drift, the resonance is practically unchanged even at this high energy. Since resonance locations are unchanged the eigenfunctions are perfectly capable of exciting a resonance at any energy, making it impossible to modify low energy distributions without also affecting high energy. Since this is a general property of the resonance condition, it makes the removal of

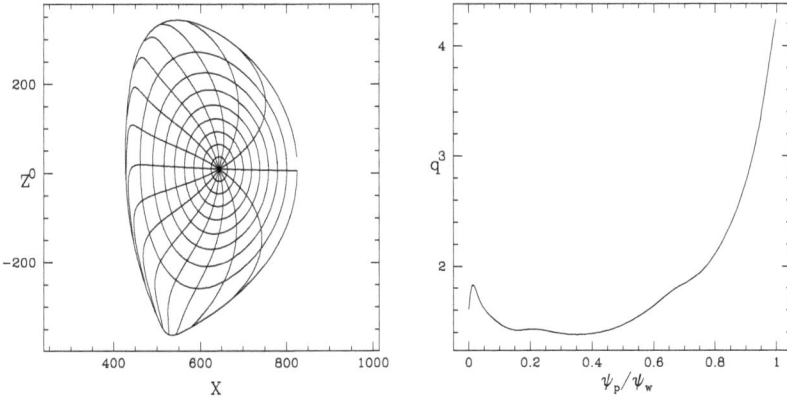

Fig. 5.6 ITER equilibrium and q profile.

cool fusion product ash not possible without also affecting the distribution of the hot alpha particles. See R.B. White et al, Phys Plasmas 28, 012503 (2021).

Other methods than simple diffusion have been suggested for ash removal, such as using chirping modes or bucket transport. However, these results suggest that any resonance intended to influence low energy particles will also affect high energy particles.

5.2.5 *JT-60U Resonances*

We examine resonances in JT-60U, replacing deuterium with hydrogen because we encountered more interesting resonances with the lighter species (albeit requiring larger perturbation amplitudes). The equilibrium and q profile are shown in Fig. 5.8. The magnetic field strength on axis was $B = 1.166$ tesla. The perturbation studied was a 50 kHz mode with $m/n = 2/1$.

In Fig. 5.9 is seen a determination of resonances caused by this mode using the method of phase vector rotation, with $\mu B = 194$ keV. There are two strong bands of resonances more or less in the center of the plasma, extending in energy almost for the full range of passing particles. There is also a resonance that exists only below 350 keV, located deep within the plasma interior. As the energy is increased, this resonance moves inward in minor radius until at 350 keV it passes through the magnetic axis and

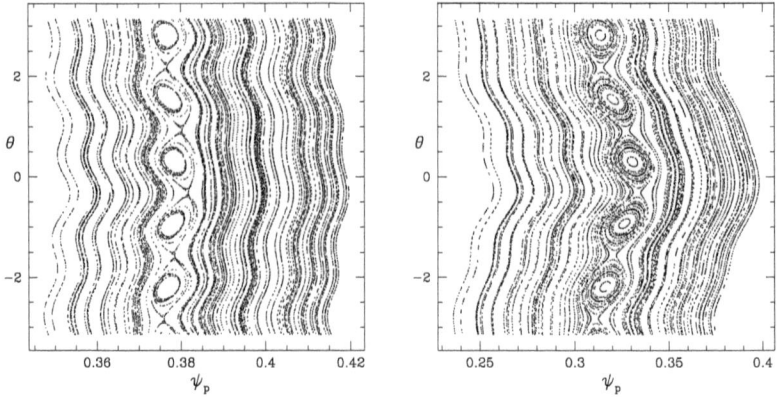

Fig. 5.7   Sample Poincare plots for the mode with a frequency of 6 kHz, with $m/n = 4/5$, with particle energy $E = 20$ keV, Left, and 1 MeV, right. High energy produces a strong shift of the radial position versus poloidal angle, but the strength of the resonance is unchanged.

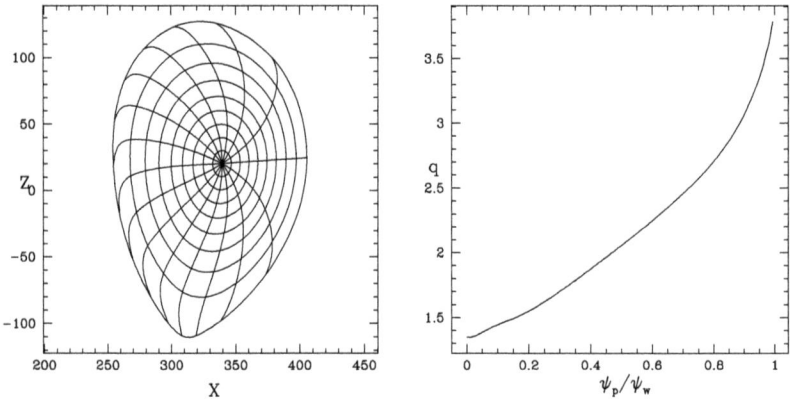

Fig. 5.8   Equilibrium and q profile, JT60-U.

thus disappears. This is a unique occurance, other resonances continue to exist at higher energies.

In Fig. 5.10 are sample Poincaré plots made along the lines shown in Fig. 5.9 along paths given by $nE - \omega P_\zeta = constant$, initiated at the plasma edge with energies of 250 keV, 320 keV, and 400 keV. The first plot with 250 keV (top left) shows a weak resonance with four poloidal elliptic points

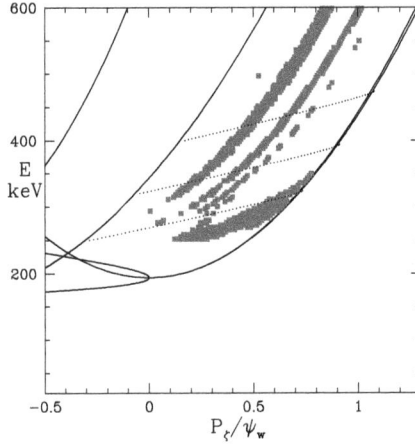

Fig. 5.9 The plane of energy and canonical momentum for the JT-60U case considered, showing the domain of co-passing ions with $\mu B = 194 keV$. Shown in red are resonance domains.

at about $P_\zeta = 0$, also barely visible in the phase vector rotation plot Fig. 5.9. There is also visible a large resonance at the axis with a single elliptic point. This is the resonance that at higher energies moves into the axis and disappears.

The plot initiated at $E = 320$ keV (top right) shows the left most resonance band with three elliptic points at $P_\zeta = 0.2$ as well as two weaker resonances, one with four elliptic points at $P_\zeta = 0.4$ and one with five at $P_\zeta = 0.55$. The third plot with E=400 keV (bottom left) shows both strong bands as well as the weaker five period resonance at $P_\zeta = 0.3$. In the left band we find $q \simeq 2.5$ and in the middle band we find $q \simeq 2.0$. The number of poloidal elliptic points is 3 at $q = 2.5$ and 4 at $q = 2$. The last plot (bottom right) shows ζ in the moving frame at $E = 400$ keV, showing that the $p = 3$ resonance has $l = 2$, the $p = 4$ resonance has $l = 3$ and that the $p = 5$ resonance has $l = 4$.

5.2.6 *JT-60U Reversed Shear*

To more fully understand the JT-60U results showing a resonance existing only for a limited range of energy, we examine a reversed shear profile, with the same perturbation, a $m/n = 2/1$ mode at 50 kHz. Also to demonstrate the weak dependence on the value of μ, we use a much lower value of the

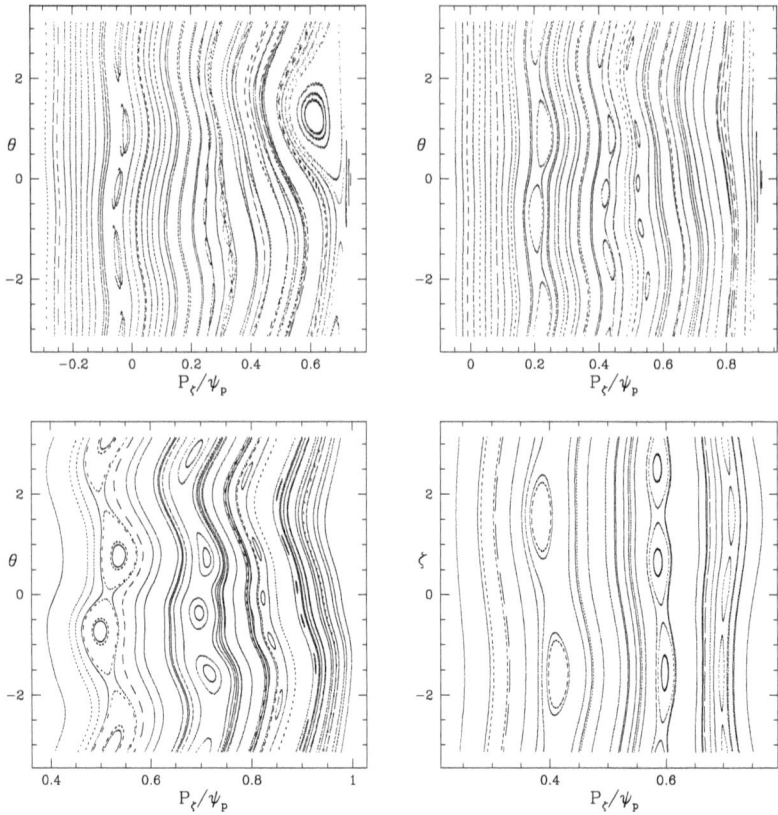

Fig. 5.10 Sample Poincaré plots made along the lines shown in Fig. 5.9 given by $nE - \omega P_\zeta = c$, initiated on the left at energies of 250, 320, and 400 keV. The last plot shows ζ in the moving frame, $E = 400$ keV, showing that the $p = 3$ resonance has $l = 2$, the $p = 4$ resonance has $l = 3$, and the $p = 5$ resonance has $l = 4$.

magnetic moment, $\mu B = 14$ keV. Resonance primarily involves the particle parallel velocities, not the total energy. In Fig. 5.11 is shown the low energy resonance that moves into the magnetic axis in the monotone shear case, now at energies of only 50 keV because of the lower value of μ, to be compared with Fig. 5.9. There is again a $p = 1$ resonance beginning at low energy in the center of the discharge and moving toward the axis at higher energy.

However, because of the reversed shear, instead of reaching the axis, at about 70 keV it is joined by an image $p = 1$ resonance exiting from

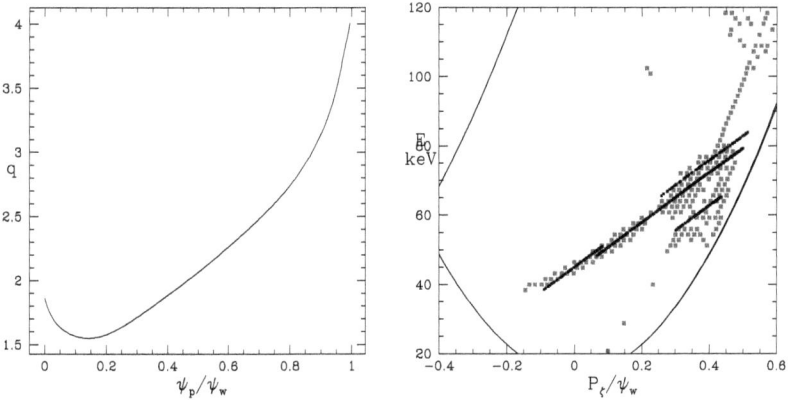

Fig. 5.11   The q profile for a reversed shear equilibrium in JT-60U, along with a plot of the plane of P_ζ, E, showing the location of resonances.

the axis, at about the point of the minimum in the q profile. This is similar to the $p = 9$ and $p = 10$ resonances merging at high energy in the DIII-D case, also an equilibrium with reversed shear, allowing two solutions for p at different radii. The two resonances merge at the point of the minimum in the q profile, and because of the local vanishing of the shear they are much larger than normal for the small amplitude $A = 10^{-5}$. At energies above 70 keV, we have $1 < p < 2$ with the unperturbed transit frequencies ω_θ and ω_ζ. Indeed, no resonance is seen for this value of A. However, an island structure does appear for $A > 10^{-4}$ and can become very large for larger amplitudes. In the following paragraphs, we will illustrate these observations with some concrete examples. After that, we discuss possible reasons for how islands can appear at large values of A.

Poincaré plots of the resonances associated with the lines in Fig. 5.11 are shown in Fig. 5.12. The initial $p = 1$ resonance approaches the $p = 1$ resonance which emerges from the magnetic axis, and these two resonances meet and annihilate. When they meet, the resonance emerging from the axis splits in two, and is pressed against the original resonance. Rotation about the elliptic point is clockwise in the right half and counterclockwise in the left half. Rotation about the elliptic point of the far left island, the original resonance coming from the plasma center, is clockwise. Since this merging occurs at the minimum of q where the shear vanishes, islands are

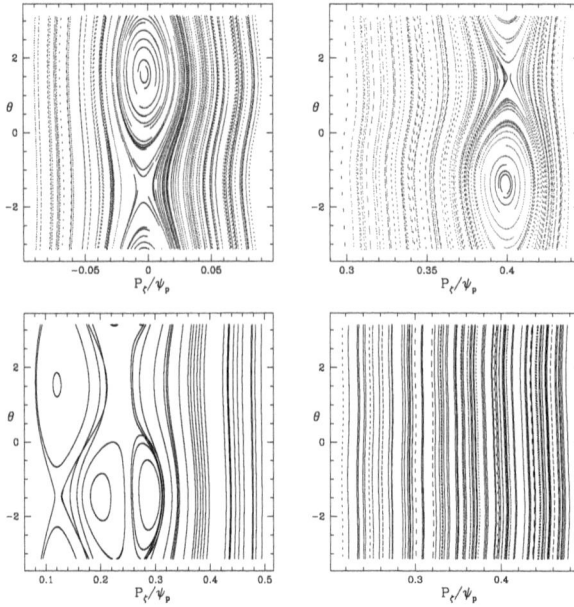

Fig. 5.12 Poincaré plots associated with the lines in Fig. 5.11. Top left, the $p = 1$ resonance, shown at $E = 45$, $P_\zeta/\psi_w = 0$, Top right, the $p = 1$ resonance, shown at $E = 60.5$, $P_\zeta/\psi_w = .37$, Bottom left, Merging of the two resonances at $E = 65$, $P_\zeta/\psi_w = .3$. Bottom right, no resonance visible at $E = 75$, $P_\zeta/\psi_w = .35$. Plots are all shown with $A = 10^{-5}$.

very large. Finally at $E = 75$, $P_\zeta/\psi_w = .39$ we see that a resonance no longer exists inside the plasma, where "inside the plasma" in the present context also means "above q_{min}". Thus this resonance is similar to that seen moving into the axis and disappearing in the last section for monatonic q. In the present reversed shear case, instead of moving into the axis the resonance is annihilated by an image resonance coming out of the axis and joining it. Although Fig. 5.11 shows a resonance continuing upward in energy after this merger, resonances in this region where $p > 1$ are created by helicity modification, only appearing with a Poincaré plot for $A > 10^{-4}$.

In Fig. 5.13 are plots of h and p for points along the trajectory of the initial low energy mode. We see $p = 1$, allowing a resonance from 40 keV to 75 keV. At and above 75 keV the resonance is at the minimum q value and as the energy increases p moves above 1, not allowing the existence of a resonance at least for unperturbed ω_θ and ω_ζ. However even though

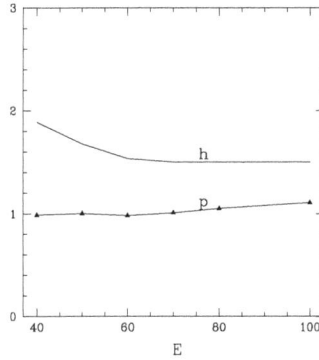

Fig. 5.13   Plots of h and p for the low frequency resonance existing from 40 keV to 75 keV. The resonance is at the minimum q value and as the energy increases above 75 keV p moves above 1, not allowing the existence of resonance at least for unperturbed ω_θ and ω_ζ.

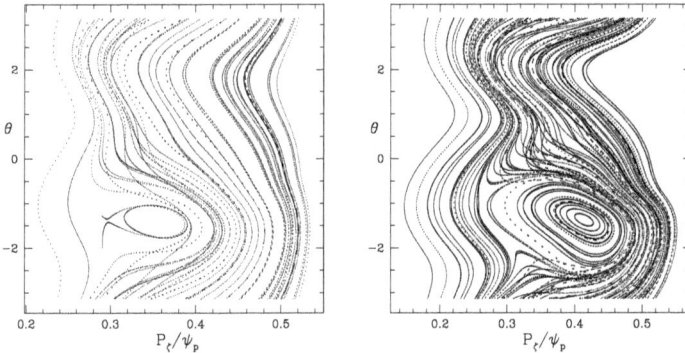

Fig. 5.14   Poincaré plots at $E = 75$, $P_\zeta/\psi_w = .35$ with increasing mode amplitude, showing the appearance of a resonance with $A = 2 \times 10^{-4}$ and $A = 5 \times 10^{-4}$.

$p \neq 1$ and no resonance is visible for small amplitude, as seen in the last plot of Fig. 5.12, at $A > 10^{-4}$ a resonance suddenly appears. In Fig. 5.14 are shown two plots at the same $E = 75$ keV location as the last plot in Fig. 5.12, but with amplitudes of $A = 2 \times 10^{-4}$ and $A = 5 \times 10^{-4}$. These resonances account for the line above 75 keV in Fig. 5.11, made with $A > 10^{-4}$. The apparent intersection of KAM surfaces is due to the fact that the orbits are very sensitive to small deviations from the line $nE - \omega P_\zeta = const$, producing small numerical errors.

5.3   STELLARATOR RESONANCES

5.3.1   *Introduction*

A resonance always provides a site for an Alfvén mode. A mode is likely unstable if there exist local high energy particle gradients. If there is toroidal
dependence of B or $\partial B/\partial\psi$ an Alfvén mode produces chaotic orbits of
particles independent of energy or pitch.

In addition, in stellarators exhibiting periodic variation of the field magnitude and with a resonance that matches the period of the equilibrium
islands appear in particle orbits without a perturbing mode, increasing in
size with energy. Confinement can be affected even without the appearance
of an unstable mode. (See R. White and S. Ethier, 1921)

5.3.2   *Resonance Determination*

The covariant expression for the magnetic field used in the guiding center
code Orbit is

$$\vec{B} = g(\psi)\nabla\zeta + I(\psi)\nabla\theta + \delta(\psi,\theta,\zeta)\nabla\psi,$$

θ and ζ are poloidal and toroidal coordinates and ψ is the toroidal flux. $g(\psi)$
is the total poloidal current outside the surface ψ, including the current in
the field coils. $I(\psi)$ is the total toroidal plasma current inside ψ. Typically
in a stellarator I is very small. The function δ is a result of non orthogonal
coordinates.

The Hamiltonian is $H = \rho_\parallel^2 B^2/2 + \mu B + \Phi$ with $\rho_\parallel = v_\parallel/B$. An Alfvén
mode is given by $\delta\vec{B} = \nabla \times \alpha\vec{B}$. The Lagrangian with an Alfvén mode
present is

$$L = (\psi + \rho_\parallel I + \alpha I)\dot{\theta} + (\rho_\parallel g + \alpha g - \psi_p)\dot{\zeta} + \mu\dot{\xi} - H(\psi_p,\rho_\parallel,\theta,\zeta),$$

For resonance a particle orbit must periodically return to experience the
same perturbation. Thus, at some time T

$$\theta(T) - \theta(0) = 2\pi n, \quad \zeta(T) - \zeta(0) = 2\pi m, \quad \psi(T) = \psi(0),$$

for integers m, n. The frequency is $\omega = 2\pi/T$.

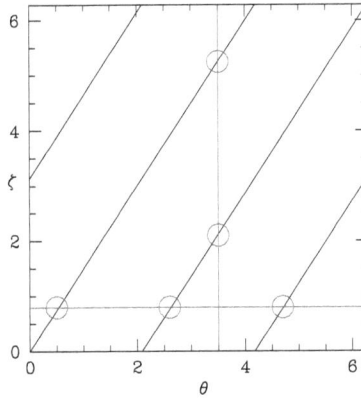

Fig. 5.15   Trajectory of a 3/2 resonant particle through multiple transits, starting at 0,0.

To find resonances launch passing particles distributed in E, μ, ψ, ζ, and θ and for every particle that returns to its initial position a resonance is given with the poloidal transits $n = mod(\theta(t) - \theta(0), 2\pi)$ and toroidal transits $m = mod(\zeta(t) - \zeta(0), 2\pi)$. The helicity m/n is equal to the field line helicity q at low energy. Drift causes a resonance to move in minor radius with energy change and thus a resonance can also enter or leave the plasma.

In Fig. 5.15 is shown the orbit associated with a 3/2 resonance. Three elliptic points are found in a Poincaré section in θ, fixed ζ. Two elliptic points are found in a Poincaré section in ζ, fixed θ. Normally it is the ζ dependence of a stellarator equilibrium that can match the resonance.

The particle motion in ψ is only drift motion, second order in $\rho_\parallel$

$$\dot{\psi} = -\frac{g}{D}(\mu + \rho_\parallel^2 B)\frac{\partial B}{\partial \theta}.$$

The particle motion in the poloidal and toroidal directions are

$$\dot{\theta} = \frac{\rho_\parallel B^2}{D}(1 - \rho_\parallel g') + \frac{g}{D}(\mu + \rho_\parallel^2 B)\frac{\partial B}{\partial \psi_p}$$

$$\dot{\zeta} = \frac{\rho_\parallel B^2}{D}(q + \rho_\parallel I'_{\psi_p}) - \frac{I}{D}(\mu + \rho_\parallel^2 B)\frac{\partial B}{\partial \psi_p}.$$

where $D = gq + I + \rho_\parallel(gI'_{\psi_p} - Ig'_{\psi_p})$.

At low energy (first order in $\rho_\parallel$) orbit helicity is $\dot{\zeta}/\dot{\theta} = q(\psi)$, the particles follow the magnetic field. As the particle energy increases orbit helicity can either increase or decrease due to drift depending on the orbit and the equilibrium.

5.3.3 *Islands caused by resonance with the equilibrium field*

The toroidal variation of the strength of B, typically ten percent in a stellarator, acts as a perturbation of the orbits, and if this field modulation matches an existing resonance it causes islands in the particle orbits. The stellarators NCSX and HSX have particle resonances that match the periodicity of the equilibrium field and thus possess islands in high energy particle orbits.*

The existing stellarators W7X and LHD, and the design for the quasisymmetric stellarator QA do not have particle resonances that match the periodicity of the equilibrium field for reasonable particle energies, and thus possess islands in particle orbits only in the presence of Alfvén instabilities.

5.3.4 *Resonances in NCSX*

The equilibrium for NCSX has a major radius of 145 cm and magnetic field of 15 kG, and the field has a toroidal period of 3. These are shown in Fig. 5.16. The q profile has a maximum of 3 near axis and falls to 1.5 at the edge. This device was designed for construction at Princeton but not built. To show the high energy resonances in this device we scale the design up to reactor size, with a major radius of 10 meters and a field of 5 Tesla.

The negative frequency resonances (counter moving particles) found in NCSX are shown in Fig. 5.17. The highest frequency resonance is a 2/1. Matching the toroidal variation of B are the 5/3, 6/3, 7/3, 8/3 and higher helicity resonances. There are also resonances 2/1, 5/2, 7/4, 8/5 and others present, not matching B. The radial location of the resonances is fairly independent of energy. The resonance frequency scales inversely as the

---

*Due to a numerical code error in the value of $\partial B/\partial \psi_p$ in equilibria, responsible for the energy modification of the orbit helicity, the islands due to resonance with the equilibria in Phys. Plasmas 28, 092503 (2021), by White and Ethier, are larger than they ought to be and in some cases enter the plasma at incorrectly low energies. Islands shown here are correct.

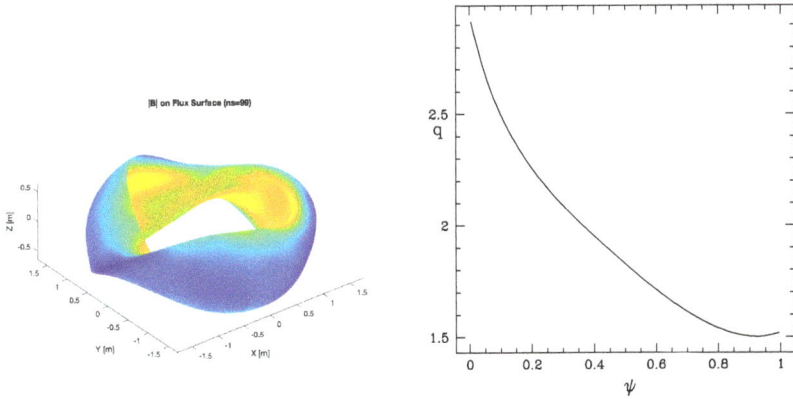

Fig. 5.16   NCSX equilibrium and q profile.

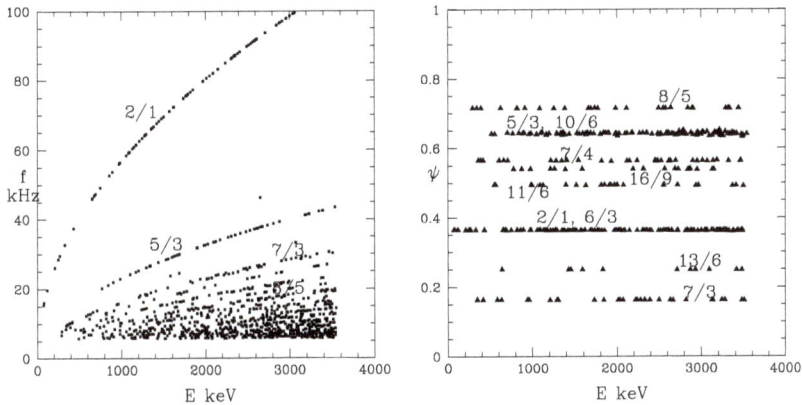

Fig. 5.17   Negative frequency resonances in NCSX vs particle energy.

number of toroidal transits of the resonance and as the square root of the particle energy.

In Fig. 5.18 is shown the 6/3 resonance in NCSX with nonzero width with no eigenmode, produced by the toroidal field variation. Islands are twice this size for barely passing orbits.

There is an adiabatic invariant $J(\alpha)$ given by an integral over the length of the low energy resonance trajectory developed by Felix Parra Diaz and

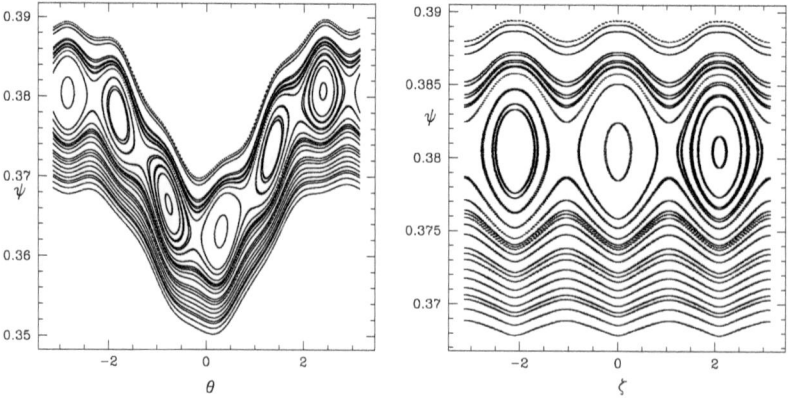

Fig. 5.18   The negative frequency 6/3 resonance in NCSX.

Thomas Foster, reproducing resonance islands produced by the equilibrium

$$J(\alpha) = \int_0^L dl \frac{\sqrt{2E - 2\mu B}(gq + I)}{Bq} - \pi n q'(\psi_r)\frac{(\psi - \psi_r)^2}{q(\psi_r)} \qquad (5.10)$$

with $\theta = \alpha + \zeta/q(\psi_r)$, $B = B(\psi_r, \theta, \zeta)$ and ψ_r the resonance flux location. The parameter α thus gives surfaces away from the resonance field line. This invariant reproduces the island widths and demonstrates that the width increases with energy as $E^{1/4}$. A numerical confirmation of this relation is shown in Fig. 5.19, along with a Poincaré section made at 100 keV, one point in the plot versus energy. The invariant 5.10 gives the island width but does not reproduce the poloidal variation of the elliptic points seen in this figure, but the addition of higher order terms can also supply this.

5.3.5   *Resonances in HSX*

The HSX stellarator in Wisconsin has a field $B = 10kG$, and a major radius of $R = 165cm$. The helicity profile q is just below $q = 1$ with a very small range of values. The equilibrium has toroidal period 4. As particle energy increases the 4/4 resonance with helicity 1 moves into the plasma from the axis at about 150 keV and moves further out as energy increases. A plot made at an energy of 200 keV is shown in Fig. 5.20. Island width depends inversely on the square root of the shear, q' and the m value of

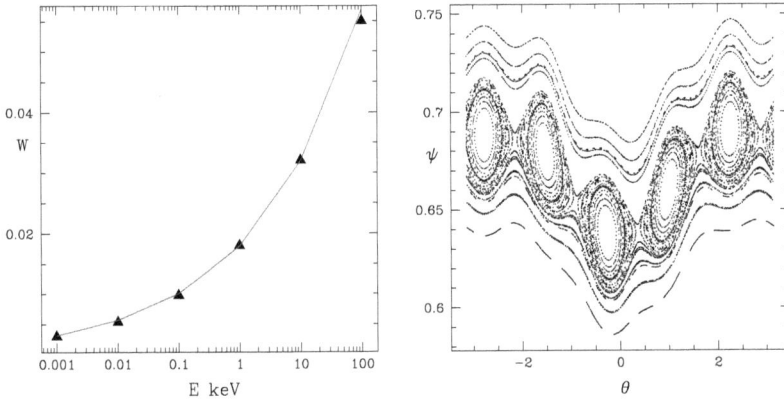

Fig. 5.19   NCSX island width vs particle energy $W = .018E^{1/4}$. The solid curve was produced by the adiabatic invariant $J(\alpha)$ and the points by Poincaré plots. One such plot is shown for particle energy of 100 keV.

the resonance, see Eq. 4.22 in Chapter 4. In this case m is not large and the shear is very small, so the islands are large. Note the wide spread of ψ values found for the resonance, caused by the large poloidal variation of the radial position of elliptic points, due to the very small range of equilibrium helicity. The drift terms modify the orbit helicity and in order for it to close on itself it must move to a different q value, meaning a large change in ψ because of the low shear.

These energies are probably too high to present a problem for the existing device. Scaling the radius to 5 meters and considering alpha particles with four times the mass moves the onset of the resonance to about 2.8 MeV, so it would still cause loss. Raising the field strength by a factor of 2 could move the onset of the resonance to above 3.5 MeV, where it would no longer present a problem.

5.3.6   *Resonances in W7X*

Wendelstein 7-X is a large fusion research stellarator located in Greiswald, Germany. The equilibrium for W7X has a major radius of 597 cm and magnetic field of 24.7 kG. In Fig. 5.21 is shown the equilibrium and the q profile. The plasma beta is 4.48%. The q profile ranges from 1 at the edge to a maximum of 1.2 near the axis, and in the plasma there are no low order rationals. While avoiding low order rationals and thus low particle energy

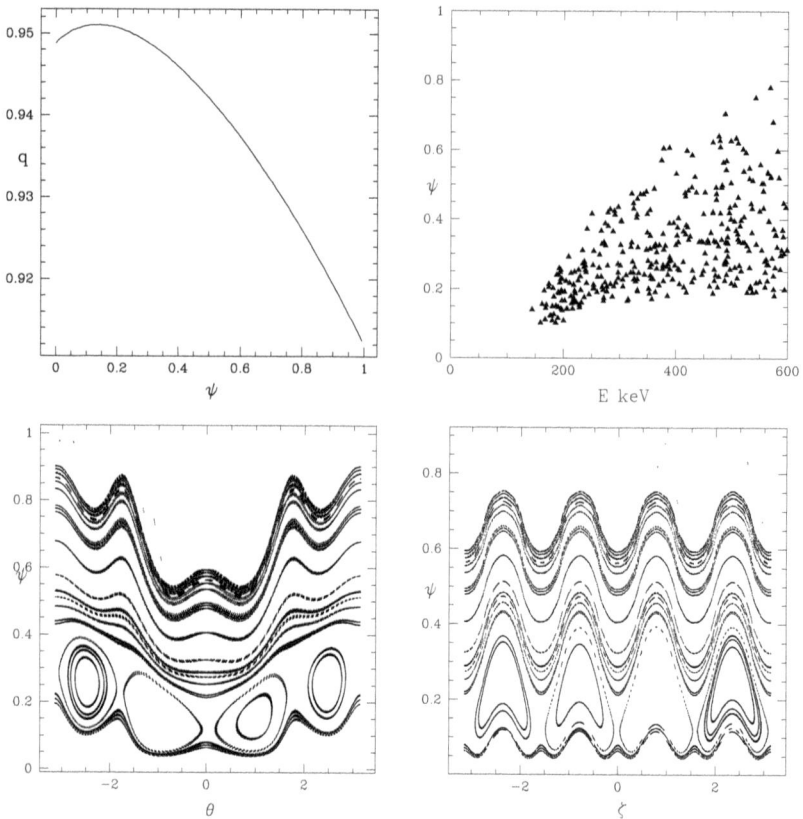

Fig. 5.20   HSX helicity profile, just below $q = 1$. At about 150 keV the $q = 1$ resonance moves into the plasma from the magnetic axis, producind a large 4/4 resonance. All orbits above those shown are lost due to the large theta dependence of the trajectories.

resonances, the plasma, while being of significant minor radius width, is very thin in terms of the range of q.

We first examine deeply passing Hydrogen orbits, small values of magnetic moment. Orbits with parallel velocity aligned with the magnetic field have resonances which enter the plasma only at energies above 2 MeV. Counter moving particles have a very different population of resonances than co-moving particles. At low energy, where both classes of orbits follow magnetic field lines they are the same, but since the drifts change the orbits in different ways, as energy increases these populations differ greatly. In Fig. 5.22 are all negative frequency resonances with $-9 < l < -4$ as

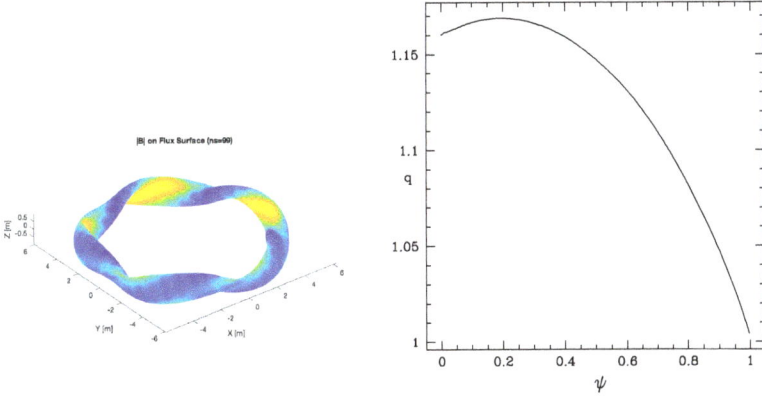

Fig. 5.21   W7X equilibrium and q profile

a function of particle energy up to 2 MeV. There are in fact resonances present with $\omega_\zeta/\omega_\theta = l/p$ equal to 5/4, 6/5, 7/6, and other rationals near the q value of the magnetic axis, and they emerge from the magnetic axis with increasing particle energy. These energies may be too high to be of practical interest for this stellarator, but we believe the resonances are of theoretical interest.

Equilibria designs should be examined for the existence of resonances, to avoid possible confinement problems.

5.3.7   *Orbital Chaos*

An Alfvén mode in a Stellarator without a symmetry does not produce a coherent Poincaré plot of a resonance, giving instead a locally chaotic image. A Poincaré frame ψ, θ is at times $n\zeta - \omega t = 2\pi k$. For a coherent Poincaré plot the step must depend only on the variables of the plot ψ, θ. Otherwise orbits are not uniquely determined in this plane and they can thus cross. But the time step depends on the value of ζ through $B(\psi, \theta, \zeta)$

$$\dot{\psi} \simeq mgB(\psi,\theta,\zeta)v_\| \alpha(\psi)cos(\Omega), \qquad \dot{\rho}_\| \simeq v_\|^2 \partial_\psi B(\psi,\theta,\zeta)\alpha(\psi)sin(\Omega).$$

for fixed ψ, θ with $\Omega = n\zeta - m\theta - \omega t$. ζ is given by ωt so there is a random step in ψ and in $\rho_\|$ given by

$$\dot{\psi} = mv_\| \delta[B]\alpha(\psi), \qquad \dot{\rho}_\| = v_\|^2 \delta[\partial_\psi B]\alpha(\psi)$$

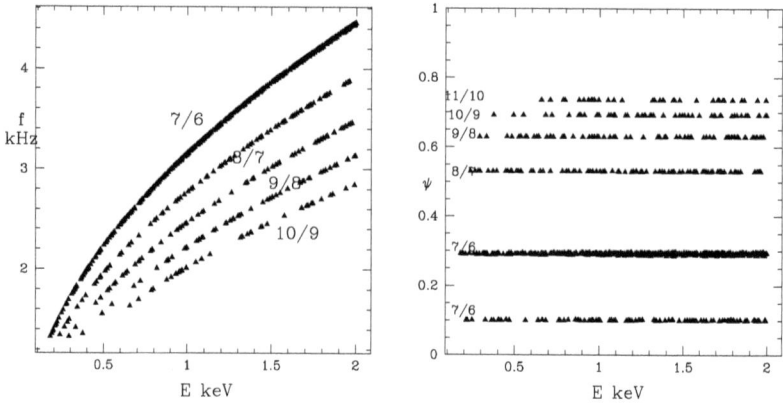

Fig. 5.22   Negative frequency (left) resonances in W7X vs particle energy with $-9 <$ $l < -4$, and position(right). The counter moving orbits have helicities slightly above the values of q on axis. At about 400 keV the $l = -5$ resonances moves out of the magnetic axis, arriving at the plasma center by 1 MeV, followed by the $l = -6$ resonance at 600 keV. Resonance frequency scales as $1/l$ and the square root of the energy.

with $\delta[X]$ equal to X minus the mean value of X over ζ. The modulation of B is typically ten percent so this introduces ten percent random noise onto the pertubation. This is enough to destroy any resonance island structure.

5.3.8   *Alfvén mode induced loss in LHD*

The Large Helical Device (LHD) is a fusion research device in Toki, Gifu, Japan, employing a heliotron field with toroidal period of 10. The shape of the equilibrium and typical q profiles during operation are shown in Fig. 5.23.

The energy and deposition location ψ of the particles injected by the three Hydrogen beams during the 2010 experiment are shown in Fig. 5.24. Two beams were counter passing, and only the beam at 185 keV was co-passing. Resonances for the hydrogen beam discharge of 2010 showing location and frequency vs particle energy are given in Fig. 5.25. These resonances exist for deeply counter passing ions. The frequency is somewhat lower for smaller pitch.

A (1/1) Alfvén instability at 70 kHz causing significant particle loss was experimentally observed. The instability was also found numerically,

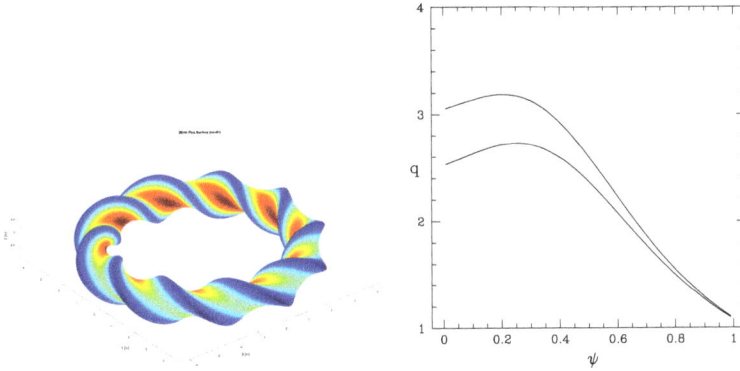

Fig. 5.23   LHD equilibrium and typical q profiles. The equilibrium has toroidal period 10.

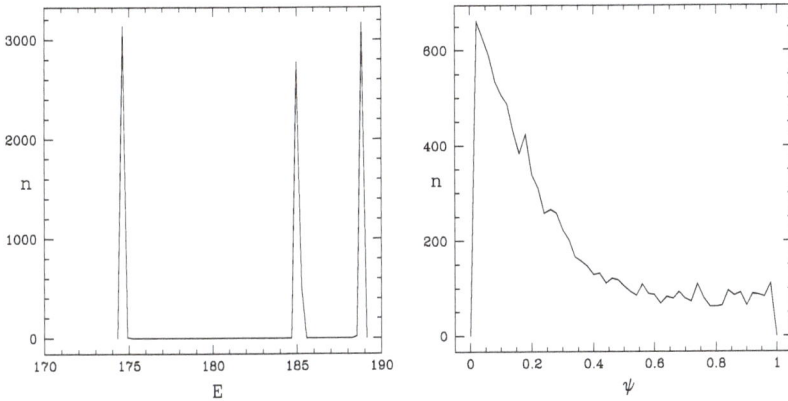

Fig. 5.24   Energy and deposition location ψ of the particles injected by the three Hydrogen beams during the 2010 experiment in LHD.

confirming the existence of the strong 1/1 resonance existing in the outer plasma.

Similarly, in the experiment of 2022 strong Alfvén instabilities were observed. Energy and deposition location ψ of the Deuterium particles injected by the two beams during the 2022 experiment are shown in Fig. 5.26. The high energy beam was deeply co-passing, and the low energy beam counter passing. Resonances for the deuterium beam discharge of

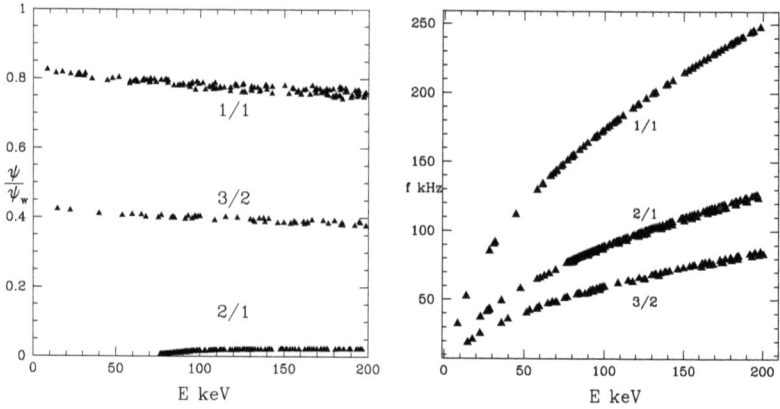

Fig. 5.25   LHD counter passing beam resonances in the 2010 experiment.

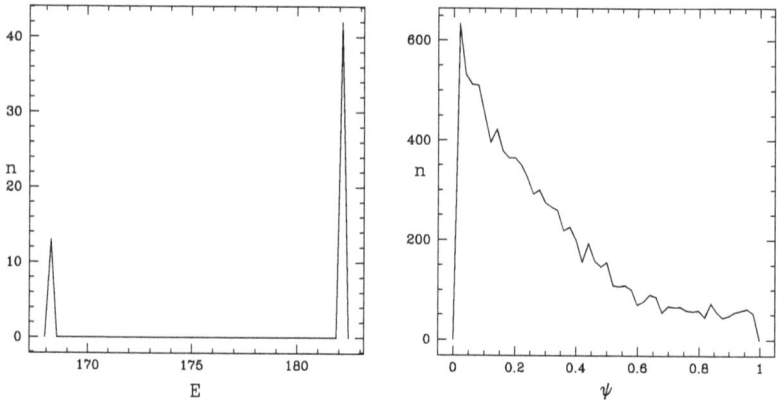

Fig. 5.26   LHD beam deposition in the experiments of 2022.

2022 showing location and frequency vs particle energy are given in Fig. 5.27. These resonances exist for deeply co- passing ions. The equilibrium toroidal period is ten, so no resonances match it. There was an experimentally observed 40 kHz avalanche with particle loss, The modes were not identified but included $n = 1$ and probably also $3/2$ and $4/3$. We conclude that the presence of a low mode number resonance and high energy particles can very well destabilize an Alfvén mode. The examination of particle res-

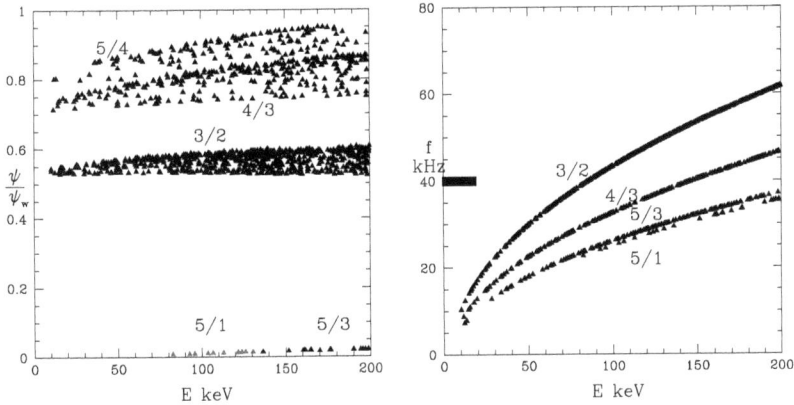

Fig. 5.27   LHD beam resonances in the experiment of 2022, for deeply co-passing particles.

onances gives a very simple means of detecting whether an unstable mode is possible.

5.3.9   *Quasisymmetry*

Exact geometric symmetry of an equilibrium is not necessary to reestablish the possibility of Poincaré resonances for Alfvén modes. It is sufficient that the magnitude of B possess a symmetry, since this is the only thing that enters into the guiding center equations. As an example we show a Poincaré section of an Alfvén mode at the 12/5 resonance existing in the quasisymmetric equilibrium QA of Landreman and Paul. The equilibrium and the helicity profile are shown in Fig. 5.28. The toroidal period of the equilibrium is two, and the helicity has a very small range, similar to HSX, but there are no lower order resonances in the equilibrium, and no resonances at all matching the periodicity of the field.

A small amplitude 12/5 Alfvén mode produces resonance islands for counter moving passing particles near the plasma edge. The island is very small, given the low shear of the equilibrium, due to the large m value, see Eq. 4.22 in Chapter 4. QA is sufficiently symmetric for a coherent Poincaré plot, shown in Fig. 5.29. The mode amplitude is $A = 3 \times 10^{-5}$, and there is no visible chaos due to the mode.

There remains some controversy regarding these equilibria. See section 2.5. The numerical code VMEC produces equilibria, but under the assump-

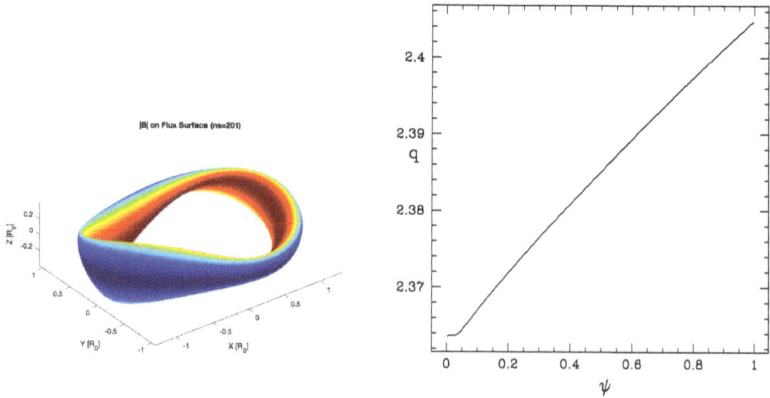

Fig. 5.28   Equilibrium and helicity profile in the quasi symmetric stellarator QA.

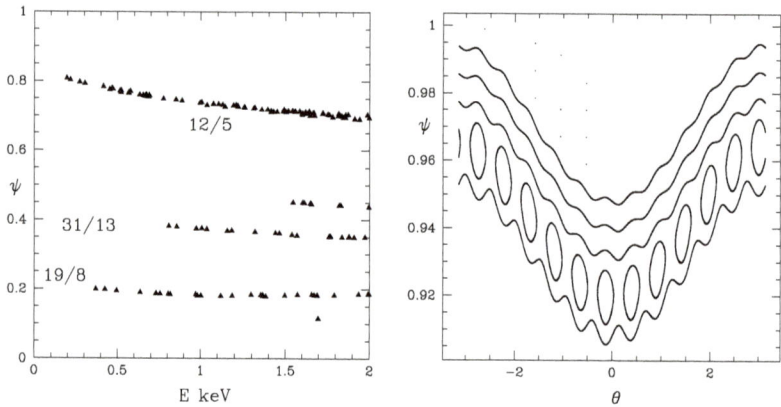

Fig. 5.29   Resonances in the quasi axisymmetric stellarator design and a Poincaré plot caused by a small amplitude 12/5 mode.

tion that good nested flux surfaces exist, whereas ideal MHD predicts that there are small current sheets at some rational surfaces. The code SPEC finds equilibria including the current sheets. The theoretical assumptions of ideal MHD are perhaps too severe to describe the real situation, since a small resistivity would change the current sheets into small localized magnetic islands. There has not yet been a complete theoretical examination of this problem. The particle gyro radii also average the field, so that small local irregularities should not affect the particle orbits. But quasisymmetric designs typically have very small shear values, leading to large island structures, so this could produce problems. The overly restrictive assumptions of ideal MHD regarding equilibria is an ancient problem in the theory of equilibria, first discussed at length by Harold Grad, 1967. However, nonsymmetric devices such as W7X and LHD have already proven to provide good confinement, and W7X also has very weak shear, so perhaps the final solution of these theoretical problems resides in the experiments.

5.4 References

- H. Grad, Phys. Fluids 10(1):1037, 1967
- J. Loizu, S. R. Hudson, A. Battacharjee, S. Lazerson and P. Helander, Phys. Plasma 22 (2015)
- T. Foster, F. Parra Diaz and R. White, to be published
- R B White, and A Bierwage, Phys. Plasmas 28, 032507 (2021)
- D. Fredrickson, and M. Podesta, Phys. of Plasmas 27, 022117 (2020)
- W.Heidbrink and R. White, Phys. Plasmas 27, 2802 (2020)
- V. N. White, R. B. Gorelenkov, N. N. Duarte, and H. L. Berk, Phys. of Plasmas 25, 1 (2018)
- R. B. White, Plasma Phys. Control. Fusion 53, (2011)
- R. B. White, Communications in Nonlinear Science and Numerical Simulation 17, 2200 (2012)
- R. B. White, Plasma Phys. Control. Fusion 57, 115008 (2015)
- Y. Gribov, ITER DATA BASE (2010)rwhite@pppl.gov
- R. B. White, Physics of Fluids B: Plasma Physics 2 4, 845 (1990)
- W. Heidbrink and R. White, Phys. Plasmas 27, 2802 (2020)
- C. S. Collins, W. W. Heidbrink, M. E. Austin, G. J. Kramer, D. C. Pace, C. C. Petty, L. Stagner, M. A. Van Zeeland, R. B. White, and Y. B. Zhu (DIII-D team), Phys. Rev. Lett 116, 095001 (2016)
- M. Podesta, M. Gorelenkova, N. N. Gorelenkov, and R. B. White, Phys. Plasmas Control Fusion 59, 095008 (2017)
- M. Podesta, M. Bardoczi, L. Collins, C. S. Gorelenkov, N. N. Heidbrink, W. W. Duarte, V. N. Kramer, V. K. Fredrickson, E. D. Gorelenkov, N. Kim, et al, Nucl. Fusion 59, 106013 (2019)
- S. P. Hirschman and J. C. Whitson, Phys. Fluids 26, 3553 (1983)
- R. B. White and S. Ethier, Phys. Plasmas 28, 092503 (2021)
- R. B. White, F. Romanelli and F. Cianfrani, Phys. Plasmas 28, 012503 (2021)
- J. M. Canik, D. T Anderson, F. S Anderson, K. M Liken, J. M Talmage, K. Zhai, Phys. Rev. Lett. 98, 85002 (2007)
- K. Toi, F. Watanabe, T. Tokuzawa, K. Ida, S. Morita, T. Ido, A. Shimizu, M. Isobe, K. Ogawa, and et al, Phys. Rev. Lett. 105, 145003 (2010)
- M. Landreman, and E. Paul, Phys. Rev. Lett. 128, 035001 (2022)

- N.N. Gorelenkov, S.D. Pinches and K. Toi, Nucl. Fusion 54, 125001 (2014)
- C.S. Collins, W.W. Heidbrink, M.E. Austin, G.E. Kramer, D.C. Pace, C.C. Petty, L. Stagner, M.A. Van Zeeland, R.B. White, Y. B. Zhu, Phys. Rev. Lett 116, 095001 (2016)
- R. B. White, N. Gorelenkov, W. W. Heidbrink, M. A. Van Zeeland, Phys. Plasmas 17, no. 5 (2010)
- N. Gorelenkov et al Nucl. Fusion 58, 082016 (2018)
- R. B. White and M. S. Chance, Phys Fluids 27, 2455 (1984)

Chapter 6

Linear Ideal Modes

6.1 INTRODUCTION

As seen in Chap. 2, an equilibrium configuration is determined by two functions, which can be taken to be the toroidal current density profile or the field helicity $q(\psi)$ and the plasma pressure profile $p(\psi)$, as well as a boundary configuration for the outermost flux surface. In addition, there may exist a vacuum region, possibly bounded by a conducting wall. The evolution of such an equilibrium state following a perturbation is determined by Eqs. 1.1–1.6. Of obvious interest is the stability of a given equilibrium to an arbitrary perturbation, and the first step is to look for exponentially growing unstable modes. There are two principle methods for examining this question. The first, straightforward method is to introduce perturbations of all quantities, linearize in the perturbation amplitudes, represent the temporal evolution in terms of an exponential $e^{-i\omega t}$, and solve the resulting partial differential equations for the eigenvalue ω. This method has the disadvantage of requiring the solution of non-trivial ordinary or partial differential equations, or even integral equations, but the advantage of yielding exact eigenmodes. The existence of discrete eigenmodes can also be complicated by the presence of continua and accumulation points in the spectrum, which are characterized by algebraic rather than exponential growth. Often the inclusion of a physical effect previously omitted in the interest of simplicity can either discretize the spectrum or otherwise clarify the transition from the stable to the unstable side of the spectrum.

Some insight into the nature of the modes to be expected in an ideal MHD plasma can be gained by examining the waves existing in a homogeneous plasma. Detailed discussions of the MHD spectrum can be found in Stix (1962), Grad (1973), Goedbloed (1975), Tataronis and Grossman (1977), and Goedbloed (1979). The linearized equations yield a dispersion relation with three branches. The highest frequency wave is the magnetoacoustic wave. It compresses both the magnetic field and the plasma pressure, and the perturbed magnetic field has components parallel and perpendicular to $\vec{B}$. For low β this wave reduces to the compressional Alfvén wave. It is nearly transverse, and describes an oscillation between perpendicular plasma kinetic energy and magnetic compressional and line bending energies. The second branch is the shear Alfvén wave, with perturbed magnetic field and velocity perpendicular to both $\vec{B}$ and the wave vector $\vec{k}$. It is an incompressible oscillation between perpendicular plasma kinetic energy and line bending magnetic energy. The lowest frequency branch is the slow magnetoacoustic wave. If the sound speed is much slower than the Alfvén speed it reduces to a sound wave. In an inhomogeneous plasma the unstable modes are combinations of these waves, but the most unstable perturbations almost always involve the shear Alfvén wave.

If considerations are restricted to ideal MHD motion a second method of finding unstable modes has been developed using the existence of a conserved energy for the system. In a perfectly conducting plasma the magnetic flux is frozen into the plasma, and thus electromagnetic and fluid perturbations are not independent, but completely described by the displacement of the fluid $\vec{\xi}$. To see this combine Eqs. 1.2 and 1.3 to find $\partial\vec{B}/\partial t = \nabla \times (\vec{v} \times \vec{B})$. Consider a surface and integrate this equation over it, giving

$$\frac{\partial \Phi}{\partial t} + \oint \vec{v} \cdot (d\vec{l} \times \vec{B}) = 0 \qquad (6.1)$$

with the flux $\Phi = \int d\vec{s} \cdot \vec{B}$ and $d\vec{l}$ an element of the surface boundary. This equation simply states that the flux convects with the fluid. Newcomb (1958) has shown rigorously that if two fluid elements are connected by a field line at one time, they remain so for all time. Thus any motion of the plasma consists of an interchange of nearby volume elements of fluid, which carry their magnetic fields along with them.

The total energy of the system is necessarily quadratic or higher order in $\vec{\xi}$ in the neighborhood of an equilibrium, and the sign of the energy change due to a fluid displacement determines the stability or instability of that equilibrium. In addition, the form of $\vec{\xi}$ minimizing the energy change

δW gives, for negative δW, the eigenfunction for a growing mode. Explicit expressions for the energy permit a convenient evaluation of the stability in many cases of interest, and Secs. 6.3 to 6.5 are devoted to the development of this method. The use of such a variational principle originated with the Rayleigh–Ritz principle in classical mechanics, and has the well-known property of allowing the determination of stability without explicit determination of the eigenfunction. Its formulation for MHD is due to Bernstein *et al.* (1958), and will be developed in detail in this chapter.

The potential energy resulting from a perturbation, derived in Sec. 6.5, allows a classification of unstable modes according to the principle driving mechanism. If the dominant destabilizing contribution is due to $j_{\parallel}$ the mode is called current driven. If the dominant contribution is due to p' the mode is called pressure driven. Modes are also classified as external or internal according to whether the displacement involves the plasma boundary. Reviews of ideal instabilities have been given by Shafranov (1966), Bateman (1978), Wesson (1978), Wesson (1987), and Freidberg (1982).

The most important current driven modes which have been identified are the kink modes, which will be discussed in Sec. 6.8. They exist in both external and internal form, and consist of a helical displacement of all or part of the plasma column, with an associated compression of the magnetic field. Stability to kink modes generally restricts the plasma current below a threshold value. The most unstable modes are those with longest wavelength, and the associated plasma motion can be large. The external kink modes can be stabilized by the close proximity of a conducting wall. The internal kink mode in a tokamak is associated with the $q = 1$ surface, and the displacement is primarily restricted to the plasma inside this surface, so a conducting wall does not affect stability. The internal modes are pressure driven and stabilized by shear, and the internal kink mode has a threshold in β unless the q profile is very flat.

For large toroidal mode numbers the internal modes are highly localized near a particular flux surface, varying slowly along the field line in such a way to maximize the displacement in the bad curvature region, and minimize it in the good curvature region. This makes the mode assume a ballooning appearance and gives it its name.

If field lines are convex toward the plasma, as is usual, their tension can make it energetically favorable for them to slip inward, exchanging places with an element of plasma at higher pressure. This interchange mode is analytically tractable and results in necessary conditions for stability, which will be derived in Sec. 6.14. It consists of perturbations in which $\delta\vec{B}_\perp = 0$ so that there is no field line bending. The perturbations are nearly constant along a field line and have a rapid variation across the field, leading to a fluting of the surface on a small scale length. The mode is also refered to as a flute mode.

The application of the energy principle began with one dimensional systems, the easiest being a cylindrically symmetric equilibrium (linear pinch) for which pressure and current depend on cylindrical radius only. Suydam (1958) found the criterion for stability against localized interchanges. Mercier (1960) generalized the Suydam criterion to axisymmetric systems. Later Mercier (1961, 1962) and Greene and Johnson (1961, 1962) found the generalization of the Suydam criterion to three dimensional systems. For a more detailed discussion of the early history of the subject see Greene and Johnson (1968). Rather than follow the historical development we defer the derivation of the stability of a plasma under local interchange until after the discussion of ballooning modes.

Much intuitive understanding of ideal modes can be gained by analytic analysis using simplified current profiles. On the other hand, some important questions such as the β limits for stability and the effect of plasma shape have not completely yielded to analytic treatment, and are only understood through numerical analysis. In this chapter we first examine those problems which have been successfully analyzed using simple models, and then the numerical techniques used for more complex problems. The Suydam condition for cylindrical equilibria is mentioned in passing in Sec. 6.6, and the general Mercier and Suydam criteria are derived in Sec. 6.14. The discussion of localized internal modes was greatly simplified by the invention of the ballooning representation in 1978. Several groups were involved, but the clearest early formulation was due to Connor *et al.* (1978). The ballooning equations are derived in Sec. 6.11 and are used to derive the Mercier criterion.

The energy principle has been extended to take into account physical effects not included in scalar pressure ideal MHD. The first extension was also by Bernstein *et al.* (1958) and treats the case of anisotropic pressure, *i.e.* a very collisionless plasma in which there is little coupling between the directions parallel and perpendicular to the magnetic field, so that longitudinal and transverse deformations take place independently. The result is called the double adiabatic or CGL energy principle after Chew, Goldberger and Low (1956), who first wrote down the double adiabatic equations of state. If an equilibrium is found to be stable in the MHD theory, then it will also be stable in the double adiabatic theory, *i.e.* $\delta W_{CGL} \geq \delta W_{MHD}$. Thus for purposes of ensuring stabililty it is sufficient to verify it with the MHD theory.

The idea of using high energy particles as a stabilizing influence in confinement devices prompted extension of the energy principle in another direction. To correctly account for heat flow along the magnetic lines it is necessary to return to kinetic theory and obtain the pressure as an appropriate velocity moment of the equilibrium distribution function. The first such kinetic energy principle is due to Kruskal and Oberman (1958). The $\vec{E} \times \vec{B}$ drift velocity of the particles is assumed to be much larger than other drifts, and the mode frequency is assumed intermediate between bounce frequency and toroidal drift frequency. The obtained energy principle is only sufficient and not necessary because they assume zero parallel equilibrium electric field and do not use charge neutrality. An extended version of this energy principle, which incorporates quasi-neutrality and electrostatic effects, was derived by Newcomb and Kulsrud in 1958 and shown to be necessary and sufficient for stability. Using Schwartz inequality arguments (Kulsrud, 1964) it can be shown that $\delta W_{CGL} > \delta W_{KO} > \delta W_{MHD}$, *i.e.* the Kruskal–Oberman energy principle predicts instability thresholds intermediate between those predicted by the double adiabatic and MHD theories. Most important, these kinetic energy principles cannot predict an instability which is not already present in the MHD formalism.

Unfortunately, this result is due to the assumptions made about mode frequency, and an important class of instabilities is overlooked by all of these formalisms. They were first found experimentally, and will be discussed in Chap. 8.

Ideal MHD instability proceeds on the rapid Alfvén time scale, and any large scale, large amplitude ideal modes are liable to be catastrophic for confinement. Thus one is normally more interested in searching for stable equilibria than in finding exact growth rates. This is particularly true if the method used is restricted to linear behavior, because nonlinear effects typically modify the mode evolution at very low amplitude.

The use of a variational energy principle allows the use of trial functions to evaluate growth rates and the form of the eigenmodes, and is the method adopted by many linear numerical codes. The method is unsatisfactory for the analysis of nonlinear behavior, for which the time evolution of the physical variables must be calculated. Appropriate formalisms for studying initial value problems and suitable for studying nonlinear evolution are discussed in Chap. 10.

6.2 PLASMA KINETIC AND POTENTIAL ENERGY

To examine stability about an equilibrium we consider small perturbations of a system in equilibrium. Take $\vec{\xi}$ to be the absolute displacement of a fluid element,

$$\vec{r} = \vec{r}_0 + \vec{\xi}(r_0, t) \tag{6.2}$$

where $\vec{r}_0$ is the equilibrium position, *i.e.* $\vec{\xi}$ is evaluated in the fixed Eulerian frame. The partial derivative of $\vec{\xi}$ with respect to time gives the velocity at a point moving with the fluid, *i.e.* in a Lagrangian frame. That is, $\partial_t \vec{\xi}(r_0, t) = \vec{v}(\vec{r}) = \vec{v}(\vec{r}_0, t) + \vec{\xi} \cdot \nabla \vec{v}(\vec{r}_0, t)$. Linearizing the equations of Sec. 1.1 about an equilibrium with $\vec{B} = \vec{B}_0 + \vec{B}_1$, $\rho = \rho_0 + \rho_1$, $p = p_0 + p_1$, we find for the equation of motion

$$\rho_0 \ddot{\vec{\xi}} = -\nabla p_1 + (\nabla \times \vec{B}_1) \times \vec{B}_0 + (\nabla \times \vec{B}_0) \times \vec{B}_1 \equiv \vec{F}(\vec{\xi}), \tag{6.3}$$

with

$$p_1 = \vec{\xi} \cdot \nabla p_0 - \gamma p_0 \nabla \cdot \vec{\xi} \tag{6.4}$$

and

$$\vec{B}_1 = \nabla \times (\vec{\xi} \times \vec{B}_0) \equiv \vec{Q}. \tag{6.5}$$

Multiply the $\ddot{\vec{\xi}}$ equation by $\dot{\vec{\xi}}$ and integrate over space, giving for the time derivative of the kinetic energy

$$\frac{1}{2} \frac{\partial}{\partial t} \int \rho_0 (\dot{\vec{\xi}})^2 d\tau = \int \dot{\vec{\xi}} \cdot \vec{F}(\vec{\xi}) d\tau. \tag{6.6}$$

This equation almost has the form of an energy conservation equation. To put it in this form it is necessary to show that the inner form defined by $\vec{F}(\vec{\xi})$ is self adjoint. There are two methods for demonstrating this. The direct method is to show explicitly that $\int \vec{\zeta} \cdot \vec{F}(\vec{\xi}) d\tau = \int \vec{\xi} \cdot \vec{F}(\vec{\zeta}) d\tau$ using the equilibrium conditions (Kadomtsev, 1966). A shorter method comes from recognizing that the self adjointness is directly related to the existence of a conserved energy integral (Bernstein *et al.*, 1958, Kulsrud, 1984). We give here the second method.

6.3 SELF ADJOINTNESS OF THE POTENTIAL ENERGY

Construct the energy of the system. Given the equation of state $p = A\rho^\gamma$ and using volume $\tau \sim 1/\rho$ we find the energy per unit volume of an adiabatic fluid by compressing it from a state of zero density, or infinite volume,

$$-\frac{1}{\tau}\int_{-\infty}^{\tau} p d\tau = \rho \int_0^{\rho} \frac{A\rho^\gamma d\rho}{\rho^2} = \frac{p}{\gamma - 1}. \tag{6.7}$$

Thus the total energy is

$$U = K + W = \int d\tau \left(\frac{\rho v^2}{2} + \frac{B^2}{2} + \frac{p}{\gamma - 1} \right). \tag{6.8}$$

It then follows from the equations of motion that U is constant in time.

Now we wish to expand U to second order in $\vec{\xi}$ and $\dot{\vec{\xi}}$. It is easy to solve for $\vec{B}$, p, and ρ to second order in $\vec{\xi}$. We find

$$\rho(\vec{r_0} + \vec{\xi}, t) = \frac{\rho(\vec{r_0}, 0)}{det(1 + \nabla\vec{\xi})}, \tag{6.9}$$

$$\frac{p(\vec{r_0} + \vec{\xi}, t)}{p(\vec{r_0}, 0)} = \left(\frac{\rho(\vec{r_0} + \vec{\xi}, t)}{\rho(\vec{r_0}, 0)} \right)^\gamma, \tag{6.10}$$

$$\frac{\vec{B}(\vec{r_0} + \vec{\xi}, t)}{\rho(\vec{r_0}, 0)} = \frac{\vec{B_0}}{\rho_0} + \left(\frac{\vec{B_0}}{\rho_0} \cdot \nabla \right) \vec{\xi}, \tag{6.11}$$

$$d\tau = d\tau_0 det(1 + \nabla\vec{\xi}). \tag{6.12}$$

These expressions are not necessary for the following except to demonstrate that U can be expanded in $\vec{\xi}$ and $\dot{\vec{\xi}}$.

Write

$$U = S(\dot{\vec{\xi}}) + T(\vec{\xi}) + K(\dot{\vec{\xi}}, \dot{\vec{\xi}}) + M(\vec{\xi}, \dot{\vec{\xi}}) + \delta W(\vec{\xi}, \vec{\xi}) \tag{6.13}$$

where S, T are linear functions of the arguments, and K, M, and δW are quadratic forms. Without loss of generality K and δW can be taken to be symmetric in the two arguments. First show that S and T are zero and find K. Consider a displacement with $\vec{\xi} = 0$ but $\dot{\vec{\xi}} = v \neq 0$. Then $\delta \vec{B} = \delta p = \delta \rho = 0$, $d\tau = d\tau_0$, and $U = \int (\rho/2) v^2 d\tau$. Thus $K = (1/2) \int \rho v^2 d\tau$ and $S = 0$. Differentiate the expression for U, giving $\dot{U} = 0 = T(\dot{\vec{\xi}}) + O(\xi^2)$, and thus $T = 0$. Also to second order in ξ we must require that $\dot{U} = 0$ and thus this gives $2K(\ddot{\vec{\xi}}, \dot{\vec{\xi}}) + M(\dot{\vec{\xi}}, \dot{\vec{\xi}}) + M(\vec{\xi}, \ddot{\vec{\xi}}) + 2\delta W(\dot{\vec{\xi}}, \vec{\xi}) = 0$. We also have $\ddot{\vec{\xi}} = \vec{F}(\vec{\xi})/\rho$ and thus

$$0 = 2K\left(\frac{\vec{F}(\vec{\xi})}{\rho}, \dot{\vec{\xi}}\right) + M(\dot{\vec{\xi}}, \dot{\vec{\xi}}) + M\left(\vec{\xi}, \frac{\vec{F}(\dot{\vec{\xi}})}{\rho}\right) + 2\delta W(\dot{\vec{\xi}}, \vec{\xi}) \quad (6.14)$$

and this must hold for all $\vec{\xi}, \dot{\vec{\xi}}$. Set $\vec{\xi} = 0$, and find $M(\dot{\vec{\xi}}, \dot{\vec{\xi}}) = 0$. Setting $\dot{\vec{\xi}} = 0$ we find $M(\vec{\xi}, \vec{F}(\vec{\xi})/\rho) = 0$ and thus

$$K(\vec{F}(\vec{\xi})/\rho, \dot{\vec{\xi}}) = -\delta W(\dot{\vec{\xi}}, \vec{\xi}). \quad (6.15)$$

But δW is symmetric so we find

$$K(\vec{F}(\dot{\vec{\xi}})/\rho, \vec{\xi}) = K(\vec{F}(\vec{\xi})/\rho, \dot{\vec{\xi}}) \quad (6.16)$$

and setting $\dot{\vec{\xi}} = \eta$ and using $K = (1/2) \int \rho \dot{\xi}^2 d\tau$ we find that $\vec{F}$ is self adjoint and further

$$\delta W(\vec{\xi}, \vec{\xi}) = -\frac{1}{2} \int \vec{\xi} \cdot \vec{F}(\vec{\xi}) d\tau, \quad (6.17)$$

so δW is the potential energy.

6.4   THE ENERGY PRINCIPLE

Using the self adjointness of $\vec{F}$ we can now write the potential energy as

$$\int \dot{\vec{\xi}} \cdot \vec{F}(\vec{\xi}) d\tau = \frac{1}{2} \frac{\partial}{\partial t} \int \vec{\xi} \cdot \vec{F}(\vec{\xi}) d\tau. \quad (6.18)$$

Thus the equation of motion integrates to

$$U = \frac{1}{2} \int \rho(\dot{\vec{\xi}})^2 d\tau + \delta W = constant \quad (6.19)$$

where $\delta W = -(1/2) \int \vec{\xi} \cdot \vec{F}(\vec{\xi}) d\tau$ is the potential energy, and $(1/2) \int \rho(\dot{\vec{\xi}})^2 d\tau$ is the kinetic energy. For the study of spontaneously occuring instabilities it is appropriate to take the constant of integration to be zero, *i.e.* at $t = 0$ we begin with $\vec{\xi}$ arbitrarily small. For a normal mode, $\xi \sim e^{-i\omega t}$, the expression for the total energy then reduces to $-(\omega^2/2) \int \rho |\vec{\xi}|^2 d\tau + \delta W = 0$ and variation of this equation with respect to $\vec{\xi}$ reproduces the equation of motion. The energy principle states that an equilibrium is stable if and only if $\delta W(\vec{\xi}, \vec{\xi}) \geq 0$ for all allowable displacments $\vec{\xi}$, *i.e.* those bounded in kinetic energy and satisfying the necessary boundary conditions.

If a real spontaneous eigenmode $\vec{\xi}$ exists use $U = 0$ to find

$$\omega^2 = \frac{\delta W}{(1/2) \int \rho |\vec{\xi}|^2 d\tau}, \tag{6.20}$$

and thus ω^2 is real and the mode is either purely growing or stable, depending on the sign of δW.

However, care must be exercised in using Eq. 6.20. As we will see, for example, in the case of the $m = 1$ internal kink mode, $\vec{\xi}$ depends on ω in such a way that $\int \rho |\vec{\xi}|^2 d\tau \sim 1/\omega$ and thus $\omega \sim \delta W$.

Since δW is self adjoint, ω^2 is extremized when $\vec{\xi}$ is a solution of the equations of motion. Thus δW must be positive for all possible $\vec{\xi}$ for the system to be stable. When the eigenfunctions of ω^2 form a complete set it is sufficient for stability that the lowest ω^2 is positive. Unfortunately this simple argument is complicated by the possible existence of continua in the spectrum, for which $\vec{\xi}$ is not normalizable. Thus it is desirable to show the equivalence of the positiveness of δW for all $\vec{\xi}$ and stability without making use of the expansion of an arbitrary $\vec{\xi}$ in terms of a complete orthonormal set of discrete normal modes. Such a proof of the energy principle has been given by Laval *et al.* (1965), making use of energy conservation. The existence of a continuous stable spectrum is a common phenomenon, often being bounded by the value $\omega^2 = 0$, but whether there exist unstable continua in ideal MHD is an unsolved problem.

6.5 CONVENIENT FORM FOR δW

Consider the expression for the potential energy

$$\delta W = -(1/2) \int \vec{\xi} \cdot \vec{F}(\vec{\xi}) d\tau \tag{6.21}$$

where the expression for the force F is

$$\vec{F}(\vec{\xi}) = \nabla(\vec{\xi} \cdot \nabla)p + \gamma\nabla(p\nabla \cdot \vec{\xi}) + (\nabla \times \vec{Q}) \times \vec{B} + \vec{j} \times \vec{Q}. \quad (6.22)$$

It is convenient to write δW in a form such that each term is separately self adjoint. To do this integrate the first three terms of $\vec{F}$ by parts, giving for the volume integral (plasma)

$$\delta W_p = \frac{1}{2}\int d\tau\gamma p(\nabla \cdot \vec{\xi})^2 + \frac{1}{2}\int d\tau \vec{Q}^2 + \frac{1}{2}\int d\tau(\vec{\xi} \cdot \nabla p)(\nabla \cdot \vec{\xi})$$
$$+\frac{1}{2}\int d\tau\vec{j} \cdot (\vec{\xi} \times \vec{Q}). \quad (6.23)$$

This expression can be further simplified. It is convenient to split up the $\xi_\parallel$ and $\vec{\xi}_\perp$ contributions. First consider terms in $\xi_\parallel$. Except for the term in $(\nabla \cdot \vec{\xi})^2$ these are only

$$\frac{\xi_\parallel}{B}\vec{j}_\perp \cdot \vec{B} \times (\nabla \times \vec{\xi}_\perp \times \vec{B}) + (\vec{\xi}_\perp \cdot \nabla p)\vec{B} \cdot \nabla\frac{\xi_\parallel}{B}. \quad (6.24)$$

But

$$\vec{j}_\perp \cdot \vec{B} \times (\nabla \times \vec{\xi}_\perp \times \vec{B}) = \nabla p \cdot (\nabla \times \vec{\xi}_\perp \times \vec{B})$$
$$= -\nabla \cdot [\nabla p \times (\vec{\xi}_\perp \times \vec{B})] = \nabla \cdot [(\vec{\xi}_\perp \cdot \nabla p)\vec{B}] \quad (6.25)$$

and thus the whole expression reduces to

$$\vec{B} \cdot \nabla\left(\frac{\xi_\parallel}{B}\vec{\xi}_\perp \cdot \nabla p\right), \quad (6.26)$$

which integrates to zero over the volume since $\vec{B} \cdot \nabla = (1/\mathcal{J})(\partial_\theta + q\partial_\zeta)$ and the volume element is $d\tau = \mathcal{J}d\zeta d\theta d\psi_p$. This leaves

$$\delta W_p = \frac{1}{2}\int d\tau[\vec{Q}^2 + \vec{j} \cdot (\vec{\xi}_\perp \times \vec{Q})$$
$$+(\vec{\xi}_\perp \cdot \nabla p)\nabla \cdot \vec{\xi}_\perp + \gamma p(\nabla \cdot \vec{\xi})^2]. \quad (6.27)$$

Now write $\vec{Q}^2 = \vec{Q}_\perp^2 + Q_\parallel^2$. First simplify $Q_\parallel$, through

$$Q_\parallel = \frac{1}{B}[-B^2\nabla \cdot \vec{\xi}_\perp + \vec{B} \cdot (\vec{B} \cdot \nabla)\vec{\xi}_\perp - \vec{B} \cdot (\vec{\xi}_\perp \cdot \nabla)\vec{B}]$$
$$= \frac{1}{B}[-B^2\nabla \cdot \vec{\xi}_\perp - \vec{\xi}_\perp \cdot \nabla p - 2\vec{B} \cdot (\vec{\xi}_\perp \cdot \nabla)\vec{B}]$$
$$= -\frac{1}{B}[\vec{\xi}_\perp \cdot \nabla(B^2 + p) + B^2\nabla \cdot \vec{\xi}_\perp] \quad (6.28)$$

where we have used $\vec{\xi}_\perp \cdot \vec{B} = 0$. Now consider the term in $\vec{j}_\perp$,

$$\vec{j}_\perp \cdot (\vec{\xi}_\perp \times \vec{Q}) = \frac{\vec{B} \times \nabla p}{B^2} \cdot \vec{\xi}_\perp \times \vec{Q} = -\vec{\xi}_\perp \cdot \nabla p \frac{Q_\parallel}{B}. \qquad (6.29)$$

Combining these terms and using Eq. 2.10 for the field line curvature

$$\delta W_p = \frac{1}{2} \int d\tau \left[\vec{Q}_\perp^{\,2} - \frac{j_\parallel}{B}(\vec{\xi}_\perp \times \vec{B}) \cdot \vec{Q}_\perp - 2(\vec{\xi}_\perp \cdot \nabla p)(\vec{\xi}_\perp \cdot \vec{\kappa}) \right.$$
$$\left. + B^2(\nabla \cdot \vec{\xi}_\perp + 2\vec{\xi}_\perp \cdot \vec{\kappa})^2 + \gamma p(\nabla \cdot \vec{\xi})^2 \right]. \qquad (6.30)$$

This expression for the potential energy was first derived by Furth *et al.* (1965), and each term has a simple physical interpretation. The first term, always stabililizing, is the magnetic energy in the Alfvén wave associated with field line bending. The second term is free energy coming from the current profile and is responsible for the kink instabilities. The third term, proportional to the pressure gradient, is the energy potential for the interchange or ballooning instabilities. The fourth term is the energy in field compression in the fast magnetosonic waves, and the final term the energy in the compressional sound waves. The interchange term is destabilizing if ∇p and $\vec{\kappa}$ are parallel, *i.e.* unfavorable curvature, and stabilizing if the curvature is favorable.

If $\vec{\xi}$ is complex we have instead

$$\delta W_p(\vec{\xi}^*, \vec{\xi}) = \frac{1}{2} \int d\tau \left[|\vec{Q}_\perp|^2 - \frac{j_\parallel}{B}(\vec{\xi}_\perp^* \times \vec{B}) \cdot \vec{Q}_\perp - 2(\vec{\xi}_\perp \cdot \nabla p)(\vec{\xi}_\perp^* \cdot \vec{\kappa}) \right.$$
$$\left. + B^2 |\nabla \cdot \vec{\xi}_\perp + 2\vec{\xi}_\perp \cdot \vec{\kappa}|^2 + \gamma p|\nabla \cdot \vec{\xi}|^2 \right] (6.31)$$

The surface terms are

$$\delta W_s = \frac{1}{2} \int \gamma(\vec{\xi} \cdot d\vec{s}) p \nabla \cdot \vec{\xi} + \frac{1}{2} \int (\vec{\xi} \cdot d\vec{s})(\vec{\xi} \cdot \nabla) p - \frac{1}{2} \int (\vec{\xi} \cdot d\vec{s}) \vec{B} \cdot \vec{Q}. (6.32)$$

Across the plasma-vacuum interface the magnetic and plasma pressures must balance, *i.e.* on the unperturbed surface $\vec{r}_0$ we have $p(\vec{r}_0) + B_i^2(\vec{r}_0)/2 = B_e^2(\vec{r}_0)/2$, where i refers to interior and e to exterior. On the perturbed surface $\vec{r} = \vec{r}_0 + \vec{\xi}$ the pressure is $p(\vec{r}) + \delta p$, density $\rho(\vec{r}) + \delta\rho$, and field $\vec{B}(\vec{r}) + \delta\vec{B}$. Pressure balance is

$$p(r) + \delta p + (\vec{B}_i(\vec{r}) + \delta\vec{B}_i)^2/2 = (\vec{B}_e(\vec{r}) + \delta\vec{B}_e)^2/2. \qquad (6.33)$$

Note that the perturbations δp, $\delta\rho$, $\delta\vec{B}$ can be evaluated either at $\vec{r}$ or at $\vec{r_0}$ since the difference is second order in $\vec{\xi}$. To evaluate δp use Eqs. 1.8 and 1.9. The convective derivative describes how quantities change during the displacement $\vec{\xi}$. Eq. 1.9 gives

$$\frac{p(\vec{r_0})}{\rho(\vec{r_0})^\gamma} = \frac{p(\vec{r}) + \delta p}{[\rho(\vec{r}) + \delta\rho]^\gamma}. \tag{6.34}$$

The convective derivative for the density, Eq. 1.8 gives $\rho(\vec{r}) + \delta\rho = \rho(\vec{r_0})(1 - \nabla \cdot \vec{\xi})$. Substituting this into Eq. 6.34 we find $\delta p = -\gamma p \nabla \cdot \vec{\xi} - \vec{\xi} \cdot \nabla p$. Thus upon using $B^2(r) = B^2(r_0) + \vec{\xi} \cdot \nabla B^2$, Eq. 6.33 becomes

$$-\gamma p \nabla \cdot \vec{\xi} + \vec{B}_i \cdot \delta\vec{B}_i + \vec{\xi} \cdot \nabla B_i^2/2 = \vec{B}_e \cdot \delta\vec{B}_e + \vec{\xi} \cdot \nabla B_e^2/2. \tag{6.35}$$

Now use Eq. 6.35 to eliminate the $\gamma p \nabla \cdot \vec{\xi}$ term in Eq. 6.32, giving

$$\delta W_s = \frac{1}{2} \int \vec{\xi} \cdot d\vec{S} \left[\xi_n \frac{dp}{dn} + \vec{\xi} \cdot \nabla(B_i^2/2 - B_e^2/2) - \vec{B}_e \cdot \delta\vec{B}_e \right] \tag{6.36}$$

with n normal to the surface.

Extend the definition of $\vec{\xi}$ by writing $\delta B = \nabla \times \vec{A} = \nabla \times (\vec{\xi} \times \vec{B})$ in the vacuum. The gauge choice is $\vec{A} \cdot \vec{B} = 0$, and the general form is $\vec{A} = \vec{\xi} \times \vec{B}$. Writing $d\vec{S} = \hat{n} ds$ note that

$$\int d\tau \nabla \cdot [\vec{A} \times (\nabla \times \vec{A})] = \int d\vec{S} \cdot [\vec{A} \times (\nabla \times \vec{A})]$$

$$= \int dS(\hat{n} \times \vec{A}) \cdot (\nabla \times \vec{A}). \tag{6.37}$$

Perfect conductivity within the surface implies that $\hat{n} \times (\vec{E} + \vec{v} \times \vec{B}) = 0$ with $\vec{v} = d\vec{\xi}/dt$. Since $\vec{E} = -\partial_t \vec{A}$ we find $\hat{n} \times [-\vec{A} + \vec{\xi} \times \vec{B}] = 0$ or $\hat{n} \times \vec{A} = -(\hat{n} \cdot \vec{\xi})\vec{B}$.

The first surface integral can be converted to a volume integral over the vacuum by using

$$\int ds(\hat{n} \cdot \vec{\xi})\vec{B} \cdot (\nabla \times \vec{A}) = -\int_V d\tau \nabla \cdot [\vec{A} \times (\nabla \times \vec{A})] = \int_V d\tau(\nabla \times \vec{A})^2 \tag{6.38}$$

where we have also used the fact that there is no current in the vacuum, $\nabla \times (\nabla \times \vec{A}) = 0$. The surface terms then reduce to

$$\delta W_s = \frac{1}{2} \int ds(\hat{n} \cdot \vec{\xi})^2 \left(\frac{dp}{dn} + \frac{1}{2}\frac{\partial B_i^2}{\partial n} - \frac{1}{2}\frac{\partial B_e^2}{\partial n} \right) \tag{6.39}$$

plus the vacuum term

$$\delta W_v = \frac{1}{2} \int d\tau (\nabla \times \vec{A})^2 \qquad (6.40)$$

with $\delta W = \delta W_p + \delta W_s + \delta W_v$. Note from Eq. 6.30 that $\xi_\parallel$ appears in δW_p only in the term $\gamma p (\nabla \cdot \vec{\xi})^2$. Minimization of δW_p over $\xi_\parallel$ is achieved by choosing $\xi_\parallel$ so that $\nabla \cdot \vec{\xi} = 0$. Thus at threshold, where the kinetic energy term can be ignored, the minimizing $\vec{\xi}$ satisfies $\nabla \cdot \vec{\xi} = 0$.

6.6 CYLINDRICAL GEOMETRY ENERGY PRINCIPLE

For most ideal modes in tokamaks the destabilizing forces are large compared to the modifications due to toroidal coupling, and cylindrical geometry gives a very good approximation to the stability problem (Shafranov, 1966). Even the nonlinear behavior of most modes is approximately described by the cylindrical treatment. The exceptions are modes with $n = 1$ and low m, for which toroidal effects can be quite strong.

The form of δW_p can be simplified considerably in cylindrical geometry (Newcomb, 1960). It is possible, because the symmetry allows the Fourier decomposition of the displacement $\vec{\xi}$, to explicitly carry out the minimization of δW_p with respect to two components of $\vec{\xi}$, leaving δW_p a function of the radial component of $\vec{\xi}$ alone. Consider δW_p, as given in Sec. 6.5. The equilibrium is independent of $z = R\phi$, and the magnetic flux surfaces are given by the radius r. Consider the Fourier decomposition of the displacement, $\vec{\xi} = \vec{\xi}(r)e^{i(m\theta + kz)}$. Without loss of generality we can consider a single Fourier mode. Note that all terms in δW are such that only combinations ξ_r, $i\xi_\theta$, and $i\xi_z$ appear. For example

$$\nabla \cdot \vec{\xi} = \frac{1}{r}\frac{d}{dr}(r\xi_r) + \frac{im}{r}\xi_\theta + ik\xi_z. \qquad (6.41)$$

Thus it is reasonable (Newcomb, 1958) to introduce the variables ξ, η, ζ with

$$\xi = \xi_r,$$
$$\eta = \nabla \cdot \vec{\xi} - \frac{1}{r}\frac{d}{dr}(r\xi_r) = \frac{im}{r}\xi_\theta + ik\xi_z,$$
$$\zeta = i(\vec{\xi} \times \vec{B})_r = i\xi_\theta B_z - i\xi_z B_\theta, \qquad (6.42)$$

and thus

$$\xi_\theta = -i\frac{kr\zeta + r\eta B_\theta}{krB_z + mB_\theta}, \tag{6.43}$$

$$\xi_z = i\frac{m\zeta - r\eta B_z}{krB_z + mB_\theta}. \tag{6.44}$$

The current can be eliminated by using $\vec{j} = \nabla \times \vec{B}$ and the pressure gradient eliminated by using force balance, $\nabla p = \vec{j} \times \vec{B}$, giving

$$\frac{dp}{dr} + B_z\frac{dB_z}{dr} + \frac{B_\theta}{r}\frac{d}{dr}(rB_\theta) = 0. \tag{6.45}$$

Substituting $\vec{\xi}$ into $\vec{Q} = \nabla \times (\vec{\xi} \times \vec{B})$ we find (drop the $e^{i(m\theta + kz)}$, which will cancel in all terms)

$$\vec{Q} = i\hat{r}\left(\frac{m\xi}{r}B_\theta + k\xi B_z\right) + \hat{\theta}\left(k\zeta - \frac{d(\xi B_\theta)}{dr}\right) - \hat{z}\left(\frac{d(r\xi B_z)}{rdr} + \frac{m\zeta}{r}\right) \tag{6.46}$$

Similarly calculating $\vec{j} \times \vec{\xi}$ we find

$$\vec{Q}^* \cdot (\vec{j} \times \vec{\xi}) = \frac{\xi^*}{r}[-(m\zeta - r\eta B_z)\partial_r B_z + \partial_r(rB_\theta)(k\zeta + \eta B_\theta)]$$

$$+\frac{\xi}{r}\partial_r(rB_\theta)[k\zeta^* - \partial_r(\xi^* B_\theta)] - \xi\partial_r B_z\left[\frac{1}{r}\partial_r(r\xi^* B_z) + \frac{m}{r}\zeta^*\right]. \tag{6.47}$$

Finally, $\delta W_p = (\pi/2)\int rdr H$, with

$$H = \gamma p\left|\eta + \frac{1}{r}\frac{d}{dr}(r\xi)\right|^2 + \frac{|\xi|^2}{r^2}(mB_\theta + krB_z)^2 + \left|k\zeta - \frac{d}{dr}(\xi B_\theta)\right|^2$$

$$+\left|\frac{1}{r}\frac{d}{dr}(r\xi B_z) + \frac{m\zeta}{r}\right|^2 - \xi^*\left[B_z\frac{dB_z}{dr} + \frac{B_\theta}{r}\frac{d}{dr}(rB_\theta)\right]\frac{1}{r}\frac{d}{dr}(r\xi)$$

$$-\frac{\xi}{r}\frac{d}{dr}(rB_\theta)\frac{d}{dr}(\xi^* B_\theta) - \frac{\xi}{r}\frac{dB_z}{dr}\frac{d}{dr}(r\xi^* B_z)$$

$$+\left[\frac{k}{r}\frac{d}{dr}(rB_\theta) - \frac{m}{r}\frac{dB_z}{dr}\right](\zeta\xi^* + \xi\zeta^*). \tag{6.48}$$

The integrand is a quadratic polynomial in ζ so δW_p can be written

$$\delta W_p = \frac{\pi}{2}\int rdr\left(\gamma p\left|\eta + \frac{1}{r}\frac{d}{dr}(r\xi)\right|^2 + \frac{k^2r^2 + m^2}{r^2}|\zeta - \zeta_0|^2 + \Lambda\right) \tag{6.49}$$

where

$$\zeta_0 = \frac{r(krB_\theta - mB_z)(d\xi/dr) - r(krB_\theta + mB_z)\xi/r}{k^2r^2 + m^2} \tag{6.50}$$

and

$$\Lambda = \frac{1}{k^2r^2 + m^2} \left| (krB_z + mB_\theta)\frac{d\xi}{dr} + (krB_z - mB_\theta)\frac{\xi}{r} \right|^2 \tag{6.51}$$

$$+ \left[(krB_z + mB_\theta)^2 - 2B_\theta\frac{d}{dr}(rB_\theta) \right]\frac{|\xi|^2}{r^2}. \tag{6.52}$$

But the minimization with respect to η and ζ is trivial, δW_p is minimized with $\eta = -(1/r)(d/dr)(r\xi)$ or $\nabla \cdot \vec{\xi} = 0$ and $\zeta = \zeta_0$. Thus δW_p reduces to

$$\delta W_p = \frac{\pi}{2} \int r dr \Lambda \tag{6.53}$$

and only the minimization with respect to ξ remains to be done.

Note that Λ is positive definite, except for the $B_\theta(d/dr)(rB_\theta)$ term. Thus for fixed boundary (internal modes) the plasma is stable for all k, m if rB_θ decreases in r, or equivalently $(1/rB_\theta)(d/dr)(rB_\theta) < 0$. This is the Rosenbluth stability criterion.

The Suydam condition, obtained in an early calculation of localized interchange stability (Suydam, 1958) is closely related to the Rosenbluth criterion. Since it is a special case of the Mercier condition it will be derived in Sec. 6.15, by use of the ballooning representation. Consider the identity $[(d/dr)ln(\mu r^4)]^2 = [(d/dr)ln\mu + (4/r)]^2$ and take $\mu = B_\theta/rB_z$. Then the quantity $S = dp/dr + (r/8)B_z^2(dln\mu/dr)^2$ can be written as

$$S = \frac{r}{8}B_z^2 \left[\frac{d}{dr}ln(\mu r^4) \right]^2 - \frac{B^2}{B_\theta r}\frac{d}{dr}(rB_\theta) + \frac{2B_z^2}{r}. \tag{6.54}$$

Thus the Rosenbluth criterion implies $S > 0$, which is the Suydam condition, which is necessary for stability. The Suydam condition shows that for internal modes the pressure gradient is normally destabilizing, but the shear is stabilizing.

The expression for $\delta W_p = (\pi/2) \int r dr \Lambda$ can be integrated by parts to eliminate the $d\xi/dr$ term, giving

$$\delta W_p = \frac{\pi}{2} \int dr \left(f\left|\frac{d\xi}{dr}\right|^2 + g|\xi|^2 \right) + \frac{\pi}{2}\frac{(k^2r^2B_z^2 - m^2B_\theta^2)|\xi|^2}{m^2 + k^2r^2}\bigg|_s \tag{6.55}$$

where the second term is a new surface term to be added to the surface terms found in Sec. 6.5, and

$$f = \frac{n^2 r^3 B_z^2}{R^2(m^2 + n^2 r^2/R^2)} \left(1 - \frac{m}{nq}\right)^2 \tag{6.56}$$

$$g = \frac{B_z^2 r(n + m/q)^2}{R^2(m^2 + n^2 r^2/R^2)} + \frac{B_z^2 r(n - m/q)^2}{R^2}$$

$$- \frac{2B_z}{R^2 q} \frac{d}{dr}\left(\frac{r^2 B_z}{q}\right) - \frac{d}{dr}\left[\left(\frac{n^2 r^2 - m^2 r^2/q^2}{m^2 + n^2 r^2/R^2}\right)\frac{B_z^2}{R^2}\right]. \tag{6.57}$$

Now consider the vacuum term, given by $\delta W_v = (1/2)\int d\tau (\nabla \times \vec{A})^2$. Write $\vec{A} = \vec{\xi} \times \vec{B}$ which is the gauge in which $\vec{A} \cdot \vec{B} = 0$. Then $\delta W_v = (1/2)\int d\tau Q^2$. But in the vacuum $\vec{j} = 0$ implies $dB_z/dr = (d/dr)(rB_\theta) = 0$ and thus δW_v is the same as δW_p, and we again obtain minimization with $\zeta = \zeta_0$. We also use the continuous limit of $\vec{\xi}$ for $p \to 0$, and for any nonzero p we have $\nabla \cdot \vec{\xi} = 0$, leaving

$$\delta W_v = \frac{\pi}{2}\int r dr \Lambda(r). \tag{6.58}$$

One can avoid refering to the pressure by noting that $\delta j = \nabla \times \vec{Q} = 0$ minimizes the vacuum energy $\delta W_v = \int |Q|^2 d\tau$. Thus the variation of the field by the perturbation cannot produce any current in the vacuum.* See problem 8 at the end of the chapter.

Letting $Y(r) = (krB_z + mB_\theta)\xi(r)$, $\Lambda(r)$ considerably simplifies in the vacuum giving

$$\delta W_v = \frac{\pi}{2}\int r dr \left[\frac{1}{m^2 + k^2 r^2}\left(\frac{dY}{dr}\right)^2 + \frac{Y^2}{r^2}\right] \tag{6.59}$$

with boundary conditions that $\xi(r)$ be continuous at the plasma-vacuum interface and $\xi = 0$ at a conducting wall.

The surface terms in cylindrical geometry are

$$\delta W_s = \frac{\pi}{2}\frac{(k^2 r^2 B_z^2 - m^2 B_\theta^2)\xi^2}{m^2 + k^2 r^2} - \frac{1}{2}\int ds(\hat{n} \cdot \vec{\xi})^2 \frac{d}{dn}\left(p + \frac{B^2}{2} - \frac{B_v^2}{2}\right). \tag{6.60}$$

*Private communication, Zhixuan Wang, 2012.

Again using force balance, Eq. 6.45, we find

$$\frac{d}{dr}\left(p + \frac{B^2}{2}\right) = -\frac{B_\theta^2}{2}. \tag{6.61}$$

The second surface term is thus nonzero only if B_θ is discontinuous at the plasma boundary, *i.e.* only if there is a surface current.

6.7 MHD INSTABILITIES IN LOW β TOKAMAKS

The lowest order, cylindrical, approximation to a tokamak is adequate for the understanding of many of the ideal MHD phenomena occurring. There are four major types of ideal instabilities predicted from the δW analysis. Kink modes consist of a helical distortion of the plasma surface, and are driven by magnetic energy. The minimized δW is of order ϵ^2, $\epsilon = r/R$. Excess pressure outside a kinked current channel can also increase the kinking. The toroidal field is stabilizing. Ballooning or internal modes, are driven by p', *i.e.* by the expansion energy. The high m modes are strongly localized radially. The low m, nonlocal modes are unstable only at high β. They have a minimized $\delta W \sim \epsilon^2$ and are unstable for $\beta_p \sim 1/\epsilon$. The internal $m = 1$ kink is unique. In cylindrical geometry it is always unstable, with $\delta W \sim \epsilon^4$, so within the lowest order tokamak approximation (δW calculated only to order ϵ^2) it is marginally stable. However, the toroidal effects in this case are not negligible, and in fact the mode is stable for $\beta_p < 0.3$ when the q profile is parabolic. The fourth type of modes are the axisymmetric modes, with $n = 0$.

The form of a lowest order, cylindrical equilibrium is discussed in Sec. 2.9. Now perturb this equilibrium with a single Fourier mode, given by $\xi(r)e^{i(m\theta - n\phi)}$, and calculate the resulting δW. Consider first the vacuum term, as given in Sec. 6.6, which becomes

$$\delta W_v = \frac{\pi}{2}\int r\, dr \left[\frac{1}{m^2}\left(\frac{dY}{dr}\right)^2 + \frac{Y^2}{r^2}\right]. \tag{6.62}$$

Minimize by varying with respect to the function Y, giving the Euler equation $(d/dr)[r(dY/dr)] = m^2Y/r$, with solution

$$Y = Cr^m + Dr^{-m}. \tag{6.63}$$

The boundary conditions are that $Y(b) = 0$ and $Y(a) = mB_\theta \xi_a[1 - nq_a/m]$. The minimized vacuum term then becomes

$$\delta W_v = \frac{\pi}{2} B_\theta^2(a) \xi_a^2 m \lambda \left(1 - \frac{nq_a}{m}\right)^2 \tag{6.64}$$

with $\lambda = [1 + (a/b)^{2m}]/[1 - (a/b)^{2m}]$. Taking an equilibrium with no surface current, the surface contribution to δW, Eq. 6.60, becomes

$$\delta W_s = \frac{\pi}{2} B_\theta^2(a) \xi_a^2 \left[\left(\frac{nq_a}{m}\right)^2 - 1\right]$$

$$= \frac{\pi}{2} B_\theta^2(a) \xi_a^2 \left[\left(1 - \frac{nq_a}{m}\right)^2 - 2\left(1 - \frac{nq_a}{m}\right)\right]. \tag{6.65}$$

Keeping only terms of lowest order in ϵ the functions f, g in δW_p reduce to

$$f = \frac{B_z^2}{R^2} r^3 \left(\frac{1}{q} - \frac{n}{m}\right)^2 \tag{6.66}$$

and

$$g = \frac{rB_z^2(m^2 - 1)}{R^2} \left(\frac{1}{q} - \frac{n}{m}\right)^2. \tag{6.67}$$

Substituting these into δW_p we find for the potential energy

$$\delta W = \frac{\pi}{2} B_\theta^2(a) |\xi_a|^2 \left[(1 + m\lambda)\left(1 - \frac{nq_a}{m}\right)^2 - 2\left(1 - \frac{nq_a}{m}\right)\right] \tag{6.68}$$

$$+ \frac{\pi}{2} \frac{1}{R^2} \int_0^a r\, dr\, B_z^2 \left(\frac{1}{q} - \frac{n}{m}\right)^2 \left[r^2 \left|\frac{d\xi}{dr}\right|^2 + (m^2 - 1)|\xi|^2\right]. \tag{6.69}$$

From this expression we note that if there is no vacuum region ($b = a$), $\xi_a = 0$ and only δW_p is nonzero. The system is always stable, except perhaps for $m = 1$. Note that $m = 1$ is qualitatively different from higher m. This is because a constant ξ, $m = 1$ displacement consists of a uniform plasma displacement with no deformation of the cross section, the only field line bending occurring toroidally. All higher m modes involve cross section deformation. Secondly, if $q(a) > m/n$ the system is stable for any b/a, *i.e.* the singular surface must lie outside the plasma for instability. Note that $\delta W \sim \epsilon^2$. Kink modes are the only instability arising in this order at low β. The instability depends entirely on the q profile, *i.e.* it is current driven. It consists of a helical distortion of the plasma surface with the same sense as the helicity of the magnetic field, just outside the plasma surface, and releases magnetic energy by relaxing stress in the field. The perturbation is such that a magnetic island is formed at the rational surface, which can thus only be in the vacuum for an ideal plasma. From the last term in δW we note that for $m > 1$ the modes must be restricted to the region where $(1/q - n/m)^2$ is small, and the localization is pronounced for larger m.

The tearing mode, or resistive kink mode, treated in Chap. 7, consists of this same perturbation with the rational surface lying in the plasma. The magnetic island formation is then slowed by the resistive reconnection, and the growth rate is smaller by the ratio $(\tau_A/\tau_R)^{2/5}$, with τ_A the Alfvén time and τ_R the resistive time.

The kink modes are similar, even in their nonlinear behavior, to resistive kink modes, which are much more common and important in practice. The toroidal effects are of order ϵ^4, so cylindrical geometry suffices for a general picture of kink mode behavior.

6.8   KINK MODE

There are several models for which the energy expressions and the growth rate can be explicitly evaluated for the kink mode. The case with a constant current profile in the plasma ($q(r) = const, r < a$) with a conducting wall at $r = b$ is analytically tractable and gives a first approximation to the stability domains of a tokamak. In this case the minimizing $\xi(r)$ in δW_p is $\xi = \xi_a(r/a)^{m-1}$. To see this let $\xi = Y/r$, and the integral takes the same form as Eq. 6.62. The minimized δW is then

$$\delta W = -\pi B_\theta^2(a)\xi_a^2 \left(1 - \frac{nq_a}{m}\right) \left[1 - \frac{m - nq_a}{1 - (a/b)^{2m}}\right]. \qquad (6.70)$$

Substituting ξ_r we find $\xi_\theta = i(r/a)^{m-1}\xi_a$ and $\xi_z = 0$, giving

$$\vec{\xi} = \xi_a(r/a)^{m-1}\hat{r} + i\xi_a(r/a)^{m-1}\hat{\theta} \qquad (6.71)$$

and

$$\frac{1}{2}\int \rho\vec{\xi}^2 d\tau = \frac{\pi\rho a^2}{m}\xi_a^2, \qquad (6.72)$$

and the growth rate is given by

$$\gamma^2\tau_A^2 = 2(m - nq_a)\left[1 - \frac{m - nq_a}{1 - (a/b)^{2m}}\right], \qquad (6.73)$$

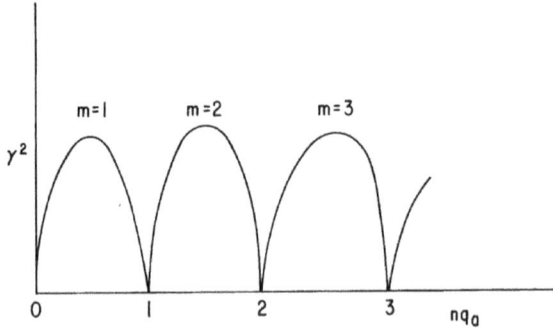

Fig. 6.1   Kink mode growth rate.

where the Alfvén time is given by $\tau_A = \sqrt{\rho}Rq/B_\phi = \sqrt{\rho}a/B_\theta$ and the growth rate is of the order $\gamma \sim 1/\tau_A$. Various characteristic times are used in the literature, associated with the shear Alfvén wave. Since this wave propagates toroidally along the field line we will use the Alfvén velocity given by the toroidal field, and q times the major radius, the connection length, to define the time τ_A.

In Fig. 6.1 is shown the square of the growth rate for the case with no conducting shell, $a/b \to 0$. It is seen that there is no stable domain in this case. The presence of a conducting shell has a small stabilizing influence, producing small windows of stability for nq(a) slightly larger than integer. The width of the stable window scales as $(a/b)^{2(m+1)}$ at the window $nq(a) \simeq m$. The shell is thus significant only for low m, and is not very helpful. Note that even with a conducting shell the mode is always unstable for $m = 1$, $n = 1$, if $q_a < 1$. This is called the Kruskal–Shafranov limit (Kruskal and Schwartschild, 1954; Shafranov, 1956) and we will see that it persists for arbitrary forms of the current profile. This limit sets a maximum value for the current for stable operation in cylindrical geometry. In toroidal geometry the ideal mode has a threshold in β.

Another analytically tractable model showing the effect of the toroidal current not extending to the plasma edge is the step model (Shafranov, 1970). The plasma is assumed to extend to $r = a$, with a conducting wall at $r = b$ as before, but the current is taken to be constant for $0 < r < r_0$, and zero beyond r_0. Thus $q = q_0$ for $0 < r < r_0$, and $q = q_a(r/a)^2$ for $r_0 < r < b$, as shown in Fig. 6.2.

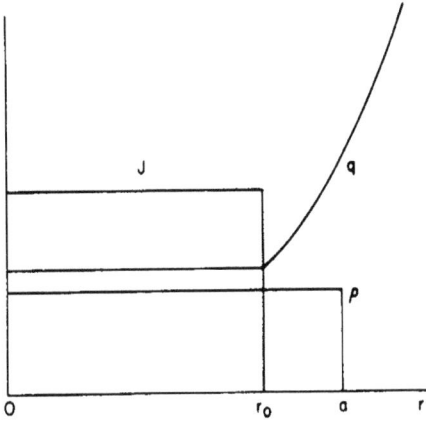

Fig. 6.2   Current density, density, and q profile for the step model.

The vacuum and surface terms are minimized as before, giving

$$\delta W_s + \delta W_v = \frac{\pi}{2} B_\theta^2(a)\xi_a^2 \left[(1 + m\lambda) \left(1 - \frac{nq_a}{m} \right)^2 - 2 \left(1 - \frac{nq_a}{m} \right) \right] . \tag{6.74}$$

For the plasma contribution, we again find for $r < r_0$ the minimizing $\xi_0(r/r_0)^{m-1} = \xi_a(r/a)^{m-1}$, and the plasma contribution to δW becomes

$$\delta W_p = \frac{\pi}{2} B_\theta^2(r_0)\xi_0^2(m - 1) \left(1 - \frac{nq_0}{m} \right)^2 + \tag{6.75}$$

$$\frac{\pi}{2} \frac{B_z^2}{R^2} \int_{r_0}^a rdr \left(\frac{1}{q} - \frac{n}{m} \right)^2 \left[r \left(\frac{d\xi}{dr} \right)^2 + (m^2 - 1)\xi^2 \right] . \tag{6.76}$$

Finally, in the range $r_0 < r < a$ we need to minimize the integral

$$\int_{r_0}^a \frac{dr}{r^3} \left(\frac{a^2}{q_a} - \frac{nr^2}{m} \right)^2 \left[r^2 \left(\frac{d\xi}{dr} \right)^2 + (m^2 - 1)\xi^2 \right] . \tag{6.77}$$

The resulting expression for $b \to \infty$ is

$$\delta W = -\pi B_\theta^2(a)\xi_a^2 \frac{m[1 - (nq_a/m)]^2[1 - m + nq_a(r_0/a)^2]}{m - 1 + (r_0/a)^{2m} - nq_a(r_0/a)^2} . \tag{6.78}$$

The stability domains are considerably larger than the constant current case, as is shown in Fig. 6.3. For a given q_a the unstable range of m-values

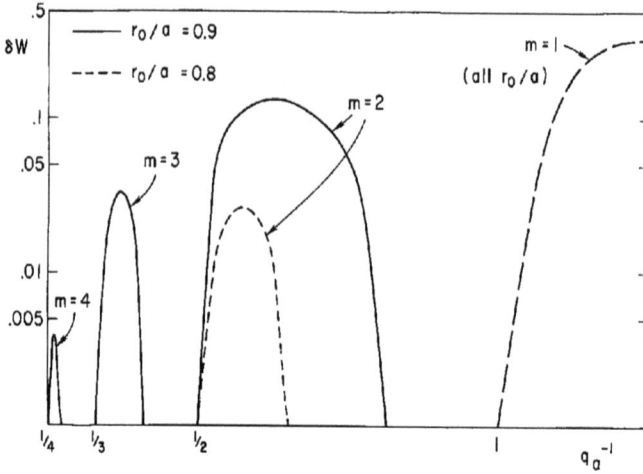

Fig. 6.3 δW for the step model.

is reduced to

$$nq_a < m < 1 + nq_a \left(\frac{r_0}{a}\right)^2. \tag{6.79}$$

For a given m, stability exists for all values of nq_a if $(r_0/a)^2 < (m-1)/m$. For example, the $m = 2$ mode is stable if $r_0/a < 0.7$. The high m modes are stabilized even for a mildly stepped profile. To stabilize the low m modes, profiles which are strongly peaked in the center are necessary. A conducting shell can help stabilize in a case which is already almost stable. For example, with $r_0/a = 0.8$ the $m = 2$ mode is stable if $b/a = 1.1$.

These windows of stability are, however, decreased again with increasing β. Pogutse and Yurchenko (1978) showed that the modes are coupled with a coupling parameter $\beta a/R$, and that for sufficiently large values of this parameter the stability windows can disappear altogether.

For realistic profiles the minimization of δW must be carried out numerically. Results have been given by Wesson (1978) for the family of profiles of the form

$$j(r) = j_0 \left(1 - \frac{r^2}{a^2}\right)^\nu. \tag{6.80}$$

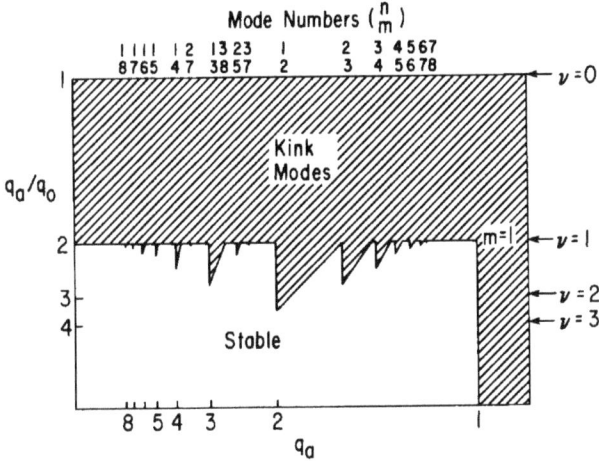

Fig. 6.4   Stability diagram for kink modes.

For these profiles $q_a = (\nu + 1)q_0$. The resulting stability diagram in the plane of q_a/q_0 and q_a is shown in Fig. 6.4. The vertical axis measures the peaking of the current profile, and the horizontal axis, q_a is inversely proportional to the total current I. The stable windows found for the step current profile are considerably larger, giving a large stable domain with $q_a > 1$. If $q_a < 1$ the $m = 1$ mode is always unstable. The large sawtooth-shaped unstable domains extending into the stable region are due to low values of m, n.

Freidberg et $al.$ (1983) have examined the stabilizing influence of limiters on the external kink mode. They find that a toroidal limiter has negligible influence on stability, but that a poloidal limiter is stabilizing.

Laval and Pellat (1973) have considered very high-m kink modes localized around the plasma-vacuum interface. These modes are of interest in two cases with the current profile more or less sharply cut off, as shown in Fig. 6.5. In case one there is instability if

$$0 < m - nq_a < j_a / \langle j \rangle , \tag{6.81}$$

and in case two there is instability if $0 < m - nq_a < me^{-2m\langle j \rangle / (aj'_a)}$, $i.e.$ within a narrow band of nq_a values just less than integers m.

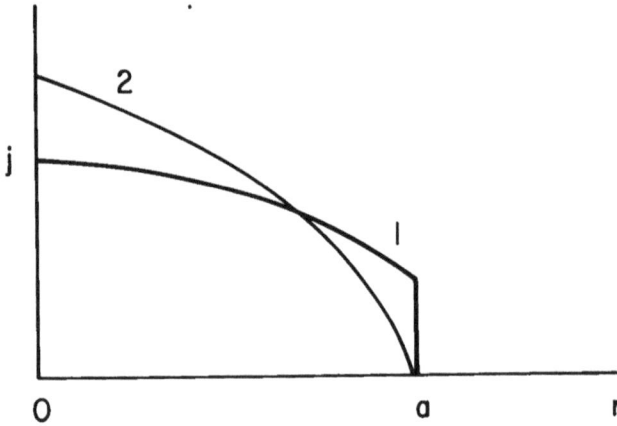

Fig. 6.5   Current profiles for high m surface kink modes.

6.9   THE M = 1 EXTERNAL KINK

Consider the case $q_a < 1$ so that the singular surface falls outside the plasma. From the expression for δW in Sec. 6.5 we see that δW_p within the plasma is minimized by $\xi = constant$, making the instability independent of the current profile within the plasma. We then find for $b \to \infty$

$$\delta W = -\pi B_\theta^2(a)\xi_a^2 nq_a(1 - nq_a). \qquad (6.82)$$

For $n = 1$ the mode is always unstable if $q_a < 1$, which is again the Kruskal–Shafranov limit. The physical mechanism is that the kinking produces a crowding of the field lines on the concave side of the plasma, which further increases the kinking. This is of course true for all m values, but for $m = 1$ there is no stabilizing effect due to the plasma interior, because of the absence of cross-sectional deformation, and the large scale lengths produce much less toroidal field line bending than for higher m.

6.10   THE INTERNAL KINK MODE

Now we examine the fixed boundary $m = 1$ mode with the singular surface inside the plasma. Consider a displacement $\xi(r) = \xi_0$ for $0 < r < r_s$, with $q(r_s) = 1/n$, $\xi(r) = \xi_0[1 - (r - r_s)/\delta]$ for $r_s < r < r_s + \delta$, and zero for $r > r_s + \delta$. The expression for δW found in Sec. 6.5 then is proportional to $\delta \epsilon^2$, and thus can be made as small as desired, tending to zero as $\xi(r)$ approaches the step function. To find δW we must return to the expression in Sec. 6.7 and evaluate it to order ϵ^4. The potential energy has the form

$$\delta W = \frac{\pi}{2} \int dr \left[f \left| \frac{d\xi}{dr} \right|^2 + g|\xi|^2 \right] \tag{6.83}$$

with

$$f = \frac{r^3 (1/q - n)^2}{1 + n^2 r^2} \tag{6.84}$$

and

$$g = \frac{n^2 r^2}{1 + n^2 r^2} [2p' - r(1/q - n)(1/q + 3n)]. \tag{6.85}$$

The step function $\xi(r)$ which minimizes δW to order ϵ^2 then gives $\delta W_{min} = (\pi/2)\xi_0^2 \int_0^{r_s} g\,dr$, which is of order $\xi_0^2 \epsilon^4$. But if $p' < 0$, $\delta W_{min} < 0$ so the mode is always unstable. To calculate the growth rate we need the kinetic energy term, $(\gamma^2/2) \int \rho \xi^2 d\tau$. But from the general cylindrical minimization, Sec. 6.7, we have that $\eta = -(1/r)(d/dr)(r\xi)$ and $\zeta = \zeta_0$. To lowest order in ϵ we then find $\zeta_0 = -(d/dr)(r\xi)$, $\xi_\theta - i(d/dr)(r\xi)$ and $\xi_z = 0$. The poloidal displacement ξ_θ is thus singular at $r = r_s$, and the inertial term is infinite. This is simply interpreted physically, a step function displacement ξ_r, given that the displacement is incompressible, $\nabla \cdot \vec{\xi} = 0$, must result in singular flows at $r = r_s$ carrying the fluid back around in θ.

To resolve this singularity some structure must be allowed to remain in ξ at $r = r_s$. An approximate expression can be obtained by a method which can be used for all values of shear, by using the ramp function $\xi = \xi_0[1 - (r - r_s)/\delta]$ as a trial function. We consider only the case $\delta \ll r_s$, otherwise δW is of order ϵ^2. Using this, f can be approximated near the singular surface, $f \sim r_s s^2 n^2 x^2$ where $s = r_s q'/q$ is the local shear, and

$x = r - r_s$. The kinetic energy term is

$$K = \frac{\pi}{2}\rho\gamma^2 \int rdr \left[\xi^2 + \left| \frac{d}{dr}(r\xi) \right|^2 + r^2 \left(\xi^* \frac{d\xi}{dr} + \xi \frac{d\xi^*}{dr} \right) \right], \qquad (6.86)$$

using $\delta \ll r_s$, is equal to $K = \pi\rho\gamma^2\xi_0^2 r_s^3/(2\delta)$ and the potential energy term reduces to $\delta W = (\pi/6)r_s\xi_0^2 s^2 n^2\delta + \delta W_{min}$, where $\delta W_{min} = \xi_0^2(\pi/2)\int_0^{r_s} gdr$ is the minimized potential obtained with the step function. Now both the kinetic and potential energies are functions of the trial function width, δ.

Note that minimizing $K + \delta W$ leads to the equation of motion for the mode. Since δW is the plasma potential energy, it is the negative of the wave energy, and this combination is the Lagrangian for the wave. Write $L = \int \rho\gamma^2\xi^2/2d\tau + \delta W$ and δW is self adjoint. Variation gives $0 = \int(\rho\gamma^2\xi - F)\delta\xi d\tau$ and thus the correct equation for ξ. Minimizing $L = K + \delta W$ with respect to δ we find $\delta_{min} = \sqrt{3}\gamma\tau_A r_s/(ns)$. Substituting δ_{min} and using the fact that the kinetic energy in the wave equals $-\delta W$ we find

$$\gamma\tau_A = -\frac{R^2}{r_s^2 sB_z^2}\int_0^{r_s} drg \simeq O\left(\frac{r^2}{R^2}\right) \qquad (6.87)$$

with $\tau_A = \sqrt{\rho}R/B_\phi$ the Alfvén time.

The displacement ξ and the growth rate can be found exactly in the case of large shear (Rosenbluth *et al.* 1973). Varying the equation for the Lagrangian we find

$$0 = \int dr \left[(2r\rho\gamma^2 + g)|\xi|^2 + (\rho r^3\gamma^2 + r^3(1/q - n)^2)\left| \frac{d\xi}{dr} \right|^2 \right]. \qquad (6.88)$$

Using $r\rho\gamma^2 \sim \epsilon^5$, since we know the order of γ to be ϵ^2 from Eq. 6.87, and $g \sim \epsilon^3$, the $2r\rho\gamma^2$ term can be neglected, giving

$$\frac{d}{dr}\left([\rho\gamma^2 + (1/q - n)^2]r^3\frac{d\xi}{dr} \right) - g\xi = 0. \qquad (6.89)$$

Solve this equation order by order in ϵ. To lowest order in ϵ, $g\xi^2$ and $\rho\gamma^2$ are negligible and the solution is the step function. To next order these terms are not negligible, and modify the solution. Consider first the exterior region, $|x| \gg \gamma\sqrt{\rho}/(ns)$. Then the $\rho\gamma^2$ term can be neglected and we find

$$\frac{d\xi}{dr} = \begin{cases} \frac{\xi_0}{r^3(1/q - n)^2}\int_0^r gdr & r < r_s \\ \frac{c}{r^3(1/q - n)^2} & r > r_s. \end{cases} \qquad (6.90)$$

Within the inertial layer $d\xi/dr$ is large and we can neglect g, and expand $(1/q - n)^2 = n^2 s^2 x^2$. The differential equation for ξ becomes

$$\frac{d\xi}{dx} = \frac{c}{\rho \gamma^2 r_s^2 + n^2 s^2 x^2}. \tag{6.91}$$

The solution which matches to ξ_0 at $x \to -\infty$ and 0 at $x \to \infty$ is

$$\xi = \frac{1}{2}\xi_0 \left[1 - \frac{2}{\pi}atan\left(\frac{nsx}{\sqrt{\rho}\gamma r_s}\right)\right]. \tag{6.92}$$

Matching $d\xi/dx$ to the exterior solution gives $\gamma\sqrt{\rho} = -\pi/(r_s^2 ns)\int_0^{r_s} g\,dr$ or

$$\gamma\tau_A = -\frac{2\delta W}{\xi_0^2 r_s^2 ns} \tag{6.93}$$

in agreement with the trial function estimate, Eq. 6.87. Notice that the interior solution, for $|x|/r_s \ll (\gamma\sqrt{\rho})/(ns)$, is a solution in slab geometry, and is valid whether the global geometry is cylindrical or toroidal. The global geometry manifests itself in the value of δW_{min}, which as we have seen is always destabilizing in cylindrical geometry.

In toroidal geometry the situation is qualitatively different. For the $m = 1$ mode, since δW is of the same order as the toroidal effects, i.e. ϵ^4, the toroidal effects cannot be expected to produce only a minor modification of the growth rate. In fact, as found by Bussac et al. (1976), the toroidal coupling to the $m = 2$ mode produces a stabilizing effect, and the mode is destabilized only for β above a threshold value. For a quadratic q profile the growth rate takes the form

$$\gamma\tau_A = -\frac{\pi\sqrt{3}}{R^2 q''}(1 - q_0)\left(\beta_p^2 - \frac{13}{144}\right) \tag{6.94}$$

and the mode is thus unstable only for $\beta_p \sim 0.3$. Thus in a toroidal device the internal kink is normally a high β phenomenon. However, the threshold depends on the q profile and for very low shear the mode can be unstable at all β. Experimentally tokamak behavior is strongly influenced by the resistive version of this mode, treated in Chap. 7, which has no pressure threshold.

The factor $\sqrt{3}$ in Eq. 6.94 is not part of the toroidal δW calculation. It is a part of $\vec{\xi}^2$ in the kinetic energy term, resulting from a toroidal component of $\vec{\xi}$, $\xi_\phi = 2i\xi_0/n$, not present in a cylindrical calculation.

6.11   BALLOONING INSTABILITIES

Ballooning instabilities are pressure driven interchanges which are localized in the region of unfavorable curvature. They have a threshold in plasma β and thus perhaps impose a limit on β. The uncertainty rests in the fact that the nonlinear consequences of the mode are to some degree unknown, and in the existence of a second stable domain for even higher β. The β limit, which we will derive by minimizing δW, can be estimated by considering the energy available from an interchange displacement in an unfavorable curvature region. The relevant contributions to δW are given by Eq. 6.30. The energy density from the pressure is $W_p \simeq (\vec{\xi}_\perp \cdot \nabla p)(\vec{\xi}_\perp \cdot \kappa) \simeq p' \xi^2 / R$. The energy density needed to bend the field lines in a distance L is

$$W_{bend} = \vec{Q}_\perp^2 \simeq \frac{B^2 \xi^2}{L^2} \tag{6.95}$$

where $L \sim qR$, the connection length, is the distance a field line spends in the bad curvature region, and hence the region where it is bent, before returning again to the good curvature region, where it is effectively tied down. The system is stable to such a displacement if $W_p < W_{bend}$ or

$$\beta \leq \frac{a}{Rq^2}. \tag{6.96}$$

The minimizing perturbation takes the form shown in Fig. 6.6, with net plasma motion into the lower field region, and the motion localized in the direction across the field lines. In fact the most strongly growing mode is very localized in this direction, and thus the most pessimistic results can be obtained by going to the limit of very rapid across-field variation of the mode. This is called the ballooning limit.

   To obtain a more exact idea of the nature of the instability, δW_p is minimized by considering a perturbation which is slowly varying along a field line and rapidly varying across it. As usual, carry out the minimization order by order. Begin with the expression for δW_p, Eq. 6.30, and use tokamak ordering, so that $q \simeq 1$ and thus $j_\parallel \simeq 1$, $B \simeq 1$. From Eq. 6.96 we order the pressure as $p \sim \epsilon$. To lowest order δW_p is dominated by $\nabla \cdot \vec{\xi}_\perp$, of order given by the inverse scale length of the perturbation. We choose a perturbation with mode number n across the field, thus $\nabla \cdot \vec{\xi}_\perp$ is ostensibly of order $n\vec{\xi}_\perp$. Expand $\xi_\perp$ in powers of $1/n$ and choose a form for $\vec{\xi}_\perp$, which to lowest order in $1/n$ is divergence free. The parameter n is regarded as

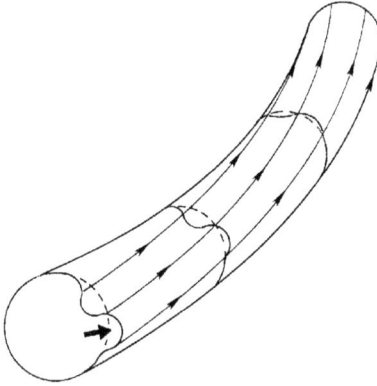

Fig. 6.6   A ballooning displacement.

very large, so ordering in n supercedes aspect ratio ordering. Write

$$\vec{\xi}_\perp = \frac{i\vec{B} \times \nabla\Phi}{nB^2} + \frac{\vec{\xi}_1}{n} \qquad (6.97)$$

where Φ is a stream function. The first term is of order one since $\nabla\Phi \sim n$, but the form is chosen so that to lowest order in $1/n$, $\nabla \cdot \vec{\xi}_\perp$ is zero. Φ is chosen to be slowly varying along a field line. The vector $\vec{\xi}_1$ is of order one, and this $1/n$ correction to $\vec{\xi}_\perp$ will be determined in the course of the minimization. Use a Clebsch representation for the field

$$\vec{B} = \nabla\beta \times \nabla\psi_p \qquad (6.98)$$

with $\beta = \zeta - q(\psi_p)\theta$, *i.e.* Eq. 2.14. The stream function is then taken to be of eikonal form

$$\Phi = A(\psi_p, \theta)e^{-in\beta} \qquad (6.99)$$

with $A(\psi_p, \theta)$ slowly varying, and n is considered large. Thus Φ is rapidly varying across field lines with mode number n and slowly varying along them where $\beta = constant$. We then have

$$\nabla\Phi = (\nabla A - inA\nabla\beta)e^{-in\beta} \qquad (6.100)$$

and substituting we find

$$\vec{\xi}_\perp = \frac{i\vec{B} \times \nabla A}{nB^2} e^{-in\beta} + \frac{\vec{B} \times \nabla \beta}{B^2} \Phi + \frac{\vec{\xi}_1}{n}. \tag{6.101}$$

It then follows that

$$\nabla \cdot \vec{\xi}_\perp = \frac{ie^{-in\beta}}{n} \left[(\vec{B} \times \nabla A) \cdot \nabla \frac{1}{B^2} + \frac{\vec{j} \cdot \nabla A}{B^2} \right]$$

$$+ \Phi(\vec{B} \times \nabla \beta) \cdot \nabla(1/B^2) + \frac{\Phi p'}{B^2} + \frac{1}{n} \nabla \cdot \vec{\xi}_1 \tag{6.102}$$

which is of order one in n. Here $p' = dp/\psi_p$ and we have used the equilibrium condition. Now calculate the components of $\vec{\xi}_\perp$, $\vec{Q}$.

$$\vec{\xi}_\perp \cdot \nabla \psi_p = \frac{i}{nB^2} (\vec{B} \times \nabla A) \cdot \nabla \psi_p e^{-in\beta} + \Phi + \frac{1}{n} \vec{\xi}_1 \cdot \nabla \psi_p \tag{6.103}$$

and similarly

$$\vec{Q} \cdot \nabla \psi_p = \nabla \psi_p \cdot \nabla \times (\vec{\xi}_\perp \times \vec{B}) = \nabla \cdot [(\vec{\xi}_\perp \times \vec{B}) \times \nabla \psi_p]$$

$$= \nabla \cdot (\vec{B} \vec{\xi}_\perp \cdot \nabla \psi_p) = \vec{B} \cdot \nabla(\vec{\xi}_\perp \cdot \nabla \psi_p) \tag{6.104}$$

which is of order unity. The component in the $\nabla\beta$ direction is

$$\vec{Q} \cdot \nabla \beta = \nabla \beta \cdot \nabla \times (\vec{\xi}_\perp \times \vec{B}) = \nabla \cdot [(\vec{\xi}_\perp \times \vec{B}) \times \nabla \beta]$$

$$= \nabla \cdot (\vec{B} \vec{\xi}_\perp \cdot \nabla \beta) = \vec{B} \cdot \nabla(\vec{\xi}_\perp \cdot \nabla \beta) \tag{6.105}$$

which is of order $1/n$. Finally

$$\vec{Q} \cdot \vec{B} = -\nabla \cdot (\vec{\xi}_\perp B^2) - \vec{\xi}_\perp \cdot \nabla p =$$

$$-B^2 \left[\nabla \cdot \vec{\xi}_\perp + 2\vec{\xi}_\perp \cdot \vec{\kappa} - \frac{1}{B^2} \vec{\xi}_\perp \cdot \nabla p \right] \tag{6.106}$$

where the curvature, given in Sec. 2.3, can be expressed as

$$\vec{\kappa} = \frac{\vec{B} \times [\nabla(p + B^2/2) \times \vec{B}]}{B^4}. \tag{6.107}$$

Now consider the ordering of the terms in δW_p, from Sec. 6.5. First consider the field compression terms, which can be written in the form

$$\vec{Q} \cdot \vec{B} - \vec{\xi}_\perp \cdot \nabla p = -B^2(\nabla \cdot \vec{\xi}_\perp + 2\vec{\xi}_\perp \cdot \vec{\kappa}). \tag{6.108}$$

To lowest order we have

$$\vec{Q} \cdot \vec{B} - \vec{\xi}_\perp \cdot \nabla p = -B^2 \left[\frac{\Phi p'}{B^2} + \frac{2\Phi}{B^2} (\vec{B} \times \nabla \beta) \cdot \vec{\kappa} + \frac{1}{n} \nabla \cdot \vec{\xi}_1 \right]$$

$$\simeq -\Phi [p' + 2(\vec{B} \times \nabla \beta) \cdot \kappa] + \frac{B^2 \nabla \cdot \vec{\xi}_1}{n}. \quad (6.109)$$

Choose $\vec{\xi}_1$ to minimize this term so that $\vec{Q} \cdot \vec{B} - \vec{\xi}_\perp \cdot \nabla p \simeq 1/n$. Now consider the kinking term, $(\vec{B} \times \vec{\xi}_\perp) \cdot (\nabla \times \vec{\xi}_\perp \times \vec{B})$. Substituting $\vec{\xi}_\perp$ we have

$$\vec{B} \times \vec{\xi}_\perp = \frac{i}{n} \left(\vec{B} \frac{\vec{B} \cdot \nabla \Phi}{B^2} - \nabla \Phi \right) + \frac{\vec{B} \times \vec{\xi}_1}{n}. \quad (6.110)$$

Note that

$$\nabla \times (\vec{B} \times \vec{\xi}_\perp) = \frac{i}{n} \left[\vec{j} \frac{\vec{B} \cdot \nabla \Phi}{B^2} + \nabla (\frac{\vec{B} \cdot \nabla \Phi}{B^2}) \times \vec{B} \right] + \frac{\nabla \times (\vec{B} \times \vec{\xi}_1)}{n} \quad (6.111)$$

is of order unity.

But write the kinking term as

$$\vec{B} \times \vec{\xi}_\perp \cdot \nabla \times (\vec{\xi}_\perp \times \vec{B}) = \frac{i}{n} \left(\frac{\vec{B} \cdot \nabla \Phi}{n} \vec{B} - \nabla \Phi \right) \cdot \nabla \times (\vec{\xi}_\perp \times \vec{B})$$

$$+ \frac{(\vec{B} \times \vec{\xi}_1) \cdot \nabla \times (\vec{\xi}_\perp \times \vec{B})}{n} \quad (6.112)$$

then note that the first term on the right hand side $\equiv iR_1/n$, can be transformed in the following way:

$$R_1 = \frac{\vec{B} \cdot \nabla \Phi}{B^2} \left[-\nabla \cdot (B^2 \vec{\xi}_\perp) - \vec{\xi}_\perp \cdot \nabla p \right] - \nabla \cdot [(\vec{\xi}_\perp \times \vec{B}) \times \nabla \Phi]$$

$$= \frac{\vec{B} \cdot \nabla \Phi}{B^2} \left[-B^2 \nabla \cdot \vec{\xi}_\perp - \vec{\xi}_\perp \cdot \nabla (B^2 + p) \right] - \nabla \cdot \left(\vec{B} \vec{\xi}_\perp \cdot \nabla \Phi - \vec{\xi}_\perp \vec{B} \cdot \nabla \Phi \right)$$

$$= \frac{-\vec{B} \cdot \nabla \Phi}{B^2} \vec{\xi}_\perp \cdot \nabla (B^2 + p) - \vec{B} \cdot \nabla (\vec{\xi}_\perp \cdot \nabla \Phi) + \vec{\xi}_\perp \cdot \nabla (\vec{B} \cdot \nabla \Phi)$$

where terms in $\nabla \cdot \vec{\xi}$ have added to zero. But $\vec{\xi}_\perp \cdot \nabla (\vec{B} \cdot \nabla \Phi) = (\vec{\xi}_\perp \cdot \nabla A) e^{-in\beta}$, and $\nabla (\vec{B} \cdot \nabla \Phi) = [\nabla (\vec{\xi}_\perp \cdot \nabla A) - in(\vec{\xi}_\perp \cdot \nabla A) \nabla \beta] e^{-in\beta}$. Substituting we find that the leading order terms cancel and the kinking term is of order $1/n$.

Now we can write the leading terms in δW_p. Using

$$\vec{\xi}_\perp^2 = \Phi^2 \frac{(\vec{B} \times \nabla\beta)^2}{B^4} = \frac{\Phi^2(\nabla\beta)^2}{B^2} \tag{6.113}$$

we find this to be of order unity. Then $Q_\perp^2 \sim (\vec{Q} \cdot \nabla\psi_p)^2(\nabla\beta)^2/B^2$ since Q_β is of order $1/n$. Thus

$$Q_\perp^2 = \frac{(\nabla\beta)^2}{B^2}[\vec{B} \cdot \nabla(\xi_\perp \cdot \nabla\psi_p)]^2 = \frac{(\nabla\beta)^2}{B^2}(\vec{B} \cdot \nabla\Phi)^2. \tag{6.114}$$

Finally $\vec{\xi} \cdot \nabla p \vec{\xi} \cdot \vec{\kappa} = \Phi^2 p'(\vec{B} \times \nabla\beta) \cdot \vec{\kappa}/B^2$ giving $\delta W_p = \frac{1}{2}\int d\tau M$ with

$$M = \frac{(\nabla\beta)^2}{B^2}(\vec{B} \cdot \nabla\Phi)^2 - 2p'\frac{(\vec{B} \times \nabla\beta) \cdot \vec{\kappa}\Phi^2}{B^2}$$

$$+B^2[\nabla \cdot \vec{\xi}_\perp + 2\vec{\xi}_\perp \cdot \vec{\kappa}]^2 + \gamma p(\nabla \cdot \vec{\xi})^2 + O(\frac{1}{n}). \tag{6.115}$$

The first term is the field line bending term, which is the major stabilizing contribution. The second term is the destabilizing term responsible for ballooning, and the third and fourth terms are weaker stabilizing terms due to field and plasma compression, respectively. We extremize the total energy keeping terms in inertia and $\xi_\parallel$, *i.e.* compressional wave effects. The kinetic energy is

$$K = -\frac{1}{2}\int d\tau \rho \omega^2 \vec{\xi}^2. \tag{6.116}$$

Write $\nabla \cdot \vec{\xi} = \nabla \cdot \vec{\xi}_\perp + \vec{B} \cdot \nabla\xi_b$, with $\xi_b = \vec{\xi} \cdot \vec{B}/B^2$ and $\vec{\xi} = \vec{\xi}_\perp + \vec{B}\xi_b$ and note that $\nabla \cdot \vec{\xi}_\perp = p'\Phi/B^2 + \nabla \cdot \vec{\xi}_1/n$ giving for the total energy $U = \frac{1}{2}\int d\tau H$ with

$$H = \frac{(\nabla\beta)^2}{B^2}(\vec{B} \cdot \nabla\Phi)^2 + B^2\left[(\frac{p'}{B^2} + 2\kappa_W)\Phi + \frac{\nabla \cdot \vec{\xi}_1}{n}\right]^2 - 2p'\kappa_W\Phi^2$$

$$+\gamma p\left[\frac{p'\Phi}{B^2} + \frac{\nabla \cdot \vec{\xi}_1}{n} + \vec{B} \cdot \nabla\xi_b\right]^2 - \rho\omega^2\left[\frac{(\nabla\beta)^2}{B^2}\Phi^2 + B^2\xi_b^2\right] \tag{6.117}$$

where $\kappa_W = (\vec{B} \times \nabla\beta) \cdot \vec{\kappa}/B^2$, fondly known as the weird component of the curvature by those in the field. Extremizing with respect to $\nabla \cdot \vec{\xi}_1/n$ gives

$$\frac{\nabla \cdot \vec{\xi}_1}{n} = -\left(\frac{p'}{B^2} + \frac{2B^2\kappa_W}{B^2 + \gamma p}\right)\Phi - \frac{\gamma p\vec{B} \cdot \nabla\xi_b}{B^2 + \gamma p}. \tag{6.118}$$

Note that then

$$\nabla \cdot \vec{\xi} = -\frac{B^2}{B^2 + \gamma p}(2\kappa_W \Phi - \vec{B} \cdot \nabla \xi_b). \tag{6.119}$$

The total energy then simplifies to

$$U = \frac{1}{2} \int d\tau \left[\begin{array}{c} \frac{(\nabla\beta)^2}{B^2}(\vec{B} \cdot \nabla\Phi)^2 + \frac{B^2\gamma p}{B^2+\gamma p}(2\kappa_W \Phi - \vec{B} \cdot \nabla \xi_b)^2 \\ -2p'\kappa_W \Phi^2 - \rho\omega^2(\frac{(\nabla\beta)^2}{B^2}\Phi^2 + B^2\xi_b^2) \end{array} \right]. \tag{6.120}$$

The destabilizing term is $2p'\kappa_W \Phi^2$, p' being negative normally, and κ_W being negative in the bad curvature region. At threshold, minimization over ξ_b is trivial with $\vec{B} \cdot \xi_b = 2\kappa_W \Phi$, which corresponds to the condition $\nabla \cdot \vec{\xi} = 0$, which we know must occur at threshold. Note the ψ_p, θ, ζ and ψ_p, θ, β systems have the same Jacobian and $d\tau = \mathcal{J}d\psi_p d\theta d\zeta$ and $\vec{B}\cdot\nabla = \vec{B}\cdot(\nabla\psi_p\partial_{\psi_p}+\nabla\theta\partial_\theta+\nabla\beta\partial_\beta) = (1/\mathcal{J})\partial_\theta$, where the partial derivative now means fixed ψ_p and fixed β. Extremizing with respect to ξ_b gives

$$\partial_\theta \left[\frac{B^2\gamma p}{B^2 + \gamma p}(\partial_\theta\xi_b/\mathcal{J} - 2\kappa_W \Phi) \right] + \rho\omega^2 B\mathcal{J}\xi_b = 0, \tag{6.121}$$

and extremizing with respect to Φ gives

$$\partial_\theta \left(\frac{(\nabla\beta)^2}{\mathcal{J}B^2}\partial_\theta\Phi \right) + 2\mathcal{J}p'\kappa_W \Phi + \rho\omega^2 \frac{\mathcal{J}(\nabla\beta)^2}{B^2}\Phi$$
$$-\frac{\mathcal{J}B^2\gamma p}{B^2 + \gamma p}2\kappa_W \left(2\kappa_W \Phi - \frac{\partial_\theta\xi_b}{\mathcal{J}} \right) = 0. \tag{6.122}$$

Using Eq. 6.118 to eliminate ξ_b we find for the fourth order set of ballooning equations

$$\frac{1}{\mathcal{J}}\partial_\theta \left[\frac{(\nabla\beta)^2}{\mathcal{J}B^2}\partial_\theta\Phi \right] + \left(2p'\kappa_W + \rho\omega^2\frac{(\nabla\beta)^2}{B^2} \right)\Phi + 2\kappa_W\gamma p\nabla \cdot \vec{\xi} = 0 \tag{6.123}$$

$$\frac{1}{\mathcal{J}}\partial_\theta \left[\frac{1}{B^2}\partial_\theta(\gamma p\nabla\cdot\vec{\xi}) \right] + \rho\omega^2 \left(\frac{1}{\gamma p} + \frac{1}{B^2} \right)(\gamma p\nabla\cdot\vec{\xi}) + \rho\omega^2 2\kappa_W \Phi = 0 \tag{6.124}$$

The boundary conditions on $\Phi(\theta)$ must be such that the solution is single valued on the toroidal surface. This does not impose conditions at $\theta = \pm\pi$ because the eikonal gives a solution following a field line, which does not return to the same point after a toroidal transit. What is necessary is that a superposition of solutions be taken such that the result is single valued on the surface. Thus take

$$\hat{\Phi}(\psi_p, \theta) = \sum \phi_m e^{-im\theta} \qquad (6.125)$$

with $\phi_m = \int_{-\infty}^{\infty} e^{im\theta} \Phi(\psi_p, \theta) d\theta$ with the integral extending over $(-\infty, \infty)$, because Φ is defined following a field line. $\hat{\Phi}(\psi_p, \theta)$ is thus single valued on the toroidal surface. The boundary condition on Φ then follows from the requirement that the integral defining δW converge. The expression for δW involves the integral $\int_0^{2\pi} d\theta |\hat{\Phi}(\psi_p, \theta)|^2$, which is equal to $\int_{-\infty}^{\infty} d\theta |\Phi(\psi_p, \theta)|^2$. Convergence of this integral requires that $|\Phi(\psi_p, \theta)|^2 < 1/\theta$ for large θ.

This representation of the function $\hat{\Phi}$ which is single valued on the toroidal surface, by a function Φ, defined in the interval $-\infty < \theta < \infty$ using the eikonal formulation, is called the ballooning representation. It may be viewed as a representation on the topological covering space of the torus, which has infinitely many sheets, with the function chosen in such a way as to yield a single valued function $\hat{\Phi}$, defined through the Fourier transform. An alternate way of viewing this representation is to recall that a Wenzel, Kramers, Brillion, Jeffries (WKBJ) solution representing one branch of a dispersion relation is never expected to satisfy the boundary conditions. Only the correct linear superposition of all solutions with the same frequency will satisfy the given boundary conditions. In this case there is an infinite degeneracy because of the periodicity of the torus and the solution satisfying the correct boundary conditions on the torus is given by the infinite sum in the ballooning representation (Dewar *et al.*, 1981).

6.12 MAGNETIC WELL

The stabilizing effect of a magnetic well can be seen by examining the destabilizing interchange term of Eq. 6.120,

$$U_I = -\int d\tau p' \kappa_W \Phi^2. \qquad (6.126)$$

Supposing $p' < 0$ and using the fact that the perturbation is localized on a single flux surface we find for each surface a necessary condition for stability,

$$\int \mathcal{J} d\theta d\beta \kappa_W > 0. \qquad (6.127)$$

This condition is referred to as average good curvature on the surface. The expression for κ_W gives

$$B^2\kappa_W = [(\nabla\beta)^2(\nabla\psi_p)^2 - (\nabla\beta \cdot \nabla\psi_p)^2]\partial_{\psi_p}(B^2/2 + p)$$
$$+[(\nabla\beta)^2\nabla\psi_p \cdot \nabla\theta - (\nabla\beta \cdot \nabla\psi_p(\nabla\beta \cdot \nabla\theta)]\partial_\theta(B^2/2 + p). \qquad (6.128)$$

For a low beta equilibrium, using leading order expressions for $\nabla\beta = \nabla(\zeta - \theta q)$, $\nabla\psi_p = r\nabla r/q$, and $\mathcal{J} = q/B^2$ we find

$$\kappa_W \simeq \frac{1}{B^2}\partial_{\psi_p}(B^2/2 + p). \qquad (6.129)$$

Terms neglected include contributions of shear and effects of equilibrium deformation due to plasma pressure. This leads to the intuitively useful but not completely rigorous concept of the stabilizing effect of an average magnetic well.

If the equilibrium is shear free, so that the field lines are closed, the condition for stability under local interchange simplifies further. This restriction is not applicable in a tokamak where the field lines ergodically cover a flux surface except for surfaces where q is rational. In this case the volume integral in Eq. 6.127 is localized to a single value of β, and neglecting the pressure we find for stability

$$-\partial_{\psi_p}\int d\zeta(1/B^2) > 0. \qquad (6.130)$$

Converting to an integration along the line element following $\vec{B}$ by using $d\vec{r}/dl = \vec{B}/B$ or

$$\vec{B} \cdot \nabla l = B\frac{d\vec{r}}{dl} \cdot \nabla l = B \qquad (6.131)$$

and $\vec{B} \cdot \nabla\zeta d\zeta = \vec{B} \cdot \nabla l dl$ we find for the stability condition

$$\partial_{\psi_p}\oint \frac{dl}{B} < 0. \qquad (6.132)$$

This condition is equivalent to the existence of a vacuum magnetic well.

6.13   BALLOONING, SIMPLE EQUILIBRIA

To illustrate this mode, consider the case of a second order equilibrium, as discussed in Sec. 2.8, and neglect compressional effects, *i.e.* $\nabla \cdot \vec{\xi} = 0$. The ballooning mode is then described by Eq. 6.123 alone, which becomes

$$\frac{1}{J}\frac{\partial}{\partial\theta}\left(\frac{(\nabla\beta)^2}{B^2}\frac{\partial\Phi}{J\partial\theta}\right) + 2p'\kappa_W\Phi + \rho\omega^2\frac{(\nabla\beta)^2\Phi}{B^2} = 0, \qquad (6.133)$$

where we recall that $p' = dp/d\psi_p$. Note that the approximation of neglecting compression is not consistent with Eq. 6.124 except very near threshold. Use Eq. 6.98 to describe the magnetic field. Neglecting terms of order r^2 we have $\beta = \phi - q(\theta - r\sin\theta - \Delta'\sin\theta)$ and the Jacobian in the ψ_p, θ, β system is given by

$$\frac{1}{J} = \nabla\phi\cdot(\nabla\psi_p\times\nabla\theta) = \nabla\beta\cdot(\nabla\psi_p\times\nabla\theta) = \frac{1}{Xq(1-\Delta'\cos\theta)}.(6.134)$$

Evaluating $\nabla\beta$ and dropping terms of higher order in ϵ and bounded in θ we find

$$\frac{(\nabla\beta)^2}{B^2} = \frac{q^2}{r^2}\left(1 + [\theta s - (M' + \frac{q\Delta'}{r})\sin\theta]^2\right)$$
$$+ \frac{q^2}{r^2}\cos\theta\left[2r - \frac{2M}{q} + \frac{2M}{q}\theta^2 s^2\right] \qquad (6.135)$$

where $M = q(\Delta'+r) = (1+l/2+\beta_\theta)qr$ and $s = rq'/q$, and we have used Eq. 2.82. The primes in these expressions refer to differentiation with respect to r.

The curvature is given by $\vec{\kappa} = -\nabla X/X - r\nabla r/q^2$, the two terms corresponding to toroidal and cylindrical contributions. In addition we will be interested in the cylindrical limit, for which $\vec{\kappa} = -r\nabla r/q^2$ and β is given by $\beta = \phi - q\theta$. Substituting these expressions we find

$$\kappa_W = \begin{cases} -1/q & \textit{cylinder} \\ -\frac{q\cos\theta}{r} + q - \frac{1}{q} - \frac{q'\Delta'}{r} + q'\theta\sin\theta + \sin^2\theta(M' - q) & \textit{torus} \end{cases}$$
$$(6.136)$$

where again $M = (1 + l/2 + \beta_\theta)qr$. Now we can understand the effect of the curvature. The interchange term in the energy integral $-2p'\kappa_W\Phi^2$ is, to leading order, $2p'(q/r)\Phi^2\cos^2\theta$, and thus to minimize the energy with $p' < 0$ the function $\Phi(\theta)$ must be large near $\theta \simeq 0$ (bad curvature region) and small near $\theta \simeq \pi$ (good curvature region). Thus the envelope of the perturbation takes on the ballooning form, with a large bulge appearing in the bad curvature region.

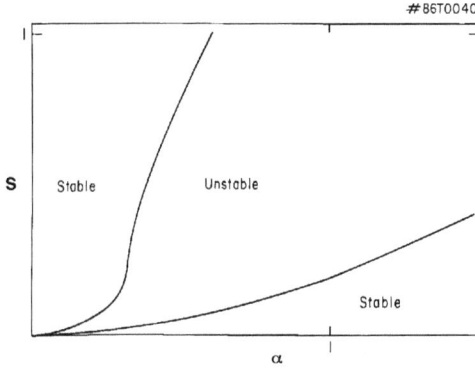

Fig. 6.7   Ballooning stability domain.

Finally, we have for the ballooning equation, keeping only leading order terms in r,

$$\frac{d}{d\theta}\left[1 + h^2(\theta)\right]\frac{d\Phi}{d\theta} + g(\theta)\Phi + \frac{\omega^2}{\omega_A^2}\left[1 + h^2(\theta)\right]\Phi = 0 \qquad (6.137)$$

with $h(\theta) = s\theta - \alpha sin\theta$, $s = rq'/q$ is the average shear, and $\alpha = rM'/q + \Delta'$. We are interested in values of the pressure near threshold, and thus from Eq. 6.96 we take $p \sim \epsilon$. From Sec. 2.8 we then find that to leading order $\alpha = -2Rq^2p'/B_\phi^2$. The curvature enters through the function $g(\theta) = 2p'r\kappa_W q$ where now $p' = dp/dr$, in keeping with the use of the prime in other expressions in the second order equilibrium. Here $\omega_A = V_A/(qR)$ with $V_A B_\phi/\sqrt{\rho}$ the Alfvén velocity. Note that in toroidal geometry, if we take $q \gg 1$, we have approximately $g(\theta) = \alpha[cos\theta + h(\theta)sin\theta]$, which is the expression used by Connor *et al.* (1978). Any relation connecting toroidal terms to the pressure gradient in this way of course makes it impossible to recover the cylindrical limit.

Numerical solution of the differential equation for $\Phi(\theta)$, subject to the boundary condition $|\Phi(r,\theta)|^2 < 1/|\theta|$ for large $|\theta|$, results in a stability regime of the form shown in Fig. 6.7. The details are dependent on the form of the pressure profile and the q profile. The boundary of the first stability region is given roughly by $\alpha = 0.8s$ for $0.5 < s < 1$.

As r varies from 0 to a, the functions s and α describe a curve in the s, α plane. For example, a profile with $p(r) = (1 - r^2/a^2)^2$ and $q(r) =$

$1 + (q_a - 1)(r^2/a^2)$ has a maximum in the pressure gradient at $r^2 = a^2/3$ and at this point $\beta \simeq (0.05)a/R$ is the limit for the onset of the ballooning instability for $q_a \sim 3$.

The second stability regime arises from the strong local shear in the bad curvature region due to the outward shift of the flux surfaces. It was discovered practically simultaneously by several different groups, and reported in 1978 first by Coppi at the Gordon Conference, and then by Mercier (1978), Sykes *et al.* (1978) and Coppi *et al.* (1978), all at the Innsbruck International Atomic Energy Agency meeting. If in fact tokamaks are able to operate in the second stability domain, it could provide a very attractive range of operation for a reactor. Access to the domain, and stability and transport within it, are of great interest both experimentally and theoretically. It has been shown (Sykes and Turner 1979) that a stable route to the second regime exists provided the safety factor q on axis can be kept above a critical value, typically between one and two, during heating. Special shaping of the plasma cross section can lead to a total disappearance of the unstable domain. Chance *et al.* (1983) showed that a bean shaped plasma (see Figs. 6.12, 6.13) provides direct stable access to the second regime.

6.14 MERCIER, SUYDAM CRITERIA

The Mercier stability criterion follows from the large θ behavior of the function $\Phi(\theta)$ in the ballooning representation, and thus is associated with the local behavior in x, *i.e.* it is a criterion for the stability of the equilibrium to local interchanges. In general the ballooning equation, at threshold, $\omega^2 = 0$, has the form

$$(f\Phi')' + g\Phi = 0 \tag{6.138}$$

with $f = a + b\theta + c\theta^2$ and $g = d + e\theta$, and a, b, c, d, e are functions periodic in θ. Later we will evaluate these terms for a second order shifted equilibrium, in order to understand their origin and their stabilizing or destabilizing effect. Examine the behavior of Φ for large θ, and recall that $\Phi < 1/\theta^{1/2}$ as $|\theta| \to \infty$. Expand Φ in terms of its large θ secular behavior, *i.e.*

$$\Phi = \theta^p (u_0 + \frac{u_1}{\theta} + \frac{u_2}{\theta^2} + ...) \tag{6.139}$$

where also the u_k are periodic functions of θ. Substitute this representation into the differential equation, giving for terms of the form θ^{p+2}, $(cu_0')' = 0$, or $cu_0' = k_1$. Integrating, we have

$$u_0 = k_1 \int \frac{d\theta}{c} + k_2, \tag{6.140}$$

which implies that $k_1 = 0$, since c has no singularities, and thus the integral is secular. Thus u_0 is constant, and we set it equal to one. The terms of the form θ^{p+1} give

$$[c(u_1' + p)]' + e = 0. \tag{6.141}$$

Integrating we find

$$cu_1' = -(cp + \hat{e}) + k_3 \tag{6.142}$$

where $\hat{e} = \int_0^\theta e \, d\theta$ and k_3 an integration constant. To find this constant, divide by c and average this equation over one period. This gives

$$u_1' = \frac{\langle \frac{\hat{e}}{c} \rangle + p}{c \langle \frac{1}{c} \rangle} - \left(p + \frac{\hat{e}}{c} \right) \tag{6.143}$$

with $\langle f \rangle = \frac{1}{2\pi} \int_0^{2\pi} f \, d\theta$. Now consider terms in θ^p. Using $u_0' = 0$ we find

$$(p+1)c(u_1' + p) + \{c[u_2' + (p-1)u_1] + b(u_1' + p)\}' + eu_1 + d = 0. \tag{6.144}$$

Averaging in θ eliminates u_2 and gives

$$(p+1)\langle cu_1' \rangle + p(p+1)\langle c \rangle + \langle eu_1 + d \rangle = 0. \tag{6.145}$$

But from the equation for u_1' we have

$$\langle cu_1' \rangle = -p \langle c \rangle - \langle \hat{e} \rangle + \frac{\langle \hat{e}/c \rangle + p}{\langle 1/c \rangle}. \tag{6.146}$$

Also note that from $(\hat{e}u_1)' = eu_1 + \hat{e}u_1'$ we have $\langle eu_1 \rangle + \langle \hat{e}u_1' \rangle = 0$. Multiply Eq. 6.143 by $\hat{e}$ and average, giving

$$\langle eu_1 \rangle = p \langle \hat{e} \rangle + \left\langle \frac{\hat{e}\hat{e}}{c} \right\rangle - \left\langle \frac{\hat{e}}{c} \right\rangle \frac{\langle \hat{e}/c \rangle + p}{\langle 1/c \rangle}. \tag{6.147}$$

Substituting these expressions and simplifying we find a quadratic equation for p with the solution

$$p = -\frac{1}{2} + \left(\frac{1}{4} - D \right)^{1/2} \tag{6.148}$$

with

$$D = \left\langle \frac{\hat{e}}{c} \right\rangle - \left\langle \frac{1}{c} \right\rangle \langle \hat{e} \rangle + \left\langle \frac{1}{c} \right\rangle \left\langle \frac{\hat{e}\hat{e}}{c} \right\rangle - \left\langle \frac{\hat{e}}{c} \right\rangle^2 + \left\langle \frac{1}{c} \right\rangle \langle d \rangle . \quad (6.149)$$

For $D < 1/4$ no well behaved solution can exist, so the system is stable. Thus the condition $D < 1/4$ sufficient for stability. If $D > 1/4$ a well behaved solution may or may not exist, the two asymptotic expressions must be joined smoothly, so one cannot conclude instability.

To understand the stability properties of the mode we evaluate this expression for the second order equilibrium of Sec. 2.8. We have

$$f = 1 + (s\theta - \alpha sin\theta)^2 = a + b\theta + c\theta^2, \quad (6.150)$$

$$g = 2p'rq\kappa_W = d + e\theta \quad (6.151)$$

and thus

$$a = 1 + \alpha^2 sin^2\theta, \qquad b = -2\alpha s \, sin\theta, \qquad c = s^2. \quad (6.152)$$

We also find that $d = -2p'r$ in cylindrical geometry and

$$d = -2p'r\left(1 - q^2 + \frac{q^2 cos\theta}{r} + \frac{qq'\Delta'}{r} - q'q sin\theta - q(M' - q)sin^2\theta\right) (6.153)$$

in the torus. Similarly

$$e = \begin{cases} 0 & cylinder \\ 2p'sq^2 sin\theta & torus. \end{cases} \quad (6.154)$$

The functions a and b do not appear in the Mercier expression. Then

$$\langle d \rangle = \begin{cases} -2p'r & cylinder \\ -2p'r(1 - q^2) - 2p'qq'\Delta' + p'rq(M' - q) & torus \end{cases} \quad (6.155)$$

and also, for the torus only, $\langle \hat{e} \rangle = 2p'q^2 s$ and $\langle \hat{e}\hat{e}/c \rangle = 6(p')^2 q^4$.

Substituting, we find for the cylindrical case $D = -2p'r/s^2$. The stablity condition $D < 1/4$ then gives

$$p'r + \frac{s^2}{8} > 0 \quad (6.156)$$

which is the Suydam condition. We see that shear is stabilizing, and the pressure gradient is destabilizing (normally $p' < 0$), and thus the shear provides a threshold for the instability.

For the toroidal case we find to leading order, with p of order ϵ

$$D = -\frac{2p'r}{s^2}(1 - q^2) \qquad (6.157)$$

and the stability condition $D < 1/4$ gives

$$p'r(1 - q^2) + \frac{s^2}{8} > 0 \qquad (6.158)$$

which is the Mercier condition. For $q^2 > 1$ the stabilizing influence of the toroidal curvature dominates and the mode is stable. This was first shown by Shafranov and Yurchenko (1967).

The expressions for D indicate that shear is stabilizing. Greene and Chance (1981) pointed out that the vanishing of the local shear in the bad curvature region is what allows the ballooning mode to become unstable. The Shafranov shift compresses the poloidal field in the outside of the torus, locally decreasing q, and the shear. The toroidal field simply slips around to the inside, and is not compressed by the shift. See Eqs. 2.86, 2.89. The safety factor in the physical coordinates ϕ, θ is given by $d\phi/d\theta$ along a field line. In a second order equilibrium we have

$$q(r, \theta) = \frac{d\phi}{d\theta} = q(r) + \frac{\partial \nu}{\partial \theta}. \qquad (6.159)$$

Substituting, we find $q(r, \theta) = q(1 - r\cos\theta - \Delta'\cos\theta)$ and thus the local shear is given by

$$s(\theta) \simeq \frac{rq'}{q} - r\cos\theta(\Delta'' + 1). \qquad (6.160)$$

Increasing β, and hence Δ'', produces vanishing local shear initially at $\theta = 0$, the center of the bad curvature region. As Δ'' increases further the point of vanishing local shear moves toward $\theta = \pi/2$, *i.e.* toward the domain of more favorable curvature. At this point the plasma is again stable to ballooning. This is the second stability regime, although of course this heuristic demonstration is not a proof of its existence.

6.15 BALLOONING EQUATION MODIFICATION

There are two distinct modifications of the ballooning equations which were derived in Sec. 6.11. The first is purely mathematical, and has to do with

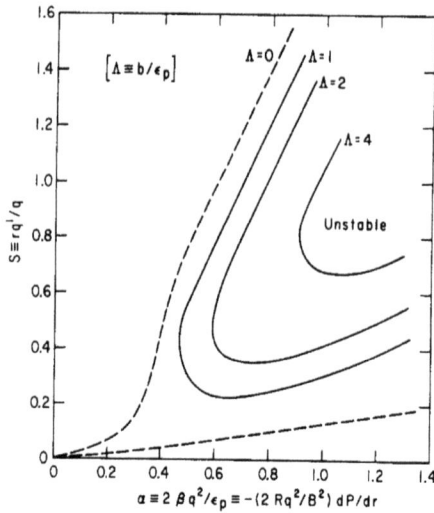

Fig. 6.8   Ballooning stability including diamagnetic effects.

the higher order terms in the expansion parameter $1/n$. The corrections due to them were examined by Connor *et al.* (1979) and Dewar *et al.* (1981). It was demonstrated using WKB analysis and numerically that the $n = \infty$ limiting equations give the most unstable mode, with the corrections taking the form $\beta_{crit} = \beta_{crit}(\infty) + cn$ with $c > 0$. Numerical results, however, showed that this first order correction is inadequate for predicting the lowest value of n giving instability, because the limiting β for ballooning stability possesses oscillations as a function of $1/n$. For low shear the oscillations in $1/n$ become so strong that they result in bands of unstable n-values, which are present even when the standard ballooning theory predicts complete stability. The occurrence of the instabilities, referred to as infernal modes (Manickam *et al.*, 1987), depends on the pressure and current profile.

The second kind of correction to the ballooning β limit is due to kinetic effects. The large n limit involves very short wave length modes, and these must be expected to be modified by nonzero gyro radius. This effect is stabilizing and significantly raises the critical β for the onset of ballooning. It is simply interpreted as defining a largest n (shortest wavelength) for which the ballooning representation is valid, and hence raising β_{crit}. A simple demonstration of this effect was given by Tang *et al.*

(1981) by the inclusion of ion diamagnetic effects in the simple circular equilibrium ballooning model introduced by Connor *et al.*, and discussed in Sec. 6.13. The inclusion of the diamagnetic effects in a large aspect ratio equilibrium modifies the ω^2 term in Eq. 6.137 to become $\omega(\omega - \omega_{*pi})$, with $\omega_{*pi} = -(c/neBr)dp_i/dr$ the pressure driven ion diamagnetic drift frequency. This modification shifts the stability boundaries for the mode, increasing the size of the stable region.

If $b = \omega_{*pi}/\omega_A$ the stability boundary is simply characterized by the value of $\Lambda = bRdln(p)/dr$. In Fig. 6.8 is shown the shrinking of the unstable domain as the parameter Λ is increased.

6.16 TAE MODES

Consider a transverse Alfvén wave in a cylindrical plasma, of the form

$$e^{i(n\phi - m\theta - \omega t)}. \tag{6.161}$$

Consider a wave propagating along a field line, $\theta = \phi/q$ so the exponent has the form $\phi(n - m/q) - \omega t$. The wave propagates with the Alfvén velocity so $R\phi/t = v_A$, giving for the frequency

$$\omega = v_A(n - m/q)/R. \tag{6.162}$$

Due to the toroidal bending in tokamaks, the differential equation describing the waves has an additional term periodic in the toroidal angle, and all poloidal harmonics m are coupled. This is a standard problem treated by Floquet theory, and very common particularly in condensed matter physics. In Fig. 6.9 is shown a typical dispersion relation for modes $m = 2, 3$, with (solid) and without (dashed) toroidal coupling. Due to this coupling there is a gap in the spectrum due to harmonics m and $m + 1$ located at $q(r) = (m + 1/2)/n$. In this gap, Cheng and Chance (1986) showed that there exists a discrete mode, called the Toroidal Alfvén Eigenmode (TAE), with a frequency given by the point where modes $m, m + 1$ intersect, $q \simeq (m + 1/2)/n$ giving $\omega \simeq V_A/(2qR)$. The mode is somewhat localized at this radius, but is fairly global, and due to the toroidal coupling possesses many m values in addition to the dominant two. Depending on the plasma density and q profile the gap in the Alfvén continuum spectrum may extend far enough that the mode is only weakly coupled to other continuum waves. Within ideal MHD the frequency is real, but this discrete

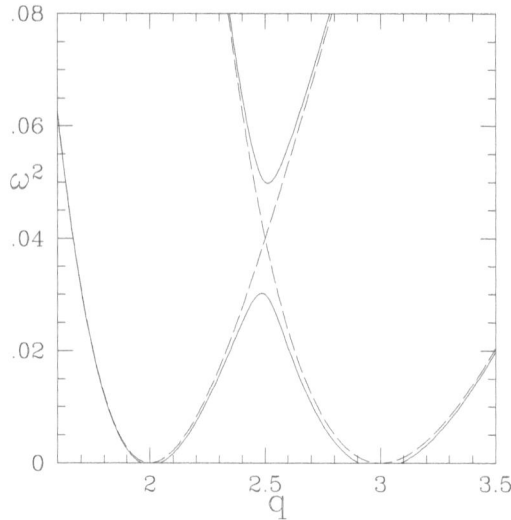

Fig. 6.9   Alfvén waves with and without toroidal coupling.

mode can be driven by resonance with a high energy particle distribution, as will be discussed in Chap. 8, and if it is only weakly coupled to the continuum it is only weakly damped.

6.17   AXISYMMETRIC MODES

Modes involving large scale, essentially rigid, plasma displacement lead to contact of the plasma with the wall of the confinement vessel, and must be eliminated by modification of the field design. Because of the dependence of these modes on plasma shape this can normally be achieved only through active feedback control systems, which must be designed to respond to the plasma on a time scale which is shorter than the instability growth time. Large scale motion of the plasma surface is achieved by the kink modes, discussed in Sec. 6.8, and by axisymmetric modes. Consider first the case of a large aspect ratio circular (cylindrical) plasma, surrounded by a perfectly conducting wall. From Sec. 6.7 we find that the potential energy for $n = 0$

reduces to

$$\delta W = \frac{\pi}{2} B_\theta^2(a) \xi_a^2 (m\lambda - 1) + \frac{\pi B_z^2}{2R^2} \int_0^a r dr \frac{[r^2(d\xi/dr)^2 + (m^2 - 1)\xi^2]}{q^2}.$$

(6.163)

This expression is minimized by taking $m = 1$ and ξ constant, giving

$$\delta W = \frac{\pi B_\theta^2(a) \xi_a^2 a^2}{b^2 - a^2},$$

(6.164)

which is of course always positive, the mode becoming only marginally stable for $b \to \infty$. We conclude that axisymmetric modes are not a problem for circular cross section plasmas. This is, however, not true for shaped cross section plasmas.

In the case of elliptically shaped plasmas, stability against axisymmetric vertical displacements can be improved by adding a quadripole field to the basic vertical field B_v, in such a way so as to increase the poloidal field moving either up or down from $Z = 0$. Thus the quadripole is oriented so that $\partial B_v/\partial R = \partial B_R/\partial Z < 0$. Defining the decay index

$$n = -\frac{R}{B_v} \frac{\partial B_v}{\partial R}$$

(6.165)

the condition for vertical stability is $n > 0$. Horizontal instability sets in for $n > 3/2$, so there is a wide stable window.

The case of an elliptical plasma cross section with a constant current profile was first studied by Laval $et\ al.$ (1972). They found a sufficient condition for stability as a function of ellipticity and wall position. Subsequently it was pointed out by Rutherford that all even m modes are stable. A closed analytic expression for the stability was found by Laval $et\ al.$ (1974) for the case of cylindrical geometry. As shown in Sec. 6.6 the displacement is incompressible. The stability condition is

$$\left(\frac{m}{1 - \alpha^{2m}} - \frac{1 + \epsilon^2}{1 - \epsilon^2}\right)^2 > \epsilon^{2m} \left(\frac{m\alpha^{2m}}{1 - \alpha^{2m}} - \frac{1 + \epsilon^2}{1 - \epsilon^2}\right)$$

(6.166)

where $\epsilon = (b - a)/(b + a)$ and $\alpha = (a + b)/(a' + b')$, with a and b the minor and major semi-axes of the elliptical plasma, and a' and b' the minor and major semi-axes of the confocal conducting wall.

Laval and Pellat (1973) found the stability criterion for a $m = 1$ mode for a shaped current profile, linear in the poloidal flux. The results for this

Linear Ideal Modes

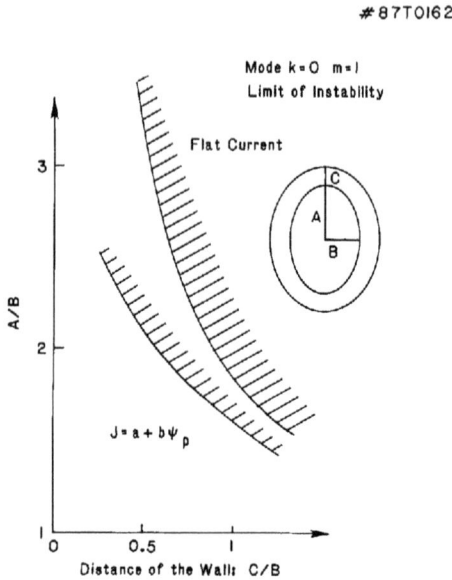

Fig. 6.10 Axisymmetric modes.

profile and for the flat profile are shown in Fig. 6.10. The stable region lies below the curves.

It was shown by Pomphrey and Jardin (1987) that a feedback system can enter into the stability considerations in an unusual manner because of the possibility of deformation of the plasma cross section. The control system consists of magnetic flux pickup loops, an amplifier with gain G, and current-carrying poloidal field coils. For some placement of the pickup loops a system unstable for gain $G = 0$ remains unstable for all values of G. This results because the plasma can find a deformation which leads to a lower energy state while avoiding the pickup loops.

6.18 NUMERICAL MHD SPECTRUM

A detailed treatment of the full spectrum of modes in a plasma is beyond the scope of this book. We consider primarily global, unstable modes which can have an important effect on the possible allowed equilibria in toroidal devices. For a general treatment of waves in a plasma see Stix (1962).

For general shaped equilibria, analytic derivation of the MHD spectrum and the form of the eigenfunctions is generally not possible. As seen in Sec. 6.4 the equation of motion results from minimizing the total energy. The stability analysis is carried out by finding numerically the stationary points for the total energy

$$T = K + \delta W. \tag{6.167}$$

To carry out the extremization, Galerkin's method of expanding the perturbations in terms of a finite subset of a complete set of expansion functions is used. The ERATO code (Gruber *et al.* 1981) uses a square grid mesh in the X, Z plane, and Fourier decomposition only in the toroidal angle. The PEST code (Grimm *et al.*, 1976) uses Fourier decomposition in both toroidal and poloidal angles, the expansion taking the form

$$\vec{\xi} = \sum_{kmn} a_{kmn}\vec{\xi}_k(\psi)e^{i(m\theta - n\phi)} \tag{6.168}$$

with $\phi = 2\pi z/L$, L the column length, and where the $\vec{\xi}_k(\psi)$ are generally taken to be piecewise linear functions of ψ, nonzero only in a small domain centered at ψ_k. The resolution in the determination of the extremizing $\vec{\xi}_k$ is then limited by the size of the domain of the $\vec{\xi}_k$ and the number of m, n values included in the analysis. The use of the finite set of basis functions then reduces the problem to a matrix inversion.

This method has been used to determine the full spectrum for a straight elliptical cross section plasma by Chance *et al.* (1977), and by Kerner *et al.* (1981). There are two continuum spectra, those of Alfvén waves and slow sound waves. In addition there are interchange modes and discrete spectra due to magnetosonic waves and kinks. The full spectrum in a circular cross section shear-free column for $m = 2$ and $L = 10\pi a$ is shown in Fig. 6.11. The square of the frequency $\Omega^2 = \omega^2\tau_A^2$ is plotted as a function of nq. There is no conducting wall. The most stable spectrum, with $\Omega^2 \simeq 10^5$, is that of the magnetosonic waves, which form a discrete spectrum.

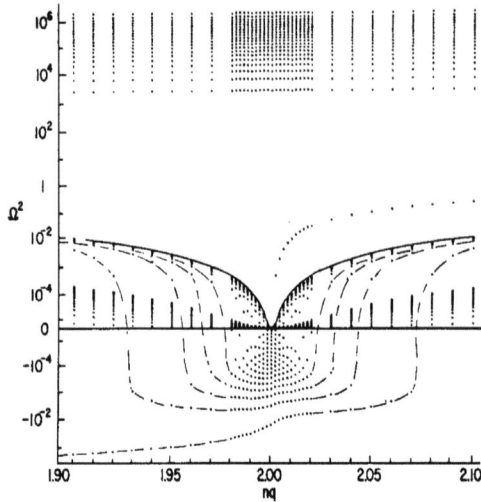

Fig. 6.11   MHD spectrum in a circular cross section cylinder.

Moving down in Ω^2, the next branch encountered is the kink mode. The dispersion relation, which is derived in Sec. 6.8, is $\Omega^2 = 2(nq - m)(nq - m + 1)$. Next is the Alfvén branch, which has a stable discrete spectrum with an accumulation point at $\Omega^2 = \gamma p(m - nq)^2/(\gamma p + B^2)$. The unstable modes, other than the single kink mode branch, are interchange modes, which appear out of the Alfvén spectrum. These codes play an important role in the design of new devices as well as in interpretation of results. An important fusion objective is the attainment of stable high β equilibrium, and the use of these codes for that purpose will be discussed in Sec. 6.19.

6.19   SHAPE AND ASPECT RATIO

As discussed in Sec. 6.14 the local shear is modified by the Shafranov shift, and the point at which the local shear vanishes affects ballooning stability. The vanishing of the local shear is also dependent on plasma shape, and thus ballooning can be inhibited by choice of a particular cross section.

For low values of n, or for strongly shaped plasma, the ballooning analysis of Sec. 6.11 is not applicable and numerical analysis is required. Extensive work has been done using the spectral codes ERATO and PEST, and

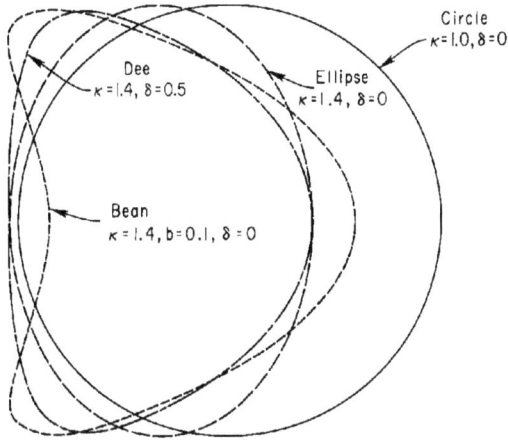

Fig. 6.12   Shaped equilibrium cross section.

some general conclusions can be drawn. The interpretation of the results is complicated by the fact that stability is dependent on the shapes of the current profile and the pressure profile, and these profiles cannot be freely adjusted in an experiment. Thus optimization over the profile shape can lead to unphysical conclusions. The object is of course to discover combinations of plasma shape, aspect ratio and profiles which lead to high β stable equilibria. The effect of plasma shape was analyzed extensively by Manickam (1984), primarily to determine the effect of introducing a large indentation in the plasma shape. Plasma shapes studied were defined by the location of the outermost surface. The representation used was

$$X = X_0 - b + (a + b\cos\theta)\cos(\theta + \delta\sin\theta), \qquad (6.169)$$

$$Z = \kappa a \sin\theta \qquad (6.170)$$

where X_0 is the major radius, a the minor radius, κ the ellipticity, δ the triangularity, and b the indentation, which yields a bean shape.

In Fig. 6.12 are shown typical plasma cross sections which have been considered, including circular, elliptic, D, and bean deformation. In Fig. 6.13 are shown the resulting stability domains for the $n = 1$ ballooning

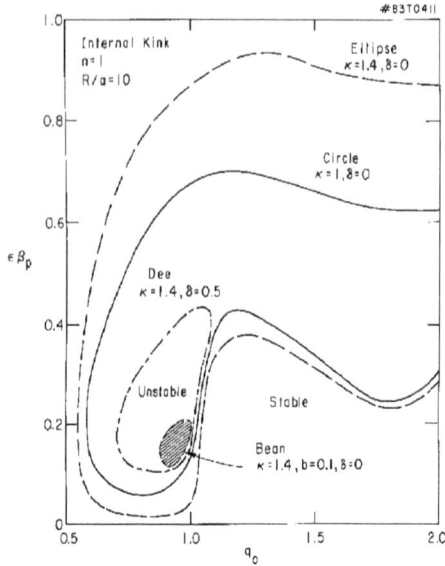

Fig. 6.13   Ballooning stability domain.

mode. The second stability domain is evident for all plasma shapes, and the extent of the unstable domain is strongly shape dependent. In addition, plasma elongation allows much more plasma current and hence better confinement for a given major radius.

A numerical study of internal and external modes was carried out for noncircular tokamaks using ERATO by Bernard *et al.* (1980). They found the external kink modes to be primarily responsible for restricting β. Troyon *et al.* (1984) carried out a pressure profile optimization for a variety of equilibria using ERATO, to find the maximum β giving stable confinement at constant current. The shape of the plasma surface was given by Eqs. 4.171 and 4.172, but with no bean shape, *i.e.* $b = 0$. The toroidal flux XB_ϕ and the pressure were taken to be polynomial functions of the poloidal flux. The limitations to β arise from the ballooning limit, the Mercier limit on axis, and the $n = 1$ free boundary kink. On axis the Mercier and the ballooning criterion coincide, so an equilibrium which is Mercier unstable also possesses an unstable ballooning region near the axis. The Mercier limit restricts $q(0)$ to be slightly greater than one. For $q(0) > 1$ the $n = 1$ external kink limits the value of β, and the highest value is reached when $q(0)$ is near the Mercier limit. For a given value of the current, the ballooning limit is given roughly by Eq. 6.96, $\beta < a/Rq^2$. Introducing the current by using $q \sim a^2 B_\phi/RI$ we find a limiting β scaling as $\beta \sim I/(aBq)$. Instead the numerical optimization over profile shapes gives $\beta = 3I/aB$ where β is in percent, I is in Mamps, B in Tesla, and a in meters. In addition to this limit in β for fixed current, there is a limitation in current, due to the external kink mode. The ballooning and Mercier limit forces q on axis to be slightly above one. Raising the current eventually produces a fairly flat current profile with q near two at the plasma edge. The $m = 2$ external kink then causes a major disruption. More will be said about this "Troyon limit", which appears to be well satisfied experimentally, in Sec. 10.10.

Goedbloed (1982) has made extensive use of conformal mapping techniques to investigate the stability of free boundary high β equilibria. He finds that the Kruskal–Shafranov limit can be surpassed with a conducting wall close to the plasma and high β.

6.20   Problems

1. Show that the normal component of the perturbed magnetic field $\vec{Q} = \nabla \times (\vec{\xi} \times \vec{B})$ satisfies the identity $\nabla p \cdot \vec{Q} = \vec{B} \cdot \nabla(\vec{\xi} \cdot \nabla p)$.

2. Calculate the local shear for the second order equilibrium of Sec. 2.8, and verify the ballooning mode threshold predicted from its vanishing.

3. Consider Eq. 6.30, δW_p. Show that if the pressure gradient is not antiparallel to κ, the interchange term is destabilizing for some $\vec{\xi}$.

4. Calculate the next order (in r) corrections to the compressionless ballooning equation, Eq. 6.137.

5. Consider a q profile with $q' = 0$ at the $q = 1$ surface. Use the trial function introduced in Sec. 6.10 to find the growth rate for the internal kink mode.

6. A flute mode, or interchange, consists of the exchange of flux tubes with no field line bending, $\vec{Q}_\perp = 0$. Also the displacement of the flux tube is directly across the original field, so $\xi_\parallel = 0$. Use Eq. 6.30 to find a condition on the pressure profile for stability to interchanges. Discuss the result for

a. the Van Allen belts in the Earth's field. Use the expression for $\vec{B}$ in Problem 5 of Chap. 1.

b. large and small aspect ratio tokamaks. Can you properly treat shear? (Liu Chen, private conversation, 2008)

7. Show that for any volume V the integral $I = \int_V |\nabla \times \vec{A}|^2 d\tau$ is extremized if, and only if, $\nabla \times (\nabla \times \vec{A}) = 0$ everywhere in V.

6.21 References

Ideal MHD energy principle

- Bernstein, I. B., E. A. Frieman, M. D. Kruskal, and R. M. Kulsrud, Proc. Roy. Soc. A244, 17 (1958).
- Chew, G. F., M. L. Goldberger, and F. E. Low, Proc. Roy. Soc. Ser. A 236, 112 (1956).
- Furth, H. P., J. Killeen, M.N. Rosenbluth, and B. Coppi, in Plasma Physics and Controlled Nuclear Fusion Research (IAEA, Vienna, 1965) Vol. 1, p. 103.
- Greene J. M., and J. L. Johnson, Plasma Physics 10, 729 (1968).
- Kadomtsev, B. B., Reviews of Plasma Physics (Consultants Bureau, New York, 1966) Vol. 2, p. 153.
- Kruskal, M. D., and C. R. Oberman, Phys. Fluids lt 275 (1958).
- Kulsrud, R. M., in Advanced Plasma Theory, in Proc. of the Enrico Fermi International School of Physics, edited by M. N. Rosenbluth (Academic, New York, 1964).
- Kulsrud, R. M., in Proc. of the Varenna Conference 1985.
- Laval, G., C. Mercier, and R. M. Pellat, Nucl. Fusion 5, 156 (1965).
- Newcomb, W. A., Ann. of Phys. 10, 232 (1960).
- Newcomb, W. A., Ann. Phys. 3, 347 (1958).
- Rosenbluth, M. N., and N. Rostoker, Phys. Fluids 2, 23 (1959).

Stability

- Freidberg, J. P., J. P. Goedbloed, and R. Rohatgi, Phys. Rev. Lett. 51, 2105 (1983).
- Goedbloed, J. P., Phys. Fluids 25, 2073 (1982).
- Greene, J. M., and J. L. Johnson, Phys. Rev. Lett. 11, 401 (1961).
- Greene, J. M., and J. L. Johnson, Physics Fluids 5, 510 (1962).
- Mercier, C., Nucl. Fusion, 1, 47 (1960).
- Mercier, C., c.r. hebd. Seanc. Acad. Sci. Paris 252, 1577 (1961).
- Mercier, C., Nucl. Fusion Suppl. Pt. 29, 801 (1962).
- Shafranov, V. D., and E. I. Yurchenko, Zh. Eksp Teor. Fiz. 539 1157 (1967) [Sov. Phys. JETP 26, 682 (1968)].
- Suydam, B. R., Peaceful Uses of Atomic Energy in Proc. of the 2nd. Int. Conf. United Nations, Geneva, 31, 157 (1958).

Large scale ideal modes

- Bateman, G., MHD Instabilities (MIT, Cambridge, Mass, 1978).
- Bussac, M., R. Pellet, D. Edery, and J. Soule, Phys. Rev. Lett. 35, 1638 (1976).
- Freidberg, J. P., Rev. Mod. Phys. 54, 801 (1982).
- Jardin, S. C., and D. A. Larrabee, Nucl. Fusion 22, 1095 (1982).
- Kruskal, M. D., and M. Shafranov, Proc. of the Royal Society of London, Ser. A 223, 348 (1954).
- Laval, G., and R. Pellat, in Controlled Fusion and Plasma Physics, in Proc. of the 6th. European Conference, Moscow (1973) 2, p. 640.
- Laval, G., R. Pellat, J. L. Soule, Controlled Fusion and Plasma Physics, in Proc. of the 5th. European Conference, Grenoble, 1972) 1, p. 25.
- Laval, G., R. Pellat, and J. L. Soule, Phys. Fluids 17, 895 (1974).
- Pogutse, O. P., and E. I. Yurchenko, Nucl. Fusion 18, 1629 (1978).
- Pomphrey, N., and S. C. Jardin, Princeton University, Plasma Physics Laboratory Report, PPPL 2468 (1987).
- Rosenbluth, M. N., R. Y. Dagazian, and P. H. Rutherford, Phys. Fluids 16, 1894 (1973).
- Shafranov, V. D., At. Energy 5, 38 (1956).
- Shafranov, V. D., Zh. Tekh. Fiz. 40, 240 (Sov. Phys. Tech. Phys. 15, 175) (1970).
- Wesson, J. A., Nucl. Fusion 18, 87 (1978). (Review Paper).
- Wesson, J., Tokamaks (Clarendon Press, Oxford, 1987).

Numerical spectrum

- Berger, D., L. C. Bernard, R. Gruber, and F. Troyon, in Proc. of the Conference on Plasma Physics and Controlled Nuclear Fusion Research, Berchtesgaden, 1976 (IAEA, Viennna, 1977) Vol. 2, p. 411.
- Bernard, L. C., D. Dobrott, F. J. Helton, R.W. Moo, Nuclear Fusion 20, 1199 (1980).
- Chance, M. S., J. M. Greene, R. C. Grimm and J. L. Johnson, Nucl. Fusion 17, 65 (1977).
- Grimm, R. C., J.M. Greene, and J.L. Johnson in Methods in Computational Physics 16, 253 (Academic, New York, 1976).

- Gruber, R., F. Troyon, D. Berger, W. C. Bernard *et al.*, Comput. Phys. Comm. 21, 323 (1981).
- Kerner, W., R. Gruber, and F. Troyon, Phys. Rev. Lett. 44, 536 (1980).
- Kerner, W. P., Goutier, K. Lackner, W. Schneider *et al.*, Nuclear Fusion 21, 1383 (1981).
- Kerner, W., K. Lerbinger, R. Gruber, and T. Tsunematsu, Computer Physics Communications 36, 225 (1985).
- Troyon, F. R., R. Gruber, H. Saurenmann, S. Semenzato *et al.*, Plasma Physics and Controlled Fusion 26, 209 (1984).

Plasma waves

- Goedbloed, J. P., Phys. Fluids 1, 1258 (1975).
- Goedbloed, J. P., Lecture notes on Ideal Magnetohydrodynamics (Fom Institut voor Plasmafysica, Nieuweigein, Netherlands, 1979).
- Grad, H., in Proc. of the National Academy of Science U.S.A. 70, 3277 (1973).
- Stix, T. H., The Theory of Plasma Waves (McGraw-Hill, New York 1962).
- Tataronis, J. A., and W. Grossmann, Courant Institute of Mathematical Science, Report COO3077-102, MF-84, unpublished (1977).

Ballooning modes

- Bateman, G., and Y-K M. Peng, Phys. Rev. Lett 38, 829 (1977).
- Chance, M. S., S. C. Jardin, and T. H. Stix, Phys. Rev. Lett. 51, 1963 (1983).
- Connor, J. W., R. J. Hastie, J. B. Taylor, Phys. Rev. Lett. 40, 396 (1978).
- Connor, J. W., R. J. Hastie, J. B. Taylor, in Proc. of the Royal Society of London A, 365 (1979).
- Coppi, B., Phys. Rev. Lett. 39, 939 (1977).
- Coppi, B., J. Filreis, and J. W. Mark in Proc. of the 1978 Conference on Plasma Physics and Controlled Nuclear Fusion Research (Innsbruck, 1978) Vol. 1, p. 793.
- Coppi, B., A. Ferreira, J. W. Mark, J. J. Ramos, Comments Plasma Phys. Controlled Fusion 51, 1 (1979).

- Dewar, R. L., J. Manickam, R. C. Grimm, and M. S. Chance, Nucl. Fusion 21, 493 (1981).
- Dobrott, D., D. B. Nelson, J. M. Greene, A. H. Glasser *et al.*, Phys. Rev. Lett. 39, 943 (1977).
- Greene, J. and M., M. S. Chance, Nucl. Fusion 21, 453 (1981)
- Lortz, D., and J. Nuhrenberg, Phys. Lett. 68A, 49 (1978).
- Lortz, D., and J. Nuhrenberg, Nucl. Fusion 19, 1207 (1979).
- Manickam, J., Nucl. Fusion 24, 595 (1984).
- Manickam, J. N. Pomphrey and A. M. M. Todd, Nucl. Fusion 27, 1461 (1987).
- Mercier, C., Proc. of the Conference on Plasma Physics and Controlled Nuclear Fusion Research, Innsbruck, 1978 (IAEA, Vienna, 1979) Vol. 1, p. 701.
- Pogutse, O. P., and E. I. Yurchenko, Sov. J. Plasma Phys. 5, 441 (1979).
- Strauss, H. R., W. Park, D. A. Monticello, R. B. White *et al.*, Nucl. Fusion 20, 638 (1980).
- Sykes, A., M. F. Turner, P. J. Fielding, and F. A. Haas, in Proc. of the Conference on Plasma Physics and Controlled Nuclear Fusion Research, Innsbruck, 1978 (IAEA, Vienna, 1979) Vol. 1, p. 625.
- Tang, W. M., J. W. Connor, and R. B. White, Nuclear Fusion 21, 891 (1981).
- Todd, A. M. M., M. S. Chance, J. M. Greene, R. C. Grimm *et al.*, Phys. Rev. Lett. 38, 826 (1977).
- Ware, A. A., and F. A. Haas, Phys. Fluids 9, 956 (1966).
- Zakharov, L. E., Nucl. Fusion 18, 355 (1978).
- Zakharov, L. E., Proc. of the Conference on Plasma Physics and Controlled Nuclear Fusion Research, Innsbruck, 1978 (IAEA, Vienna, 1979) Vol. 1, p. 689.

TAE modes

- Cheng C. Z. and M. S. Chance, Phys Fluids 29, 3695 (1986).
- Cheng C. Z., Phys Reports 211, 1 (1992).

Chapter 7

Linear Resistive Modes

7.1 INTRODUCTION

As seen in Sec. 1.7, in a system with well defined flux surfaces, magnetic islands are produced by resonant perturbations of $\vec{B}$ directed across the flux surfaces. These perturbations are automatically excluded from ideal MHD, since the magnetic perturbation has the form $\delta\vec{B} = \nabla \times (\vec{\xi} \times \vec{B})$. Fourier decomposing $\vec{\xi} = \vec{\xi}(r)e^{i(m\theta - n\zeta)}$ and using $\vec{B} = \nabla\zeta \times \nabla\psi_p + q(\psi)(\nabla\psi_p \times \nabla\theta)$ we find $\delta\vec{B}\cdot\nabla\psi \sim (\vec{\xi}\cdot\nabla\psi)(m/q - n)$, which vanishes on the resonant surface $q = m/n$, producing no magnetic island.

As shown in Chap. 6 (see Eq. 6.1), in the absence of plasma resistivity the plasma remains attached to the magnetic field lines. The mechanism by which it does so for small values of η is easily seen. Consider a field B_y and attempt to drive plasma across it by inducing a flow v_x. Then $\eta\vec{j} = \vec{v} \times \vec{B}$ gives $j_z = v_x B_y/\eta$, which induces a force $\vec{F} = \vec{j} \times \vec{B}$, which has an x-component $F_x = -v_x B_y^2/\eta$ opposing the induced flow. This force is infinite in the limit $\eta \to 0$ provided $B_y \neq 0$. However, if B_y vanishes for some value of x the opposing force also vanishes locally, and a field configuration can form about this x which gives rise to rapid diffusion of the plasma through the field. Now suppose $B_y(x) = B_0$ for $x \gg L$, $-B_0$ for $x \ll L$, and vanishes along the surface $x = 0$. Consider the magnetic flux passing through the x, z plane $\Phi(x) = \int^x B_y dx$. From Maxwell's equations and Ohm's law, we find the local flux anihilation rate to be $\partial_t \Phi = \eta\partial_x B_y$, where we have ignored structure in the y direction, which we will see is generally of larger scale than that in x. At $x = 0$, depending on the nature of the driving forces, the field strength gradient $\partial_x B_y$ can be made arbitrarily large, and thus an arbitrarily large reconnection rate attained for any nonzero η. A self-consistent solution shows that B_y is modified to produce a sharp gradient in a narrow range of x where the opposing force is small, $|x| < x_T$, referred to as the tearing layer. The width of this layer and the gradient of B_y depend on the nature of the driving forces.

There is, everywhere in a plasma of nonzero resistivity, a slow diffusion of the plasma across the field lines, characterized by the resistive time. Combining Eqs. 1.5 and 1.7 we find, for constant resistivity and neglecting fluid velocity,

$$\partial_t \vec{B} = \eta \nabla^2 \vec{B}, \tag{7.1}$$

which is the diffusion equation. However, rapid reconnection can occur only in thin resistive layers along magnetic surfaces where the perturbation is resonant and produces a magnetic island configuration. In most of the plasma, where the reconnection is negligible, the ideal MHD equations are valid, and the form of the perturbation derived by minimizing δW is correct. These ideal solutions must be joined across the thin resistive layers by an analysis taking into account the fast resistive diffusion in the layer. If the distant plasma is at rest the reconnection is driven by the magnetic energy difference between the initial and final states, and the time scale for the mode growth is determined by a combination of the resistive and Alfvén times. Even so, the growth rate can be substantially different for different geometries. An inkling of this can be seen in the nonlinear behaviour of certain ideal modes. The internal kink mode ($m = 1$) nonlinearly evolves to a state with an infinite gradient of B across the singular surface. There is no similar phenomenon for $m \neq 1$. This tendency of the mode to develop a steep gradient in B leads to a much faster resistive growth rate in this case than for the case with $m > 1$.

To discuss this more quantitatively, represent the $\vec{B}$ field through the flux function $\psi(x, y)$ with

$$\vec{B} = \nabla \psi \times \hat{z} \tag{7.2}$$

and assume steady state geometry given by Fig. 7.1. At large x an incompressible plasma is driven toward the X-point with velocity $\vec{u}$, imbedded in a field $\vec{B} = B_{y0}\hat{y}$. The configuration is symmetric in both x and y, with reconnection occuring in the neighborhood of the X-point and the plasma flowing away in the y direction, with velocity v and field $\vec{B} = B_{x0}\hat{x}$. Since

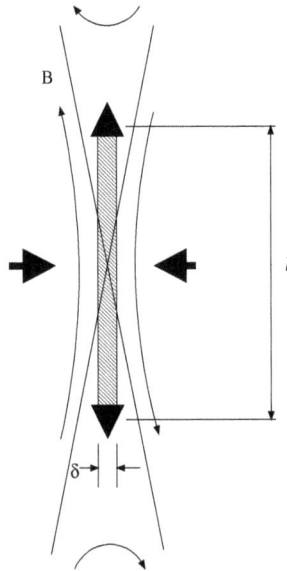

Fig. 7.1   Geometry for resistive reconnection.

the configuration is steady state, ψ has the form $\psi = \psi_0(x, y) - at$, and $\psi_0(x, y)$ has the shape of a saddle facing the y axis. The plasma pressure is $B_{y0}^2/2$, larger in the reconnection region than for $|x| > \delta$. This is balanced by the acceleration of the plasma along the y axis, up to kinetic energy $(1/2)\rho v^2 = B_{y0}^2/2$, giving $v_y \simeq V_a$, the Alfvén speed. Equations 1.6 and 1.7 give

$$\partial_t \psi + (\vec{v} \cdot \nabla)\psi = \eta \nabla^2 \psi + E. \tag{7.3}$$

Outside the reconnection region the rate of change of ψ is given by convection. Along the x and y axes this gives

$$\partial_t \psi = v_{x0} B_{y0} = -V_A B_{x0}. \tag{7.4}$$

Mass conservation gives $v_{x0} l = v_A \delta$ and thus $B_{x0} = (\delta/l) B_{y0}$. At the X-point, where there is no flow, $\partial_t \psi$ is given entirely by $\eta j = \eta \nabla^2 \psi \simeq 2\eta B_{y0}/\delta$, and the width of the resistive layer δ must adjust until this value equals that determined by the boundary conditions. A dimensionless re-connection rate is defined as $M = \partial_t \psi / B_{y0} V_A$ and thus $M = 2\eta / V_A \delta = \delta / l$ or $M = (\eta / V_A l)^{1/2}$.

There are two major models for steady state reconnection, having to do with the size of the reconnection region, and, in particular, depending on the scaling of l with η. Writing $l = \eta^r$ we find $M \tau_A \sim \eta^{(1-r)/2}$. Sweet (1956) and Parker (1963) assumed that l had global dimensions ($r = 0$), giving $M \tau_A \sim \eta^{1/2}$. The current in the resistive layer, $j = v_A BM/\eta \sim \eta^{-1/2}$, is singular in the limit $\eta \to 0$. The model proposed by Petschek (1964) attaches shock fronts to the diffusion region, which is similar to that in the Sweet–Parker model, but the length of the diffusion layer l is proportional to η, giving M independent of η. The current is more singular than in the Sweet–Parker model, going as $1/\eta$ for $\eta \to 0$.

The rate at which work is done on the fluid is given by

$$P = v_x F_x = v_x^2 (B'\delta)^2 / \eta \tag{7.5}$$

where $B'\delta = B_y$ and δ is the width of the reconnection region.

In general, the instability wavelength is much larger than δ and the fluid kinetic energy along $\vec{k}$ is dominant. The rate of change of kinetic energy for a growing mode is then

$$\frac{dK}{dt} = \gamma \rho v_x^2 / (k^2 \delta^2). \tag{7.6}$$

Equating these two gives

$$\delta = \frac{(\gamma \rho \eta)^{1/4}}{(kB')^{1/2}}. \tag{7.7}$$

Thus the width of the reconnection region is related to the mode growth rate.

However, these analytical treatments leave many questions unanswered. Aside from the problem of smoothly matching the exterior solution to the diffusive region, the accessibility and stability of the steady state patterns are not addressed by the analysis. This problem has been examined through numerical simulation with initial value codes. All simulations employing a resistivity which was a function of position only led to Sweet–Parker type reconnection, with l independent of η. (Waddell *et al.*, 1976; Syrovatskii, 1980; Biskamp, 1980, 1982; Priest, 1983 and Park *et al.*, 1984).

The behavior of the current at the singular layer as $\eta \to 0$ reveals the nature of the forces driving the instability. In both the Sweet–Parker and Petschek models the plasma is asymptotically driven toward the resistive layer, producing a singular j at this point. This is also the case for the internal kink mode (Park *et al.*, 1980) and ballooning mode (Monticello *et al.*, 1981). When forces exist, driving the plasma toward the singular surface, j will be singular and the reconnection will proceed at a rate faster than that given by the resistive time. An example of nonsingular behavior is provided by the nonlinear tearing mode for $m > 1$, which is stable in the limit $\eta \to 0$. The absence of driving forces means that j is finite for $\eta \to 0$, and reconnection proceeds at the resistive rate.

Much of the history of a typical tokamak discharge is known, or conjectured, to be due to tearing-mode activity. In the initial stages of a discharge, the double tearing mode is thought to play a role in producing rapid current penetration into the plasma. During the discharge, there occur sawtooth oscillations and Mirnov oscillations, which are understood to be due to tearing modes, and the observed anomalous electron thermal transport is possibly due to small magnetic islands and resulting stochastic field behavior produced by microinstabilities. Finally, abrupt termination of a discharge, through what is referred to as a major disruption, is understood to be due primarily to tearing-mode activity. In addition, magnetic reconnection plays an important, but perhaps less well-understood role, in stellar and interstellar phenomena such as solar flares and the interaction of the solar wind with the earth's magnetic field.

The possible importance of the tearing mode for tokamak plasmas was first shown by Furth *et al.* (1963). The growth rate for the mode was shown to depend on the global current profile, and to involve the plasma resistivity only in a narrow tearing layer. It was found to be strongest for modes of large wave length, indicating possible large scale effects on plasma equilibrium. In 1973 the growth rate was evaluated for several current profile shapes by Furth *et al.* By this time, the telltale magnetic signals indicating the presence of the mode had already been observed by Mirnov and Semenov (1971). The special case of the $m = 1$ mode was analyzed by Coppi *et al.* in 1976.

However, these models of resistive reconnection fail to explain some very fast reconnection rates seen both in fusion experiments and in astrophysical observations, and an explanation of these rates requires an extension of resistive magnetohydrodynamics to include the coupling to dispersive whistler waves and kinetic Alfvén waves. These effects can be included by a modification of Ohm's law to include the electron inertia and the Hall term,

$$d_e^2 \frac{d\vec{j}}{dt} = \vec{E} + \vec{v} \times \vec{B} - \eta\vec{j} - \frac{1}{n}\vec{j} \times \vec{B} + \frac{1}{n}\nabla p \qquad (7.8)$$

with $d_e = c/\omega_{pe}$ the collisionless skin depth, and the last term is the Hall term. The initial inclusion of kinetic effects was done by Drake and Lee (1977) and this work has been extended and modifed by other authors. See Cowley *et al.* (1986). Wesson (1990) was the first to introduce the collisionless skin depth as the characteristic scale of the reconnection layer. This reduced the discrepancy between theory and experiment for the reconnection time for the $m = 1$ mode in tokamaks by an order of magnitude, leaving only a factor of three or four to explain. Zakharov and Rogers (1992) then eliminated this factor by showing that in a typical tokamak discharge the ion-sound Larmor radius takes the place of the characteristic reconnection width. The onset of the fast reconnection was described by Waelbroeck (1989), using a set of nonlinear equations valid in the early nonlinear stage of the reconnection, and these equations were further investigated by Zakharov *et al.* (1993), again finding the ion-sound Larmor radius to be the relevant scale. This understanding of reconnection was compared to sawtooth stabilization in TFTR by Levinton *et al.* (1994). See also articles on fast reconnection by Aydemir *et al.* (1992), Ottaviani and Porcelli (1993), X. Wang and Battacharjee (1993), Biskamp and Drake (1994), Cafaro *et al.* (1998). A full analysis of parameter domains (the value of the resistivity, the plasma beta, the magnitude of the transverse field, and the electron-ion mass ratio) describing when the reconnection is dominated by the resistivity, the electron inertia, ion kinetic effects, or the whistler dynamics has yet to be completed, and we will not attempt to describe the state of the present analysis here, most of the following sections giving results obtained by using the equations of resistive MHD alone.

7.2 THE TEARING MODE

As in the case of the external kink mode, the forces driving the tearing mode are large compared to toroidal effects, and this mode can be examined in cylindrical geometry. Plasma equilibria are discussed in Sec. 2.9.

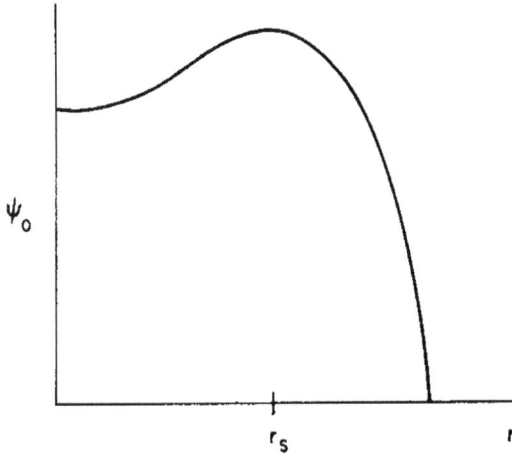

Fig. 7.2   Equilibrium helical flux.

As seen in Sec. 7.1, resistive effects play a role only in a thin layer at a rational surface. Outside this layer the form of the field perturbation is found by solving the ideal MHD equations, *i.e.* minimizing δW. As in the case of ideal MHD, we Fourier decompose the perturbation and consider a perturbation of a single helicity, $e^{i(m\theta - n\phi)}$ with the rational surface $q(r_s) = m/n$ in the plasma. A perturbation of this form, as shown in Sec. 1.7, produces a topological change in the magnetic field, in the form of magnetic islands, only at the resonant surface $q = m/n$. The form of the equilibrium helical flux $\psi_0(r)$ in a typical tokamak configuration, with $q(r)$ monotonically increasing and the resonant surface $q = m/n$ within the plasma, is shown in Fig. 7.2. The helical flux (see Eq. 2.109) is arbitrary within an additive constant, and the convention used is to normalize it to zero on the cylindrical boundary.

Introduce a perturbation of a single helicity through $\psi = \psi_0(r) + \psi_1(r)e^{i(m\theta - n\phi)}$. Thus the perturbation wave vector is $\vec{k} = \nabla(m\theta - n\phi)$ and $\vec{k} \cdot \vec{B}_0 = (m/r)\psi_0'$. The perturbation ψ_1 is simply related to the ideal displacement, $\vec{\xi}$. To lowest order in ϵ, *i.e.* the cylindrical geometry tokamak approximation, $\vec{\xi}$ is given by $\xi_r = \xi$, $\xi_\theta = id(r\xi)/(mdr)$, $\xi_\phi = 0$, and thus

$$\vec{\xi} \times \vec{B} = \left[\hat{r}\frac{id(r\xi)}{mdr} - \hat{\theta}\xi + \hat{\phi}\xi\left(\psi_0' + \frac{nr}{m}\right) \right] e^{i(m\theta - n\phi)}. \tag{7.9}$$

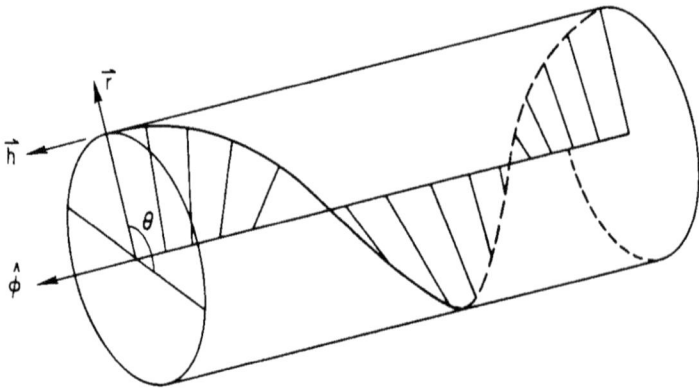

Fig. 7.3 Helical surface.

Also, from Eq. 2.110, to lowest order the perturbed field is given by $\delta\vec{B} = -\nabla \times (\psi_1\hat{\phi})$.

Thus writing $-\psi_1\hat{\phi} = \vec{\xi}\times\vec{B}+\nabla G$, with G the difference in gauge between the two representations, we find $G = -(i/m)r\xi e^{i(m\theta-n\phi)}$ and $\psi_1(r) = \xi\psi_0'$, or $\psi_1(r) = \xi r(1/q - n/m)$. Thus for finite ξ, $\psi_1(r)$ vanishes at the rational surface. The presence of a magnetic island corresponds to taking ξ infinite. This does not mean that plasma displacement becomes infinite, but rather that because of nonzero η, and the slipping of the plasma through the field, the ideal displacement no longer correctly describes the plasma motion in the tearing layer.

Consider the helical ribbon defined by the vector $\vec{h} = \hat{\phi} + nr\hat{\theta}/m$, as shown in Fig. 7.3. At the rational surface, $\psi_0' = 0$, the $\vec{B}$ field lies in the surface defined by $\vec{h}$. The vectors $\vec{k}, \vec{h}$, and $\vec{r}$ are mutually orthogonal, and in the coordinate system defined by them the component of $\vec{B}$ in the direction of $\vec{k}$ vanishes at $r = r_s$, as shown in Fig. 7.4.

The surfaces of constant $\psi = \psi_0 + \psi_1(r)cos(m\theta - n\phi)$ at fixed ϕ define magnetic islands, with width $w = 4(-\psi_1/\psi_0'')^{1/2}$, as shown in Fig. 7.5. In the exterior region ψ_1 can be found by minimizing δW. However, directly substituting ψ_1 into δW and carrying out the variation leads to infinite contributions at the rational surface if ψ_1 is finite there (not allowed in ideal MHD) when the necessary integration by parts is attempted, as does

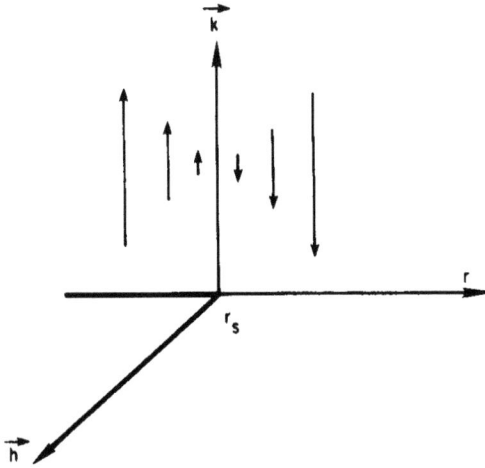

Fig. 7.4 Magnetic field near $r = r_s$.

variation using the function ξ itself. Instead, let $\chi = \xi F = \psi_1/r$, with $F = (1/q - n/m)$ and carry out the variation of δW using χ. Integrating by parts we find

$$\delta W = -\frac{\pi r^3}{2} \chi \frac{d\chi}{dr} \Big|_-^+$$

$$+ \frac{\pi}{2} \int dr \left\{ -\chi \frac{d}{dr} \left(r^3 \frac{d\chi}{dr} \right) + \left[(m^2 - 1)r + \frac{1}{F} \frac{d}{dr} \left(r^3 \frac{dF}{dr} \right) \right] \chi^2 \right\} . (7.10)$$

Thus the Euler–Lagrange equation is given by setting the integrand to zero, and

$$\delta W_{min} = -\frac{\pi}{2} r^3 \chi \frac{d\chi}{dr} \Big|_-^+ . \tag{7.11}$$

Substituting $\psi_1 = r\chi$ and using the fact that ψ_1 is continuous across the resistive layer, but not necessarily $d\psi_1/dr$, we find

$$\delta W_{min} = -\frac{\pi}{2} r_s \psi_1^2 \Delta' \tag{7.12}$$

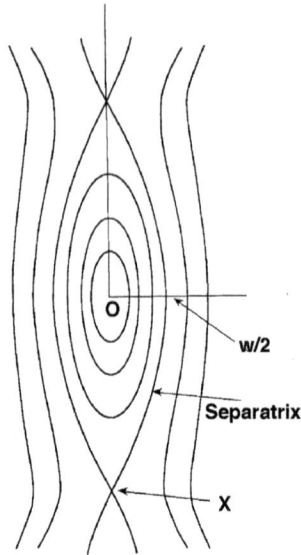

Fig. 7.5   Magnetic island.

where

$$\Delta' = \frac{1}{\psi_1} \frac{d\psi_1}{dr} \Big|_{-}^{+} \tag{7.13}$$

is the discontinuity of ψ_1' across the tearing layer, and is a measure of the magnetic energy to be gained by a perturbation producing magnetic islands at $r = r_s$. Substituting ψ_1 into the integrand and using $F'' = j_0'/r - 3F'/r$ we find the Euler–Lagrange equation for ψ_1,

$$\nabla_\perp^2 \psi_1 = \frac{dj_0}{d\psi_0} \psi_1 \tag{7.14}$$

where $dj_0/d\psi_0 = dj_0/(\psi_0' dr)$ and $\nabla_\perp$ is the gradient in the r, θ plane.

The solution of this equation for ψ_1 in the exterior region determines Δ'. The energy differential driving the mode and proportional to Δ' is a function of the whole current profile, not a quantity determined locally at the rational surface.

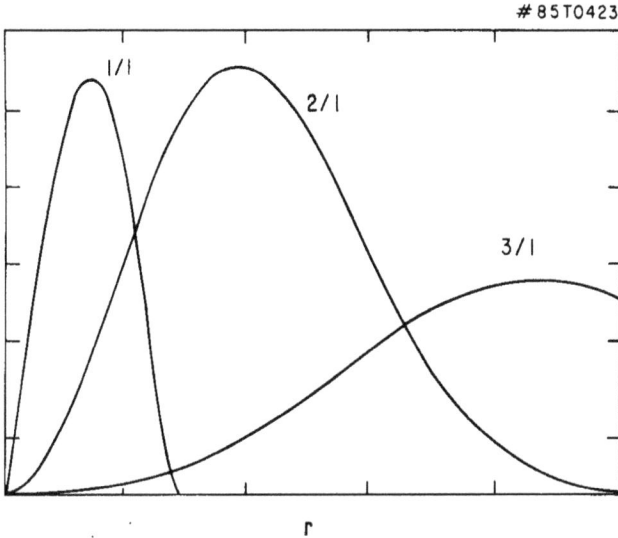

Fig. 7.6   Perturbations of the helical flux.

Shown in Fig.   7.6 are typical solutions ψ_1 obtained in the exterior region. They behave as r^m for small r, vanish at the conducting wall, and have discontinuities in $d\psi_1/dr$ at $r = r_s$. The $m = 1$ solution is zero for $r > r_s$ and is therefore rapidly changing in the vicinity of $r = r_s$, but for $m > 1$ ψ_1 is essentially constant near $r = r_s$.

The change in magnetic energy produced by the formation of an island is not due to the exterior region alone; the magnetic energy is also modified in the interior region by the island itself.   Nevertheless, a careful analysis of the interior and exterior regions shows (Adler *et al.*, 1980) that the integrated energy change is correctly given by the exterior ideal analysis, $\delta W \sim -\Delta'\psi_1^2$.

It should be kept in mind that these are not calculations of energy dissipation during mode growth, which would involve also Ohmic heating, but only considerations of the magnetic potential energy available to drive the mode.

Now consider the full resistive equations, which are valid in the ideal domain as well as in the tearing layer. The energy principle cannot be used, and we return to the original equations describing resistive MHD.

The poloidal flux and the helical flux satisfy very similar equations, as expected from the simple way they are related (see Sec. 1.7). First write $\vec{B}$

in terms of the poloidal flux, $\vec{B} = \nabla\phi - \nabla \times \psi_p \nabla\phi$, and substitute it into Faraday's law and Ohm's law, Eqs. 1.6 and 1.7, giving

$$\nabla \times [\partial_t \psi_p \nabla\phi] = \nabla \times (\eta\vec{j} - \vec{v} \times \vec{B}). \tag{7.15}$$

Thus

$$\partial_t \psi_p \nabla\phi = \eta\vec{j} - \vec{v} \times \vec{B}, \tag{7.16}$$

except for a gauge term ∇G, which can be taken to be zero.

Dotting Eq. 7.16 with $\nabla\phi$ we find

$$\partial_t \psi_p + \vec{v} \cdot \nabla_\perp \psi_p = \eta j_\phi. \tag{7.17}$$

We also have, from Eq. 2.106,

$$j_\phi = \nabla_\perp^2 \psi_p. \tag{7.18}$$

The helical flux satisfies a similar equation. Dot Eq. 7.16 with $\vec{h}$ and substitute Eq. 2.110, $\vec{B} = \vec{h} + \nabla\phi \times \nabla_\perp \psi_h$, and Eq. 2.111 for $\vec{j}$. Using also $\psi_h = \psi_h(r, \tau)$ so that $\vec{h} \cdot \psi_h = 0$ we find

$$\partial_t \psi_h + \vec{v} \cdot \nabla_\perp \psi_h = \eta j_\phi \tag{7.19}$$

and

$$j_\phi = \nabla_\perp^2 \psi_h + \frac{2n}{m}. \tag{7.20}$$

The velocity is governed by the equation of motion

$$\rho\frac{d\vec{v}}{dt} = \vec{j} \times \vec{B} - \nabla p \tag{7.21}$$

where $d/dt = \partial_t + \vec{v} \cdot \nabla$ is the convective derivative.

The large aspect ratio approximation, $R/r \gg 1$, and tokamak ordering with $\beta_\theta \simeq 1$, $q \simeq 1$, can be used to give a consistent ordering for the quantities in Eqs. 1.5–1.10 in the parameter $\epsilon = r/R$. The result is that B_ϕ, ρ, $\nabla_\perp$, ∂_ϕ are of order one, and B_θ, B_r, j_ϕ, $v_\perp$, ∂_t are of order ϵ, while δB_ϕ, p, $j_\perp$, v_ϕ are of order ϵ^2 and $\nabla \cdot \vec{v}_\perp$ is of order ϵ^3. The derivation of the reduced equations is discussed in detail in Sec. 10.2. Thus in cylindrical approximation $\nabla \cdot \vec{v}_\perp \simeq 0$, which implies that the velocity can be represented by a stream function U with $\vec{v} = \hat{\phi} \times \nabla_\perp U$, where $\perp$ refers to the r, θ plane. Operating with $\hat{\phi} \cdot \nabla \times$ we find

$$\frac{d}{dt}\rho \nabla^2 U = \hat{\phi} \cdot \nabla \times (\vec{j} \times \vec{B}). \tag{7.22}$$

The equation resulting from taking the divergence of the equation of motion relates the order ϵ^2 quantities $\nabla \cdot \vec{v}$ and δB_ϕ, which remain unknown when the equations are solved to lowest order in ϵ.

In the remainder of this chapter we will use only the helical flux ψ_h and omit the subscript. Substituting the expression for $\vec{v}$ into Eq. 7.19, and the expression for $\vec{B}$ into the equation for U, we have two coupled nonlinear differential equations for ψ, U

$$\frac{\partial \psi}{\partial t} + \hat{\phi} \cdot (\nabla_\perp U \times \nabla_\perp \psi) = \eta \left(\nabla_\perp^2 \psi + \frac{2n}{m} \right), \tag{7.23}$$

$$\frac{d}{dt}\rho \nabla_\perp^2 U = \hat{\phi} \cdot [\nabla_\perp \psi \times \nabla_\perp (\nabla_\perp^2 \psi)]. \tag{7.24}$$

These equations give the full nonlinear evolution of a single helicity mode, as is discussed in Chap. 10. To study the linear behavior, introduce the perturbations

$$\psi = \psi_0(r) + \psi_1(r)cos(m\theta - n\phi), \tag{7.25}$$

$$U = \frac{r\gamma\xi(r)}{m}sin(m\theta - n\phi). \tag{7.26}$$

Substituting these expressions then results in two coupled, second order ordinary differential equations for ξ, ψ_1:

$$\psi_1 - \psi_0'\xi = \frac{r_s^2}{\gamma\tau_R}\nabla_\perp^2 \psi_1 \tag{7.27}$$

$$\gamma^2\tau_A^2\nabla_\perp^2 r\xi = \frac{m^2\psi_0'}{r}\left(\frac{dj_0}{d\psi_0}\psi_1 - \nabla_\perp^2\psi_1 \right) \tag{7.28}$$

where now $\nabla_\perp^2 = \frac{1}{r}\frac{d}{dr}r\frac{d}{dr} - \frac{m^2}{r^2}$ and $\tau_A = R\sqrt{\rho}/B_z$ is the characteristic Alfvén time for these equations, equal to the poloidal Alfvén time divided by q, and $\tau_R = r_s^2/\eta$ is the resistive time given by the rational surface. In the following it is convenient to be able to refer to dimensional arguments as a check on algebra, and this convenience has not been sacrificed by our units. Recall that $\vec{B}, \vec{j}$ are dimensionless, ξ has the dimensions of length, and flux (ψ) the dimensions of length squared.

Outside the tearing layer the operator $r_s^2 \nabla_\perp^2$ is of order unity and using $\gamma \tau_R \gg 1$, $\gamma \tau_A \ll 1$ we find the solutions in the external region satisfy

$$\nabla_\perp^2 \psi_1 = \frac{dj_0}{d\psi_0} \psi_1 \tag{7.29}$$

and

$$\psi_1 = \xi \psi_0', \tag{7.30}$$

which is Eq. 7.14, *i.e.* these solutions are those derived by minimizing the ideal δW.

To complete the analysis, solutions must be obtained in the interior region, and matched to the exterior ideal solutions. Consider the interior region, $|x| \ll r_s$ with $x = r - r_s$. In this region ψ_1' is rapidly changing. Thus we can take $\nabla_\perp^2 \psi_1 \simeq \psi_1''$, and $\psi_0' = -snx/m$ with $s = r_s q'/q$. We then find for $|x| \ll r_s$, $\vec{j} \simeq -\psi_1'' \hat{\phi}$, and

$$\psi_1(x) + \frac{snx\xi(x)}{m} = \frac{r_s^2 \psi_1''(x)}{\gamma \tau_R} \tag{7.31}$$

and

$$\gamma^2 \tau_A^2 \xi''(x) = \frac{smnx}{r^2} \psi_1''(x). \tag{7.32}$$

The nature of the boundary layer is found using a Kruskal–Newton diagram. For a given linear, homogeneous differential equation of the form

$$\sum_{q,n,m} C_{q,m,n} \epsilon^q x^m \frac{d^n y}{dx^n} = 0, \tag{7.33}$$

construct a Kruskal–Newton diagram with terms given by $x^p \epsilon^q$. The layer is at $x = 0$ so count powers of x as well as powers of dx giving $p = m - n$. The function y and the numbers $C_{q,m,n}$ are all assumed to be of order one. Each term in the sum can be represented as a point in the p, q plane. All points above a given line represent terms smaller than points on the line, and all points below a line represent terms larger than those on the line.

To find all possible combinations of dominant balance for a given equation, find all possible placements of a line so that it includes two or more terms of the equation, with all other points lying above the line. Graphically this may be understood as bringing the line up from below until it makes contact with a point, and then rotating it one way or the other until it makes contact with a second point. The scaling of x is quickly determined by the slope of the line in the diagram. If the terms defining the line are given by $x^p \epsilon^q$ and $x^r \epsilon^s$ then balancing them gives the scaling $x \sim \epsilon^{(s-q)/(p-r)}$, the power of ϵ being minus the slope of the line. Such a plot is known as a Kruskal–Newton diagram. It was first used by Newton in considering infinitesimal displacements for the development of differential calculus and subsequently further developed by Kruskal. See also White, (2010).

7.2.1 The tearing mode, $m \neq 1$

Because of the quite different form of the solutions for $m = 1$, it must be treated as a special case. First examine $m \neq 1$. From the nature of the exterior solution for $m \neq 1$ we can approximate $\psi_1(x)$ as constant. Substituting we then obtain a single second order differential equation

$$\psi_1 + \frac{snx\xi(x)}{m} = \frac{r_s^4 \gamma^2 \tau_A^2 \xi''(x)}{smnx\gamma\tau_R}. \tag{7.34}$$

The Kruskal–Newton diagram is shown in Fig. 7.7, where the point a is the ξ'' term, b is the ξ term, and c the constant ψ_1, and $\gamma^2 \tau_A^2/(\gamma\tau_R) \sim \epsilon^3$. The other points above the line connecting a and b are terms neglected in approximating $\nabla_\perp^2 \psi_1$ as ψ_1'', and ψ_0' as $-snx/m$.

We now use Kruskal's asymptotic principle of maximal balance (Kruskal, 1963), according to which "the most informative ordering is that which simplifies the least, maintaining a maximal set of comparable terms". We obtain this by renormalizing ξ through $\xi = m\psi_1(\gamma\tau_R/(n^2 s^2 \gamma^2 \tau_A^2))^{1/4}\chi(z)/r_s$ to bring all points to lie on the same line. The slope of the line gives the scaling of the layer as $x \sim (\gamma^2 \tau_A^2/\gamma\tau_R)^{1/4}$, and we introduce the independent variable z through $x = (\gamma^2 \tau_A^2/\gamma\tau_R n^2 s^2)^{1/4} r_s z$, giving

$$\chi'' - z^2 \chi = z, \tag{7.35}$$

with boundary condition $\chi \to -1/z$ for $z \to \infty$. As written, the equation involves z^2, which would introduce second order derivatives under Fourier Laplace transformation, not leading to simplification. A simple change of variables can remedy this. Let $\chi = zg(x)$ with $x = z^2/2$. Substituting, find

$$2xg'' + 3g' - 2xg = 1, \tag{7.36}$$

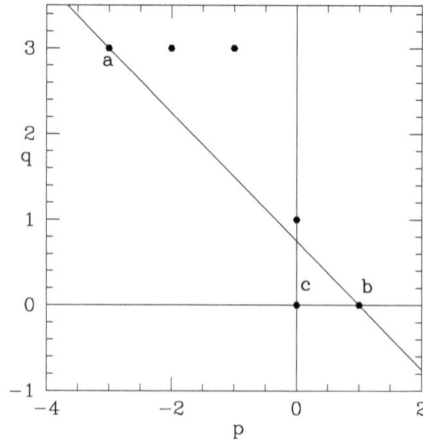

Fig. 7.7   Kruskal–Newton diagram for $m \geq 2$.

an equation containing only x and thus leading to a first order differential equation under transformation. Now make a Fourier–Laplace transformation of the function g,

$$g(x) = \int_C e^{-tx} f(t) dt, \tag{7.37}$$

and substitute this into Eq. 7.36. Integrating by parts we find

$$\int_C dt e^{tx} \left[tf(t) + 2(t^2 - 1)f'(t) \right] - 2f(t)(t^2 - 1)e^{tx}|_a^b = 1, \tag{7.38}$$

with the ends of the integration contour C at a, b. We then have a solution provided

$$f(t) = c(1 - t^2)^{-1/4}, \qquad 2f(t)(1 - t^2)e^{tx}|_a^b = 1. \tag{7.39}$$

There is only one end point making $f(t)(1-t^2)e^{tx}$ independent of x, namely $t = 0$. We thus find the two contours $a = 0$, $b = \pm 1$, giving $c = -1/2$. The two solutions have different asymptotic behavior for $z \to \infty$. Choosing the solution which tends to $-1/z$ (given by $b = 1$) we find

$$\chi = -\frac{z}{2} \int_0^1 d\mu e^{-z^2 \mu/2} (1 - \mu^2)^{-1/4}. \tag{7.40}$$

For the necessary matching to the exterior solution we have

$$\frac{r^2\gamma^2\tau_A^2}{msn\psi_1(r_s)}\int_{-\infty}^{\infty}\frac{\xi''}{x}dx = \Delta'. \tag{7.41}$$

The matching condition becomes

$$\gamma^{5/4}\tau_A^{5/4}S^{3/4}(ns)^{-1/2}\int_{-\infty}^{\infty}\frac{dz}{z}\chi'' = \Delta'r \tag{7.42}$$

with $S = \tau_R/\tau_A$ and $\gamma\tau_R \gg 1$ and $\gamma\tau_A \ll 1$, as assumed.

To evaluate the integral $I = \int_{-\infty}^{\infty}(dz/z)\chi''$, substitute χ and integrate over z, giving

$$I = \sqrt{\pi/2}\int_0^1 d\mu\frac{\mu^{1/2}}{(1-\mu^2)^{1/4}} = \pi\Gamma(3/4)/\Gamma(1/4). \tag{7.43}$$

Thus

$$\gamma\tau_A = \left(\frac{\Gamma(1/4)\Delta'r}{\pi\Gamma(3/4)}\right)^{4/5}S^{-3/5}(ns)^{2/5}. \tag{7.44}$$

We thus find $\gamma \sim \tau_R^{-3/5}\tau_H^{-2/5}$, where the hydromagnetic time $\tau_H = \tau_A/(sn)$ is the time for a shear Alfvén wave to cross the plasma.

Because the differential equation for $\chi(z)$ is parameter free, the variation of $\chi(z)$ takes place in a scale of $\Delta z \simeq 1$. Substituting the expression for γ we then find the tearing layer width, for $\Delta' \simeq 1$,

$$\frac{W_T}{r} = \left(\frac{\tau_A}{\tau_R}\right)^{2/5}. \tag{7.45}$$

Shafranov (1970) calculated Δ' for a constant current profile of finite extent. This derivation will be given in Sec. 7.5. Furth *et al.*, (1973) carried out a detailed investigation of the dependence of Δ' on the profile shape. They introduced the peaked, rounded, and flat profiles described in Sec. 2.9. They included lowest order toroidal corrections to Eq. 7.29, which are of order $(r/R)^2$ but used an aspect ratio of 10, in which case these corrections are not visible in plots of Δ', such as Figs. 7.8–7.10.

The peaked model is found to be unstable to $m = 2$ and $m = 3$ tearing modes. In Fig. 7.8, Δ' is shown as a function of the position of the singular surface, r_s/r_0 for various values of wall position, $X_b = b/r_0$. The rounded model is generally more unstable, with $m = 2, 3$ existing over a wider range of r_s and $m = 4$ also becoming unstable, as is shown in Fig. 7.9.

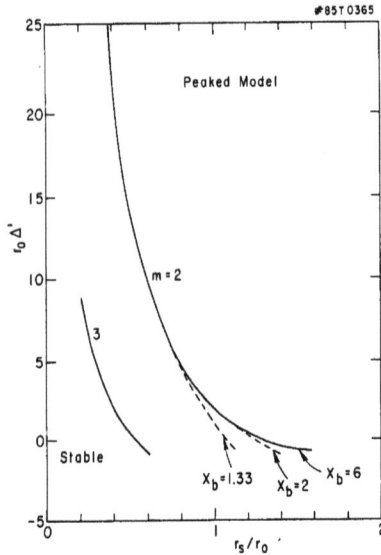

Fig. 7.8 Δ' for the peaked current profile.

The flat profile is even more unstable, as shown in Fig. 7.10. Since all tokamaks tend to operate with $q_0 \simeq 1$, the most important modes are $m/n = 1/1$, $2/1$, $3/2$. As we will see, the 1/1 mode is always unstable if $q_0 < 1$. For flat current profiles all three of these modes can be unstable for q_0 slightly below one. This is not possible for peaked and rounded profiles.

7.2.2   *The $m = 1$ tearing mode*

For the $m = 1$ mode, recall that the exterior displacement ξ is to lowest order a step function (see Sec. 6.10), and ψ_1 is also very rapidly changing at the rational surface. Thus the constant ψ approximation used to obtain Eq. 7.40 is not valid. Return to the coupled Eqs. 7.32 for the interior region and substitute $\psi_1 = xZ$, giving equations in order one quantities

$$xZ + \frac{snx\xi(x)}{m} = \frac{r_s^2(2Z' + xZ'')}{\gamma\tau_R}, \qquad \gamma^2\tau_A^2\xi''(x) = \frac{smn}{r^2}Z. \quad (7.46)$$

The Kruskal–Newton diagrams for these equations are shown in Fig. 7.11. In the left plot $\epsilon = 1/(\gamma\tau_R)$ and point a is the derivative term, point b the

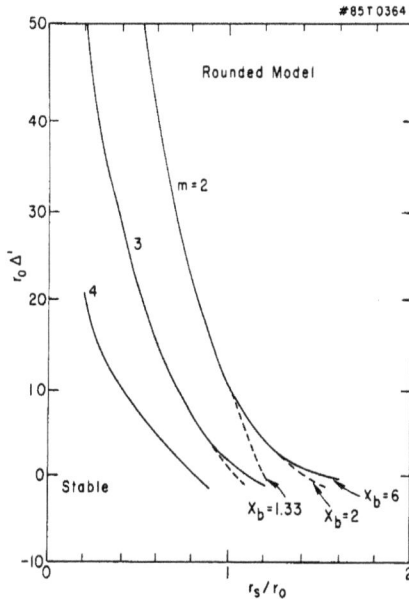

Fig. 7.9 Δ' for the rounded current profile.

xZ and $x\xi$ terms. This plot gives a layer scaling of $x \sim 1/\sqrt{\gamma\tau_R}$. In the right plot point a is the ξ'' term, and point b the Z term. This plot gives a layer scaling of $x \sim \gamma\tau_A$. But of course these equations are coupled and there is only one layer, and equating these we find $\gamma \sim \tau_A^{-2/3}\tau_R^{-1/3}$.

Introducing the layer variable $z = (\gamma\tau_R)^{1/2}x/r_s$ we find an exact solution which satisfies the boundary conditions $\psi_1 \to 0, x \to +\infty$ and $\psi_1 \to -\xi_0 snx, x \to -\infty$

$$\xi = \frac{\xi_0}{\sqrt{2\pi}}\int_z^\infty e^{-z^2/2}dz, \quad \psi_1 = \frac{\xi_0 r_s}{\sqrt{2\pi}}\frac{sn}{(\gamma\tau_R)^{1/2}}\left[e^{-z^2/2} - z\int_z^\infty e^{-z^2/2}dz\right] \quad (7.47)$$

and

$$\gamma = \tau_R^{-1/3}\tau_H^{-2/3}, \quad (7.48)$$

where again $\tau_H = \tau_A/(sn)$, the hydromagnetic time, is the time for a shear Alfvén wave to cross the plasma. This growth rate is much larger than the growth rate for $m > 2$, and further the $m = 1$ mode is unstable to all

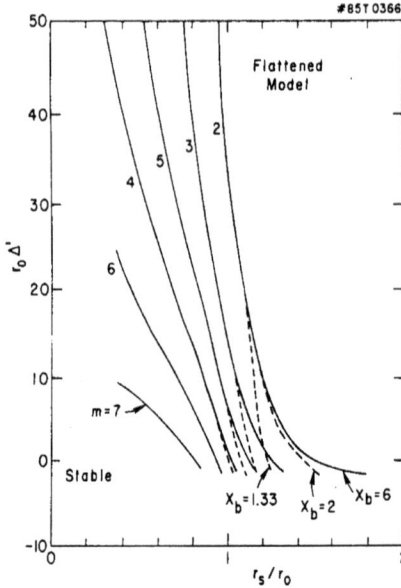

Fig. 7.10 Δ' for the flattened current profile.

profiles with $q_0 < 1$. The very different behavior is due to existence of the ideal $m = 1$ mode, which causes a steepening of the field gradients near the rational surface and drives the reconnection. This is true even though we are treating the case of a low β approximately cylindrical equilibrium, in which the ideal mode is only weakly unstable. Substituting we find for the width of the tearing layer

$$W_T = (snS)^{-1/3} r_s. \qquad (7.49)$$

From the form of the solution we can immediately find the magnitude of the current in the resistive layer. Using Eq. 7.47, and $j_1 \sim \psi_1''$, we find that

$$j \sim \left(\frac{\tau_R}{\tau_A}\right)^{1/3}. \qquad (7.50)$$

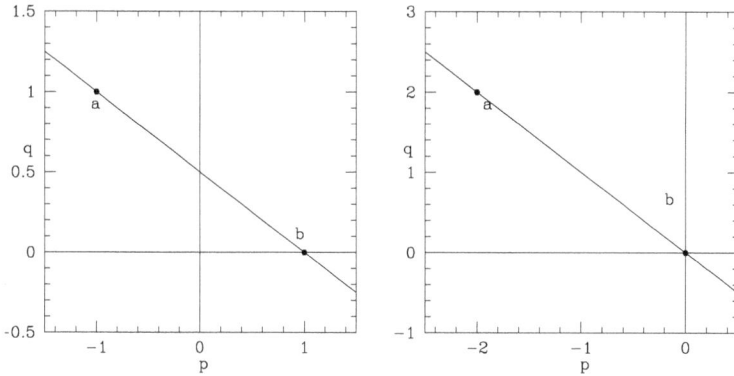

Fig. 7.11   Kruskal–Newton diagrams for $m = 1$.

Thus the current sheet at the X-point becomes increasingly more singular with increasing temperature.

In Fig. 7.12 is shown the helical flux surfaces and a cross section of the current density through the midplane for a developing $m = 1$ magnetic island. The original plasma center and current profile has shifted to the left. The strong negative current sheet at the X-point, and the positive current in the island O-point are both intuitively understandable as inductive response of the plasma to this shift of current profile. Namely, the negative current sheet at the X-point and the positive current flowing down the island O-point both partially cancel the leftward shift of the original current column, and oppose the island growth. Of course all these structures have helical $m = 1$, $n = 1$ symmetry.

Because of the effect of the ideal $m = 1$ mode to steepen the field gradients in the vicinity of the $q = 1$ surface, the $m = 1$ tearing mode can be expected to have quite different behavior depending on whether the mode is ideally stable, neutral stable, or unstable. The linear growth rate can be calculated for all three cases in a single analysis (Coppi $et\ al.$, 1976).

Since the interior region is narrow the analysis there is the same for cylindrical or toroidal geometry. Further, as seen in Sec. 6.10, the ideal growth rate is determined by the asymptotic form of the inner solution, obtained by matching to the exterior solution, which depends on toroidicity,

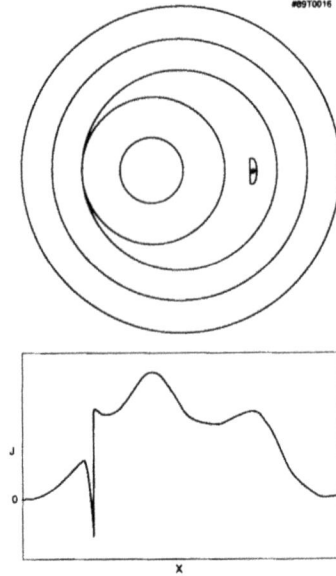

Fig. 7.12　Flux surfaces and current density during $m = 1$ tearing mode.

plasma pressure, etc.. From Eq. 6.92 we have for $x \to -\infty$

$$\frac{1}{\xi_0}\frac{d\xi}{dx} \to -\frac{\gamma_I \tau_A}{\pi x^2 n s} \tag{7.51}$$

with γ_I the ideal growth rate, given by Eq. 6.87, and $\tau_A = r_s\sqrt{\rho}/B_\phi$ the Alfvén time. Now we require the resistive layer solution to have this same asymptotic form, and to thus match onto the external ideal solution. Substitute $\chi = x\psi_1' - \psi_1$ into Eqs. 7.31 and 7.32 and note that $\chi' = x\psi_1'' = \Gamma^2\xi''$ with $\Gamma^2 = \gamma^2\tau_A^2 r^2/(sn)$ so that

$$\chi = \Gamma^2\xi' + \chi_\infty, \tag{7.52}$$

and $\chi \to \chi_\infty$ for $x \to \pm\infty$.

Now integrate $\psi_1'' = \chi'/x$ to find ψ_1'. The integration constant is evaluated by noting that in the ideal domain $\psi_1 = \xi\psi_0'$ so that, asymptotically for $x \to -\infty$, $\psi_1 \to -sxn\xi_0$ and for $x \to \infty$, $\psi_1 \to -\gamma_I\tau_A\xi_0/\pi$. Thus we find that at $x \to \infty$, $\chi = -\psi_1$, and

$$\psi_1 = -\chi - x \int_x^\infty \frac{dx}{x} \frac{d\chi}{dx}. \tag{7.53}$$

Using $\psi_1 \to -sxn\xi(x)$ for $x \to -\infty$ then gives

$$\xi_0 = \int_{-\infty}^\infty \frac{dx}{x} \frac{d\chi}{dx} \simeq 2 \int_0^\infty \frac{dx}{x} \frac{d\chi}{dx} \tag{7.54}$$

and

$$\frac{d\xi}{dx} = -\chi_\infty/(snx^2). \tag{7.55}$$

Thus Eq. 7.51 reduces to

$$\xi_\infty = \frac{2\gamma_I \tau_A}{\pi} \int_0^\infty \frac{dx}{x} \frac{d\chi}{dx}. \tag{7.56}$$

Divide Eq. 7.31 by x, substitute ψ_1, substitute $\psi_1'' = \chi'/x$, and differentiate. Then change the independent variable to $x = (\gamma\tau_A/s^2 n^2 S)^{1/4} r_s z$, giving

$$\frac{d^2\chi}{dx^2} - \frac{2}{z}\frac{d\chi}{dz} - (z^2 + \lambda^{3/2})\chi = -z^2\chi_\infty \tag{7.57}$$

with $\lambda = \gamma\tau_R^{1/3}\tau_H^{2/3}$, and with $\tau_H = \tau_A/(sn)$ the hydromagnetic time. Thus from Eq. 7.48, $\lambda = 1$ if the growth rate is given by the usual cylindrical result.

This equation again involves z^2 and thus would lead to a second order differential equation using the Fourier–Laplace transform. Change variables using $\chi = g(x)$ with $x = z^2/2$, giving

$$2xg''(x) - g'(x) - (2x + \lambda^{3/2})g(x) = -2xg_\infty, \tag{7.58}$$

an equation containing only x and thus leading to a first order differential equation under transformation. The x dependence of the inhomogeneous term can be removed by the Fourier–Laplace transformation

$$g(x) = g_\infty + \int_C e^{tx} f(t)dt. \tag{7.59}$$

Integrating by parts we find

$$\int_C dt e^{tx}[(\lambda^{3/2} + 5t)f(t) + 2(1 - t^2)f'] - 2f(t)e^{tx}|_a^b = \lambda^{3/2}g_\infty, \tag{7.60}$$

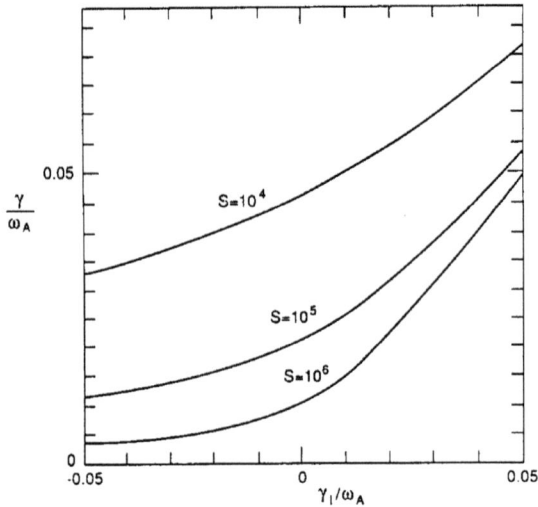

Fig. 7.13   Growth rate for the $m = 1$ mode.

with the ends of the integration contour C at a, b. We then have a solution provided,

$$f(t) = c(1 - t)^{-(\lambda^{3/2}+5)/4}(1 + t)^{(\lambda^{3/2}-5)/4}, \quad 2f(t)e^{tx}|_a^b = \lambda^{3/2}g_\infty,$$
(7.61)

giving $a = 0$, $b = \pm 1$, $c = \lambda^{3/2}/2$. We must choose $b = -1$ to prevent y from diverging at $z \to \pm\infty$. Changing variable from t to $-t$ we then find

$$\frac{\chi}{\chi_\infty} = 1 - \frac{\lambda^{3/2}}{2} \int_0^1 dt(1 - t)^{(\lambda^{3/2}-5)/4}(1 + t)^{-(\lambda^{3/2}+5)/4}e^{-tz^2/2}. \quad (7.62)$$

The asymptotic matching condition, Eq. 7.56, can be evaluated by integration, giving

$$\gamma_I - \frac{8\gamma\Gamma[(\lambda^{3/2} + 5)/4]}{\lambda^{9/4}\Gamma[\lambda^{3/2} - 1)/4]} = 0. \quad (7.63)$$

Three limiting cases are of interest. If $\gamma_I = 0$, the solution is $\lambda = 1$, or

$$\gamma = \tau_R^{-1/3}\tau_H^{-2/3}, \quad (7.64)$$

i.e. the usual cylindrical result. In the ideal MHD limit, with $\tau_R \to \infty$, we can then take the large argument limit of the Γ functions $\Gamma(z+a)/\Gamma(z) \simeq z^a$ to find

$$\gamma = \gamma_I. \tag{7.65}$$

Finally, in the limit of large negative γ_I, the solution is given by $\lambda \ll 1$, and we readily obtain

$$\gamma = \tau_R^{-3/5} \tau_H^{-2/5} \left[\frac{\xi_0 r_s n^2 s^2 \Gamma(1/4)}{4\delta W \Gamma(3/4)} \right]^{4/5} \tag{7.66}$$

where we have used Eq. 6.93 for the ideal growth rate γ_I. In this limit the growth rate scales with resistivity in the same way that modes with $m \geq 2$ do (see Eq. 7.44), with $\xi_0^2/\delta W$ playing the role of Δ'. Thus as marginal stability is approached from the stable side ($\delta W \to 0+$), $\Delta' \to \infty$ and the mode changes character from a $m \geq 2$ tearing mode to an ideally driven mode.

Notice that for all values of γ_I the mode is purely growing and there is no threshold. The introduction of diamagnetic effects and the associated real frequencies, discussed in Sec. 7.10, changes this considerably, introducing complex frequencies and thresholds. Shown in Fig. 7.13 is the growth rate of the $m = 1$ mode as a function of the ideal growth rate γ_I.

7.3 THE SKIN CURRENT PROFILE

In the early stages of a discharge, a hollow current profile frequently occurs, and can lead to the existence of two singular surfaces of the same $q = m/n$. This leads to the double tearing mode, in which a single helical perturbation produces magnetic islands at both surfaces. The external Euler–Lagrange equation takes the form

$$\nabla_\perp^2 \psi_1 = \left(\frac{3F'}{rF} + \frac{F''}{F} \right) \psi_1. \tag{7.67}$$

Suppose the resonant surfaces S_1, S_2 are close to one another, with $S_2 - S_1 = \Delta \ll a$. Then $F'' \simeq F/\Delta^2$ is large and an appropriate solution is $\psi_1 = F$ for $S_1 < r < S_2$ and zero outside. Then we have

$$\delta W = \frac{\pi}{2} \int_{S_1}^{S_2} dr r^3 \frac{F''}{F} < 0 \tag{7.68}$$

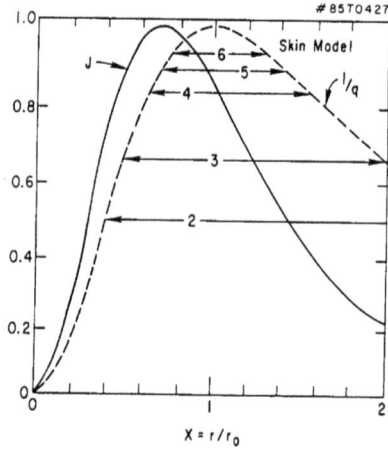

Fig. 7.14   Unstable domains for a hollow profile.

and thus this configuration is always unstable (see Eq. 7.12). If the surfaces are not close together it is necessary to solve the Euler–Lagrange equation to find $\Delta'_1(\gamma_1)$ and $\Delta'_2(\gamma_2)$ through the analysis of Sec. 7.2, and choose that combination of solutions which makes $\gamma_1 = \gamma_2$. If one surface (S_2) falls in a very conducting region then $\psi(S_2) = 0$ and stability is determined by Δ'_1 alone.

A hollow current profile model, $j \sim (r^2/r_0^2)/(1 + r^2/r_0^2)^3$, has been investigated in detail (Carreras *et al.*, 1979), and the unstable domains for various modes are shown in Fig. 7.14.

7.4   TOROIDAL AND SHAPING EFFECTS

The effect of toroidicity was first examined by Glasser *et al.* (1976). They found that the toroidal curvature, in the presence of plasma compressibility, can stabilize the tearing mode, producing a threshold value for the value of Δ' which appears in the cylindrical theory. The dispersion relation takes the form

$$\Delta = f \left(\frac{\gamma}{\eta^{3/5}} \right)^{5/4} \left[1 - d \left(\frac{\eta^{1/3}}{\gamma} \right)^{3/2} \right] \qquad (7.69)$$

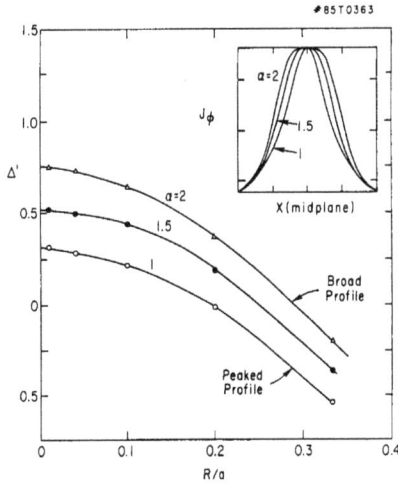

Fig. 7.15   The stabilizing effect of aspect ratio.

where Δ is a generalization of the cylindrical Δ' for toroidal geometry, stabilizing for $d < 0$. For sufficiently large Δ, γ is real and there are two roots, the large one being the usual tearing mode. As Δ decreases, the two roots coalesce with $\gamma > 0$ and then form a complex conjugate pair. The magnitude of the real frequency (Im γ) increases and the magnitude of the growth rate Re γ decreases until at some point $\Delta_c > 0$, Re $\gamma = 0$ and the mode becomes stable. The value of Δ_c depends on the magnitude of the resistivity, and the stabilizing toroidal effect is most pronounced at high temperatures (low η).

The effect of toroidal coupling, and noncircular plasma cross section has been investigated numerically, through a minimization of δW to determine the Δ' at the rational surfaces. Typically the helicities $m/n = 2/1,\ 1/1,\ 3/1$ are considered, and the Δ' must be found at all the rational surfaces. For the simple case given by $q(\psi) = 1.1 + 1.8(\psi/\psi_a)^\alpha$, only a single rational surface occurs, at $q = 2$, and the analysis is relatively simple. The toroidal effects are found to be stabilizing, as shown in Fig. 7.15. The effect of ellipticity is shown in Fig. 7.16.

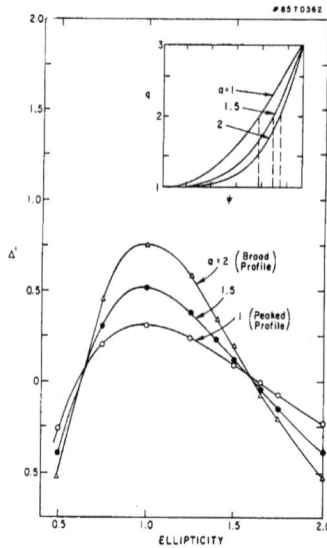

Fig. 7.16   The stabilizing effect of ellipticity.

7.5   THE RESISTIVE SURFACE KINK MODE

The ideal surface kink, and the resistive tearing mode, can be viewed as
two different limits of the same instability. In the case of the ideal kink, the
magnetic island must be located outside the plasma in the surrounding vacuum region, and is felt only as a distortion of the plasma-vacuum boundary.
If a resistive plasma is introduced in the vacuum region, the magnetic reconnection is resistively limited and the growth rate of the mode accordingly
modified. A simple derivation of the linear growth rate for the resistive
kink mode is presented here, which makes obvious this close connection
(Pogutse and Yurchenko, 1977).

Consider an equilibrium with a dense plasma and a constant current
density extending to $r = a$, and a much less dense, resistive plasma, and
zero current, for $a < r < b$ with a conducting wall at $r = b$. The safety
factor is then constant for $r < a$ and parabolic for $r > a$. The singular
surface is assumed to lie in the low-density resistive plasma. For a constant
current profile equilibrium the solution to the Euler–Lagrange equation
(Sec. 7.2) is $\psi_1 = cr^{\pm m}$. Thus, solving for $\psi_1(r)$ reduces to matching these

solutions in the three regions $r < a$, $a < r < r_s$, and $r_s < r < b$. To find the discontinuity in ψ_1' at $r = a$ integrate Eq. 7.28 across the surface $r = a$, relating the discontinuity in ξ' to that in ψ_1',

$$\gamma^2 \tau_A^2 (r\xi)' |_-^+ = \frac{m^2}{a}(-\psi_0' \psi_1'|_-^+ + \psi_1 j_0|_-^+).\tag{7.70}$$

Away from the singular surface we have $\psi_1 - \xi r F = 0$, with $F = 1/q - n/m$. Thus also

$$
\begin{aligned}
\psi_1' - (\xi r)' F = 0 \qquad & r < a \\
\psi_1' - (\xi r)' - \xi r F = 0 \qquad & r > a
\end{aligned}\tag{7.71}
$$

and $F' = -2/(rq)$ everywhere outside the current column. The discontinuity in ψ_1' at the singular surface is given by Δ'. Writing the solution as $(r/a)^m$ for the interval $r < a$, $c_1(r/a)^m + c_2(r/a)^{-m}$ for the interval $a < r < r_s$ and $c_3(r/r_s)^m + c_4(r/r_s)^{-m}$ for $r_s < r < b$, the matching conditions at $r = a$ give

$$c_2 = (m - nq_a)^{-1}\left(1 + \frac{\gamma^2(\tau_+^2 - \tau_-^2)}{2F} + \frac{\gamma^2\tau_+^2}{q_a^2 F^2}\right)\left(1 + \frac{\gamma^2\tau_+^2}{q_a^2 F^2}\right)^{-1}\tag{7.72}$$

where $\tau_\pm$ are the characteristic Alfvén times of the inner and outer regions. Similarly from matching at $r = r_s$ we find

$$\Delta' = \frac{2m}{r_s}\left[\frac{1}{1 - (\frac{b}{r_s})^{2m}} - \frac{1}{1 + \frac{c_2}{1 - c_2}(\frac{a}{r_s})^{2m}}\right].\tag{7.73}$$

The growth rate is given by Δ' from the analysis of Sec. 7.2. The two limits are readily obtained. First consider the ideal limit with no plasma in the exterior region $r > a$. Letting $\Delta' = 0$ and $\tau_+ = 0$, and using $\tau_- = \tau_A$, we find

$$\gamma^2 \tau_A^2 = 2(m - nq_a) - \frac{2(m - nq_a)^2}{1 - (a/b)^{2m}},\tag{7.74}$$

which is the growth rate of the kink mode, Sec. 6.8. In this limit the island growth is impeded by the inertia of the central core rather than the resistive tearing.

The tearing-mode limit is found by setting the density discontinuity equal to zero, giving $C_2 = (m - nq_a)^{-1}$, independent of γ. Substituting this expression into the equation for Δ' then gives Δ' for the step current profile,

$$\Delta' = \frac{2m}{r_s} \left[\frac{1}{(a/r_s)^{2m}\frac{1}{1-m+nq_a} - 1} - \frac{1}{(b/r_s)^{2m} - 1} \right]. \tag{7.75}$$

From this expression it is clear that the conducting wall is stabilizing also for the tearing mode, Δ' decreases with decreasing b, and as b approaches r_s the stabilizing term becomes infinite. For no conducting wall, $b \to \infty$, the mode is unstable only for

$$m - 1 < nq_a < m - 1 + \left(\frac{a}{r_s}\right)^{2m}. \tag{7.76}$$

The domains of instability are similar to those for the kink mode found in Sec. 6.8, see Figs. 6.2, 6.4. Note that for nq_a slightly larger than $m - 1$ the mode is unstable even with a conducting wall; the conducting wall shifts the right hand edge of the stability domain.

In Fig. 7.17 is shown the growth rate γ for the resistive surface kink mode as a function of $S = \tau_R/\tau_A$. Note that the resistive growth rate can be either larger or smaller than the ideal kink growth rate, but that the mode growth rate is smaller than either.

The close relation between the surface kink mode and the resistive tearing mode is intuitively useful as it indicates that, for fairly sharply limited current profiles and with rather large resistivity in the region outside the current channel, the nonlinear behavior of the tearing mode can be expected to be similar to that of the kink mode (Rosenbluth *et al.*, 1976).

7.6 OPTIMIZED PROFILES

The magnitude of the destabilizing force driving the tearing mode can be effectively modified by changing the current profile locally, in the vicinity of the rational surface. To see this, consider the effect of locally flattening the current profile at a point $r = r_0$, for a width $\Delta r = W_f$. Calculate the modification to Δ' produced by this flattening. The perturbation satisfies

$$\nabla_\perp^2 \psi_1 = \frac{1}{\psi_0'} \frac{dj_0}{dr} \psi_1. \tag{7.77}$$

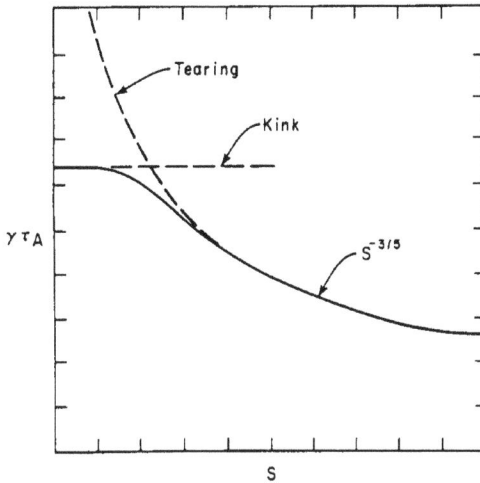

Fig. 7.17   Growth rate for the kink-tearing mode.

Integrating this equation across Δr, and using the constant ψ approximation, we find that the change in ψ_1' due to the flattening is

$$\delta\psi_1' = -\left(\frac{1}{\psi_0'}\frac{dj_0}{dr}\right)W_f. \tag{7.78}$$

For a normal profile $(dj_0/dr < 0)$ we see that Δ' is decreased if the flattening occurs at $r_0 < r_s$ where $\psi_0' > 0$, and increased if $r_0 > r_s$. Note that for $r_0 < r_s$ only ψ_-' is changed, and for $r > r_s$ only ψ_+' is changed, with $\Delta' = (\psi_+' - \psi_-')/\psi_1$, and that the maximum effect is obtained if the flattening is adjacent to the singular surface, where ψ_0' is smallest. Unfortunately, if the value of the current density on axis is held fixed, introducing a flattening just inside one rational surface will introduce steepening elsewhere. Nevertheless, since only the lowest m, n values give strongly unstable modes, it is possible to find profiles stable to all tearing modes (Glasser *et al.*, 1977). Shown in Fig. 7.18 is an example of a stable profile with $q = 2.7$ at the plasma edge.

It is tempting to argue that stable discharges consist of profiles the shape of which is dominated by the Δ' values, and the tearing modes undoubtedly play some role in the "profile consistency" hypothesis (Coppi, 1980). However, for the tearing modes to remain innocuous it is not neces-

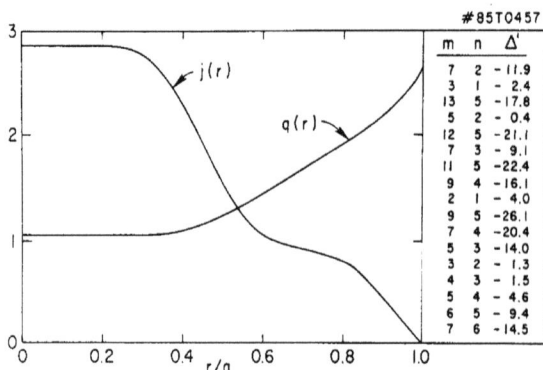

Fig. 7.18 A tearing mode stable profile.

sary that they be stable, but only that they saturate at small island width, and this requirement is much weaker than that of stability, leading to a much broader class of allowed profiles. This saturation mechanism will be discussed in Chap. 10.

7.7 THE RIPPLING MODE

This mode is caused by the local channeling of the current through ripples in the resistivity. Ohm's law is modified by the resistivity change to

$$\eta_0 j_1 = -\eta_1 j_0 + \vec{v} \times \vec{B} \tag{7.79}$$

where η_1 is given by the convection, $\eta_1 = (\vec{v}/\gamma) \cdot \nabla \eta_0$. Within the reconnection layer $\dot{E} = 0$ and the η_1 term gives rise to a force

$$\vec{F} = \vec{j}_1 \times \vec{B} = (\vec{j}_0 \times \vec{B}) \vec{v} \cdot \nabla \eta_0 / (\gamma \eta_0), \tag{7.80}$$

which changes sign at the singular surface. Thus F is stabilizing on the side of higher resistivity and destabilizing on the side of lower resistivity. An unstable mode results if the region of decoupled flow lies on the lower resistivity side and has a width δ such that the driving power

$$vF \simeq v^2 \eta_0' (B')^2 \delta / \eta_0 \gamma \tag{7.81}$$

is the rate of work done on the fluid.

Equating this to the expression in Sec. 7.1 we find $\delta \simeq \eta_0'/\gamma$ for the reconnection domain width. Substituting this into the relation between δ, γ derived in Sec. 7.1 we find

$$\gamma \sim \left[\frac{(\eta')^2 kB'}{(\eta\rho)^{1/2}}\right]^{2/5} \sim S^{2/5}\eta. \tag{7.82}$$

This mode is thus most rapidly growing for small wavelengths. It is not responsible for large scale plasma motion but may be of relevance for the understanding of anomalous transport.

7.8 THE RESISTIVE INTERCHANGE MODE

This mode occurs in the presence of a mass density gradient, and a vertical, y-directed "gravitational" field, which can also be due to the field line curvature. The driving force is of the form

$$F = v_y \rho' g/\gamma, \tag{7.83}$$

which is destabilizing if g points toward decreasing density. Comparing this with Sec. 7.1 gives

$$\delta = \frac{(\rho' g \eta)^{1/2}}{B'\sqrt{\gamma}} \tag{7.84}$$

and substituting we find

$$\gamma = \left[\frac{kg\rho'}{B'}\left(\frac{\eta}{\rho}\right)^{1/2}\right]^{2/3} \sim S^{-1/3}/\tau_A. \tag{7.85}$$

This mode is also most unstable for small wavelengths. As seen in Sec. 6.14, localized interchange is a limiting case of the ballooning mode, which also has a resistive version. This mode is discussed from the point of view of the ballooning representation in Sec. 7.9.

7.9 RESISTIVE BALLOONING

As seen in Sec. 7.1, resistive effects are enhanced by large gradients in a singular layer about a rational surface. The behavior of the ballooning mode eigenfunction at small x is reflected in its behavior at large θ in the ballooning representation, as noted in Sec. 6.14. The large gradients are associated with the twisting of the field lines in the presence of shear and enter the problem through the secularities in $|\nabla\beta|^2$ and κ_W. They introduce a new class of pressure driven resistive ballooning modes, with growth rates scaling as fractional powers of the resistivity, which are unstable at lower values of β than ideal ballooning modes. Thus the introduction of resistive effects modifies the ballooning equation for large θ, with the transition from ideal to resistive occuring at $\theta \simeq (S/n^2)^{1/3}$ with $S = \tau_R/\tau_A$, and n the toroidal eigenvalue. The solutions in the large θ resistive domain are exponentially decreasing as $\theta \to \infty$ and for θ near the transition value match to the large θ limit of the ideal ballooning equations, $\Phi \to \theta^p$ with

$$p = -\frac{1}{2} \pm \left(\frac{1}{4} - D\right)^{1/2}. \tag{7.86}$$

For a description of the resistive modifications see Chance *et al.* (1978), Glasser *et al.* (1980), and Connor *et al.* (1985). The modification of the usual ideal ballooning mode due to resistivity is to make it more unstable, and to lower the threshold in β. Shown in Fig. 7.19 is the modification of the ideal stability boundary in the $s - \alpha$ plane due to resistivity (Sykes *et al.*, 1987). The stability boundary for the resistive mode is seen to be very close to the ideal boundary in the second stability domain. Recall that Δ' behaves as $1/\delta W$ near ideal marginal stability (see Eq. 7.66). In the second stability domain Δ' very rapidly drops from ∞ to zero. This does not occur in the first stability domain, and resistive effects make the entire first domain unstable. However, the shifted circular flux surface equilibrium model used in the analysis does not possess a magnetic well, so this work omits an important stabilizing effect. In fact, in a realistic equilibrium, the resistive ballooning modes become unstable only as the first ideal stability boundary is approached.

The resistive ballooning modes are a resistive version of an ideal mode, for which the perturbed helical flux $\psi_{m,n}$ must vanish at a rational surface m/n. Thus the modes are distinguished from tearing modes by having odd ballooning mode parity about the rational surface, as opposed to even, or tearing mode parity. In addition, new unstable branches appear which are, however, highly localized, and although of interest for the modification of transport, are probably not as important in determining a hard β limit to tokamak operation as are the ideal modes.

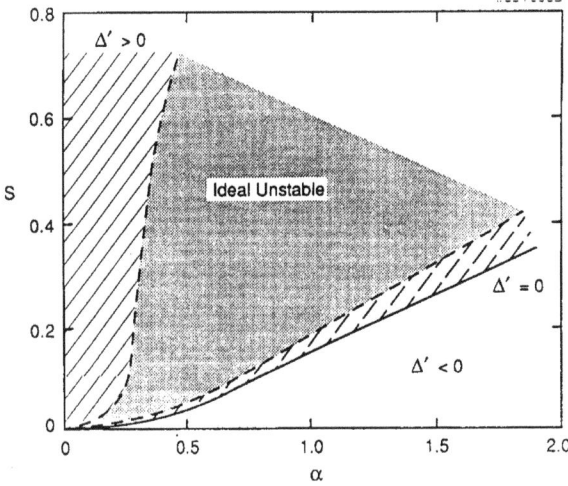

Fig. 7.19   Stability domain for the resitive ballooning mode.

The fastest growing mode, in the vicinity of the ballooning threshold, $\beta_p \simeq 1$ has a growth rate (Strauss, 1981)

$$\gamma = \left(\frac{n^2}{S}\right)^{1/3} \left[\frac{R\beta q^2}{r\rho L_p}\right]^{2/3} \frac{1}{\tau_A} \tag{7.87}$$

with L_p the pressure scale length, $L_p^{-1} = -[dp/d\psi]/p$, and ψ is a flux surface label $0 < \psi < 1$.

7.10   DIAMAGNETIC ROTATION

Inclusion of the effect of ion gyro radius and diamagnetic frequencies, as in the case of resistivity, modifies the analysis only in the singular layer, where $q \simeq m/n$. Away from this domain the ideal equations apply. In the singular layer a two fluid model must be used, and the electron and ion velocities are Doppler shifted by their diamagnetic frequencies, which can be written

$$\omega_{*i} = \frac{-c}{neBr} \frac{dp_i}{dr} = \frac{-1}{m_i nr\omega_{ci}} \frac{dp_i}{dr}$$

$$\omega_{*e} = \frac{c}{neBr} \frac{dp_e}{dr} = \frac{1}{m_e nr\omega_{ce}} \frac{dp_e}{dr}. \tag{7.88}$$

Thus Eq. 7.26 becomes, for electrons and ions,

$$u_{e,i} = \frac{r\xi(\vec{r})}{m}(\omega - \omega_{*e,i})e^{i(m\theta - n\phi - \omega t)}. \tag{7.89}$$

First consider the case of the $m = 1$ mode. The inclusion of the diamagnetic effects on the $m = 1$ mode was carried out by Bussac *et al.* (1977) and Ara *et al.* (1978) generalizing the work of Coppi *et al.* (1976). The analysis proceeds by again introducing $\chi = x\psi' - \psi$, resulting in a second order equation similar to Eq. 7.57, for $\chi(z)$ with

$$x = \left[\frac{-i\omega\tau_A}{s^2n^2S}\left(\frac{\omega - \omega_{*i}}{\omega - \hat{\omega}_{*e}}\right)\right]^{1/4} r_s z, \tag{7.90}$$

$$\frac{d^2\chi}{dz^2} - \frac{2}{z}\frac{d\chi}{dz} - (z^2 + \Lambda^{3/2})\chi = -z^2\chi_\infty, \tag{7.91}$$

with $\hat{\omega}_{*e} = \omega_{*e} + [.71c/(eBr)]dT_e/dr$ and

$$\Lambda = \frac{-i[\omega(\omega - \hat{\omega}_{*e})(\omega - \omega_{*i})]^{1/3}}{\gamma_R}, \tag{7.92}$$

with γ_R the cylindrical geometry $m = 1$ growth rate, $\gamma_R = \tau_R^{-1/3}\tau_H^{-2/3}$. The eigenvalue equation then follows from Eq. 7.56 and gives

$$\gamma_I + \frac{8i[\omega(\omega - \omega_{*i})]^{1/2}\Gamma((\Lambda^{3/2} + 5)/4)}{\Lambda^{9/4}\Gamma((\Lambda^{3/2} - 1)/4)} = 0 \tag{7.93}$$

which reduces to Eq. 7.63 in the case $\hat{\omega}_{*e} = \omega_{*i} = 0.$*

This dispersion relation thus depends parametrically on the four frequencies $\hat{\omega}_{*e}, \omega_{*i}, \gamma_I, \gamma_R$. Typically $\hat{\omega}_{*e} < 0$, and $\omega_{*i} > 0$ are of the same order.

First consider the ideal limit, $\gamma_R \to 0$, and use $\Lambda \gg 1$ to take the large argument limit of the Γ functions. This gives

$$i\gamma_I - [\omega(\omega - \omega_{*i})]^{1/2} = 0 \tag{7.94}$$

and thus the frequency is $\omega = \omega_{*i}/2 \pm (\omega_{*i}^2/4 - \gamma_I^2)^{1/2}$, with the minus sign for $\gamma_I > 0$ and the plus sign for $\gamma_I < 0$. For large positive γ_I the mode is growing with $\gamma \simeq \gamma_I$ (kink mode branch) and as $\gamma_I \to 0+$ the solution enters the cut at $\omega = 0$. For large negative γ_I the mode is damped (ion

---

*For the solution to the differential equation see Sec. 16.5 in White, (2010).

branch) and for $\gamma_I \to 0-$ the solution enters the cut at $\omega = \omega_{*i}$. In the ideal limit these two branches coalesce at $\omega = \omega_{*i}/2$ for $\gamma_I = \omega_{*i}/2$, and the addition of a small amount of resistivity is necessary to separate them. Note that for any value of $\gamma_R > 0$ the solution near $\gamma_I = 0$ is not valid since $\Lambda \to 0$, and the large argument expansion of the Γ functions cannot be used. Now consider the resistive limit, $\gamma_R > 0$. First take $\gamma_R \gg \omega_{*i}, \omega_{*e}$. In this case the solution is not significantly different from that found in Sec. 7.2.2, with the growth rate given by $\gamma = \gamma_R$ if $\gamma_R \gg \gamma_I$. The modification of the frequency due to $\hat{\omega}_{*e}, \omega_{*i}, \gamma_I$ can be calculated perturbatively, giving

$$\omega = \frac{\hat{\omega}_{*e} + \omega_{*i}}{3} + i \left[\gamma_R + \frac{4\gamma_I}{3\sqrt{\pi}} \right]. \tag{7.95}$$

If $\gamma_I \gg \gamma_R$ the solution reverts to the ideal case with $\Lambda \gg 1$ and the growth rate is γ_I. If γ_I is large and negative the solution is given by $\Lambda \ll 1$ and reduces to Eq. 7.66, provided $\gamma > \omega_{*i}, |\hat{\omega}_{*e}|$.

Now consider $\gamma_R \ll \omega_{*i}, \hat{\omega}_{*e}$. First take $\gamma_I = 0$. In this case the solution is $\Lambda = 1$ and the only possiblities are $\omega = \omega_{*i} + \delta$, $\hat{\omega}_{*e} + \delta$, δ with $\delta \ll \omega_{*i}, \hat{\omega}_{*e}$. Refer to these as the ion, electron, and kink mode branches, respectively. Substituting we find that only the kink branch is unstable, with growth rate $\gamma = \gamma_R^3/|\omega_{*i}\hat{\omega}_{*e}|$.

To find the range of γ_I in which each branch is unstable, look for threshold solutions, again with $\omega = \omega_{*i} + \delta$, $\hat{\omega}_{*e} + \delta$, δ. For the kink mode branch one threshold is found, with the mode being unstable for

$$-3\sqrt{2/\pi} \frac{\gamma_R^{3/2}}{\sqrt{|\hat{\omega}_{*e}|}} < \gamma_I, \tag{7.96}$$

and the frequency at the threshold is

$$\omega = -\frac{\pi^2}{64} \frac{\gamma_R^3}{|\omega_{*i}\hat{\omega}_{*e}|}. \tag{7.97}$$

There is no threshold for the ion branch, it is thus always stable. For the electron branch there is again a single threshold, the mode unstable for

$$\gamma_I < \frac{-3\sqrt{3\pi}}{2} [\hat{\omega}_{*e}(\hat{\omega}_{*e} - \omega_{*i})]^{1/2}, \tag{7.98}$$

and the threshold frequency is

$$\omega = \hat{\omega}_{*e} - \frac{\gamma_R^3}{9\hat{\omega}_{*e}(\hat{\omega}_{*e} - \omega_{*i})}. \tag{7.99}$$

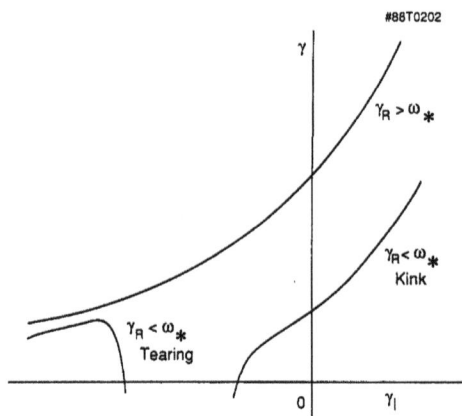

Fig. 7.20   Growth rate of the kink and tearing mode branches.

Thus for the case $\gamma_R \ll \omega_{*i}$, $\hat{\omega}_{*e}$, there is an unstable electron branch for negative γ_I, a stable gap, then the zero frequency branch near $\gamma_I \sim 0$. In Fig. 7.20 is shown the growth rate as a function of γ_I for the kink and electron (tearing) branches, for $\gamma_R > \omega_*$ and $\gamma_R < \omega_*$.

Inclusion of the ion diamagnetic effects for the tearing mode ($m \geq 2$) was first carried out by Coppi (1964). See also the treatment by Hazeltine and Ross (1978) for an analysis including both ion and electron diamagnetic effects. The dispersion relation obtained is

$$\omega(\omega - \omega_{*T})^3(\omega - \omega_{*i}) = i\gamma_t^5 \qquad (7.100)$$

where $\omega_{*T} = -n'mc(T_i+T_e)/(nreB)$, ω_{*i} is the ion diamagnetic frequency, and γ_t is the tearing mode growth rate calculated in Sec. 7.2. For $\gamma_t \gg \omega_{*T}$, ω_{*i} the solution reduces to $\omega \simeq i\gamma_t+(3\omega_{*T}+\omega_{*i})/5$. For $\gamma_t \ll \omega_{*T}$, ω_{*i} the mode is also growing with $\omega \simeq i\gamma_t^5/(\omega_{*T}^3\omega_{*i})$. Thus the diamagnetic terms are stabilizing, although there is no threshold for finite values of ω_{*T}, ω_{*i}.

7.11   Problems

Consider the tearing stability of a slab model with an equilibrium magnetic field $\vec{B}_0 = B_z \hat{z} + b \tanh(x/L)\hat{y}$, where B_z and b are constants with $b \ll B_z$. (Private communication, Allan Reiman, 2005)

1. What is the equilibrium current density? What is the tearing mode equation for the exterior region?

2. The exterior tearing mode equation can be solved analytically in this case. The solutions take the form

$$\psi_1(x) = \exp(\pm kx)[1 + \alpha \tanh(x/L)],$$

where

$$\psi = \psi_0 - \epsilon \psi_1 \cos(kz)$$

and α is a constant that must be chosen appropriately. Find the appropriate values of α for which this expression gives the two solutions to the equation.

3. Calculate Δ', determining the linear stability of the tearing mode, applying the appropriate boundary conditions at $|x| \to \infty$. Apply periodic boundary conditions at $y = \pm l_y/2$ (i.e. $\vec{B}(x, y + l_y) = \vec{B}(x, y)$). Under what conditions are the allowable perturbations unstable? Is the mode most unstable for perturbations of short wavelength or long wavelength? Is the mode most unstable for sharp current gradients or gentle current gradients? What is the maximum value of Δ'?

4. Show that in the large m limit the tearing mode is stable for any current profile. Calculate Δ'.

5. Profile consistency has been suggested to result from an energy principle. Extremize the total magnetic energy using the constraints of fixed toroidal current and fixed total poloidal flux, *i.e.* minimize

$$\int (B^2/2 + \alpha \vec{j} \cdot \nabla\phi + \lambda \vec{B} \cdot \nabla\theta)d^3x.$$

Use $\vec{B} = \nabla\phi - \nabla \times (\psi_p \nabla\phi)$, let $\vec{j} \cdot \nabla\phi = \mu(\psi_p)$ and solve for the toroidal current profile $\mu(\psi_p)$

6. Consider the dispersion relation for the $m = 1$ mode, Eq. 7.93. For $\gamma_R \gg \omega_{*i}, \omega_{*e}, \gamma_I$ the mode is purely growing with $\gamma \simeq \gamma_R$. Find the effect of nonzero $\omega_{*i}, \omega_{*e}, \gamma_I$ on this solution perturbatively.

7. Calculate the frequencies and values of γ_I at the thresholds for the electron and kink mode branches of the $m = 1$ mode for the high Reynolds number case, $\gamma_R \ll \omega_{*i}, \omega_{*e}$.

7.12   References

Basic

- Biskamp, D., Phys. Fluids 29, 1520 (1986).
- Bondeson A., and J. R. Sobel, Phys. Fluids 27, 2028 (1984).
- Carreras, B., H. R. Hicks, and B. V. Waddell, Nucl. Fusion 19, 583 (1979).
- Coppi, B., J. M. Greene, J. L. Johnson, Nucl. Fusion 6, 101 (1966).
- Furth, H. P., J. Killeen, and M. N. Rosenbluth, Phys. Fluids, 6, 459 (1963).
- Furth, H. P., P. H. Rutherford, and H. Selberg, Phys. Fluids 16, 1054 (1973).
- Glasser, A. H., H. P. Furth, and P. H. Rutherford, Phys. Rev. Lett. 38, 234 (1977).
- Ivanov, N. V., A. M. Kakurin, and A. N. Chudnovskii, Sov. J. Plasma Phys. 10, 38 (1984).
- Mirnov, S. V., I. B. Semenov, Sov. Phys. JETP 33, 1134 (1971).
- Park, W., D. A. Monticello, and R. B. White, Phys. Fluids 27, 137 (1984).
- Parker, E. N., Astrophys. Journal, Suppl. Ser. 77, 8:177 (1963).
- Petschek, H. E., in Symposium on Physics of Solar Flares, edited by W. N. Hess, SP-50 (1964).
- Shafranov, V. D., Zh. Tekh. Fiz. 40, 240 (Sov. Phys. Tech. Phys. 15, 175) (1970).
- Sweet, P. A., in Proc. of the International Astronomical Union Symposium on Electromagnetic Phenomena in Cosmical Physics, No. 6 (Stockholm, 1956) p. 123.
- Wesson, J. A., Nuclear Fusion, 18, 87 (1978). (Review paper).
- White, R. B., Handbook of Plasma Physics, edited by M. N. Rosenbluth and R. Z. Sagdeev (North Holland, Amsterdam 1983) Vol. 1, p. 612. (Review paper).

Profile consistency

- Biskamp, D., Comments on Plasma Physics and Controlled Nuclear Fusion 10, 165 (1986).
- Coppi, B., Comments on Plasma Physics and Controlled Nuclear Fusion 5, 261 (1980).

M = 1 mode

- Ara, G., B. Basu, B. Coppi, G. Laval, *et al.*, Ann. Phys. 112, 443 (1978).
- Bussac, M. N., R. Pellat, D. Edery and J. L. Soule, Phys. Rev. Lett., 35, 1638 (1975).
- Bussac, M. N., D. Edery, R. Pellat, and J. L. Soule, Proc. of the 6th. International Conference on Plasma Physics and Controlled Nuclear Fusion Research, Berchtesgaden (IAEA, Vienna, 1977) Vol. 1, p. 607.
- Coppi, B., R. Galvao, R. Pellat, M. N. Rosenbluth *et al.*, Sov. J. Plasma Phys. 2, 533 (1976).

Toroidal

- Connor, J. W. , S. C. Cowley, R. J. Hastie, T. C. Hender *et al.*, Phys. Fluids 31, 577 (1988).
- Glasser, A. H., J. M. Greene, and J. L. Johnson, Phys. Fluids 19, 567 (1976).
- Grimm, R. C., R. L. Dewar, J. Manickam, in Proc. of the Ninth International Conference on Plasma Physics and Controlled Nuclear Fusion Research, Baltimore, 1982 (IAEA, Vienna, 1983).

Kinetic Effects and Fast Reconnection

- Aydemir, A., Phys. Fluids B 4, 3469 (1992).
- Biskamp, D. and J. Drake, Phys. Rev. Lett., 73, 971 (1994).
- Coppi, B., Phys. Fluids 7, 1501 (1964).
- Cowley, S., R. M. Kulsrud, and T. S. Hahm, Phys. Fluids 29, 3230 (1986).
- Cafaro, E., D. Grasso, F. Pegoraro, F. Porcelli *et al.*, Phys. Rev. Lett. 80, 4430 (1998).
- Drake, J. F., and Y. C. Lee, Phys. Rev. Lett. 39, 453 (1977).
- Hazeltine, R. D., D. Dobrott, T. S. Wang, Phys. Fluids 18, 1778 (1975).
- Hazeltine, R. D., and D. W. Ross, Phys. Fluids 21, 1140 (1978).
- Hazeltine, R. D., C. T. Hsu, and P. J. Morrison, Phys. Fluids 30, 3204 (1993).
- Kleva, R., J. Drake, and F. L. Waelbroeck, Phys. Plasmas 2, 23 (1995).

- Levinton, F., L. Zakharov, S. H. Batha, J. Manickam, *et al.*, Phys. Rev. Lett. 72, 2895 (1994).
- Monticello, D. A., and R. B. White, Phys. Fluids 23, 366 (1980).
- Ottaviani, M., and F. Porcelli, Phys. Rev. Lett. 71, 3802 (1993).
- Waelbroeck, F. L., Phys. Fluids B 1, 2372 (1989).
- Wang, X. and A. Battacharjee, Phys. Rev. Lett. 70, 1627 (1993).
- Wesson, J. A., Nuclear Fusion 30, 2545 (1990).
- Zakharov, L., and B. Rogers, Phys. Fluids B 4, 3285 (1992).
- Zakharov, L., B. Rogers, and S. Migliuolo, Phys. Fluids B 5, 2498 (1993).

Resistive ballooning

- Chance, M. S., R. L. Dewar, E. A. Frieman, A. H. Glasser *et al.*, in Plasma Physics and Controlled Nuclear Fusion Research, in Proc. of the 7th. International Conference, Innsbruck (1978).
- Connor, J. W., L. Chen, Phys. Fluids 28, 2201 (1985).
- Connor, J. W., R. J. Hastie, and T. J. Martin, Plasma Physics and Controlled Fusion 27, 1509 (1985).
- Glasser, A. H., M. S. Chance, and R. L. Dewar, Nucl. Fusion 20, 105 (1980).
- Gribkov, V. M., D. Kh. Morozov, O. P. Pogutse and E. I. Yurchenko, Plasma Physics and Controlled Nuclear Fusion Research, (IAEA, 1981).
- Strauss, H. R., Phys. Fluids 24, 2004 (1981).
- Sykes, A., C. M. Bishop and R. J. Hastie, Plasma Physics and Controlled Fusion 29, 719 (1987).

Kruskal–Newton diagrams

- Newton, Isaac, Methods of series and fluxions, in The Mathematical Papers of Isaac Newton, edited by D. T. Whiteside, Volume III 1670–1673, pp 50–71, (Cambridge University Press, 1969).
- Kruskal M., Asymptotology, in Mathematical Models in Physical Sciences, edited by S. Drobot and P. A. Viebrock, Prentice-Hall, New Jersey (1963).
- White, R. B., Asymptotic Analysis of Differential Equations, (Imperial College Press, London 2010).

Chapter 8

Mode-Particle Interaction

8.1 INTRODUCTION

High energy particles injected into a confined plasma have been considered for some time as a stabilizing influence. When the hot particles precess much faster than the frequency of the mode in question, they average the potentials associated with the perturbation, and also form a diamagnetic well, giving rigidity to the plasma, and stabilizing the equilibrium. Deliberate introduction of hot trapped particles was used in mirror machines (Berk, 1976), field reversed hot electron rings (Sudan and Rosenbluth, 1976), ion rings (Fleischmann, 1975), astron (Lovelace, 1976), and the Elmo Bumpy Torus (Antonson *et al.*, 1983). Conner *et al.* (1982) found that an isotropic population of energetic ions would have a stabilizing effect on ballooning modes in the limit of rapid precession frequency, larger than the bounce frequency. Stabilization by a population of trapped particles in the opposite limit, with the precession frequency less than the bounce frequency, was analyzed by Rosenbluth *et al.* (1983). Both calculations, however, assumed the precession frequency large compared to the mode frequency.

The first experiments involving large scale near perpendicular neutral beam injection into a high β plasma on PDX provided a surprise (PDX Group, 1983). A very large amplitude $n = 1$ mode, with a real frequency near 20 KHz, and a few low m values, appeared in bursts with a repetition rate of a few milliseconds. The mode was called the fishbone because of the structure of the magnetic signals observed, as shown in Fig. 8.1. Subsequently, it was recognized that the telltale high frequency signals had been observed, although they were less pronounced, on other tokamaks (ISX-B

271

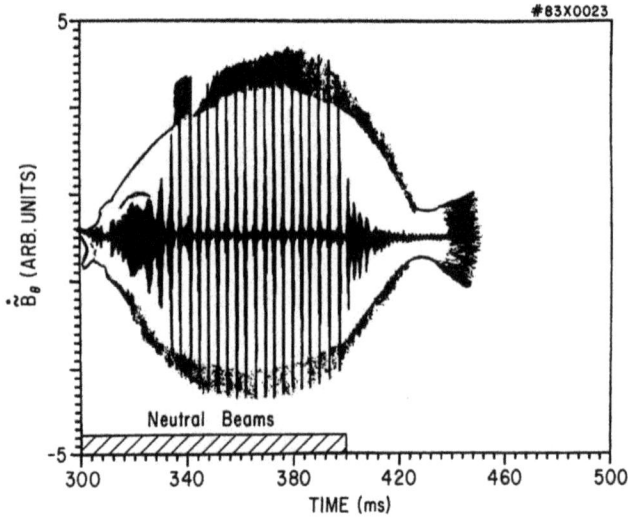

Fig. 8.1 Mirnov signal during fishbone mode, outline added by artist.

1982, DIII 1984, JFT 1981). During the bursts of mode activity a sharp drop in neutron emission indicated a loss of beam particles from the plasma, and charge exchange measurements on PDX subsequently confirmed a loss of as much as one third of the beam particles in each burst.

An analytical calculation and Monte Carlo simulation by White *et al.* (1983) showed that a rotating mode with $n = 1$ and $m = 1, 2, 3$ was capable of resonant ejection of trapped particles, provided that the mode frequency matched their toroidal precession rate. As seen in Fig. 8.2, it was noted that the plasma β and q profile were such that the plasma was marginally stable to the internal kink mode, although this mode, being purely growing, could offer no explanation for the real frequency, which appeared to be associated with the trapped particles. The mode structure obtained by resistive MHD simulations of an unstable internal kink mode agreed well with the Mirnov and soft X-ray signals associated with the mode, provided the experimental frequency was used in the simulations. The internal kink mode naturally has higher m values in a high β plasma because of the Shafranov shift, and the observed in-phase rotation of all m values indicated that one fundamental mode was driving the whole coupled system. Finally,

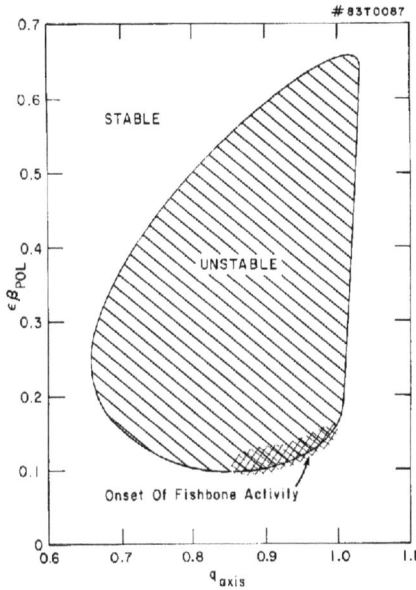

Fig. 8.2   Internal kink mode stability diagram.

it was shown by Chen *et al.* (1984) that a trapped particle population could destabilize the internal kink mode provided it was near its ideal MHD threshold, and that the excited mode would have a real frequency and growth rate determined by the trapped particle precession rate.

There are many fascinating details of this mode still under analysis. It is of particular interest because of the strongly coupled kinetic theory and MHD components, giving diagnostic tools for investigating both theories.

With the understanding of the fishbone mode, it was realized that there probably existed many other possibilities of exciting MHD modes which are normally stable by means of a high energy particle population. The most important are toroidal Alfvén modes, and there are several different branches, the most relevant being the shear Alfvén waves. In slab geometry the waves have a continuous spectrum, but due to toroidal bending there is also a spectrum of discrete modes existing inside the gaps in the continuum, as described in Sec. 6.16. In the continuum the modes are damped by dispersion, but the gap modes may be undamped except for Landau damping with the plasma, and hence are easily excited by particles with a velocity resonant with the wave. Another branch is the Kinetic Toroidal Alfvén Eigenmodes (KTAE). These modes have a frequency within the Alfvén continuum and would normally have a singular mode structure, but due to the interaction with high energy particles the mode can be discretized. The situation is further complicated by the fact that there can exist modes which are not present at all in the absence of a high energy particle population, and thus cannot be calculated perturbatively. There are refered to as Energetic Particle Modes (EPM), and are characterized by the frequency associated with the particle motion. For a description of these modes see Zonca and Chen (1996), and the description of a numerical code HINST (Gorelenkov, 1998), which gives a non-perturbative construction of the form of the wave functions.

Other symmetry breaking produces further gaps in the Alfvén continuum, for example an elliptically shaped plasma cross section (Betti and Freidberg, 1991) produces a gap and the modes are called the Elliptic Alfvén Eigenmodes (EAE). All such modes can be important, depending on their frequencies, especially if they are weakly damped by dispersion through the continuum and by Landau damping.

Important interaction between plasma waves and particles involves resonances. This follows since the guiding center motion is Hamiltonian, and as seen in Chap. 3, the orbits are closed in a poloidal cross section in the limit of axisymmetry. Without resonances there are only small perturbations of the existing KAM surfaces, which produce no topological changes in them, and thus there is no large effect on particle orbits. The resonance effects are most important for high energy particles, including fusion alpha particles, injected beam particles, and particles heated through resonance with waves launched from antennae. Even outside the approximations of guiding center theory, important modifications of particle orbits are by resonance effects, the most important of these being through waves which resonate with the cyclotron motion or its harmonics.

Consider an electromagnetic wave with a frequency much smaller than the gyro frequency and a wave length long compared to the gyro radius, so that the guiding center equations are a good approximation to the particle motion. Perturb the equilibrium field with a perturbation of the form $\delta \vec{B} = \nabla \times \alpha \vec{B}$ and also introduce an electric perturbation Φ. The particle motion is then described by Eqs. 3.93–3.96. We are primarily interested in motion across the flux surface, which can lead to particle loss, and changes in the particle energy. Consider a particular harmonic of the wave, of the form

$$\alpha = \alpha_{mn}e^{i\chi} \qquad \Phi = \Phi_{mn}e^{i\chi}. \tag{8.1}$$

with $\chi = n\zeta - m\theta - \omega t$. We then find

$$\frac{dE}{dt} = i[\rho_\parallel B^2 \alpha_{mn} - \Phi_{mn}]\omega e^{i\chi}, \tag{8.2}$$

$$\frac{d\psi_p}{dt} = -i\frac{mg + nI}{D}[\rho_\parallel B^2 \alpha_{mn} - \Phi_{mn}]e^{i\chi} - \frac{g(\rho_\parallel^2 B + \mu)}{D}\partial_\theta B. \tag{8.3}$$

Now carry out an average over one orbit transit (passing) or one bounce (trapped). To zero order in α and Φ the orbit closes and the $\partial_\theta B$ term integrates to zero. With orbit modifications due to alpha this term produces a correction of order $\rho^2 \alpha_{mn}$, which we neglect. We thus find that radial motion is directly related to energy change,

$$\frac{d\psi_p}{dt} = -\frac{(mg + nI)}{\omega D}\frac{dE}{dt} \tag{8.4}$$

and thus for positive frequency outward motion is coupled to energy loss. Use the large aspect ratio approximation to illustrate the results; the equations can easily be generalized to an arbitrary equilibrium. In a large aspect ratio circular device $D \simeq gq$, $g \simeq 1$, $I \ll 1$ giving

$$rdr \simeq -\frac{m}{\omega}dE. \tag{8.5}$$

Thus to move a distance of the order of the minor radius a, the energy change is (using $E = \rho^2/2$)

$$\frac{\Delta E}{E} \simeq -\frac{\omega}{\omega_0}\frac{a^2}{m\rho^2}. \tag{8.6}$$

For typical machine parameters the energy change produced in moving across the minor radius is small, and thus it is not practical to extract energy from alpha particles using waves with frequency small compared to the gyro frequency. Such a process would cause loss of alpha particles of large energy. For such a scheme, referred to as alpha channeling, it is necessary to also introduce waves with frequencies near the gyro frequency to attempt to reduce the energy without producing radial motion.

8.2 IDEAL DISPLACEMENT $\vec{\xi}$, α, $\delta\vec{B}$, AND POTENTIAL

The most rapidly growing instabilities in a plasma are the ideal modes, including interchange, kink, and ballooning modes. For this reason they are associated with some of the most violent events occuring in plasmas, both in laboratory fusion experiments and in astrophysical plasmas. There is a long history of theoretical analysis of stability to, and the calculation of growth rates for, ideal MHD modes. One of the most important properties of MHD plasmas is the condition that the magnetic field be frozen in the plasma, *i.e.* that there can be no changes in the topology of the magnetic flux surfaces. Instabilities are studied with two major methods, that of the use of an energy principle, whereby stability of a system is guaranteed provided the energy is increased by any arbitrary plasma displacement, and the initial value approach, whereby an initial perturbation is introduced and equations of motion developed to ascertain whether the instability will grow. In fact in both these methods using the perturbation to lowest order destroys the frozen in condition, the magnetic topology of the initial state is changed. To ensure topological invariance of the original flux surfaces the effect of the perturbation must be calculated to all orders, or a full Lagrangian treatment used to find the perturbed field. But normally in instability calculations or in the analysis of the effect of ideal modes on particle trajectories only the lowest order perturbed field is used.

There are two representations of ideal MHD modes used to describe instabilities regularly observed in toroidal magnetic confinement devices, to carry out theoretical analyses of stability and growth rates, and to analyze the effect of these modes on high energy particle populations. The first is $\delta\vec{B} = \nabla \times (\vec{\xi} \times \vec{B})$ with $\vec{\xi}$ the plasma displacement, and the second $\delta\vec{B} = \nabla \times \alpha\vec{B}$ with α a scalar function of position and time, and where $\vec{B}$ is the equilibrium field. The second representation is particularly useful for the analysis of guiding center equations for charged particles in a perturbed field. These two representations are related and they produce equivalent changes of the cross field perturbation, the most important for the analysis of resonances, but to lowest order both destroy the initial flux surface topology of the equilibrium.

The magnetic perturbation $\delta\vec{B} = \nabla \times \alpha\vec{B}$ with $\vec{B}$ the equilibrium field is convenient for representing modes in low β plasmas. For resistive modes it exactly represents the cross field magnitude of the perturbation, responsible for producing magnetic islands and most important for the production of resonances between high energy particles and MHD perturbations, and the guiding center equations for charged particles in this field are derived in Chap. 3. The function α is also simply related to the ideal MHD displacement $\vec{\xi}$. Using a time dependent α to describe the perturbation introduces an electric field parallel to $\vec{B}$ proportional to the mode frequency. But the rapid mobility of the electrons shorts out the parallel electric field felt by the ions; nonzero parallel electric field is forbidden in ideal MHD, so a potential must be introduced to cancel this field if the α form is to represent an ideal perturbation. The equilibrium field in a toroidal axisymmetric equilibrium has covariant and contravariant representations, given by $\vec{B} = (\nabla\zeta \times \nabla\psi)/q(\psi) + \nabla\psi \times \nabla\theta = g\nabla\zeta + I\nabla\theta + \delta\nabla\psi_p$, with $q(\psi)$ the field line helicity, ψ the toroidal flux, ψ_p the poloidal flux, θ and ζ poloidal and toroidal coordinates and ψ_p, θ, and ζ forming a right-handed coordinate system with Jacobian $1/\mathcal{J} = \nabla\psi \cdot (\nabla\theta \times \nabla\zeta)$. Contravariant bases for the coordinate system are given by $\vec{e}^\beta = \nabla\beta$ with $\beta = \psi$, θ, and ζ.

Expand α in a Fourier series of the form

$$\alpha = \sum_{m,n} \alpha_{m,n}(\psi)sin(n\zeta - m\theta - \omega t), \qquad (8.7)$$

with ω the mode frequency, giving

$$\delta\vec{B} \cdot \nabla\psi_p = \sum_{m,n} \frac{mg + nI}{\mathcal{J}} \alpha_{m,n} cos(n\zeta - m\theta - \omega t). \qquad (8.8)$$

Now add an electric potential Φ to cancel the parallel electric field induced by $d\vec{B}/dt$, with

$$\sum_{m,n} \omega B \alpha_{m,n} cos(n\zeta - m\theta - \omega t) - \vec{B} \cdot \nabla\Phi/B = 0, \qquad (8.9)$$

and make a similar Fourier expansion of the potential Φ, giving a very simple expression if one uses Boozer coordinates with I independent of θ,

$$(gq + I)\omega\alpha_{mn} = (nq - m)\Phi_{mn}. \qquad (8.10)$$

If coordinates are used in which I is a function of θ there is instead a coupling of different harmonics in this expression.

Now relate α to the ideal displacement $\vec{\xi}$. We require that the cross field component of the perturbation is exactly given by α,

$$\nabla\psi \cdot [\nabla \times (\vec{\xi} \times \vec{B})] = \nabla\psi \cdot [\nabla \times \alpha\vec{B}]. \tag{8.11}$$

Use the identity $\nabla \times (\vec{\xi} \times \vec{B}) = \vec{\xi}\nabla \cdot \vec{B} - \vec{B}\nabla \cdot \vec{\xi} + (\vec{B} \cdot \nabla)\vec{\xi} - (\vec{\xi} \cdot \nabla)\vec{B}$ and the fact that $\nabla \cdot \vec{B} = 0$. Now expand $\vec{B}$ and $\vec{\xi}$ in the covariant basis $\vec{\xi} = \xi^\alpha \vec{e}_\alpha$, $\vec{B} = B^\alpha \vec{e}_\alpha$, giving

$$\vec{e}^\psi \cdot \nabla \times (\vec{\xi} \times \vec{B}) = \vec{e}^\psi \cdot [B^\alpha \partial_\alpha(\xi^\beta \vec{e}_\beta) - \xi^\alpha \partial_\alpha(B^\beta \vec{e}_\beta)]. \tag{8.12}$$

All terms in $\partial_\alpha \vec{e}_\beta$ cancel. Since $\vec{e}_\beta = \partial_\beta \vec{r}$ these terms become

$$[B^\alpha \xi^\beta - B^\beta \xi^\alpha]\partial_\alpha \partial_\beta \vec{r} \tag{8.13}$$

which is zero by symmetry.

We then have

$$\vec{e}^\psi \cdot \nabla \times (\vec{\xi} \times \vec{B}) = \vec{e}^\psi \cdot \vec{e}_\beta[B^\alpha \partial_\alpha \xi^\beta - \xi^\alpha \partial_\alpha B^\beta], \tag{8.14}$$

and using $\vec{e}^\psi \cdot \vec{e}_\beta = \delta^\psi_\beta$ and $B^\psi = 0$ we have $\vec{e}^\psi \cdot \nabla \times (\vec{\xi} \times \vec{B}) = B^\alpha \partial_\alpha \xi^\psi$. The right hand side of Eq. 8.11 is simplified using $\nabla \times \vec{B} = \vec{j}$ and $\vec{j} \cdot \nabla\psi = 0$, giving finally

$$B^\alpha \partial_\alpha \xi^\psi = \nabla\alpha \cdot (\vec{B} \times \nabla\psi). \tag{8.15}$$

Again using Eq. 8.7 and $\vec{B} \cdot \nabla = \frac{1}{qJ}(\partial_\theta + q\partial_\zeta)$ and the covariant representation for $\vec{B}$, we have

$$\alpha_{mn} = \frac{(m/q - n)}{(mg + nI)}\xi^\psi_{mn}. \tag{8.16}$$

Thus alpha is simply related to the cross field component of the ideal displacement, and is zero at the rational surface $q = m/n$.

Magnetic islands are produced by resonant perturbations of $\vec{B}$ directed across the flux surfaces of an equilibrium, so the vanishing of α_{mn} at the rational surface and Eq. 8.8 is thought to exclude island formation. But the condition $\delta\vec{B}\cdot\nabla\psi = 0$ is a necessary but not a sufficient condition to exclude the formation of islands. In Fig. 8.3 is an example of the modification of magnetic flux surfaces by a single mode, showing islands appearing adjacent to the rational surface $r/a = 0.7$. We have used a circular equilibrium, with major radius $R = 1$, minor radius $a = 1/4$, and $B = 1$ at the magnetic

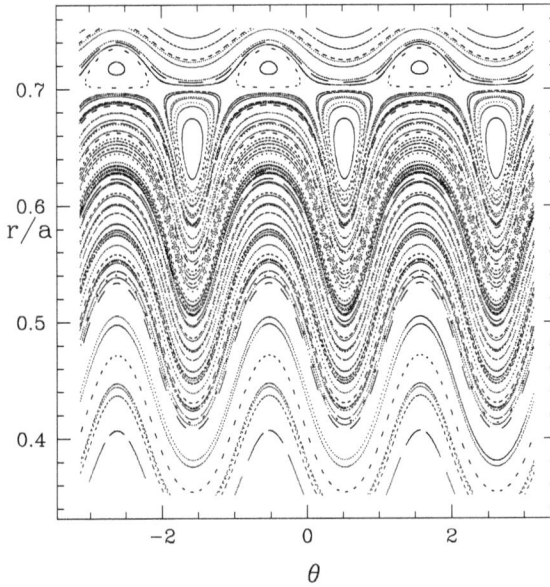

Fig. 8.3   A Poincaré plot of field lines for an odd parity perturbation with $m/n = 3/2$, with $\psi_{m,n}(r_s) = 0$ at the rational surface $r_s/a = 0.7$ where $q = 3/2$. No island elliptic points form at the rational surface, but hyperbolic points do, and islands form on each side of the surface.

axis, so the toroidal flux is $\psi = r^2/2$. The q profile is quadratic in r. The perturbation has $m/n = 3/2$, and $\alpha_{32}(\psi)$ is zero at the rational surface.

We can find the nature of the magnetic field lines analytically. Let $Q = n\zeta - m\theta$, use the same circular equilibrium for which $g = 1$, and neglect I, which is second order in r/R. Then the field line equations $d\psi/d\zeta = \vec{B} \cdot \nabla\psi/\vec{B} \cdot \nabla\zeta$, and $d\theta/d\zeta = \vec{B} \cdot \nabla\theta/\vec{B} \cdot \nabla\zeta$, give

$$\frac{d\psi}{d\zeta} = -m\alpha_{mn}cosQ, \qquad \frac{dQ}{d\zeta} = n - m/q + m\alpha'_{mn}sinQ. \qquad (8.17)$$

with $\alpha'_{mn} = d\alpha_{mn}/d\psi$. Fixed points are given by $d\psi = 0$, $dQ = 0$, or

$$\alpha_{mn}cosQ = 0, \qquad n - m/q + m\alpha'_{mn}sinQ = 0 \qquad (8.18)$$

The first equation has two solutions. First consider $\alpha_{mn} = 0$, i.e. $\psi = \psi_r$, the rational surface. The second equation then gives $sinQ = 0$. Expand

$\psi = \psi_r + x$, and $sinQ = \pm y$, $cosQ = \pm(1 - y^2/2)$, with $x, y << 1$ and drop terms of third order in Eq. 8.17. We then find

$$ydx + xdy \mp \frac{q'}{q^2\alpha'_{mn}}xdx = 0, \tag{8.19}$$

giving

$$\frac{q'x^2}{q^2\alpha'_{mn}} \mp xy = constant, \tag{8.20}$$

so these are hyperbolic points, located on the rational surface at $m\theta = k\pi$ with k integer, and with one separatrix lying along the x axis.

Similarly, if $cosQ = 0$ we have $Q = \pm\pi/2$, $sinQ = \pm 1$, and $n - m/q = \pm m\alpha'_{mn}$ defines surfaces $\psi_\pm$ located away from the rational surface. Again expanding $\psi = \psi_\pm + x$ and $Q = \pm\pi/2 + y$ gives

$$xdx + \frac{\alpha_{mn}(\psi_\pm)q^2}{q'}ydy = 0, \tag{8.21}$$

giving

$$x^2 + \frac{\alpha_{mn}(\psi_\pm)q^2}{q'}y^2 = constant, \tag{8.22}$$

so these are elliptic points located at $\psi_\pm$ and $m\theta = \pm\pi/2$, clearly seen in Fig. 8.3. Since one separatrix lies along the rational surface we can estimate the island width to be twice the distance from the separatrix to the O-point, giving

$$\Delta\psi \simeq \frac{2\alpha'_{mn}q^2}{q'}, \tag{8.23}$$

and thus these islands have a width proportional to the perturbation amplitude, not the square root of the amplitude as do even parity modes, so are much smaller. Ideal MHD does not allow reconnection to occur, so evolution of a mode from an infinitesimal initial state should not allow islands to evolve, but this is apparently not equivalent to adding a nonzero amplitude ideal perturbation to an equilibrium state. Note that this analysis is linear in the perturbation; the obtained structure scales with the magnitude of the perturbation, but is essentially unchanged. The equilibrium field has a continuum of fixed points along the rational surface. Every point on this line satisfies $dQ = d\psi = 0$.

Now consider the field lines defined by the full MHD perturbation $\delta\vec{B} = \nabla \times (\vec{\xi} \times \vec{B})$, giving from Eq. 8.14

$$\vec{B} \cdot \nabla\psi = (\vec{B} \cdot \nabla)\xi^\psi,$$
$$\vec{B} \cdot \nabla\theta = 1/(q\mathcal{J}) + (\vec{B} \cdot \nabla)\xi^\theta - (\vec{\xi} \cdot \nabla)B^\theta,$$
$$\vec{B} \cdot \nabla\zeta = 1/\mathcal{J} + (\vec{B} \cdot \nabla)\xi^\zeta - (\vec{\xi} \cdot \nabla)B^\zeta. \tag{8.24}$$

Again use the same circular equilibrium with $R = 1$, so $\mathcal{J} = 1$, $B^\zeta = 1$, $B^\theta = 1/q$, and $g = 1$, and I is second order in r/R and negligible.

Write $\xi^\psi = \xi^\psi_{mn} sin(Q)$, $\xi^\theta = \xi^\theta_{mn} cos(Q)$, $\xi^\zeta = \xi^\zeta_{mn} cos(Q)$, and $\nabla \cdot \vec{\xi} = 0$ gives

$$m\xi^\theta_{mn} - n\xi^\zeta_{mn} + \xi'^\psi_{mn} = 0. \tag{8.25}$$

Then the field lines satisfy, to lowest order in ξ

$$\frac{d\psi}{d\zeta} = -(m/q - n)\xi^\psi_{mn} cosQ,$$
$$\frac{dQ}{d\zeta} = n - m/q + [(m/q - n)\xi^\psi_{mn}]' sinQ, \tag{8.26}$$

but using $\alpha_{mn} = (1/q - n/m)\xi^\psi_{mn}$ we again obtain Eq. 8.17, leading to the islands located away from the rational surface.

To demonstrate how the topology is maintained, provided one goes to all orders in the perturbation magnitude, consider a fluid perturbation of velocity $\vec{v}$, constant in time. Then

$$\frac{\partial\vec{B}}{\partial t} = \nabla \times (\vec{v} \times \vec{B}), \tag{8.27}$$

but one must iterate this equation to find successive orders (in ξ) of the perturbed field with

$$\frac{\partial\vec{B}_{n+1}}{\partial t} = \nabla \times (\vec{v} \times \vec{B}_n), \tag{8.28}$$

and at each order the fixed points for the map move closer and closer to the resonance surface, and the resonance surface itself is perturbed in a sinusoidal fashion.

A better way to envision what happens is to use the ideal MHD condition that the field is frozen into the plasma. Use the Lundquist identity

(Lundquist, 1951; Roberts, 1967)

$$\vec{B}(\vec{r},t) = \vec{B}_0(\vec{r}_0,0) + (\vec{B}_0(\vec{r}_0,0) \cdot \nabla_0)\vec{\xi}(\vec{r}_0,t), \tag{8.29}$$

where $\vec{r} = \vec{r}_0 + \vec{\xi}$. Use a simple circular large aspect ratio equilibrium field

$$\vec{B}_0(\vec{r}_0,0) = \hat{\phi} + \frac{r_0}{q(r_0)}\hat{\theta} \tag{8.30}$$

and take $\vec{\xi} = \vec{\xi}_0 sinQ$ with $\vec{\xi}_0$ a constant and $Q = n\phi - m\theta$. We then have

$$\vec{B}(\vec{r}_0,t) = \vec{B}_0(\vec{r}_0 - \vec{\xi},0) + [\vec{B}_0(\vec{r}_0 - \vec{\xi},0) \cdot \nabla_0]\vec{\xi}(\vec{r}_0,t) \tag{8.31}$$

or

$$\vec{B}(\vec{r}_0,t) = \hat{\phi} + \frac{(r_0 - \xi_0^r sinQ)}{q(r_0 - \xi_0^r sinQ)}\hat{\theta} + \left(n - \frac{m}{q(r_0 - \xi_0^r sinQ)}\right)\vec{\xi}_0 cosQ. \tag{8.32}$$

We then have for the field line equations, using also $\nabla_0\theta = (\hat{\theta}/r_0)d/d\theta$,

$$\frac{dr}{d\phi} = \left(n - \frac{m}{q(r_0 - \xi_0^r sinQ)}\right)\xi_0^r cosQ$$

$$\frac{dQ}{d\phi} = n - \frac{m}{q(r_0 - \xi_0^r sinQ)} \tag{8.33}$$

where we have used Eq. 8.25 and $\vec{\xi}_0$ constant. The only fixed points are the degenerate ones, lying on the displaced rational surface $q(r_0 - \xi_0^r sinQ) = m/n$; there are no additional fixed points in the field.

Two things are notable about the field. First, expanding Eq. 8.29 in powers of ξ one obtains to lowest order the expressions leading to breaking of the topology. Secondly, if $\vec{\xi}$ is divergence free the field line equations involve only ξ^r, and thus can also be expressed in terms of α, meaning that these representations reproduce the same field to all orders.

Undoubtedly the small islands produced by the indiscriminate use of a perturbed field in either representation do not have a significant effect on either instability analysis or on the effect of ideal modes on high energy particles, but these common representations of ideal MHD perturbations in fact change the initial field line topology. Instability analysis involves the energy integral, proportional to ξ^2, and since the island has a width proportional to ξ, the modification of δW should be of order ξ^3 and hence negligible unless the instability eigenfunction is very localized near the rational surface.

8.3 RESONANCE

Only through resonance can waves modify a particle distribution. Unless the topology of the phase space orbits is changed, when a perturbation subsides a distribution will relax back to exactly the same form it had before the waves were introduced. This is because without resonance the trajectories in phase space occupy KAM surfaces that are topologically equivalent to nested nonintersecting surfaces. The interaction of particle distributions and MHD modes with frequencies well below the cyclotron frequency can be studied in the guiding center approximation. A method for determining resonance location, along with examples of resonances both in tokamaks and stellarators is given in Chapter 5.

The magnetic moment μ is conserved by the interaction of a particle with a mode with frequency much smaller than the cyclotron frequency, so only P_ζ and E are modified by interaction with it. For a perturbation of a single n the Hamiltonian is a function of the combination $n\zeta - \omega t$. Then from $\dot{P}_\zeta = -\frac{\partial H}{\partial \zeta}$ and $\frac{dE}{dt} = \frac{\partial H}{\partial t}$, we find that for fixed n we have $E - P_\zeta \omega/n = constant$ in time. This is really just a statement of energy conservation in the mode frame, and constrains particle motion to a line in the E, P_ζ plane.

Some progress can be made analytically to determine the location of resonances. For resonance the particle motion through the wave must be such that the modification of the orbit is reinforced each time the particle passes through the wave. This requirement is very similar to the quantum mechanical Bohr condition for the existence of a bound state in an atom: the electron wave function must reinforce itself in one circuit of the nucleus. For particle motion this condition requires the existence of fixed points in the map of the orbit through one period of the wave.

A Poincaré point, indicating one passage through the wave, which is a function of $n\zeta - m\theta - \omega_n t$, occurs when $n\zeta - \omega_n t = 2\pi k$, with k integer. For there to be m' periodic fixed points in θ, we also require $\Delta\theta = 2\pi l/m'$ between successive points with l integer. Here m' is the number of islands in a poloidal cross section Poincaré plot. Thus the particle can return to receive the same impulse from the wave by passing through the fixed finite set of θ values, for those values of m in the perturbation that are also consistent with the number of islands poloidally. The helicity of the resonance is then

$$h(P_\zeta, E, \mu) = \frac{\Delta\zeta - \omega_n \Delta t/n}{\Delta\theta} = \frac{m'}{nl}, \qquad (8.34)$$

which must be rational, where $\Delta\zeta$, $\Delta\theta$, Δt refer to one transit. In one Poincaré transit $\Delta\theta = 2\pi l/m'$, so $l = 1$ implies that in one such transit a particle moves from one island in the chain to the adjacent one. Note that the poloidal mode number m does not appear in this expression. For a resonance to appear there must exist integers m', l such that this relation can be satisfied.

The modification of these quantities due to the perturbation α can be neglected. Converting integrals over time to integrals over θ, we have

$$\Delta\zeta = \int \frac{\dot\zeta}{\dot\theta} d\theta, \qquad \Delta t = \int \frac{1}{\dot\theta} d\theta \qquad (8.35)$$

where the integrands must be evaluated following an unperturbed closed particle orbit, and for passing particles $\Delta\theta = 2\pi$, but for trapped particles the integrals must be between the bounce points.

For fixed values of E, μ we scan the range of P_ζ and carry out the integrals, looking for values of P_ζ for which $h(P_\zeta, E, \mu)$ is a low order rational. This determination of the existence of fixed points is not sufficient for the formation of an island, since it is also necessary that the perturbation be nonzero along the orbit and also that it be in resonance with the fixed point period, either directly or through toroidal coupling, which normally means that m' is not far removed from m values existing in the perturbation. For qualitative understanding only, use a low energy approximation for passing particles with $\Delta\zeta = q\Delta\theta$. Further, using a large aspect ratio approximation, find $R\Delta\zeta = v_\parallel \Delta t$, giving

$$h = q\left[1 - \frac{\omega_n R}{n v_\parallel}\right]. \qquad (8.36)$$

Note that Eq. 8.34 only indicates the possibility of an island existing at a particular value, but gives no information concerning island size, and Eq. 8.36 neglects particle drift motion.

There is, however, significant drift modification from a simple field line following analysis, in which $\dot\zeta = q\dot\theta$, and the perturbation is fixed in time. Toroidal precession greatly modifies $\dot\zeta$ at high energy. Particle resonances

are determined not by the q profile of the magnetic field, but by an analogous "kinetic q factor" determined by the particle motion, including the
effects of precession and the large shift of the drift surfaces. For low particle
energy, zero frequency, and pitch $\lambda = v_\parallel/v = 1$, the "kinetic q factor" approaches the magnetic q profile, but precession and shift of the drift surface
strongly depend on frequency, energy and pitch (Gobbin et $al.$ 2008).

Consider passing particle resonance. Use Eqs. 3.95 and 3.96 giving

$$\dot\zeta = \omega_t + \dot\zeta_d, \qquad \dot\theta = \frac{\omega_t}{q} + \dot\theta_d, \tag{8.37}$$

where $\omega_t = \rho_\parallel B^2 = v_\parallel/R$ is the transit frequency, and $\dot\zeta_d$, $\dot\theta_d$ are the second
order drift terms.

A qualitative understanding of resonance can be gained by examining
the large aspect ratio circular equilibrium case. Using $B = 1 - r\cos\theta$ gives
the large aspect ratio expressions $\dot\zeta_d \simeq (\rho_\parallel B^2 + \mu)r\cos\theta$, $\dot\theta_d \simeq -(\rho_\parallel B^2 + \mu)\cos\theta/r$, the $\dot\theta_d$ term dominates the energy evolution and one finds for the
particle energy evolution in the presence of a single harmonic

$$\frac{d\mathcal{E}}{dt} \sim \omega\alpha_{mn}[\cos(Q_{m+1}) + \cos(Q_{m-1})], \tag{8.38}$$

with $Q_m = n\zeta - m\theta - \omega t$. Thus the harmonic m produces two surfaces at the
points where Q_{m-1} and Q_{m+1} are resonant due to the $\cos(\theta)$ dependence
of the drift terms.

Consider the resonance $\dot Q_m = 0$, and note that $\dot Q_m$ is a function of
energy. Expand it around the resonant energy $\mathcal{E}_0$ where $\dot Q_m = 0$. Assuming
that the resonances are well separated in the vicinity of $\mathcal{E}_0$ we can drop the
nonresonant $\cos(Q_{m'})$ terms for $m' \neq m$ and taking for simplicity a far
passing particle, $i.e.$ $\rho_\parallel B^2 \gg \mu$, find

$$dQ_m = \partial_E \dot Q(\mathcal{E} - \mathcal{E}_0)dt, \tag{8.39}$$

$$d\mathcal{E} = -\frac{m\omega_t^2 \Phi_{mn}}{2r}\cos(Q_m)dt, \tag{8.40}$$

with $\partial_E \dot Q$ evaluated at $\mathcal{E}_0$.

These equations have the same form as the equation for a magnetic
island, Eq. 1.43. Far from the resonance they integrate to give $\mathcal{E} \sim \sin(Q_m)/(\mathcal{E} - \mathcal{E}_0)$. Near resonance we find an island given by

$$(\mathcal{E} - \mathcal{E}_0)^2 = k\sin(Q_m) + c, \tag{8.41}$$

with $k = -m\omega_t^2\Phi_{mn}/(r\partial_E\dot{Q})$, and c and integration constant. Now use $\partial_E\omega_t = 1/\rho_{\parallel}$ and Eq. 8.5, and introduce the shear $s = rq'/q$, giving at resonance

$$\partial_E\dot{Q}_m = \frac{\omega}{\omega_t^2} - \frac{m^2 s\omega_t}{qr^2\omega}, \tag{8.42}$$

and thus $k = \omega_t^4(\delta B_r/B)[\omega/\omega_t - m^2sR^2\omega_t^2/(qr^2\omega\omega_0)]^{-1}$. The island width is given by

$$\frac{\Delta\mathcal{E}}{\mathcal{E}} = \frac{4R^2}{\rho^2\omega_0^2}\sqrt{2|k|}, \tag{8.43}$$

and using Eq. 8.5 the width in r is

$$\Delta r = \frac{2mR^2}{r\omega\omega_0}\sqrt{2|k|}. \tag{8.44}$$

To find the frequency of rotation about the island O-point use Eq. 8.39

$$\frac{d^2Q_m}{dt^2} = (\partial_E\dot{Q})\frac{d\mathcal{E}}{dt}, \tag{8.45}$$

and using Eq. 8.40 we find the pendulum equation

$$\frac{d^2Q_m}{dt^2} = \omega_T^2 cos Q_m. \tag{8.46}$$

Using $\delta\vec{B} = \nabla \times \alpha\vec{B}$ we have $\delta B_r/B = m\alpha_{mn}/r$, and reinserting the major radius R and the gyro frequency ω_0 to restore dimensions, the trapping frequency is given by

$$\omega_T^2 \simeq \frac{\omega_t\omega}{2}\left[\frac{m^2sR^2\omega_t^3}{qr^2\omega^2\omega_0} - 1\right]\frac{\delta B_r}{B}. \tag{8.47}$$

Note that if $\omega_T^2 < 0$ we can simply redefine Q_m through $Q_m + \pi$ and again recover the pendulum equation, so the absolute value can be understood for this expression.

8.4 THE FIBONACCI SEQUENCE

The nonlinear coupling of two perturbations with poloidal and toroidal mode numbers of m/n and m'/n' give rise to a sequence of fractions studied by Fibonacci (1170–1250). Simply multiplying the terms $\alpha e^{i(m\theta - n\zeta)}$ and

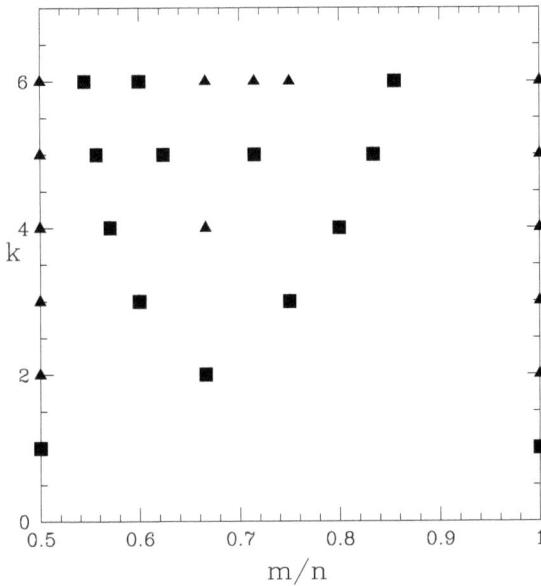

Fig. 8.4   Fibonacci sequence for the standard map, showing in triangles the resonances and in large squares those resonances that destroy new KAM surfaces at each level up to $k = 6$.

$\alpha e^{i(m'\theta - n'\varsigma)}$ gives $\alpha^2 e^{i(m+m')\theta - i(n+n')\varsigma}$. Note that the higher order fraction $(m+m')/(n+n')$ is always bounded by the parent fractions. By continuing to multiply perturbations thus produced there results an infinite number of islands produced by any pair. Nevertheless, the KAM theorem guarantees that if the original perturbations are small enough, the sum of all these island widths remains small, so that there are domains in which the original KAM surfaces are distorted but retain their original topology. To see this in a nonrigorous manner, note that the island width is proportional to the square root of the perturbation, and the new islands produced at order k are given by the fractions $[lm + (k-l)m']/[ln + (k-l)n']$ for $0 \leq l \leq k$, so there are $k+1$ of them. The total island width is then bounded by $\sum_k (k+1)\alpha^{k/2}$, which converges for $\alpha < 1$, giving a finite width occupied by islands. As long as this width is small compared to the whole domain, some KAM surfaces still exist. The vanishing of the last KAM surface implies overall stochasticity; orbits not trapped in residual islands can wander over the whole space. Even for large amplitude perturbations there will exist some trapped orbits, but there also exist trajectories that completely cross the domain.

Trying to improve the Chirikov criterion using the Fibonacci sequence is nontrivial. Note that some of the higher order islands given by the sequence exist on top of islands already accounted for, not destroying additional KAM surfaces, so the series actually contains many fewer terms, consisting only of those fractions for which the numerator and denominator are relatively prime. For example, consider the standard map, discussed in section 1.8. The simple Chirikov criterion gives a stochastic threshold of $\epsilon = 2.47$, whereas the actual value is $\epsilon = 0.9716...$ The $m = 1$ island has width given approximately by $4\sqrt{\epsilon}$. Because of the periodicity of the map, the value of n can be chosen to be any integer at $v = 0$, it then changes by integer values for v changing by 2π. We take $m/n = 1/1$ at $v = 0$, then $m/n = 1/2$ at $v = 2\pi$. In Fig. 8.4 are shown the first few relatively prime fractions in the Fibonacci series up to $k = 6$, with those fractions destroying new KAM surfaces shown in large dark squares. The resonances produced at level k are the fractions $m/n = k/n$ with $n = k, k + 1, ...2k$.

The 2/3 island for $k = 2$ is shown in Fig. 8.5 at $v = \pi$ with width approximately ϵ. The $k = 3$ islands 3/5 and 3/4 are at $v \simeq 2$ and at $v \simeq 4$ with width approximately $\epsilon^{3/2}/2$, and the $k = 4$ islands 4/5 and 4/7 are located at $v \simeq 1.5$ and $v \simeq 4.8$ with widths approximately $\epsilon^2/4$. The $k = 5$ islands are also visible in the plot but have very small widths. These widths were determined numerically for the value of $\epsilon = 0.5$ shown in the plot. The scaling of the widths with ϵ is known, but it is nontrivial to calculate the actual widths. Adding these together gives a width occupied by islands of $w = 4\sqrt{\epsilon} + \epsilon + \epsilon^{3/2} + \epsilon^2/2$. This function reaches 2π for $\epsilon \simeq 0.95$, a much more accurate estimate of stochastic threshold than that given by simple Chirikov overlap.

8.5   LANDAU PHASE MIXING

In Fig. 8.6 is shown a kinetic Poincaré plot of the action of a single harmonic of a toroidal Alfvén eigenmode on a particle distribution of 25 keV energy, with $\mu B/E \simeq 0.6$. The harmonic has $m/n = 9/2$ and a frequency of 50 kHz. One of the island resonances is shown in detail. In addition to circling the island O-point the trapped orbits are also following the helical structure of the perturbation around the torus. The original KAM surfaces are broken

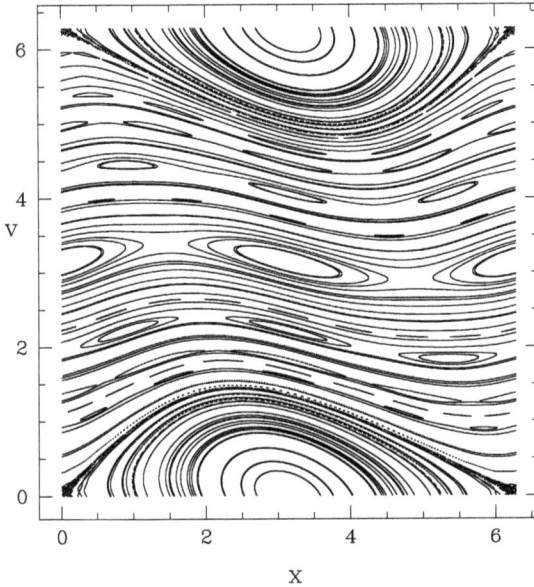

Fig. 8.5   The standard map for $\epsilon = 0.5$, showing the large $m = 1$ islands and the first few Fibonacci islands for $m = 2, 3, 4, 5$.

by the island. This breaking of topology causes particles at different drift surfaces to mix with one another, as they circle around the island O-point with frequencies that depend on the distance from the O-point. Also shown is a plot giving the time evolution of points placed initially at $\theta = -0.2$, for a period of 0.34 msec, showing the dependence of the rotation rate on distance from the O-point, with rapid rotation occuring close to it. Near the O-point a full period is approximately 0.5 msec. The particles are completing an orbit around the device in about two microseconds, so many toroidal transits are completed while rotating around the island O-point. The rotation frequency is an increasing function of perturbation strength, or island size, zero in the limit of zero perturbation, but it is typically very slow compared to particle transit times. See Fig. 1.8 for an example of the rotation frequency (internal q_I) versus mode amplitude.

The fact that the frequency depends on distance from the O-point means that the mixing occcurs on smaller and smaller scales, until finally at some scale small collisionality or other perturbative effects produce large scale

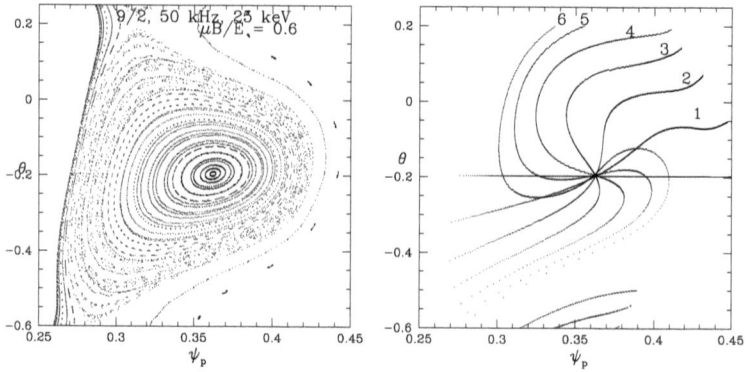

Fig. 8.6   Circulation of particles within a resonance. At left is shown the Poincaré plot given by very many toroidal transits, and at right snapshots taken at fixed time intervals of the particles initially at $\theta = -.2$, showing the variation of the rotation rate as a function of distance from the O-point.

irreversible changes in the distribution and the entropy of the system is increased.

A simple model can be constructed to show how this breaking of topology transfers energy between a mode and a particle distribution possessing a gradient in density of particles either in space or in energy. A Poincaré plot can be made in the variables of θ and ψ_p, as shown in Fig. 8.7, or variables θ and E. Locally one can also use θ and v with v the particle velocity, and we consider using these variables. Introduce at $t = 0$ a gradient in particle density in velocity space,

$$n(v, \phi) = 1 + \delta v = 1 + \delta r \sin \phi_0, \qquad (8.48)$$

where ϕ refers to the angle about the O-point and r is the "distance" from the O-point, as shown in Fig. 8.7. The initial particle location is ϕ_0 and δ is the inverse scale length for the velocity gradient. In time the particles rotate about the O-point, as seen in Fig. 8.6, with $\phi(t) = \phi_0 + \omega(r)t$.

Now calculate the average energy of particles trapped in the island given by the integral $< E > \simeq \int r dr d\phi n v^2$.

$$<v^2(t)> = \frac{1}{\pi V^2} \int_0^V r dr \int_0^{2\pi} d\phi [v_r + r \sin \phi]^2 [1 + \delta r \sin \phi \cos(\omega t) - \delta r \cos \phi \sin(\omega t)],$$

$$(8.49)$$

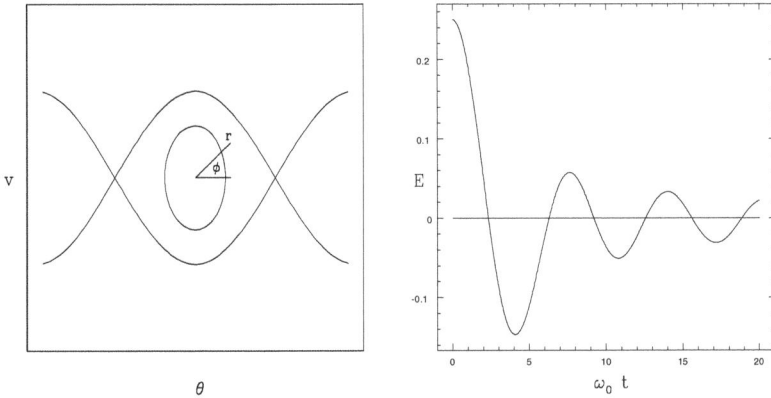

Fig. 8.7   Landau mixing model, showing resonance coordinates and mean particle energy vs time.

where V is the island half width in velocity space and v_r is the velocity at the O-point, the resonance velocity. Integration over ϕ gives

$$< v^2(t) >= v_r^2 + \frac{V^2}{4} + \frac{\delta v_r}{V^2} \int_0^V r^3 dr \cos(\omega t), \qquad (8.50)$$

and the initial mean energy is $< v^2(0) >= v_r^2 + V^2/4 + \delta v_r V^2/4$. To obtain a simple model approximate the frequency of rotation about the O-point as a parabolic function going to zero at the edge,

$$\omega(r) = \omega_0(1 - r^2/V^2). \qquad (8.51)$$

Then

$$< v^2(t) >= v_r^2 + \frac{V^2}{4} + \frac{\delta v_r^2}{2\omega_0 t} \left[\sin \omega_0 t + \frac{(\cos \omega_0 t - 1)}{\omega_0 t} \right], \qquad (8.52)$$

as shown in Fig. 8.7. Note that for small t the change is $\sim t^2$, there is no instantaneous rate of energy transfer at $t = 0$. For large time the mean energy in the resonance island is $v_r^2 + V^2/4$, the energy $\delta v_r V^2/4$ is transfered to the wave. There is a delay from the time of particle deposition to energy transfer to the mode, due to the finite mixing time in the resonance, given by the period of rotation of particles in the resonance island.

8.6   PHASE VECTOR ROTATION

A general method for numerically determining the existence of, or the destruction of, good KAM surfaces can be obtained using the method of phase vector rotation. Consider following two orbits located nearby one another. Examine a Poincaré section in P_ζ, θ and define the angle χ to give the orientation of the vector joining them in this plane. If good KAM surfaces exist χ can change by at most an angle of π, due to their relative velocity in the angular coordinate.* However, two orbits within an island rotate around one another with χ increasing with the rotation about the island O-point, as seen in Fig. 8.6, also referred to as the bounce frequency of a particle trapped in the wave, which increases with the size of the island. The rate of change of χ is a function of distance from the island O-point, dropping to zero at the separatrix. See Fig. 1.8 for an example of the dependence of the bounce frequency at the island elliptic point on the perturbation strength. Phase vector rotation is illustrated in Fig. 8.8, showing vectors between nearby points in the P_ζ, θ plane on good KAM surfaces and in a resonance.

Thus we determine the nonexistence of good KAM surfaces by examining nearby pairs of orbits for phase vector rotation χ exceeding π. An example of such a determination is given in Figure 8.9. A simple circular equilibrium was used, with the q profile equal to one on axis and to five at the plasma edge. Shown is a Poincaré plot of the field, produced by three zero frequency tearing modes with $m/n = 2/1$, $m/n = 3/2$ and $m/n = 1/1$. First order Fibonacci resonances ($\sim \alpha^2$) are thus produced at $4/3$ and $5/3$ and the next order ($\sim \alpha^3$) gives islands at $5/4$, $7/5$, $8/5$, and $7/4$. The Poincaré plot shows the three lowest order resonances and also the $5/3$ resonance at $P_\zeta = -0.5$. Barely visible is the $7/5$ resonance at $P_\zeta = -0.39$ caused by coupling of $4/3$ and $3/2$, and other high order resonances are not visible with this resolution.

In addition we show the result of the phase vector rotation criterion. Superimposed on a Poincaré plot are points obtained by launching pairs of particles distributed in P_ζ, θ, and recording only those initial values participating in phase vector rotation. The major resonances at $2/1$, $3/2$ and $1/1$ as well as the higher order resonances at $5/3$ and $4/3$ are clearly

---

*This method was used in the first edition of this book (1989) to illustrate island internal
   q values, Fig. 1.6, but I have since discovered that it was also employed in calculations
   of the stability of the solar system, see Froeschle *et. al.* Science 289, 2108 (2000).

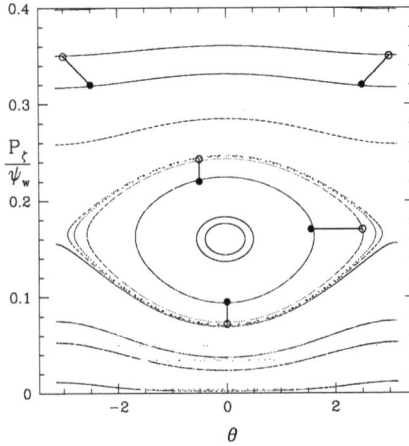

Fig. 8.8 The P_ζ, θ plane showing a single $m = 1$ resonance island, and vectors between nearby points on good KAM surfaces and in the island. On nearby KAM surfaces the phase vector can rotate by at most π, whereas a phase vector in an island rotates through 2π with a period given by the trapping bounce time.

seen. The 4/3 is more visible in the phase vector rotation plot than in the Poincaré plot. Also shown with large triangles are the locations of the resonances for 2/1, 5/3, 3/2, 4/3, and 1/1 from Eq. 8.34. The phase vector rotation also shows scattered points between the 4/3 and the 1/1 resonances. As we will see, this criterion is highly accurate, and indicates islands often not visible except in a high resolution Poincaré plot.

For this plot the particle energy was taken very small and all particles have pitch equal to one, so that the orbits simply follow field lines. Since for very small islands the phase vector rotation is very slow, the particles must be followed for a longer time to detect smaller islands, and the length of the simulation is determined by the desired resolution. In addition, islands smaller than the separation between the pairs of orbits are of course not detectable. The separation cannot be chosen too small or it will result in false positive island detection due to numerical error. Thus there are three parameters to adjust for an optimization of this procedure: the initial orbit pair separation, the number of toroidal transits followed, and the critical value of rotation to indicate KAM destruction. Typical values are a separation of 2×10^{-4} times the minor radius, a run time of 500 toroidal transits, and a critical value of rotation of $|d\chi| = 4$.

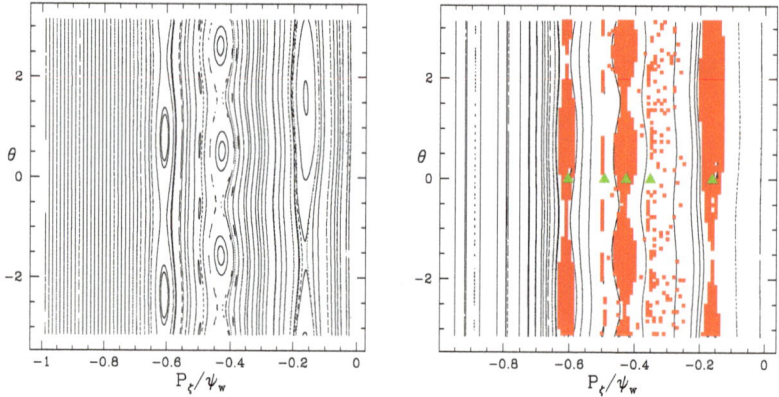

Fig. 8.9 Kinetic Poincaré plot for 2/1, 3/2, and 1/1 zero frequency tearing modes, and amplitudes $\alpha = 10^{-5}R$, and phase vector rotation indicator. Resonant surfaces from Eq. 8.34 are also shown with large triangles for surfaces 2/1, 5/3, 3/2, 4/3, and 1/1, listed in order as they appear from the left in the plot.

If the mode amplitudes are increased so that a stochastic region is produced, even more KAM surfaces are destroyed. In this case the islands overlap, and the Chirikov criterion is well satisfied. The criterion of looking for phase vector rotation also reproduces the stochastic region, that is the lack of rotation is a necessary condition for the existence of a good KAM surface. We wish to define domains in the space of P_ζ, E using this method, close to the maximum extent of the island or stochastic region, so a few particle pairs are initiated in each domain with random phase with respect to the perturbations, the rotation of any pair indicating broken KAM surfaces for the domain.

A test of this procedure was carried out using the DIII-D reversed shear equilibrium for shot 122117 (White *et al.*, 2010a, 2010b). We consider one of the eleven TAE and reversed shear Alfvén eigenmodes modes present in this discharge.

The poloidal harmonics of the TAE mode are shown in Fig. 8.10, and the E, P_ζ plane showing the result of the phase vector rotation determination for that part of the distribution with $\mu B_0 = 14$ keV. The domains in the E, P_ζ plane are shaded where phase vector rotation indicated broken KAM surfaces. It is interesting to note that the large resonances on the right edge of the domain (near the magnetic axis) consist of a sequence with the number of islands decreasing as the energy decreases, beginning with a ten

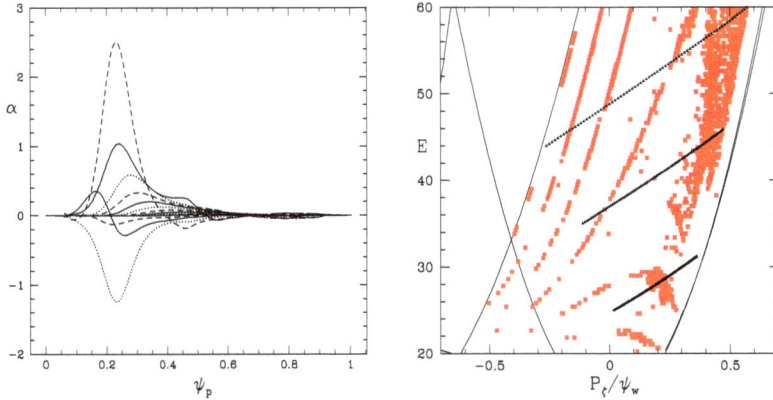

Fig. 8.10   Poloidal harmonics ($\times 10^{-6}$) of an 81 kHz $n = 3$ TAE mode with $10 \leq m \leq 23$, observed in DIII-D in shot 122117 and the plane of P_ζ, E with $\mu B_0 = 14$ keV, with paths for kinetic Poincaré plots.

island resonance at 60 keV, a gap near 30 keV followed by a nine island resonance, another gap near 25 keV followed by an eight island resonance, and so on. The phase vector rotation determination was performed using 100 domains in energy, from 20 to 60 keV and also 100 domains in canonical momentum. The initial pair separation was $\Delta\psi_p = 2 \times 10^{-4}\psi_w$, where ψ_w is the magnitude of the poloidal flux at the last flux surface. Two particle pairs were deposited in each domain, initiated at $\theta = 0$, but with random ζ so that each pair has a different phase relation with the mode. The run time was chosen to be 500 toroidal transits. Any orbit pair exhibiting phase vector rotation greater than $|\chi| = 4$ causes that domain to be labeled non-KAM. Only positive pitch was used, so that this analysis is restricted to trapped and co-passing particles, neglecting the counter-passing population.

In Fig. 8.11 are two kinetic Poincaré plots showing the nature of the resonances along lines with $\omega P_\zeta - nE = constant$, with energies at the left (end point of the Poincaré plot) of 44 and 25 keV. The phase vector rotation plot is seen to accurately indicate the location and size of resonances.

8.7   MODE INDUCED AVALANCHE

Now we wish to investigate the approach to avalanche conditions due to several resonance island chains. For this we again use a simple circular

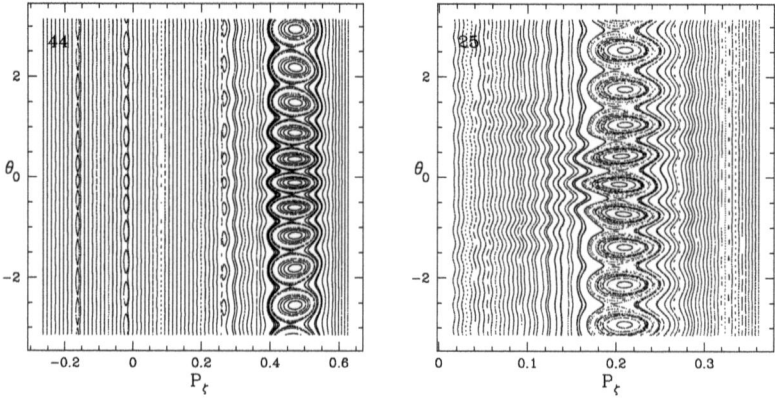

Fig. 8.11   Poincaré plots associated with lines originating at 44 and 25 keV in Fig. 8.10.

equilibrium with $0.8 < q < 4$, $R = 100$ cm, $a = 25$ cm, and a field of 20 kG on axis. In Fig. 8.12 are shown the results of the phase vector rotation determination for a single broad 10 kHz mode with $m/n = 6/5$, with amplitude of $\alpha = 5 \times 10^{-6} R$ and a kinetic Poincaré plot showing the nature of the resonances. Note that this result was obtained with a simple circular equilibrium and particles of moderately high energy. Conventional wisdom, see Eq. 8.38, has it that the outward shift of an orbit, which is essentially $m = 1$, couples to the poloidal mode number m to produce resonances at $m \pm 1$, but we see that there are island chains with the number of islands ranging from five to nine. It is not a simple matter to determine analytically the location and extent of resonances.

Also resonance locations from Eq. 8.34 are shown for $m'/n = 9/5$, $8/5$, $7/5$, $6/5$, and $5/5$. Island widths are comparable for $m' = m, m \pm 1$ and somewhat smaller for $m' = m + 2, m + 3$. The surfaces for resonances with $m' = m - 2, m - 3$ are outside the plasma. In the plane of P_ζ, E the major resonances are clearly visible as well as the first order Fibonacci sequence, appearing as thinner, partly broken lines in between the major broader resonances, existing at $m'/n = 13/10$, and at $m'/n = 11/10$. The locations are shown as triangles in the Poincaré plot, but the islands are visible only in a higher resolution plot. Because of the shift and distortion of flux surfaces caused by the largest islands, the location of the higher order Fibonacci resonances are not accurately given, being off by several island widths.

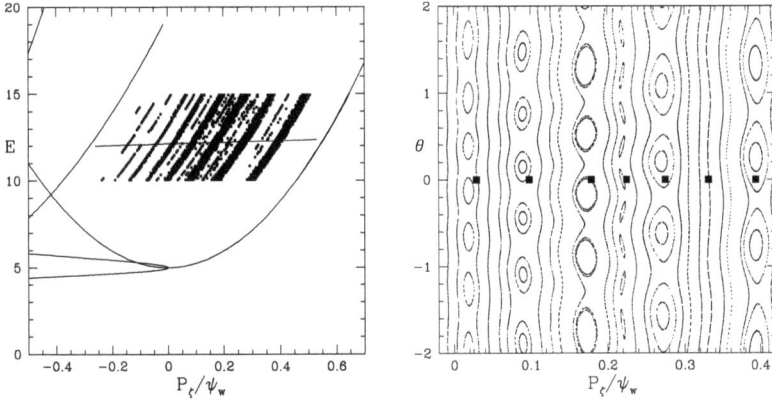

Fig. 8.12   Plane of P_ζ, E with $\mu B_0 = 5$ keV, for a broad 10 kHz mode with $m/n = 6/5$ and amplitude $\alpha = 5 \times 10^{-6} R$ and a kinetic Poincaré plot made along the line originating at the wall at 12 keV. Resonance locations from Eq. 8.34 are shown for $m'/n = 9/5$, 8/5, 7/5, 6/5, and 5/5, and also, at points where no islands are visible in the Poincaré plot, for $m'/n = 13/10$, and $m'/n = 11/10$.

In Fig. 8.13, with the mode amplitude ten times larger, a continuous stochastic domain has appeared between $P_\zeta = 0.05$ and $P_\zeta = 0.42$, containing remnant islands from the major resonances. On the left are still visible a $m'/n = 9/5$ and a $m'/n = 10/5$ island chain, also visible in the P_ζ, E plane. Note that good KAM surfaces exist both near the plasma edge and near the axis, so this mode should produce only local profile flattening, but no particle loss.

For the most part the phase vector rotation appears to give a reasonable description of the non-KAM domains. It has the advantage over methods that simply determine resonance location through integrals, such as Eq. 8.34, that the resonance is shown only if the mode is sufficiently large at the requisite surface, and in addition the full width of the resonance is displayed, not only the location. Also there can be no innacuracy regarding the location. It thus makes evident which perturbations and nonlinear produced islands of the Fibonacci sequence should be taken into consideration. Besides this, as is evident from examples shown, resonance location from Eq. 8.34 is only approximate. Primary resonances are reasonably well given, but higher order resonances are often displaced significantly.

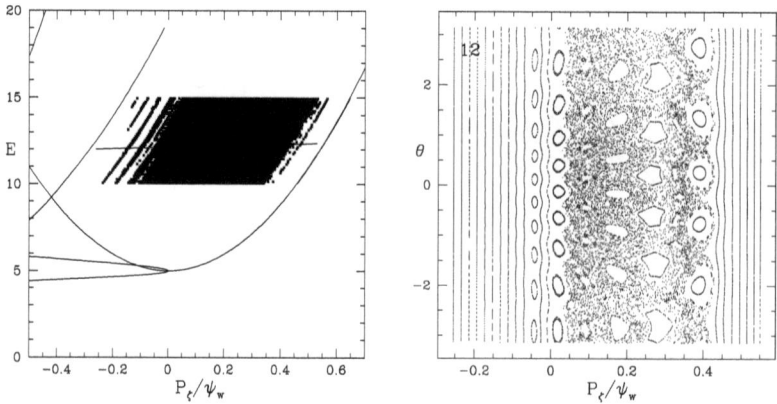

Fig. 8.13   Plane of P_ζ, E with $\mu B_0 = 5$ keV, for a 10 kHz local mode with $m/n = 6/5$ and a kinetic Poincaré plot for amplitude $\alpha = 5 \times 10^{-5} R$ made along the line originating at the wall at 12 keV.

We wish to use the determination of domains in the P_ζ, E plane with destroyed KAM surfaces to find the final state of a given particle distribution under the action of the mode spectrum. Construct a numerical method of producing the final state of a given particle distribution under the action of a given spectrum of modes. In the final state the density should be constant for a particular mode with frequency ω and toroidal mode number n, in all adjacent island and stochastic domains along lines given by $\omega P_\zeta - nE = constant$, since this combination is conserved and annealing can happen only along these lines. However, repeated annealing for multiple modes, with different values of ω/n, produces diffusive motion in the combined non-KAM domains of the modes involved. Thus the necessary algorithm must be an iterative annealing process, one mode at a time, but repeated so as to capture the effect of the combination of modes present.

Examine a high energy particle distribution as predicted by a neutral beam deposition calculation or an alpha particle birth profile calculation, and make a number of domains in the magnetic moment μ sufficient to give a good representation of the distribution. For each μ, divide the space of confined particles in the P_ζ, E plane into small domains, with size determined by the desired resolution of small islands. Then find the domains of broken KAM surfaces for that part of the plane which is occupied by the distribution by following pairs of orbits and looking for phase vector rotation, noting whether each domain is stochastic or consists of good KAM surfaces. This is the only computationally demanding part of the calculation, depending on the desired resolution for island size. Reintroduce the original distribution and distribute it into the μ, P_ζ, E domains. Then carry out an equilibration of densities in stochastic domains which are in contact along lines $\omega P_\zeta - nE = c$ for each mode, iterating this process until a final state is achieved.

To carry this out, note that the differential volume is given by $dV \sim J(\psi_p, \theta)d\theta d\psi_p$, where in Boozer coordinates $JB^2 = gq+I$. Thus the domain at μ, P_ζ, E with range dP_ζ has volume

$$dV \sim dP_\zeta \int d\theta \left(\frac{d\psi_p}{dP_\zeta} \right)_{E,\mu} J(\psi_p, \theta), \qquad (8.53)$$

the integration being over the particle orbit. Using $P_\zeta = g\rho_\| - \psi_p$ and $E = \rho_\| B^2/2 + \mu B$ we find

$$dV = dP_\zeta \int d\theta \frac{(gq+I)}{B^2} \frac{1}{1 + g(\rho_\|^2 B + \mu)\partial_{\psi_p} B/\rho_\| B^2 - g'\rho_\|}, \qquad (8.54)$$

where the integral is taken over a constant μ, P_ζ, E surface, *i.e.* along the particle orbit, and $g' = \partial_{\psi_p} g$. Thus in neighboring stochastic domains along the lines with $\omega P_\zeta - nE = c$ with initial particle numbers in the domains n_1, n_2 and final particle numbers n_1', n_2', particle conservation and density equilibration gives

$$n_1' = n_1 + \frac{n_2 dV_1 - n_1 dV_2}{dV_1 + dV_2}, \qquad n_2' = n_2 - \frac{n_2 dV_1 - n_1 dV_2}{dV_1 + dV_2}. \qquad (8.55)$$

In addition to replacing the particle numbers in adjacent stochastic domains with the modified values for the two domains, stochastic domains in contact with the outer wall are emptied of particles, they being counted as lost. This process must be repeated many times using the stochastic domain template for each mode. After the annealing the particle distribution can be reconstructed.

To test this procedure (White, 2011, 2012) a simple circular equilibrium with $q = 0.8$ on axis and $q = 4$ at the last closed flux surface was used, with a simple distribution with a single value of $\mu B_0 = 5$ KeV and with energy ranging from 10 to 15 keV, but with a steep radial density profile. The single mode of Fig. 8.13 was used, consisting of a large amplitude $m/n = 6/5$ localized 10 kHz perturbation. Also shown is the result of the phase vector

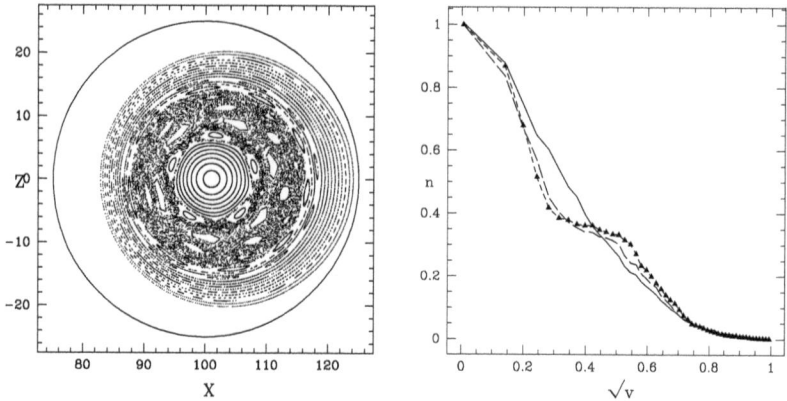

Fig. 8.14 The kinetic Poincare plot of Fig. 8.13, shown in the X, Z plane, showing the extent of the stochastic domain, with $\mu B_0 = 5$ keV, for a local 10 kHz mode with $m/n = 6/5$, and the result of annealing. Shown is the initial distribution, obtained using the annealing process, with the points marked by triangles, and a distribution obtained with a direct guiding center simulation of three milliseconds, as a dashed line, plotted vs the square root of the volume, approximately equal to the minor radius.

rotation indicator. Twenty domains in energy were used, since the range of energy of the distribution was chosen to be small, and 200 domains in canonical momentum P_ζ to provide high resolution. The orbit pairs were separated by $\Delta\psi_p = 2 \times 10^{-4}\psi_w$, with ψ_w the value of poloidal flux at the plasma edge. In the P_ζ, E plot a broad stochastic band is seen, as well as three narrow partly broken bands to the left of it. Note that the location of resonances in θ is at constant P_ζ, but not on a flux surface. The good KAM surfaces shown in the plot of the poloidal cross section in Fig. 8.14 do not coincide with flux surfaces, they are shifted outboard. The resonance location in the poloidal plane depends on poloidal angle, mode frequency, and particle energy. The magnitude and location of the perturbation was chosen so that there are bands of good KAM surfaces both in the plasma center and near the outer edge. Thus the expected result is local flattening at the location of the stochastic domain with no induced loss.

The annealing process is carrried out repeatedly in the stochastic domains, equalizing densities along the lines $\omega P_\zeta - nE = constant$ until the particle distribution has stabilized, this process taking a very small amount of computing time. The second plot of Fig. 8.14 shows the initial and modified profiles of the density, as well as the profile of a distribution obtained with a direct guiding center simulation of three milliseconds. The plots were obtained by binning particles using equal size bins in the volume inscribed by a given flux surface, plotted versus the square root of the volume, approximately equal to the minor radius, showing the canonical near flattening of the distribution in the stochastic domain. The flattening is not perfect because the boundaries of the stochastic domain do not coincide with flux surfaces.

8.8 MODE-PARTICLE ENERGY TRANSFER

Now consider net energy transfer between a particle distribution and a wave. If the particles merely oscillate with their mean position equal to the initial one, there is no net transfer. However, at a resonance the particle motion forms an island (both in energy and in minor radius). In this case particles trapped in the island have a mean value of energy equal to the resonance value, not their initial value. If the particle distribution possesses a gradient (either in energy or in radius) the flattening of the distribution function by the island produces a net energy transfer to the wave

$$\Delta E \sim \omega \partial_E F - (m/r)\partial_r F. \tag{8.56}$$

Taking $\partial_r F \sim F/r$ and $\partial_E F \sim F/\rho^2$, the ratio of the first to the second term is $\omega r^2/(m\omega_0\rho^2)$, typically small for high energy particles of interest, and in this case only the radial gradient of the distribution function is relevant.

For the net energy transfer between mode and particle we have Eq. 8.2. Adding the electric potential Φ to cancel the parallel electric field induced by $d\vec{B}/dt$ we find Eq. 8.10. Remaining with Boozer coordinates we have

$$\frac{dE}{dt} = \sum_{m,n} i \left[\rho_\parallel B^2 \frac{nq - m}{gq + I} - \omega\right] \Phi_{mn}(\psi_p)e^{i(n\zeta - m\theta - \omega t)}, \tag{8.57}$$

which becomes

$$\frac{dE}{dt} = \sum_{m,n} i \left[n\dot{\zeta} - n\dot{\zeta}_d - m\dot{\theta} + m\dot{\theta}_d - \omega\right] \Phi_{mn}(\psi_p)e^{i(n\zeta - m\theta - \omega t)}, \tag{8.58}$$

with $\dot{\zeta}_d$, $\dot{\theta}_d$ the drift terms from Eqs. 3.95 and 3.96.

But $i[n\dot{\zeta} - m\dot{\theta} - \omega]\Phi_{mn}(\psi_p)e^{i(n\zeta - m\theta - \omega t)} = (d/dt)\Phi_{mn}e^{i(m\theta - n\zeta - \omega t)} - \Phi'_{mn}\dot{\psi}_p e^{i(m\theta - n\zeta - \omega t)}$. Then subtracting an exact time derivative of the form $(d/dt)\Phi_{mn}e^{i(m\theta - n\zeta - \omega t)}$ we have

$$\frac{d\mathcal{E}}{dt} = \sum_{m,n} i\left[-n\dot{\zeta}_d + m\dot{\theta}_d\right]\Phi_{mn}e^{i(n\zeta - m\theta - \omega t)} - \sum_{m,n}\Phi'_{mn}\dot{\psi}_p e^{i(n\zeta - m\theta - \omega t)}$$

(8.59)

with $\mathcal{E} = E - \sum_{m,n}\Phi_{mn}e^{i(n\zeta - m\theta - \omega t)}$. The nonresonant exact derivative is simply the periodic energy change of the particle following the locally changing potential, whereas $\mathcal{E}$, entirely due to the drift motion, can be changed resonantly. Thus the transverse Alfvén wave can resonantly exchange energy with a passing particle only through the cross field drift motion, because there is no electric field aligned with the magnetic field.

Note from Eq. 8.57 that the contribution to energy change and to motion across flux surfaces coming from α is proportional to ρ/R and that from the potential is proportional to ω/ω_0. For high frequency modes it is essential to include this potential, but it can be safely ignored for many low frequency applications.

Another means of finding the ability of a particular mode to produce change in the particle distribution is to look for time averaged energy transfer to or from the particles, which can only happen if a particle is trapped in an island produced by the perturbation. In Fig. 8.15 are shown plots of the P_ζ, μ plane for an energy of 23 keV in a numerical equilibrium. See Sec. 3.3 for a description of this plot of orbit domains. Initially a distribution consisting of all confined co-passing orbits is launched, and then a mode is allowed to act on the distribution. Shown are results using two $n = 3$ reversed shear Alfvén eigenmodes (RSAE) with amplitudes of $dB_r/B \simeq 2 \times 10^{-4}$ and m values between 5 and 13. A time average is taken, and only particles showing a significant time averaged energy change (here a 20 percent loss in energy) are plotted. These plots show bands of resonance, indicating the particular orbits that are involved. Only co-passing orbits are resonant for these mode frequencies. The plots show resonance occuring for deeply and for barely passing particles, giving also the flux surface location. Whereas the Poincaré plots of islands for fixed values of μ, shown in Sec. 3.9.4 show resonance only for a particular value of pitch, these plots indicate all pitch values giving resonance, appearing as separate bands for different values of $\mu B_0/E$. Orbits that pass through the magnetic axis contact the far right

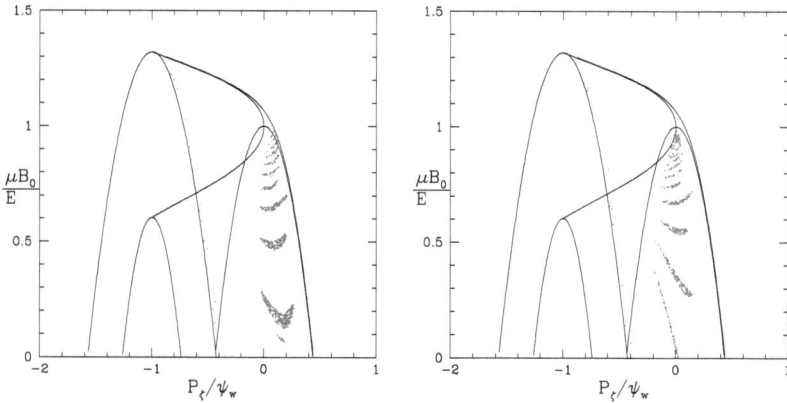

Fig. 8.15   A plot of the P_ζ, μ plane, showing only 23 keV particle orbits with 20 percent energy change due to the mode, for a $n = 3$ flute mode at 102 kHz (left) and a $n = 4$ flute mode at 96 kHz (right).

parabola symmetric about $P_\zeta = 0$. All points within this parabola are co-passing confined orbits. At a frequency of 96 kHz there are no resonances near the magnetic axis, but at 102 kHz resonances with particles of small pitch $(\mu B_0/E \geq 0.7)$ are very near the magnetic axis.

8.9   MODE EVOLUTION

There are two possible scenarios for modes driven by a high energy particle distribution. The mode frequency may be determined only by equilibrium parameters, with growth and mode phase modified by the particle distribution. This is the case with many Alfvén gap modes such as the TAE in a tokamak. The second case, where the mode frequency itself is determined by the particle distribution, is more complicated, and requires the solution of a dispersion relation involving the particle distribution. In the following sections we show how the mode evolution can be evaluated numerically with a guiding center code using Monte Carlo techniques.

8.9.1   Frequency determined by equilibrium

In the case of modes for which the frequency is determined by equilibrium parameters both the amplitude and the phase of the mode are changed by the interaction with the particle distribution. We consider a background ideal conducting plasma plus a hot component. Charge neutrality requires that the charge densities of the hot component and the background are opposite, $q_b = -q_h$. The bulk plasma current is given by

$$\vec{J}_b = \nabla \times \vec{B} - \vec{J}_h \qquad (8.60)$$

and the equation of motion for the bulk is

$$\rho_b \frac{d\vec{V}_b}{dt} = -\nabla p_b + \vec{J}_b \times \vec{B} + q_b \vec{E} = -\nabla p_b + (\nabla \times \vec{B} - \vec{J}_h) \times \vec{B} + q_b \vec{E} \qquad (8.61)$$

with ρ_b the bulk mass density. Substituting the fact that the bulk plasma is perfectly conducting, $\vec{E} + \vec{V}_b \times \vec{B} = 0$, we have

$$\rho_b \frac{d\vec{V}_b}{dt} = -\nabla p_b + (\nabla \times \vec{B} - \vec{J}_h) \times \vec{B} + q_h \vec{V}_b \times \vec{B} \qquad (8.62)$$

and the velocity is related to the displacement, $\vec{V}_b = \partial_t \vec{\xi}$. Linearizing Eq. 8.62 we have

$$\rho_b \partial_t^2 \vec{\xi} = \vec{F}(\vec{\xi}) + \vec{S} \qquad (8.63)$$

where

$$\vec{F}(\vec{\xi}) = (\nabla \times \vec{Q}) \times \vec{B} + (\nabla \times \vec{B}) \times \vec{Q} + \nabla(\vec{\xi} \cdot \nabla p + \gamma p \nabla \cdot \vec{\xi}) \qquad (8.64)$$

with $\vec{Q} = \nabla \times (\vec{\xi} \times \vec{B})$ and $\vec{S} = -(\vec{J}_h - q_h \partial_t \vec{\xi}) \times \vec{B}$.

Write the perturbation as

$$\vec{\xi} = \sum_{m,n} A_n \vec{\xi}_{m,n}(\psi_p) sin(n\zeta - m\theta - \omega_n t - \phi_n). \qquad (8.65)$$

The amplitude and phase are slowly varying, describing the evolution of the unstable mode due to the energetic particle drive. Substituting and neglecting terms in $\partial_t^2 A_n$, $\partial_t^2 \phi_n$, $\partial_t \phi_n \partial_t A_n$ we find

$$\sum_{mn} 2\omega_n \vec{\xi}_{mn}[\dot{A}_n cos(\Omega_{mn}) + A_n \dot{\phi}_n sin(\Omega_{mn})] = -\nu_A^2 \vec{S} \qquad (8.66)$$

with ν_A the Alfvén frequency, $\nu_A = V_A/R = B/(R\sqrt{4\pi n_i m_i})$, arising from the factor of ρ_b in Eq. 8.62 and $\Omega_{mn} = n\zeta - m\theta - \omega_n t - \phi_n$. Now take dot products with $\vec{\xi}_{mn} sin(\Omega_{mn})$ and $\vec{\xi}_{mn} cos(\Omega_{mn})$ and integrate over ψ_p, θ, ζ

and the fast time scale using $\int \xi_{mn}^2 d\psi_p = 1$, giving

$$\omega_n \dot{A}_n = -\nu_A^2 \left\langle \int \vec{S} \cdot \vec{\xi}_{mn} \cos(\Omega_{mn}) d\vec{x} \right\rangle,$$

$$\omega_n A_n \dot{\phi}_n = -\nu_A^2 \left\langle \int \vec{S} \cdot \vec{\xi}_{mn} \sin(\Omega_{mn}) d\vec{x} \right\rangle, \qquad (8.67)$$

where brackets indicate time averages over the rapid frequency time scale.

But the electric field corresponding to the displacement $\vec{\xi}$ is $-\partial_t \vec{\xi} \times \vec{B}$ so we have

$$\vec{E}_n = A_n \omega_n (\vec{\xi} \times \vec{B}) \cos(\Omega_{mn}), \qquad (8.68)$$

plus terms smaller in the ratio of the slow to fast time scales. Similarly

$$\partial_t \vec{E}_n = A_n \omega_n^2 (\vec{\xi} \times \vec{B}) \sin(\Omega_{mn}), \qquad (8.69)$$

and using the fact that $\vec{J}_h$ is the high energy particle velocity $\vec{v}$ we have

$$\vec{S} \cdot \vec{\xi} = -(\vec{v} \times \vec{B}) \cdot \vec{\xi} + q_h (\partial_t \vec{\xi} \times \vec{B}) \cdot \vec{\xi}, \qquad (8.70)$$

so

$$\vec{v} \cdot \vec{E}_n = A_n \omega_n (\vec{S} \cdot \vec{\xi}) \cos\Omega_{mn}, \qquad (8.71)$$

and

$$\vec{v} \cdot \partial_t \vec{E}_n = A_n \omega_n^2 (\vec{S} \cdot \vec{\xi}) \sin\Omega_{mn}, \qquad (8.72)$$

giving

$$\frac{dA_n}{dt} = -\frac{\nu_A^2}{\omega_n^2 A_n} \left\langle \int \vec{v} \cdot \vec{E}_n d^3 x \right\rangle - \gamma_d A_n, \qquad (8.73)$$

$$\frac{d\phi_n}{dt} = -\frac{\nu_A^2}{\omega_n^3 A_n^2} \left\langle \int \vec{v} \cdot \partial_t \vec{E}_n d^3 x \right\rangle, \qquad (8.74)$$

where γ_d is the sum of all damping mechanisms not produced by the high energy particle population.

But $\vec{v} \cdot \vec{E}_n$ is just the energy transfer between mode and particle and this is $dH/dt = \partial_t H$.

$$H = (\rho_c - \alpha)^2 B^2 / 2 + \mu B + \Phi \qquad (8.75)$$

and so

$$\frac{dH}{dt} = -\rho_\| B^2 \partial_t \alpha + \partial_t \Phi.$$ (8.76)

The relations between ξ, α, and Φ are given in Sec. 8.2.

The energy transfer becomes

$$\vec{v} \cdot \vec{E}_n = w_n A_n \sum_m \left[\rho_\| B^2 \alpha_{mn} - \Phi_{mn}(\psi_p) \right] cos(\Omega_{mn})$$ (8.77)

and

$$\vec{v} \cdot \partial_t \vec{E}_n = w_n^2 A_n \sum_m \left[\rho_\| B^2 \alpha_{mn} - \Phi_{mn}(\psi_p) \right] sin(\Omega_{mn}).$$ (8.78)

For the full distribution we write $f(\vec{x}) = \sum_k \delta(\vec{x} - \vec{x}_k)/N$ to do the integrals. Note that if $x = g(y)$,

$$\int \delta(x)dx = \int \delta(g(y))g'dy = \int \frac{\delta(y - y_k)}{g'(y_k)}g'dy = \int \delta(y - y_k)dy, \text{(8.79)}$$

so the Klimontovich representation is independent of the coordinate system.

So $f(\psi_p, \theta, \zeta) = \sum_k \delta(\psi_p - \psi_{pk})\delta(\theta - \theta_k)\delta(\zeta - \zeta_k)/(N)$.

$$\frac{dA_n}{dt} = \frac{-\nu_A^2}{w_n} \sum_{k,m} \left\langle \left[\rho_\| B^2 \alpha_{mn} - \Phi_{mn}(\psi_p) \right] cos(\Omega_{mn}) \right\rangle - \gamma_d A_n,$$ (8.80)

$$\frac{d\phi_n}{dt} = \frac{-\nu_A^2}{w_n A_n} \sum_{k,m} \left\langle \left[\rho_\| B^2 \alpha_{mn} - \Phi_{mn}(\psi_p) \right] sin(\Omega_{mn}) \right\rangle,$$ (8.81)

all terms in the sum evaluated at the position of particle k, and frequencies are normalized to the on-axis cyclotron frequency. Note that the particle orbits in the presence of the mode must be followed to perform these integrals. Otherwise only a set of particles of measure zero, those on the resonance surface, contribute. By following particles interacting with the mode, all those that are resonant, *i.e.* trapped in the wave, contribute to the integrals. The A_n in the denominator of the second equation can be a problem numerically. We can also write

$$\alpha = X \sum_m \alpha_{mn}(\psi_p)cos(n\zeta - m\theta - w_n t)$$

$$-Y \sum_m \alpha_{mn}(\psi_p)sin(n\zeta - m\theta - w_n t).$$ (8.82)

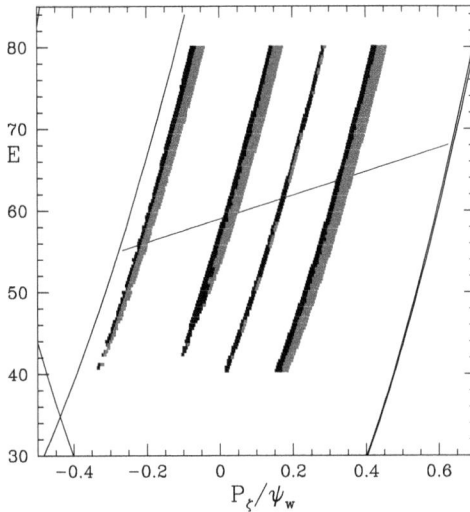

Fig. 8.16 Energy transfer for fixed μ due to a 60 kHz mode with $m/n = 4/3$ and amplitude of 10^{-4}. Domains of energy loss are shown in red, and energy gain in black.

with $X = A_n cos(\phi_n)$, $Y = A_n sin(\phi_n)$. Then

$$\left| \begin{array}{c} dX/dt \\ dY/dt \end{array} \right| = \left| \begin{array}{cc} cos(\phi_n) & -A_n sin(\phi_n) \\ sin(\phi_n) & A_n cos(\phi_n) \end{array} \right| \left| \begin{array}{c} dA/dt \\ d\phi_n/dt \end{array} \right|. \tag{8.83}$$

These equations have been used to investigate the saturation of the TAE mode, see Sec. 8.10.

Carrying out such an analysis for mode development is nontrivial, because the domains of secular energy transfer, islands, and stochastic regions can be very small, requiring high resolution in the particle distribution. In Fig. 8.16 is shown an example of the energy transfer due to a 60 kHz mode with $m/n = 4/3$ and amplitude of 10^{-4}. Domains of energy loss are shown in red, and energy gain in black. These domains are adjacent to one another, they correspond to two sides of a resonance island. Also shown with a black line is the slope $E - P_\zeta \omega/n = constant$ along which the particle motion occurs. To detect the energy flowing from the particle distribution to the mode the spatial or energy gradient in density must be sufficiently large that the black and red domains do not simply cancel within noise levels. With no gradients the net energy transfer to the mode is zero.

8.9.2 *Frequency determined by particle distribution*

As an example we examine the fishbone and sawtooth instabilities, which demonstrate the case where the mode frequency is wholly determined or significantly modified by the particle distribution. Excessive computing requirements are avoided by considering the plasma on a time scale which is short with respect to changes in the equilibrium profiles. In addition, mode-mode coupling effects are ignored. In the case of the modes considered this is a very reasonable approximation, the dominant destabilization and saturation mechanisms are due to the mode-particle interaction. Thus the MHD state of the plasma is given by parameters describing the current and pressure profiles, and by the solution of the Grad–Shafranov equation for the equilibrium to be considered. For some modes (sawtooth, fishbone, toroidal Alfvén eigenmode) the ordering of the high energy particle effects guarantees that the form of the mode is not greatly modified by the high energy particles, and an ideal MHD code can be used to determine the eigenfunctions. But for other modes, the kinetic terms are important in determining the shape of the mode and a code such as NOVA-K must be used to determine the mode spectrum and shape.

Were it not for the presence of the singular layer (either inertial or resistive, depending on plasma parameters) it would be possible to simulate the coupled system through energy conservation. Energy transferred to and from the particles through resonance with the mode is readily observed in guiding center simulations, and energy carried by the fluid with a given displacement can be evaluated numerically, and the resulting rate of energy transfer could be used to calculate the mode growth rate. The presence of the singular layer, which can transfer energy to the electrons through kinetic modes not treated in a simple MHD analysis, makes such a simple treatment incorrect, and requires that one instead model these processes or solve a dispersion relation which includes mode-particle energy exchange, the dissipation of the layer, and the MHD energy of the mode in order to find the frequency and growth rate of the mode.

The kinetic contributions to the variational dispersion relation are expressed in general magnetic coordinates. The contribution of the interaction of the particles with higher harmonics is retained. The effects of plasma shape and β (*i.e.* the modification of plasma shape from concentric circular flux surfaces and the outward shift of the surfaces due to plasma pressure) are all properly included. Next, the integrals over the particle distribution are carried out using a Monte Carlo simulation, which allows a reasonable approximation of the actual distribution functions. Finally, the bounce orbit averaging for trapped and passing particles is carried out using a Hamiltonian guiding center code using either analytical or numerically generated equilibria. This correctly gives the toroidal precession rates in the finite beta, shaped equilibria as well as including nonlinear particle dynamics. In addition the averaging over drift orbits takes into account the drift excursions of the particles. A radial electric field and suitable modification of the dispersion relation can be used to give the effects of plasma rotation.

The dispersion relation is then solved using the MHD parameters associated with the equilibrium and the kinetic contributions obtained from the guiding center averages. The results can be plotted following the variation of any parameter which does not involve a significant modification of the equilibrium, *i.e.* variation of the high energy particle density, plasma resistivity, diamagnetic drift frequencies, etc. If the amplitude of the mode is kept small and fixed, one obtains an analysis of the stabilizing-destabilizing properties of the fixed particle distribution used. If instead an unstable mode is allowed to grow to an amplitude sufficiently large to modify the particle distribution, a self-consistent time dependent solution can be obtained with the instantaneous particle distribution determining the mode frequency and growth rate.

In the remainder of this section we sketch the derivation of the dispersion relation, describe the Monte Carlo procedure using the guiding center equations, and show some generic results for the internal kink and fishbone modes. The same techniques can be applied to analyze the interaction of high energy particles with any MHD mode.

Begin with the contravariant and covariant expressions for the equilibrium field. The volume element is given by $d^3x = \mathcal{J} d\Psi_p d\theta d\zeta$. The equilibrium is obtained numerically for the plasma under consideration. The high energy particle contribution to the dispersion relation is obtained by following a Monte Carlo distribution of particles. The dispersion relation takes the form $D(\omega) = 0$ where

$$D(\omega) = I + \delta W_{MHD} + \delta W_r + \delta W_n, \qquad (8.84)$$

with I the contribution from the inertial (resistive) layer,

$$\hat{I} = -\frac{8i\Gamma((\Lambda^{3/2}+5)/4)\sqrt{\omega(\omega-\omega_{*i})}}{\Lambda^{9/4}\Gamma((\Lambda^{3/2}-1)/4)\omega_A}, \tag{8.85}$$

and

$$\Lambda = -i(\omega(\omega-\omega_{*e})(\omega-\omega_{*i}))^{1/3}/\gamma_R, \tag{8.86}$$

and the resistive growth rate $\gamma_R = S^{-1/3}\omega_A$. The diamagnetic frequencies are given by ω_{*e} and ω_{*i} and S is the ratio of the resistive time to the Alfvén time, all evaluated at the q = 1 surface. The term δW_{MHD} is the usual fluid MHD contribution $\delta\hat{W}_{MHD} = -\gamma_I/\omega_A$ with γ_I referring to the ideal growth rate, and ω_A the shear Alfvén frequency, $\omega_A = V_A/(\sqrt{3}Rr_sq')$ with V_A the Alfvén velocity and r_s the radius of the q = 1 surface. In our final dispersion relation, in agreement with the inertial term $\hat{I}$, all terms are normalized by the factor $2R/[\pi(Br_s\xi_{r0})^2]$ with R the major radius, B the on-axis magnetic field, and the hat indicates a normalized quantity. All terms in $\delta\hat{W}$ are thus dimensionless and frequencies are normalized to the shear Alfvén frequency.

The quantities δW_r, δW_n are the resonant and nonresonant contributions to $D(\omega)$ due to the high energy particles. The nonresonant contribution is

$$\delta W_n = -\int (\vec{\xi}_\perp \cdot \vec{\nabla}p)(\vec{\xi}_\perp \cdot \vec{\kappa})\, d^3x \tag{8.87}$$

where $\vec{\xi}_\perp$ is the usual displacement introduced in MHD stability theory, $\vec{\kappa}$ is the curvature and $p = \int EF\, d^3v$ is the pressure due to the high energy particles,

$$p = \sum_\sigma \int \frac{2\pi BEF}{\sqrt{2E(1-\mu B/E)}}\, d\mu dE \tag{8.88}$$

with E the energy, $F(\Psi_p, E, \mu)$ the particle distribution function, and σ the sign of $v_\parallel$.

Using these expressions the nonresonant part of the kinetic contribution becomes

$$\delta W_n = -2^{3/2}\pi^2 \sum_\sigma \int \frac{BE^{1/2}}{\sqrt{1-\mu B/E}} H\partial_{\Psi_p}F(\Psi_p, E, \mu)\mathcal{J}\, d\theta d\Psi_p d\mu dE \tag{8.89}$$

where

$$H(\Psi_p, \theta) = (\vec{\xi}_\perp \cdot \vec{\kappa})(\vec{\xi}_\perp \cdot \vec{\nabla}\Psi_p). \tag{8.90}$$

We sketch a heuristic derivation of the resonant part $\delta\hat{W}_r$. For more detail regarding the method see Catto *et al.* (1981), Chen *et al.* (1984) and Antonson *et al.* (1983). The nonadiabatic part of the perturbed distribution function is given by the solution of

$$(-\omega - i\vec{v} \cdot \vec{\nabla})g = 2EQ(1 - \mu B/2E)\vec{\xi}_\perp \cdot \vec{\kappa}. \tag{8.91}$$

This equation is solved by Fourier expanding in the variable θ. From axisymmetry we can choose perturbations of a single toroidal harmonic n. First note that $\vec{v} \cdot \vec{\nabla}f = df/dt$ along an orbit for any function f of the coordinates (not explicitly time dependent). Write g in the form

$$g = \sum_l \frac{g_l(\Psi_p)e^{i(n\zeta - l\theta - \omega t)}}{\omega_d - \omega + (q - l)\Omega_t}, \tag{8.92}$$

with $\Omega_t = \overline{d\theta/dt}$, the average poloidal frequency. Now write the left hand side as a time derivative and perform a time average,

$$\frac{1}{T}\int dt \sum_l \frac{g_l}{\omega_d - \omega + (q - l)\Omega_t} \frac{1}{i}\frac{d}{dt}e^{i(n\zeta - l\theta - \omega t)} = 2EQ\overline{J} \tag{8.93}$$

where $\overline{J}$ is the time average of J. Use an average over one transit (bounce) $t_0 - T < t < t_0 + T$ with $\theta(t_0) = 0$, $\theta(t_0 \pm T) = \pm\pi$, and $\pi = \Omega_t T$. We also have $\zeta(t_0 \pm T) = \zeta_0 \pm q\pi \pm \omega_d$. We thus find

$$e^{i(n\zeta_0 - \omega t_0)}\sum_l g_l \frac{sin((\omega_d - \omega + (q - l)\Omega_t)T)}{(\omega_d - \omega + (q - l)\Omega_t)T} \frac{d}{idt}e^{i(n\zeta - l\theta - \omega t)} = 2EQ\overline{J} \tag{8.94}$$

and thus for particles within resonance, the only relevant ones for the calculation of J, within an irrelevant phase factor $g_l = 2EQ\overline{J}$. For trapped particles instead T is taken to be half the bounce time and $\theta(t_0 - T) = \theta(t_0 + T)$, $\Omega_t = 0$, and there is no l dependence. Thus finally

$$g = \sum_l \frac{2EQ\overline{J}e^{i(n\zeta - l\theta - \omega t)}}{\omega_d - \omega + (q - l)\Omega_t} \tag{8.95}$$

for passing particles, and

$$g = \frac{2EQ\bar{J}e^{i(n\zeta - \omega t)}}{\omega_d - \omega} \tag{8.96}$$

for trapped particles. We then have

$$\delta W_r = 2^{7/2}\pi^2 \sum_\sigma \int \sum_l \frac{E^{3/2}\bar{J}^* Q\bar{J}}{B\sqrt{1 - \mu B/E}} \frac{\mathcal{J}d\theta d\Psi_p d\mu dE}{(\omega_d - \omega + (q - l)\Omega_t)} \tag{8.97}$$

where Ω_t is the poloidal transit frequency for passing particles and zero for trapped particles, and

$$J(\Psi_p, \theta) = (1 - \mu B/2E)(\vec{\xi}_\perp \cdot \vec{\kappa}) \tag{8.98}$$

with n the toroidal mode number of the mode under consideration. The bar refers to orbit averaging

$$\bar{f} = \frac{1}{2\pi K_b} \int_{-\pi}^{\pi} \frac{f}{B\sqrt{1 - \mu B/E}} d\theta \tag{8.99}$$

with

$$K_b = \frac{1}{2\pi} \int_{-\pi}^{\pi} \frac{1}{B\sqrt{1 - \mu B/E}} d\theta \tag{8.100}$$

and K_2 is defined similarly, with a factor of $cos(\theta)$ in the numerator of the integrand (see Sec. 8.9.3). If necessary (trapped particles), the range of θ is limited to the domain where $1 - \mu B/E$ is positive. In large aspect ratio approximation in a circular equilibrium, where B has a simple $cos(\theta)$ dependence, these functions reduce to lowest order to the usual elliptic integrals, but in a general equilibrium they may be significantly different.

The operator $Q = [\omega\partial_E - (\vec{B}\times\vec{\nabla}S) \cdot \vec{\nabla}]F$ with $S = n(\zeta - q\theta)$ reduces to

$$Q = \omega\partial_E F - \partial_{\Psi_p} F. \tag{8.101}$$

Introducing the particle density

$$dN = 4\pi^2 \frac{B}{\sqrt{2E(1 - \mu B/E)}} F(\Psi_p, E, \mu)\mathcal{J}\,d\theta d\Psi_p d\mu dE, \tag{8.102}$$

the nonresonant part of the kinetic contribution becomes

$$\delta W_n = -\int E\overline{H}\frac{\partial_{\Psi_p} F}{F} dN \tag{8.103}$$

and the resonant part

$$\delta W_r = 4 \int \sum_l E^2 \frac{\overline{J}^*(Q/F)\overline{J}}{(\omega_d - \omega + (q - l)\Omega_t)} dN. \tag{8.104}$$

In these expressions, the particle distribution dN is an invariant, *i.e.* equal to the time averaged distribution, so H has been replaced with its time average. This proceedure is useful for improving the statistics in the Monte Carlo evaluation, *i.e.* all the expressions in the dispersion relation; $\overline{J}, \overline{H}, \omega_d$, Ω_t are time averaged quantities. These equations complete the derivation of the dispersion relation.

The functions $\vec{\xi}_\perp \cdot \vec{\kappa}$ and $\vec{\xi}_\perp \cdot \vec{\nabla}\Psi_p$ must be supplied either analytically or from a linear MHD code for the equilibrium being considered. Write the perturbed magnetic field in the form $\delta\vec{B} = \vec{\nabla} \times \alpha\vec{B}$ (see Sec. 8.2). Then $\alpha\vec{B} = (\vec{\xi} \times \vec{B}) + i\vec{\nabla}G$ with G a gauge. We then find the two functions of interest to be given by

$$\vec{\xi} \cdot \vec{\nabla}\Psi_p = \frac{i}{JB^2}[g\partial_\theta G - I\partial_\zeta G] \tag{8.105}$$

and

$$\vec{\xi} \cdot \vec{\nabla}\theta = \frac{i}{JB^2}[-g\partial_{\Psi_p}G + \delta\partial_\zeta G]. \tag{8.106}$$

Introduce the Fourier expansions $\alpha = \sum_m \alpha_m(\Psi_p)e^{i(n\zeta - m\theta)}$ and $G = \sum_m G_m(\Psi_p)e^{i(n\zeta - m\theta)}$. We then find, on using $\vec{B} \cdot \vec{\nabla} = (1/J)(\partial_\theta + q\partial_\zeta)$,

$$G_m = \frac{(gq + I)\alpha_m}{m - nq}, \tag{8.107}$$

and finally,

$$\vec{\xi} \cdot \vec{\nabla}\Psi_p = \frac{1}{gq + I} \sum_m [gm + nI]G_m(\Psi_p)e^{i(n\zeta - m\theta)}, \tag{8.108}$$

$$\vec{\xi} \cdot \vec{\nabla}\theta = -\frac{1}{gq + I} \sum_m [igm\partial_{\Psi_p}G_m + n\delta G_m]e^{i(n\zeta - m\theta)}, \tag{8.109}$$

and

$$\vec{\xi} \cdot \vec{\kappa} = \vec{\xi} \cdot \vec{\nabla}\Psi_p[\frac{\partial_{\Psi_p}B}{B} + \frac{\partial_{\Psi_p}p}{B^2}] + \vec{\xi} \cdot \vec{\nabla}\theta\frac{\partial_\theta B}{B}. \tag{8.110}$$

Recall that these Fourier expansions are in terms of the straight field line coordinate θ. For use of the cylindrical coordinate θ_c, see Sec. 8.9.3.

It can be appropriate to use the large aspect ratio expansion while not neglecting the effects of plasma shape and pressure, which can produce very nonorthogonal coordinates. To lowest order in inverse aspect ratio the internal kink mode consists of a rigid displacement (no field line bending in the large aspect ratio limit) with no compression. For $n = 1$

$$\vec{\xi}_\perp = \hat{X}cos(\zeta) + \hat{Z}sin(\zeta) \tag{8.111}$$

for $r < r_s$, and zero for $r > r_s$. Note that this is not a $m = 1$ mode except in the cylindrical limit. Also for large aspect ratio the field line curvature is $\vec{\kappa} = -\hat{X}/X$. Thus

$$\vec{\xi}_\perp \cdot \vec{\kappa} = -cos(\zeta)/X. \tag{8.112}$$

Similarly

$$\vec{\xi}_\perp \cdot \vec{\nabla}\Psi_p = \partial_X\Psi_p cos(\zeta) + \partial_Z\Psi_p sin(\zeta), \tag{8.113}$$

giving

$$H = -\frac{cos(\zeta)}{X}(\partial_X\Psi_p cos(\zeta) + \partial_Z\Psi_p sin(\zeta)) \tag{8.114}$$

and

$$J = -(1 - \mu B/2E)\frac{cos(\zeta)}{X}. \tag{8.115}$$

For a circular cylindrical equilibrium H reduces to

$$H = -\frac{cos(\zeta)}{qX}cos((q-1)\theta) \tag{8.116}$$

and of course, neglecting drifts, one also has $\zeta = q\theta$.

Hamiltonian guiding center equations are used to evaluate δW_r and δW_n. Bounce averages are conveniently converted to time integrals. Substituting $d\theta/dt$ and neglecting terms of higher order in gyroradius it is straightforward to convert a bounce average to a time average,

$$\overline{f} = \frac{1}{T}\int f\,dt \tag{8.117}$$

with T a bounce (transit) period, or any period much longer than a bounce (transit) period.

Write the particle density in terms of the distribution function

$$dN = F(\Psi_p, E, \mu)d^3xd^3v \qquad (8.118)$$

and rewrite the volume element in terms of the magnetic coordinates through

$$d^3xd^3v = 4\pi^2\sqrt{2E}\mathcal{J}d\Psi_pd\theta d\lambda dE, \qquad (8.119)$$

with the pitch $\lambda = \pm\sqrt{1 - \mu B/E}$, the sign given by $v_\parallel$. Normalize through

$$\beta_h = \frac{8\pi}{B_0^2 V} \int E\, dN \qquad (8.120)$$

with V the plasma volume, and use the Monte Carlo representation for the density (integrated over ζ),

$$dN = c\sum_k \delta(\Psi_p - \Psi_{pk})\delta(\lambda - \lambda_k)\delta(E - E_k)\delta(\theta - \theta_k)d\Psi_p d\lambda dE d\theta \quad(8.121)$$

Fixed values of $\Psi_p, \lambda, E, \theta$ completely define the orbits. Bounce averaging is carried out numerically, *i.e.* particles are initiated at particular values of Ψ_p, θ, and ζ, and the orbits followed.

For studies of stability with a fixed distribution function it is convenient to directly evaluate $\partial_{\Psi_p}F$ in terms of F rather than integrating by parts. Substituting dN and normalizing $D(\omega)$ through the factor $2R/[\pi(B_0 r_s \xi_{r0})^2]$ with ξ the amplitude of the $m = 1$ displacement, to simplify the layer term, we find

$$\delta\hat{W} = \frac{-V\beta_h}{4\pi^2(r_s\xi)^2\sum_j E_j}\sum_k\left[\frac{\partial_{\Psi_p}F}{F}EH - \frac{4E^2Q}{F}\frac{\overline{J}^*\overline{J}}{\omega_d - \omega + (q-1)\Omega_t}\right]_k \quad(8.122)$$

where we have kept only the $l = 1$ term, since the large value of Ω_t for passing particles makes higher order terms negligible.

The use of Monte Carlo methods to represent the distribution function leads to a technical difficulty when ω is in the vicinity of the real axis. The necessary modification of the analysis in this region, a rapid means of generating Monte Carlo distributions, and the evaluation of the partial derivatives of F with respect to energy and flux using Monte Carlo distributions is discussed in Sec. 8.9.4.

Finally, it should be noted that the adiabatic term due to H is a purely MHD contribution, and should not be counted twice. That is, if the high energy particles have been taken into account in calculating the plasma equilibrium, and are therefore incorporated into the evaluation of the MHD growth rate, γ_I, this term should not be included in the dispersion relation. If, on the other hand, the high energy particles are injected into or otherwise added to an equilibrium calculated without taking them into account, then H should be included in the dispersion relation.

It is convenient to use simple reference examples to serve as numerical test cases, and to illustrate the basic properties of the solutions. For this purpose we use the circular large aspect ratio equilibrium, neglect resistive modifications, and restrict the perturbation to consist of a single poloidal harmonic. In the nonresistive limit the dispersion relation reduces to

$$-i\sqrt{\omega(\omega - \omega_{*i})} - \gamma_I + \omega_A\delta\hat{W} = 0. \tag{8.123}$$

In circular flux surface approximation the ideal MHD growth rate γ_I can be evaluated analytically, and for a quadratic q profile the result is

$$\gamma_I = \frac{3\pi\omega_A r_s^2}{R^2}[1 - q(0)]\left(\beta_p^2 - \frac{13}{144}\right), \tag{8.124}$$

with ω_A the shear Alfvén frequency, and

$$\beta_p = -\frac{2}{B_\theta^2}\int_0^{r_s}\left(\frac{r}{r_s}\right)^2\frac{dp}{dr}dr, \tag{8.125}$$

where p is the pressure of the background MHD plasma, not that of the high energy particles.

A trapped particle distribution can be produced by initiating all particles at $\theta = \pm\pi/4$ with zero parallel velocity. For the equilibrium used $\alpha = \mu/E$ is not exactly constant over the minor radius, varying from 1.01 to 1.1. If, on the other hand, α is kept constant, then all particles inside some radius are passing rather than trapped. The analytic approximations made in evaluating the dispersion relation are not self-consistent but can be closely if not exactly duplicated. Parameters used are similar to those of previous works, but with very high aspect ratio. The q profile was taken to be quadratic in r, ranging from 0.8 to 1., *i.e.* the entire plasma volume is taken to be inside the $q = 1$ surface, as is consistent with previous calculations, and there is a single $m = 1$ harmonic. The ideal limit (large S) is used.

First, consider stabilization of the internal kink mode . The initial effect of a high energy population can be understood with a perturbation analysis. The solutions of the dispersion relation have qualitatively different features according to the relative magnitudes of γ_I and $\omega_{*i}/2$. Either term can be larger in a tokamak discharge. First, consider $\gamma_I > \omega_{*i}/2$. Writing

$$\omega = \omega_{*i}/2 + i\sqrt{\gamma_I^2 - \omega_{*i}^2/4} + \Delta \tag{8.126}$$

we find to lowest order

$$\frac{\Delta}{\omega_A} = \frac{-i\gamma_I \delta\hat{W}}{\sqrt{\gamma_I^2 - \omega_{*i}^2/4}} \tag{8.127}$$

and thus $Imag\ \delta\hat{W}$ positive causes the mode frequency to increase and $real\ \delta\hat{W}$ positive causes stabilization. Now consider the various contributions to $\delta\hat{W}$. First consider the nonresonant contribution, which is real. For this equilibrium we find (Sec. 8.9.3)

$$\delta\hat{W}_n = \frac{2^{5/2}\pi}{r_s^2} \sum_\sigma \int \frac{E^{1/2}cos(\theta)}{\sqrt{1 - \mu B/E}} \partial_r F\, r dr d\theta d\mu dE. \tag{8.128}$$

Normally $\partial_r F$ is negative, so $\delta\hat{W}_n$ is destabilizing for deeply trapped particles (increased pressure gradient in the bad curvature region) and stabilizing for barely trapped and passing particles (increased pressure gradient in the good curvature region), with the largest contribution coming from particles near the trapped-passing boundary, which spend a larger fraction of their time in the good curvature region.

Now consider the resonant contribution. Neglecting the $\omega \partial_E F$ term in Q, we find

$$\delta\hat{W}_r = -\frac{2^{9/2}\pi}{r_s^2} \sum_\sigma \int \frac{E^{3/2}\overline{J}^*\overline{J}\partial_r F}{(\omega_d - \omega + (q-1)\Omega_t)\sqrt{1 - \mu B/E}}\, dr d\theta d\mu dE. \tag{8.129}$$

Real $\delta\hat{W}_r$ is positive, (stabilizing) if $\omega_d + (q-1)\Omega_t - \omega_r$ is positive, and negative (destabilizing) if negative, where $\omega = \omega_r + i\gamma$. Typically, for deeply trapped high energy particles, $\omega_d >> \omega$, so they are stabilizing. Low energy trapped particles, barely trapped particles, and passing particles act to destabilize the mode, but note that $\overline{J}$ is much larger for trapped than for passing particles. *Imag* $\delta\hat{W}_r$ is positive for $\gamma > 0$ and thus this contribution causes an increase in the frequency of the mode for trapped or passing particles. This frequency change can play an important role in the stabilization process.

A similar analysis, for $\gamma_I < \omega_{*i}/2$, shows there are initially two marginally stable modes with frequencies $\omega_{*i}/2 \pm \sqrt{\omega_{*i}^2/4 - \gamma_I^2}$, and the frequency change is given by

$$\frac{\Delta}{\omega_A} = \pm \frac{\gamma_I \delta\hat{W}}{\sqrt{\omega_{*i}^2/4 - \gamma_I^2}}, \qquad (8.130)$$

and thus *Imag* $\delta\hat{W}$ positive causes destabilization of the high frequency branch and stabilization of the low frequency branch. *Real* $\delta\hat{W}$ positive causes a frequency increase of the high frequency branch and a frequency decrease of the low frequency branch.

To go beyond a perturbative analysis analytically it is necessary to make further assumptions. First, consider a deeply trapped slowing-down particle distribution. Model a particle distribution in r, E, and μ (*i.e.* integrated over ϕ) by $dN = cE^{-1}f(r)\delta(\mu - \alpha E)dErdrd\mu d\theta$, for $E_0 < E < E_m$, with $f(r)$ the radial density profile normalized through $\int f(r)\,rdr = 1$. Then

$$\sum_\sigma F = \frac{B_0^2 V \beta_h \sqrt{1 - \alpha B}}{2^{9/2}\pi^4(E_m - E_0)} E^{-1/2}f(r)\delta(\mu - \alpha B). \qquad (8.131)$$

Neglecting the $\omega\partial_E F$ term in Q, and also using $\omega_d = EqK_2/(rK_b)$ and approximating q as one everywhere inside the rational surface, we find

$$\delta\hat{W} = \frac{\beta_h V c_f}{4\pi^2 r_s^2 a} \frac{\omega}{\omega_{dm}} \ln\left(\frac{\omega - \omega_{dm}}{\omega - \omega_{d0}}\right) \qquad (8.132)$$

with ω_{dm}, ω_{d0} the maximum and minimum precession frequencies in the distribution, a the minor radius, and c_f a shape dependent constant of order unity.

$$c_f = -\frac{1}{a}\int_0^a \frac{df}{dr} rdr. \qquad (8.133)$$

For a parabolic density profile, $f(r) \sim 1 - r^2/a^2$, $c_f = 8/3$. For a Gaussian profile, $f(r) \sim e^{-(r/h)^2}$, $c_f = \sqrt{\pi}a/h$. Here we have approximated the deeply trapped precession frequency $\omega_d = Eq/rR$ as independent of r. This approximation is necessary to obtain simple analytic expressions, but it can lead to errors in thresholds of as much as 50 percent. Using $V = 2\pi^2 a^2 R$

gives in the nonresistive limit the dispersion relation

$$-i\sqrt{\omega(\omega - \omega_{*i})} - \gamma_I + \frac{\beta_h \omega}{\pi \beta_c} ln\left(\frac{\omega - \omega_{dm}}{\omega - \omega_{d0}}\right) = 0 \qquad (8.134)$$

with the maximum precession frequency ω_{dm} corresponding to energy E_m, and the critical β given by

$$\beta_c = \frac{4r_s^2 \overline{\omega}_d}{\pi c_f a R \omega_A}, \qquad (8.135)$$

with $\overline{\omega}_d$ a weighted average of the precession frequency. The value is normally taken to be $\omega_{dm}/2$, but this is subject to the approximations made to perform the radial integrals.

In the absence of high energy particles, the frequency is given by $\omega = \omega_{*i}/2 + i\sqrt{\gamma_I^2 - \omega_{*i}^2/4}$. This is of course significantly modified if resistive effects are important.

Again, first consider $\gamma_I > \omega_{*i}/2$. Neglecting ω_{*i}, ω_{d0}, and γ_I and using the total plasma volume to be that contained within the $q = 1$ surface, (*i.e.* take $a = r_s$) this dispersion relation possesses two thresholds, one at $\omega = 0$ and $\beta_h = 0$ (the kink) and the second (the fishbone mode) at $\beta_h = \beta_c$, and mode frequency given by the average precession frequency $\overline{\omega}_d$. Note that since ω_d is proportional to particle energy the threshold condition is actually a requirement on hot particle density, not on beta.

For $0 < \gamma_I < .18\overline{\omega}_d$ and $\omega_{*i} > 0$ there are two thresholds, which correspond to the stabilization of the internal kink mode, and the fishbone threshold. Depending on the values of ω_{d0} and ω_{*i} the internal kink stabilization threshold may take on two very different characters. If stabilization occurs with ω greater than ω_{d0} the logarithm term is complex and ω must be greater than ω_{*i}. In this case the mode becomes damped for further increase in β_h. If instead stabilization occurs with ω less than ω_{d0}, the logarithm term is real and thus ω must be less than ω_{*i}. For further increase in β_h, the frequency remains real.

For $\omega > \omega_{d0}$ we have $\omega > \omega_{*i}$ and the mode frequencies are given by the solution of

$$\sqrt{\omega(\omega - \omega_{*i})} \; ln\left(\frac{\omega_{dm} - \omega}{\omega - \omega_0}\right) = \pi\gamma_I. \qquad (8.136)$$

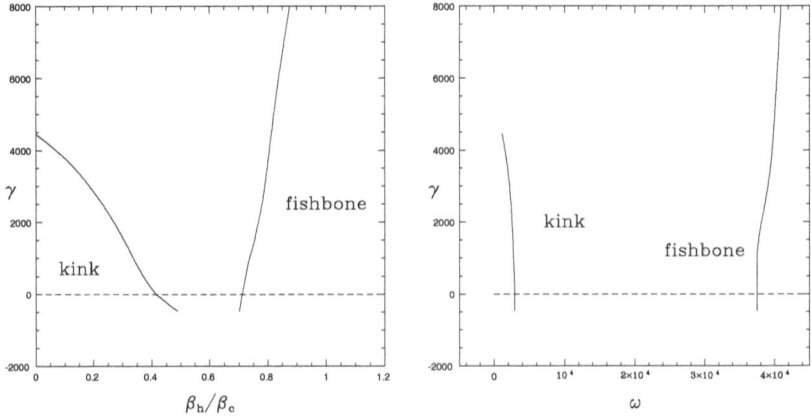

Fig. 8.17 Growth rate vs hot particle beta and the complex frequency plane for a trapped particle distribution with $\gamma_I > \omega_{*i}/2$.

The corresponding values of β_h are given by

$$\beta_h = \sqrt{1 - \frac{\omega_{*i}}{\omega}}\beta_c,\qquad(8.137)$$

a monotone increasing function of frequency. If, as is usual, $\omega_{dm} \gg \omega_{*i}$, γ_I, the two solutions can be found perturbatively with $\omega = \omega_{*i} + \Delta$, and $\omega = \omega_{dm}/2 - \Delta$.

If instead $\omega_{dm} \gg \omega_{d0} \gg \omega$ at threshold, then $\omega < \omega_{*i}$ and threshold is given by the solution of

$$\sqrt{\omega(\omega_{*i} - \omega)} - \gamma_I + \frac{\beta_h}{\pi\beta_c}\omega ln\left(\frac{\omega_{dm}}{\omega_{d0}}\right) = 0\qquad(8.138)$$

and

$$\frac{\omega_{*i}/2 - \omega}{\sqrt{\omega(\omega_{*i} - \omega)}} + \frac{\beta_h}{\pi\beta_c}\left[\frac{\omega}{\omega_{d0}} + ln\left(\frac{\omega_{dm}}{\omega_{d0}}\right)\right] = 0.\qquad(8.139)$$

The evolution of the real frequency as β_h increases further is determined by the first equation alone, and the growth rate remains zero.

For $\gamma_I > .18\overline{\omega_d}$ there is no threshold. In this case the kink mode branch frequency and growth rate increase as β_h increases. The fishbone branch instead remains stable.

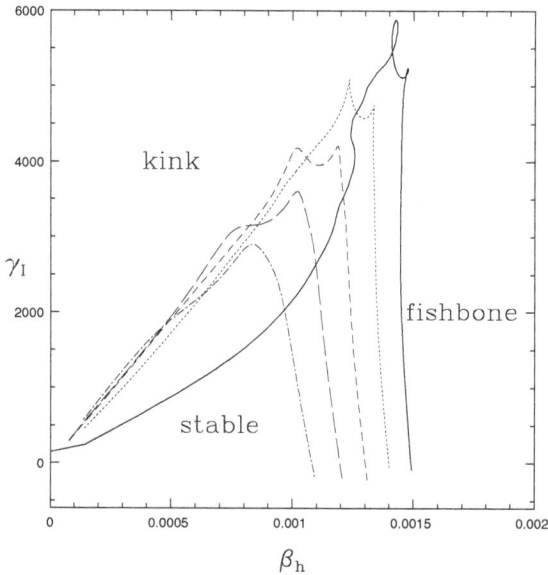

Fig. 8.18   Stable domain vs ω_{*i}, trapped particle distribution.

First consider the case $\gamma_I > \omega_{*i}/2$. We use in this case $\gamma_I = 4.5 \times 10^3$ /sec, R = 2.96 M, and very high aspect ratio, $\epsilon = .1$ at the $q = 1$ surface, $E_m = 100$ kev, $E_0 = 2$ kev, $\omega_A = 2 \times 10^6$ /sec, $\omega_{*i} = 2.3 \times 10^3/$ sec, the maximum precession frequency is 1.3×10^5 /sec, $h = a/2$. The q profile was taken to be quadratic in r, ranging from 0.8 to 1. In Fig. 8.17 are shown the growth rates as a function of hot particle beta, for both the kink branch and the fishbone branch, and the trajectories in the complex ω plane. There is a domain of β_h in which both modes are stable.

In Fig. 8.18 is shown the stable domain for a trapped particle distribution with a slowing-down energy distribution, for different values of ω_{*i}. The solid curve gives the stable domain (s) for small ω_{*i}. As the diamagnetic frequency increases, the stability domain contracts, primarily through a lowering of the fishbone threshold. The threshold curve is parameterized by the real frequency, and the vertical section of the boundary, on the right of the stable domain, is the high frequency, or fishbone threshold, part of the boundary, and the diagonal, left boundary of the stable domain is the low frequency, or kink stability, boundary. The fishbone onset and kink stabilization coincide for $\gamma_I = 0$.

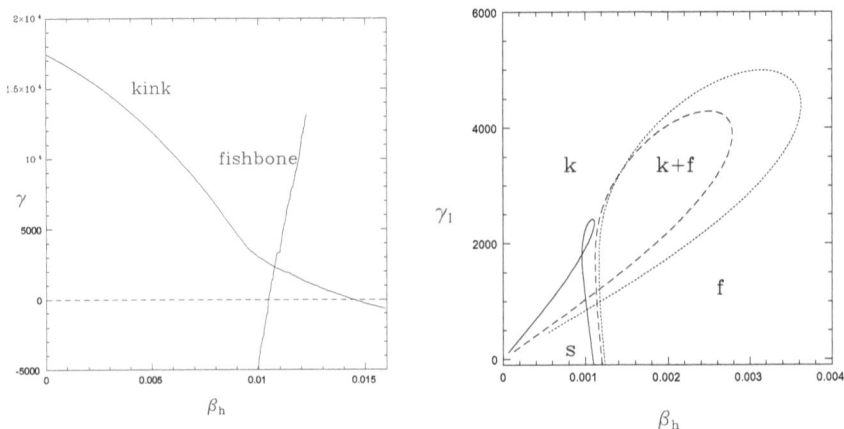

Fig. 8.19   Growth rate vs hot particle beta for a trapped particle distribution (left), and domains of stability (s), kink (k), and fishbone (f) versus hot particle beta and ideal growth rate (right).

In Fig. 8.19 are shown results obtained with somewhat different parameters. The hot particle distribution is a slowing-down distribution, but with the minimum energy only 1/3 of the maximum. In this case the fishbone is destabilized before the kink is stabilized. Also shown is the stable domain for this truncated slowing-down distribution, for different values of ω_{*i}. The solid curve gives the stable domain (s) for large ω_{*i}. As the diamagnetic frequency decreases, the domain with both kink mode and fishbone (f+k) increases. For positive values of γ_I there is always a range of β_h with both branches unstable.

Now consider $\gamma_I < \omega_{*i}/2$. In this case with $\beta_h = 0$ there are two marginally stable branches of the internal kink mode, with frequencies $\omega_{*i}/2 \pm \sqrt{\gamma_I^2 - \omega_{*i}^2/4}$. The low frequency branch is stabilized by the high energy particles, and the high frequency branch is destabilized, and has also been suggested as the cause of the fishbone in some discharges. However, this branch of the dispersion relation exists even in the absense of the high energy particles, and we will refer to it as the high frequency or ion diamagnetic frequency branch of the internal kink mode. The branch with a threshold at a nonzero value of β_h and at a frequency given by the trapped particle precession frequency we will refer to as the fishbone branch.

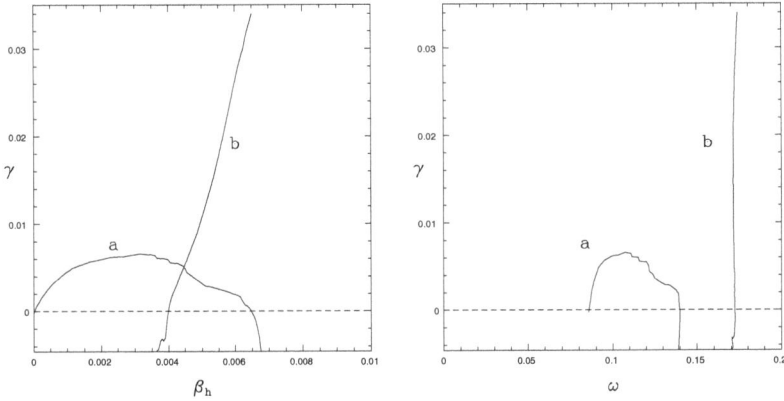

Fig. 8.20   Growth rate vs β_h and the complex frequency plane for a trapped particle distribution with $\gamma_I < \omega_{*i}/2$. The kink mode branch (a) and the fishbone branch (b) are shown.

In Fig. 8.20 are shown results with $\gamma_I < \omega_{*i}/2$. We show the growth rates and the trajectories in the complex frequency plane for the destabilized high frequency branch of the kink mode (a) and the precession frequency (fishbone) branch (b). For $\beta_h = 0$ the kink mode branch solution is marginally stable with real frequency very near the diamagnetic frequency. The fishbone branch is at $-i\infty$. As β_h increases the kink mode branch is immediately destabilized and its frequency increases. Above threshold the fishbone branch quickly dominates over the kink mode branch, which eventually stabilizes.

In this case the diamagnetic and precession frequencies are well separated, so the modes are easily distinguished, but if these frequencies are very close, it may be difficult to distinguish them in an experiment.

Also very important is a distribution uniform both poloidally and in pitch, for which the nonresonant contribution to $\delta \hat{W}$ can be evaluated explicitly (see Sec. 8.9.3). The results are independent of the form of the energy distribution. Writing the distribution in the form $F = f(r)w(E)$ we find for the trapped particle contribution

$$\delta \hat{W}_{nt} = \frac{\sqrt{2}\beta_h}{3\pi} \left(\frac{a}{r_s} \right)^2 \frac{\int_0^{r_s} r^{3/2}[1 + \frac{21r}{15}]\frac{df}{dr}\, dr}{\int_0^a rf\, dr} \tag{8.140}$$

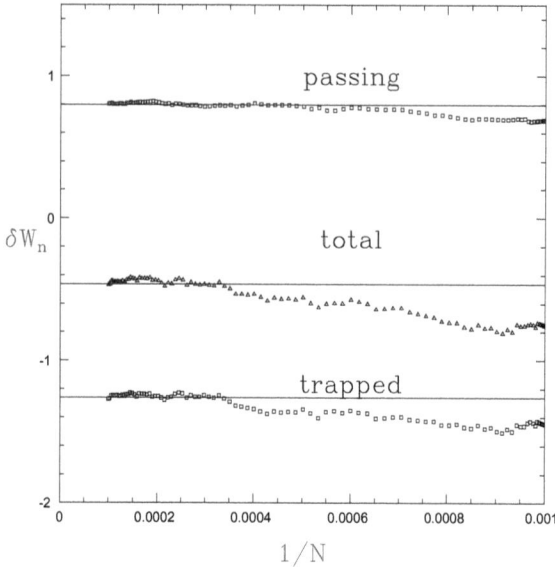

Fig. 8.21   Convergence of δW_n in particle number, $\beta_h = 1$.

where r_s is the $q = 1$ radius, and a the plasma radius. The total of the trapped and passing particle contributions $\delta \hat{W}_n = \delta \hat{W}_{nt} + \delta \hat{W}_{np}$ is

$$\delta \hat{W}_n = \frac{\beta_h}{4} \left(\frac{a}{r_s} \right)^2 \frac{\int_0^{r_s} r^2 \frac{df}{dr}\, dr}{\int_0^a rf\, dr}. \tag{8.141}$$

The numerical scheme employed involves convergence in the number of Monte Carlo particles used, which must be controlled to ensure that the desired accuracy has been obtained. This example presents more problems numerically than most cases, because in this case the leading order trapped and passing contributions to $\delta \hat{W}_n$, of order $\sqrt{R/r}$, have opposite sign and cancel, with the total $\delta \hat{W}_n$, of order unity, being significantly smaller than either. Furthermore, much of the contribution of the passing particles is due to particles near the trapped-passing boundary.

We use a large aspect ratio, $\epsilon = .1$, with a large magnetic field strength so that particle excursions away from the flux surface are negligible. In Fig. 8.21 are shown the convergence of the trapped, passing and total contributions to δW_n as a function of the number of particles used. The values shown are for $\beta_h = 1$. Both the trapped and passing contributions

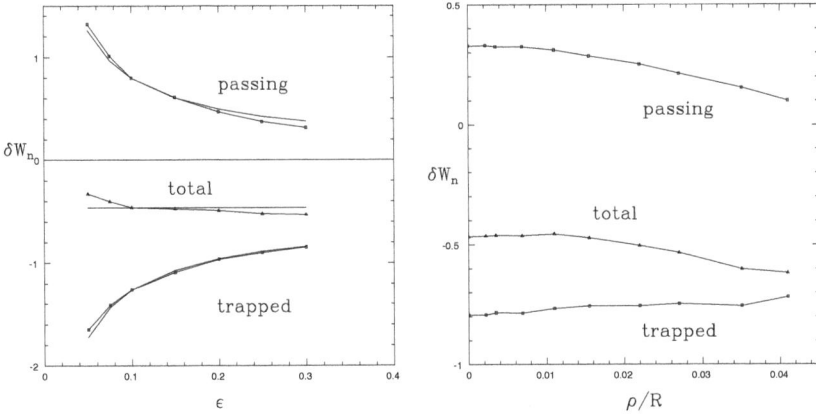

Fig. 8.22 Contributions to the dispersion relation term δW_n vs ϵ (left) and δW_n vs ρ/R, (right) for $\beta_h = 1$.

are seen to converge reasonably well to the analytic values (shown as lines) when the number of particles is a few thousand. Statistical fluctuations in $\delta \hat{W}_n$ are expected of order $(R/r)^{1/2}/\sqrt{N}$ because of the partial cancellation of the trapped and passing contributions. Here, the time step was set to require 20 steps per toroidal transit time ($2\pi R/v_{\parallel}$) and the orbit averages were done using 30 such toroidal transit times. For distributions other than uniform in pitch the numerical problems are much less severe, and we conclude that a few thousand particles are generally sufficient. For this example a Gaussian distribution $f(r) \simeq e^{-r^2/h^2}$ was used, for which

$$\delta \hat{W}_{nt} = -\frac{2^{5/2}\beta_h a^2 \int_0^{r_s/h} z^{5/2}[1 + \frac{21hz}{15}]e^{-z^2}dz}{3\pi r_s^2 \sqrt{h}[1 - e^{-a^2/h^2}]} \tag{8.142}$$

and

$$\delta \hat{W}_n = -\frac{\beta_h a^2[1 - e^{-r_s^2/h^2}(1 + r_s^2/h^2)]}{2r_s^2[1 - e^{-a^2/h^2}]}. \tag{8.143}$$

The q profile is quadratic in r with $.8 < q < 1$, and $r_s = a$, and $h = a/2$. Since the results for the nonresonant contribution are independent of the particle energy distribution, a monoenergetic distribution was used to restrict the particle gyro radius and drift excursions to a single value.

In Fig. 8.22 is shown the dependence of these terms as a function of the aspect ratio as well as analytic results, and also the dependence of these

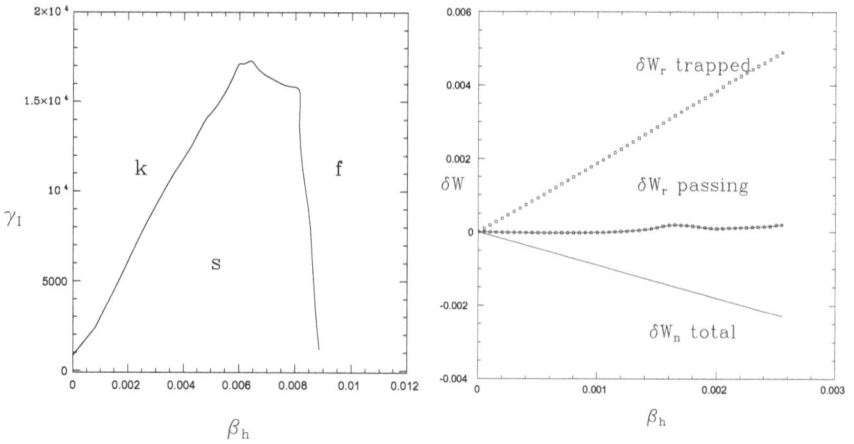

Fig. 8.23   Stable domain $\delta\hat{W}$ for a uniform pitch Maxwellian distribution. The left figure shows stable (s), kink (k) and fishbone (f) domains. On the right are shown the dependence of the contributions on β_h.

terms as the gyro radius is increased for $\epsilon = .25$. Since the aspect ratio is large, these results extend to values of the gyro radius equal to one fifth of the minor radius, and significant particle loss is encountered. The passing particle contribution is modified the most, as barely passing particles are not well confined, and they contribute the most. The trapped particle term is modified little, since barely trapped particles are lost, and they contribute the least.

In Fig. 8.23 is shown the stable domain for a uniform pitch distribution, Maxwellian in energy, with a temperature of 50 keV. In this case the fishbone threshold is at $\beta_h \simeq .01$, thirty times larger than the trapped particle case shown in Fig. 8.17. The factors producing this large increase are the small trapped particle fraction, $f_t = .16$, the larger value of $\overline{\omega}_d$ (the temperature is equal to the maximum energy of the trapped particle case), and the effect of the adiabatic contribution as a function of β_h during sawtooth stabilization. The resonant passing contribution is negligible, and the trapped resonant contribution is stabilizing. The combination of the resonant and adiabatic terms is only weakly stabilizing, and large β_h is required to reach the stable domain. Similarly destabilization of the fishbone also occurs at high β_h.

8.9.3 *Calculation of* δW_n

Recall that θ is a straight field line coordinate with a particular choice of Jacobian, not the usual cylindrical coordinate θ_c. It is defined so that $\vec{\nabla}\Psi_p \cdot (\vec{\nabla}\theta \times \vec{\nabla}\zeta) = B^2/(gq + I)$. In cylindrical coordinates instead $\vec{\nabla}\Psi_p \cdot (\vec{\nabla}\theta_c \times \vec{\nabla}\zeta) = 1/Xq$. Directly from the Hamiltonian equations, in these straight field line coordinates the integral for the bounce average contains a factor $B\sqrt{1 - \mu B/E}$ in the denominator rather than simply $\sqrt{1 - \mu B/E}$. Bounce averages of $H(\Psi_p, \theta)$, $J(\Psi_p, \theta)$, and the toroidal precession rate are needed to evaluate $\delta\hat{W}$. In evaluating analytic models it is convenient to perform these bounce averages in the cylindrical coordinate θ_c. If J is expressed in cylindrical coordinates the time integral of J in these coordinates is

$$\int J dt \simeq \int J(\theta_c)\frac{d\theta_c}{\sqrt{1 - \mu B/E}} = \int J(\theta_c)B\frac{d\theta_c}{B\sqrt{1 - \mu B/E}}. \quad (8.144)$$

But in this last expression θ_c is a dummy integration variable, and the integral is the time integral of BJ in straight field line coordinates, provided the numerical evaluation is done with $B(\Psi_p, \theta)$ taken to be the same function as was used in the analytic model, e.g. if in the analytic model $B(\Psi_p, \theta_c) = 1 - cos(\theta_c)$, then in the numerical evaluation $B(\Psi_p, \theta) = 1 - cos(\theta)$. This can be a more convenient way to evaluate analytic models with quantities given in cylindrical coordinates than re-expressing everything in terms of straight field line coordinates.

Note that the particle distribution, if uniform spatially, is not uniform in either θ_c or θ, but must be weighted by the appropriate power of B in the Jacobian. In many cases this is a small correction, but in the case of a distribution which is uniform poloidally and also in pitch (a Maxwellian or an alpha particle distribution) the leading order contribution to $\delta\hat{W}_n$ is zero, and only the second order survives.

For a distribution uniform poloidally and in pitch $\lambda = \sqrt{1 - \mu B/E}$, and using the cylindrical coordinate θ_c and neglecting $q - 1$ corrections, J and H become

$$J = -(1 - \mu B/2E)\frac{cos(q\theta)}{X}, \qquad H = -\frac{cos(q\theta)}{qX}. \quad (8.145)$$

The nonresonant contribution $\delta \hat{W}_n$ then takes the form

$$\delta \hat{W}_n = \frac{2^{7/2}\pi}{r_s^2 B_0^2} \int \frac{E^{3/2}cos(\theta_c)}{B} \partial_r F\, r dr d\theta_c d\lambda dE \qquad (8.146)$$

with F a function of E and r only. Write F in the form $F = w(E)f(r)$ and normalize through β_h, giving to two orders in inverse aspect ratio

$$\delta \hat{W}_n = \frac{\beta_h a^2}{8\pi r_s^2} \frac{\int \frac{cos(\theta_c)}{B} \partial_r f\, r dr d\theta_c d\lambda}{\int f r dr}. \qquad (8.147)$$

The trapped particle contribution is given by the restriction $|\lambda| < \sqrt{1 - B(r,\theta)/B(r,\pi)}$ and the total contribution by $|\lambda| < 1$. The total then becomes

$$\delta \hat{W}_n = \frac{\beta_h a^2}{4r_s^2} \frac{\int \partial_r f\, r^2 dr}{\int f r dr} \qquad (8.148)$$

and, substituting $w = sin(\theta/2)$ to facilitate the integration, the trapped particle contribution is

$$\delta \hat{W}_{nt} = \frac{\beta_h a^2}{3\pi r_s^2} \frac{\int \sqrt{2r}[1 + \frac{21r}{15}]\partial_r f\, r dr}{\int f r dr} \qquad (8.149)$$

with $\delta \hat{W}_n = \delta \hat{W}_{nt} + \delta \hat{W}_{np}$ and the subscripts refer to nonresonant trapped and nonresonant passing respectively.

8.9.4 *Monte Carlo evaluation*

If $\omega = \omega_r + i\gamma$ is in the vicinity of the real axis and there are Monte Carlo points near ω_r, the discreteness of the sum in the representation of the resonant part of the dispersion relation can give large errors. Consider the discrete representation of an integral

$$W = \sum \frac{g(\omega_{dk})}{\omega_{dk} - \omega} = \int \frac{dn}{d\omega_d} \frac{g(\omega_d)}{(\omega_d - \omega)} d\omega_d \qquad (8.150)$$

with $dn/d\omega_d$ the (smooth probability) density of points in the interval $d\omega_d$. The discrete sum clearly fails to give an accurate representation of the integral when $\gamma < d\omega_d/dn$, the discrete particle spacing on the real axis in the vicinity of ω_r. Since we are interested in mode stabilization and destabilization it is necessary to modify the numerical algorithm for the evaluation of the sum when ω is near the real axis. Use a domain D near the resonance chosen sufficiently large to contain many Monte Carlo points, $|\omega_r - \omega_d| < D$. Then evaluate the principal parts and the delta function contributions through

$$P\sum_k \frac{g(\omega_{dk})}{\omega_{dk} - \omega} = \sum_{k \neq D} \frac{g(\omega_{dk})}{\omega_{dk} - \omega_r} \qquad (8.151)$$

and

$$d\sum_k \frac{g(\omega_{dk})}{\omega_{dk} - \omega} = \sum_{k=D} g(\omega_{dk}). \qquad (8.152)$$

Then for $0 < \gamma < D$, use a linear combination of $W(D)$ and $P + i\pi d$, and for $-D < \gamma < 0$, use a linear combination of $W(-D) + 2i\pi d +$ and $P + i\pi d$, and for $\gamma < -D$ use $W(-D) + 2i\pi d$. Since we are only interested in threshold phenomena, and not in following roots of the dispersion relation far below threshold, this procedure is quite adequate.

Introduce the pitch $\lambda = v_\parallel / v = \sqrt{1 - \mu B / E}$. The particle density is then

$$dn = 4\pi^2 \sqrt{2E} F(E, \mu, r, \theta) \mathcal{J} d\lambda dE d\theta d\Psi_p. \qquad (8.153)$$

A Monte Carlo distribution can be constructed by selecting points uniformly distributed in $\lambda, E, \theta, \Psi_p$ and accepting them as part of the distribution with probability $4\pi^2 \sqrt{2E} F(E, \mu, r, \theta) \mathcal{J}$. (Directly calculating the distribution using the variable μ is not satisfactory because of the resulting factor of $\sqrt{1 - \mu B / E}$ appearing in the denominator of the probability, which is unbounded.) For most plasma physics applications the distribution function has exponential or near exponential form in both energy and radius. This need not be a large aspect ratio approximation. The radius can refer to the flux value on the midplane $\theta = 0$. It is then convenient to perform this part of the Monte Carlo generation analytically. Suppose the distribution to have the form

$$F = g(E, \mu, r, \theta) e^{-r^2/h^2} e^{-E/T} \qquad (8.154)$$

with $g(E, \mu, r, \theta)$ a relatively slowly varying function of its arguments. The particle density is then

$$dn \simeq \sqrt{E} g(E, \mu, r, \theta) e^{-r^2/h^2} e^{-E/T} \mathcal{J} \frac{d\Psi_p}{dr^2} d\lambda dE d\theta r dr. \qquad (8.155)$$

Introduce the variables $1 - z = e^{-r^2/h^2}$ and $1 - w = e^{-E/T}$, giving

$$dn \simeq \sqrt{E}g(E, \mu, r, \theta)\mathcal{J}(d\Psi_p/dr^2)d\lambda dz d\theta dw, \qquad (8.156)$$

and thus the distribution is formed by selecting points uniformly distributed in θ, λ z, w and then making the corresponding $E = -Tlog(1 - w)$, $r^2 = -h^2log(1 - z)$, $\mu = E(1 - \lambda^2)/B$ part of the distribution with probability $\sqrt{E}g(E, \mu, r, \theta)\mathcal{J}d\Psi_p/dr^2$. The weak dependence of this function ensures a Monte Carlo generation of the distribution with a small amount of computing.

Typically, an analysis of a given equilibrium and particle distribution requires a few thousand Monte Carlo particles and bounce averaging for 20 or 30 transit times $(2\pi R/v_{\parallel})$. Thus it is relatively easy to perform scans of parameter values.

In order to calculate $\delta \hat{W}_k$ we need to know both $\partial_{\psi_p} \log F$ and $\partial_E \log F$. This cannot be done directly if each particle is represented as a point in phase space. Either phase space must be partitioned and local densities calculated, or equivalently each particle can be assigned a finite size in both real and velocity space.

Rewrite the Monte Carlo distribution function in the form

$$F = \sum_k^N \frac{1}{(\psi_p - \psi_{pk})^2 + a^2} \frac{1}{(E - E_k)^2 + b^2} \qquad (8.157)$$

where a and b are the particle widths in real and energy space respectively. We ignore the normalization constant since we are only interested in $\partial_E F/F$ and $\partial_{\psi_p} F/F$. Partial derivatives with respect to ψ_p and E are given by

$$\partial_{\psi_p} F = \sum_k^N \frac{-2(\psi_p - \psi_{pk})}{[(\psi_p - \psi_{pk})^2 + a^2]^2} \frac{1}{(E - E_k)^2 + a^2} \qquad (8.158)$$

and

$$\partial_E F = \sum_k^N \frac{-2(E - E_k)}{[(E - E_k)^2 + b^2]^2} \frac{1}{(\psi_p - \psi_{pk})^2 + a^2}. \qquad (8.159)$$

This method gives a very accurate evaluation in test cases using known distribution functions.

8.10   TAE MODE DRIVE AND SATURATION

For the most part, wave particle interaction is adiabatic, with energy passing back and forth between the particles and the wave. However, some particles through the formation of phase space islands become trapped in the wave. This modification of the particle distribution drives the wave if the distribution has a density gradient, since, as we have seen, outward motion implies energy loss. In Fig. 8.24 is shown the particle distribution modification produced by the presence of the phase space island produced by a TAE mode. The final distribution (solid line) shows a net outward motion of part of the distribution, and hence a positive energy transfer from the distribution to the mode. This figure is only schematic, as not all particles at the resonant surface participate in the resonance, and the degree of broadening of the profile depends on the details of the distribution function and the resonance. This process produces the driving mechanism for the TAE mode, which in the absence of the particle drive has only a real frequency.

The calculation of the growth rate depends on the details of the equilibrium and the particle distribution, and must be carried out numerically. For typical equilibria the TAE mode is destabilized by α particles when the α particle pressure is large enough to overcome Landau damping, typically when $\omega_{\alpha*}/\omega_A > 1.5$ with $\omega_{\alpha*}$ the diamagnetic drift frequency of the alphas, and increases approximately linearly with $\omega_{\alpha*}$ for larger values, but γ/ω_A is typically less than 10^{-2} for $\omega_{\alpha*}/\omega_A = 10$. For the details of the numerical calculation of the growth rate, see Cheng (1992).

Due to the large alpha particle pressure in an ignited reactor-size tokamak, the TAE could grow to a sufficient level to cause energetic particles loss. Thus it is of great importance to know the TAE saturation level. For a single TAE mode, resonant particles exchange energy with the wave. If they lose sufficient parallel energy they can become trapped poloidally. If the resulting trapped orbit has a sufficiently large excursion this process can lead to particle loss, as shown in Fig. 8.25. Examination of the particle energy as a function of time shows that the resonance occurs only over a small part of the orbit, due to the variation of q caused by the drift motion, and that the energy is modified randomly up and down, until on the final transit sufficient parallel energy is lost to cause transition to a trapped orbit.

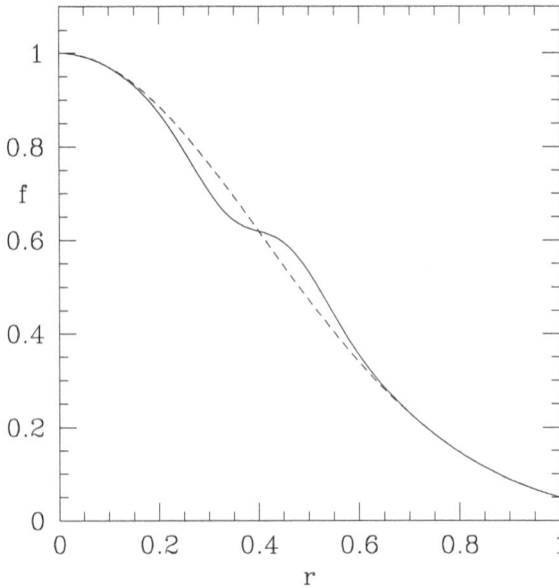

Fig. 8.24   Particle distribution modification by TAE mode.

In addition, the flattening leads to mode saturation through the classical bump-on-tail mechanism (O'Neil, 1965). If the width of the phase space island is less than the spacing between nearby resonances produced by other poloidal harmonics, as is normal for typical reactor parameters, then there is no loss, but if the mode saturation amplitude is sufficiently large the overlap of nearby islands can lead to catastrophic avalanche loss. Thus examination of single mode saturation levels can provide sufficient information for the avoidance of large scale TAE induced loss. In Fig. 8.26 is shown a phase space plot of a single resonance produced by a saturated TAE mode.

For simplicity we thus consider a single toroidal eigenvalue, n. Due to the toroidal bending in tokamaks, all poloidal harmonics m are coupled. Due to this coupling, there is a gap in the spectrum due to the cylindrical harmonics m and $m+1$. In Fig. 6.9 is shown the dispersion relation for a pair of coupled Alfvén waves. In this gap, located at the point r_0 with $q(r_0)$ between m/n and $(m+1)/n$, there exists a discrete TAE mode with a frequency given by the mode frequency at $q \simeq (m+.5)/n$, or $\omega \simeq V_A/(2qR)$,

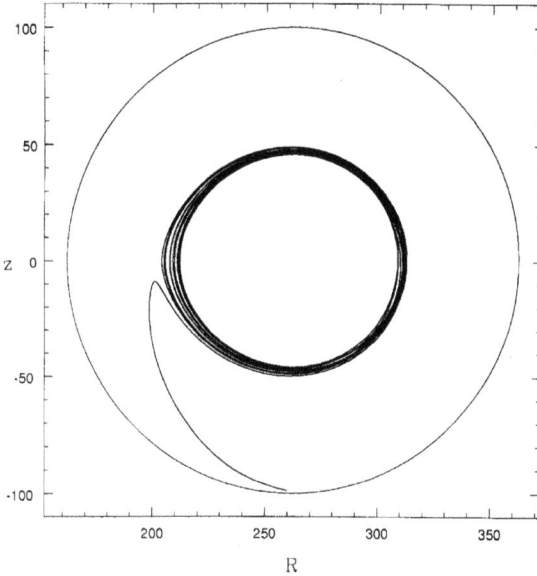

Fig. 8.25   Particle loss by TAE resonance through transition from a counter-passing orbit to a banana loss orbit.

consisting primarily of the two poloidal harmonics $m, m+1$. The resonance condition between the wave and the alpha particles is $(n - m/q)\omega_t = \omega$ and thus $\omega_t = 2q\omega$.

When the trapping frequency, given by Eq. 8.47, exceeds the growth rate, trapped particles remove energy too rapidly to permit growth (O'Neil, 1965). The TAE saturation level can be estimated by setting $\omega_T = C\gamma_L$, where C is a numerical factor and γ_L is the linear growth rate. Thus saturation occurs for

$$\frac{\delta B_r}{B} \simeq \frac{2C^2\gamma_L^2}{\omega_t\omega\left[(8m^2sR^2q^2\omega)/(r^2\omega_0) - 1\right]}. \tag{8.160}$$

Once we have found the TAE saturation level, the particle energy and radial excursions in the island formed by the drift orbit trajectories can be found readily using Eq. 8.43 and Eq. 8.44. In the reactor design device ITER typical values are $\rho \simeq 0.2$ cm, the alpha pressure is large at a minor radius of $r \simeq 50$ cm, the gyro frequency is $\omega_0 \simeq 3 \times 10^8$/sec, the alpha particle transit frequency is $\omega \simeq 5 \times 10^4$/sec, $R \simeq 900$ cm, in which case

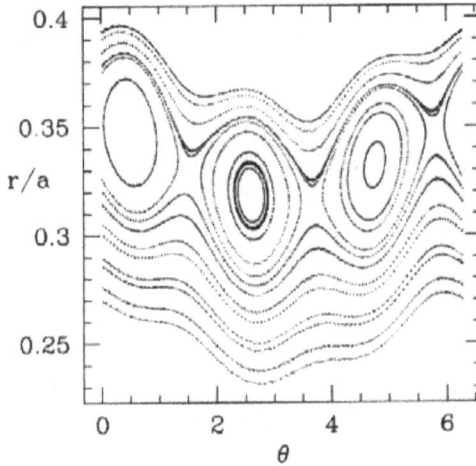

Fig. 8.26   TAE induced drift island vs minor radius, showing the resonance island as well as the outward shift of the orbits.

the denominator in Eq. 8.160 is dominated by the shear term for $s > 1$, giving

$$\frac{\Delta \mathcal{E}}{\mathcal{E}} \simeq \frac{4Cr^2\gamma_L}{sm^2\omega_0\rho^2}. \tag{8.161}$$

The linear growth rate for the TAE mode is typically $\gamma_L \simeq 10^{-2}\omega$, giving $\Delta \mathcal{E}/\mathcal{E}$ of order 10^{-1}, and using Eq. 8.5 the width in r is

$$\frac{\Delta r}{r} \simeq \frac{2C\gamma_L}{m\omega s}, \tag{8.162}$$

which is of order 10^{-2}. Using the typical resonance surface separation $\Delta_m = 1/nq'$, we find the Chirikov stochastic parameter S (island width divided by mode separation) is

$$S \simeq \frac{C\gamma_L(2 + \frac{1}{m})}{\omega}. \tag{8.163}$$

As is found numerically (Wu and White, 1994), using a guding center code with the mode particle coupling given by Sec. 8.9, the constant C is about 4 for typical distributions and profiles. Since there are many resonant surfaces with somewhat different saturation characteristics, the mode saturation involves an averaging process, and will depend somewhat on the form of the distribution and profiles. However, $\gamma_L/\omega \simeq 10^{-2}$ so a single n TAE mode is very far from producing stochastic loss.

8.11 TRAPPED PARTICLE RESONANCE

For trapped particles the orbits are too complex for complete resonance in the sense that $n\dot\zeta - m\dot\theta - \omega = 0$ for all time, but if the phase change is cyclic in the bounce period then some quantities will not average to zero, giving secular orbit changes. Thus we demand only that after one bounce and time T the phase $Q = n\zeta - m\theta - \omega t$ returns to its initial value within $2\pi l$ with l integer. The time T is given by $T = 2\pi/\omega_b$ with ω_b the bounce frequency, and we have $\theta(T) = \theta(0)$, giving $n\omega_d T - \omega T + 2\pi l = 0$, or

$$n\omega_d + l\omega_b = \omega, \qquad (8.164)$$

for some integer l. These are called bounce-precession resonances. Rewrite Eq. 8.59 as

$$\frac{d\mathcal{E}}{dt} = i\left[-n\dot\zeta_d + m\dot\theta_d\right]\Phi_{mn}(\psi_p)e^{iQ} - \Phi'_{mn}\dot\psi_p e^{iQ}. \qquad (8.165)$$

Although the bounce frequency changes significantly with increasing bounce angle, the bounce motion continues to be dominated by the fundamental. For a pendulum the leading correction to the harmonic content is $\theta = (\theta_b + \delta)sin\omega_b t + \delta sin3\omega_b t$ with $\delta = \theta_b^3/192$. See Problem 1 at the end of the chapter. Typically for a trapped orbit, even for bounce angle of $\theta_b = 2.5$, the higher harmonics are an order of magnitude smaller than the fundamental. Thus write $\theta = \theta_b sin(\omega_b t)$, $\zeta = \omega_d t + q\theta_b sin\omega_b t$, $\psi_p = \psi_0 + \rho_b e^{i\omega_b t}$ with ψ_0 the banana center and ρ_b the banana width, note $<\dot\zeta_d> = \omega_d$ with $<>$ indicating bounce averaging, and use the usual Bessel expansion $e^{ia\theta} =$

$\sum_l J_l(a\theta_b)e^{il\omega_b t}$ to find for the energy change of a trapped particle

$$\frac{d\mathcal{E}}{dt} = -i\omega_d\Phi_{mn}(\psi_0)\sum_l J_l((nq-m)\theta_b)e^{iQ_l}$$

$$-i\omega_d\rho_b\Phi'_{mn}(\psi_0)\sum_l J_l((nq-m)\theta_b)e^{iQ_{l+1}}$$

$$-i\omega_b\rho_b\Phi'_{mn}(\psi_0)\sum_l J_l((nq-m)\theta_b)e^{Q_{l+1}}, \qquad (8.166)$$

$$\frac{dQ_l}{dt} = n\omega_d + l\omega_b - \omega, \qquad (8.167)$$

with now $Q_l = n\omega_d t + l\omega_b t - \omega t$.

Expand $dQ = \partial_E Q(E - E_0)dt$ about the resonance $n\omega_d + l\omega_b = \omega$, and use $\partial_E\omega_d = \omega_d/E$, $\partial_E\omega_b = \omega_b/2E$. The ω_b term dominates, giving an island in energy of the form $(E - E_0)^2/2 = k\sin Q + c$, all quantities evaluated at the resonance $n\omega_d + l\omega_b = \omega$, with k depending on the magnitude of $\Phi'\rho_b\omega_b/\Phi\omega_d$. If Φ' is small $k = 4nE\Phi_{mn}\omega_d J_l(m - nq)/q\omega_b$. The island width in energy is $4\sqrt{k}$. This resonance island, existing in the energy variable, the radial variable, and the frequency, causes non reversible energy transfer between the wave and the particle distribution, provided either energy or radial gradients exist in the distribution.

The dominant contribution in this energy exchange is determined by the particle distribution, the radial extent of the mode Φ_{mn} and the radial dependencies of q, and the bounce and precession frequencies. In the first observed cases the fishbone, which is primarily an $n = 1, m = 1$ mode, has been dominated by the $l = 0$ resonance with $\omega \simeq \omega_d$, the precession frequency. However, if the trapped particle population has a large mean bounce angle, θ_b, the mean value of J_1 can be comparable to J_0. Depending on thresholds and growth rates, the bounce averaged energy transfer due to the $l = 1$ term can dominate the mode-particle energy exchange, producing a mode with a frequency given approximately by the mean bounce frequency of the distribution.

Assuming $\omega_b \gg \omega$ only the $l = 0$ term survives. For a deeply trapped particle $\omega_d = \mathcal{E}q/r$, so using Eq. 8.5, and using $\omega_d \simeq \omega$, we have

$$\partial_{\mathcal{E}}\omega_d \simeq \frac{\omega_d}{\mathcal{E}}\left[1 + (1 - s)\frac{mR}{qr}\right]. \qquad (8.168)$$

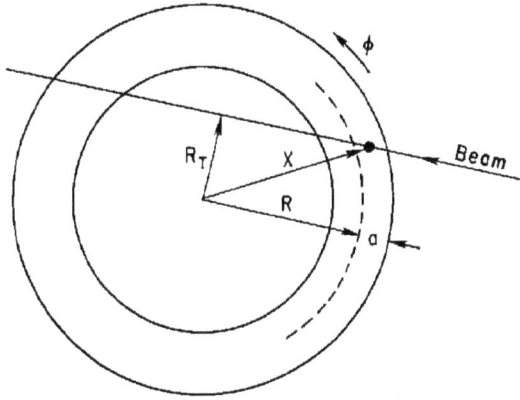

Fig. 8.27   Neutral beam injection geometry.

For less deeply trapped particles ω_d is smaller than the deeply trapped value, see Fig. 3.10. The island width in energy is $4\sqrt{k}$, and using Eq. 8.5 we find the radial extent of the island to be

$$\frac{\Delta r}{R} = 4\sqrt{\frac{qJ_0\delta B_r/B}{[1 + (1 - s)mR/(qr)](nq - m)}}. \tag{8.169}$$

The fishbone mode can attain $\delta B_r/B \sim 10^{-3}$ giving $\Delta r/R \simeq 10^{-1}$, a very large excursion. In these conditions the mode can produce extensive coherent loss.

Normally the precession rate is a monotone function of radius, and thus there is a single large island in phase space, with the position depending on particle energy.

If the particles are not deeply trapped, so that the average value of θ_b for the distribution is not small, then the particle resonance can be dominated by the $l = 1$ term in Eq. 8.166, leading to bounce frequency fishbones.

8.12   FISHBONE INDUCED LOSS

The geometry for neutral beam injection is shown in Fig. 8.27. On most
tokamaks the beam is injected along the horizontal midplane, although this
is not always the case. The particle's initial pitch $(v_\parallel/v)$ is determined at
the moment of ionization, with $\lambda = v_\parallel/v = \vec{v} \cdot \vec{B}/vB$. If the injection is
in the horizontal midplane, then $\lambda = cos\phi$ to lowest order in ϵ, with ϕ
measured from the perpendicular distance drawn between the vertical axis
and the beam line, R_T. If R_T is small compared to R, the initial velocity
is mostly across the field, and almost all of the particles ionized before
crossing the magnetic axis are trapped. For a deeply trapped particle the
toroidal precession frequency, as derived in Sec. 3.5, is $\omega_d = Eq/(mrR\omega_0)$.
In addition, particles with smaller r have a higher value of initial pitch,
and ω_d is smaller than the deeply trapped value. Shown in Fig. 8.28 is the
precession frequency ω_d as a function of r for different values of R_T. The
parameters used correspond to 50 keV deuterium ions in a 10 kG low q
discharge in PDX. For the beam injection used, $R_T = 36$ cm, the trapped
particles are seen to give a very well-defined precession frequency at 20 kHz.

Now consider the effect of a rotating mode on the particle population.
Consider δB to be rotating in the laboratory frame. In the frame moving
with the perturbation there is an electric field $\delta \vec{E} = \vec{v} \times \delta \vec{B}$, which gives
rise to a δE_θ. This in turn produces an $\vec{E} \times \vec{B}$ drift of the particles which
is radial. Particles moving in phase with the perturbation experience a
secular radial drift. This results in the island given by Eq. 8.169.

The bounce averaged radial motion is given by Eqs. 8.5 and 8.166,

$$\frac{dr}{dt} = \frac{\omega}{(n-m/q)}\frac{\delta B_r}{B}cos(Q)J_0(m\theta_b). \tag{8.170}$$

For typical PDX parameters this loss mechanism is very effective, capable
of ejecting a particle from deep within the plasma in 100 microseconds, as
shown in Fig. 8.30. The sign of dr/dt depends on the particle location with
respect to the mode phase $Q = n\zeta - \omega t$, and thus the ejected particles
form a rotating beacon, rotating toroidally with the mode frequency.

With the improved diagnostic tools developed since the first fishbone
loss simulations (White, 1983), more can be said about the loss resonance.
The particle distributions given by the geometry of Fig. 8.27 for two different values of R_T are shown in Fig. 8.29. For $R_T = 36$ cm the particles are
not trapped, they are all in stagnation orbits. For $R_T = 55$ cm almost half

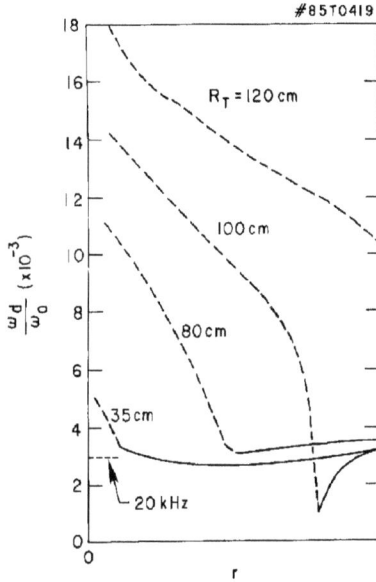

Fig. 8.28   Injected beam precession rate vs minor radius.

of the particles are in stagnation orbits and half are trapped. For R_T larger than this orbits are approximately half trapped and half co-passing. The plot of Fig. 8.30 shows the loss orbit of a stagnation particle, not a banana orbit. It is difficult to display the beam resonance, since a Poincaré plot requires all particles to have the same values of $P_\zeta = \omega E/n$ and the injected beam particles do not satisfy this. However, choosing a value approximating the distribution, we find a large resonance with stochastic domains near the plasma border, indicating induced loss, even using values of the mode perturbations smaller than the experimental ones. See Fig. 8.30.

8.13   FISHBONE DESTABILIZATION

Resonant energy transfer between a mode and a trapped particle is treated in Sec. 8.11. Monte Carlo simulations of particle loss carried out by White *et al.* (1983) showed that the total energy transfer to the trapped particles was approximately equal to the MHD energy in the mode, so that it was

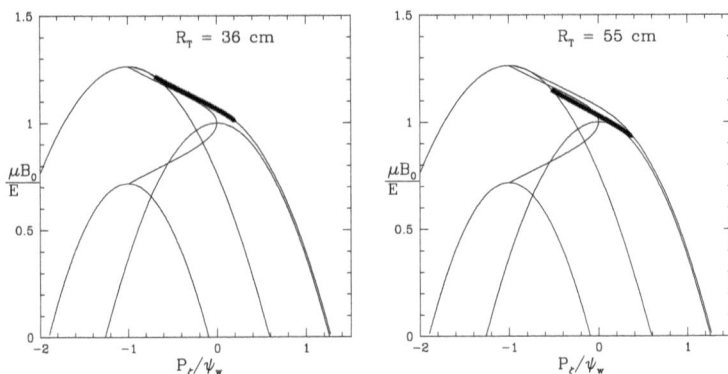

Fig. 8.29 PDX beam deposition. For near perpendicular injection ($R_T = 36$ cm) the particles are in stagnation orbits, and only for more parallel injection are there trapped and co-passing orbits.

energetically feasible that the particles were destabilizing the mode. The resonant interaction implies that the energy expressions in a δW formalism are non self-adjoint and give rise to complex frequencies. Numerical determination of the spectra of such non Hermitian operators is considerably more complicated than the usual MHD self-adjoint case, but codes have been developed for this purpose by Kerner *et al.* (1985) and Cheng and Chance (1987).

An analysis of the mode destabilization was carried out by Chen *et al.* (1984). The plasma is imagined to consist of two components, a relatively cold MHD part and a hot trapped particle component, treated with a gyrokinetic description. In the gyrokinetic description each particle is characterized by an energy E, magnetic moment μ, and minor radius r. Finite banana width corrections are ignored. A variational dispersion function $D(\xi)$ is found of the form

$$D(\xi) = \delta W_{MHD} + \delta W_k + \delta I, \qquad (8.171)$$

where δW_{MHD} is the usual MHD expression, derived in Sec. 6.5, δW_k the energy change in the hot trapped particles due to the mode, and δI the inertia term. This functional is not self-adjoint, because of the resonant form of the mode-particle interaction. However, its variation leads to the equations of motion for the coupled MHD kinetic system.

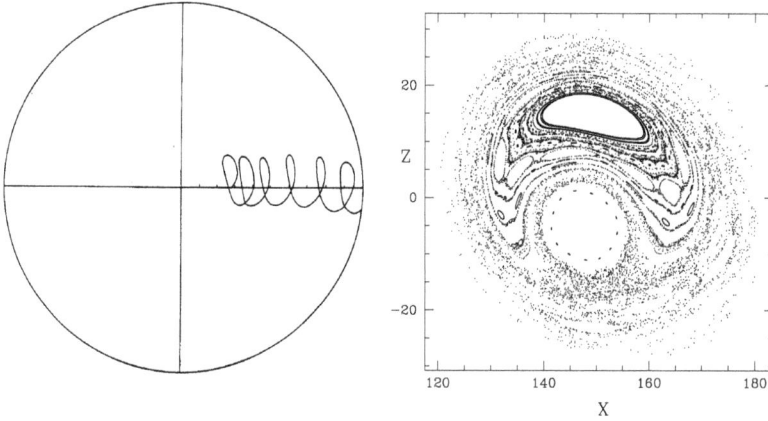

Fig. 8.30 Trajectory of an ejected beam ion, initially a stagnation orbit, and a fishbone resonance showing the large $m = 1$ island produced in the particle orbits.

The ideal MHD energy is given by Eq. 6.30,

$$\delta W_{MHD} = \frac{1}{2}\int d\tau \left[\begin{array}{c} \vec{Q}_\perp^2 - \frac{j_\parallel}{B}(\vec{\xi}_\perp \times \vec{B}) \cdot \vec{Q}_\perp \\ -2(\vec{\xi}_\perp \cdot \nabla p)(\vec{\xi}_\perp \cdot \vec{\kappa}) + B^2(\nabla \cdot \vec{\xi}_\perp + 2\vec{\xi}_\perp \cdot \vec{\kappa})^2 \\ +\gamma p(\nabla \cdot \vec{\xi})^2 \end{array} \right] \quad (8.172)$$

Now consider the kinetic energy of the particles. The full particle distribution is described by the density

$$d^6 N = d^3 v d^3 x F. \qquad (8.173)$$

The spatial volume is $d^3 x = X d\phi r d\theta dr$. It is convenient to express the volume in velocity space $d^3 v = 2\pi v_\perp dv_\perp dv_\parallel$ in terms of the particle energy and magnetic moment. The change of variables gives

$$d^3 v = \sum_\sigma \frac{2\pi B dE d\mu}{[2E(1 - \alpha B)]^{1/2}}, \qquad (8.174)$$

where $\alpha = \mu/E$, and $\sigma = sign(v_\parallel)$. The particle distribution is then

$$d^3 N = 2^{7/2}\pi^3 dE d\alpha dr dr E^{1/2} K_b F \qquad (8.175)$$

with

$$K_b = \frac{1}{\pi} \int_0^{\theta_b} \frac{d\theta}{(1 - \alpha B)^{1/2}} \tag{8.176}$$

an elliptic integral arising from the bounce averaging. The hot trapped particle beta, defined by $\beta_h = (2/B^2 V) \int d^3 N E$ with the volume $V = 2\pi^2 R r_s^2$, is given by

$$\beta_h = C_\beta \int r dr dE d\alpha E^{3/2} K_b F \tag{8.177}$$

with $C_\beta = 2^{7/2} N \pi / r_s^2$, where N is the number of particles and we normalize F to one using Eq. 8.175, or, as is normally more convenient, use the value of β_h as the normalization describing the number of particles.

In a toroidally uniform hot trapped particle distribution the total energy transfer between the mode and trapped particles integrates to zero. The energy transfer is thus second order in ξ, coming from the product of the adiabatic particle distribution response $\xi \delta F$ and the energy transfer. Thus the energy has a resonance structure of the form

$$\delta W_k = \int d^3 v \frac{\xi^2 \delta F}{\omega - \omega_d}. \tag{8.178}$$

The calculation of δF follows from gyrokinetic theory (Catto *et al.* 1981), see also Sec. 8.9.2. The result is

$$\delta W_k = 2^{9/2} \pi^3 \int r dr d\mu dE E^{3/2} \frac{K_b \vec{J}^* Q \overline{J}}{\omega_d - \omega}, \tag{8.179}$$

where

$$Q = \omega \frac{\partial F}{\partial E} - \frac{1}{\omega_0 r} \frac{\partial F}{\partial r}, \tag{8.180}$$

$$\vec{J} = \frac{\alpha B}{2} \nabla \cdot \vec{\xi}_\perp - \left(1 - \frac{3\alpha B}{2}\right) \vec{\xi}_\perp \cdot \vec{\kappa}, \tag{8.181}$$

and the bar indicates bounce averaging. Here and in the following only the lowest order in ϵ is retained, and the usual normalization of B to the on-axis value and distances to the major radius are used. Since $\omega_d = \omega_d(E, \alpha, r)$, δW_k is defined by this integral in the upper half frequency plane, and must be analytically continued into the lower half plane.

The contributions to Q are understandable in terms of the two basic effects of the mode on individual particles, as discussed in Sec. 8.1 and given by Eq. 8.6. The $\omega \partial F / \partial E$ term comes from the induced energy change of the particle and is negative and stabilizing for $\partial F / \partial E < 0$, which is normally the case. The relevant frequencies are the order of the precession frequency, $\omega \sim E/(rR\omega_0)$. Using $\partial F/\partial E \sim F/E$ this term is of order $F/(\omega_0 Ra)$. The second term, due to the radial displacement, is of the order $F/(\omega_0 a^2)$ and dominates. This term is destabilizing for $\partial F/\partial r < 0$. The driving mechanism for the instability is thus the trapped particle pressure gradient. Note that this term is also destabilizing for an electron distribution. The particle precession changes sign, but also the sign of Q in Eq. 8.180 changes sign because of ω_0.

The contribution to D from the inertia takes the form

$$\delta I = -\frac{1}{2}\omega^2 \int d^3x \rho \xi^2. \tag{8.182}$$

Just as in the case of the internal kink mode, the dominant contribution to this integral comes from the inertial layer, where $\xi_\theta \sim 1/\omega$, giving a contribution which is linear in ω. From Sec. 6.10 we have

$$\delta I = -\frac{\pi^2}{2} \frac{r_s^2 \xi_0^2 i\omega}{\omega_A}, \tag{8.183}$$

with $\omega_A = v_A/(\sqrt{3}Rrq')$ the effective Alfvén frequency.

If a general particle distribution is used, the kinetic term δW_k must be evaluated numerically, as in Sec. 8.9.2. However, model distribution functions can be used which allow analytical evaluation of δW_k, giving tractable solutions to the dispersion relation which still display all the qualitative features of a full numerical solution. The essential approximations to achieve this are to eliminate the integrations over r, μ by approximating the integrand with a mean value, or in the case of μ, assuming a δ-function distribution in μ/E, and choosing an energy dependence for F which allows δW_k to be evaluated analytically. In this section we also neglect diamagnetic effects in order to arrive at simple analytic expressions that display the essential physics of the mode. For a more complete but numerical analysis see Sec. 8.9.2.

Fortunately, two experimental situations produce distributions which are well described by this procedure, neutral beam injection experiments and minority species ion cyclotron heating. Ion cyclotron heating gives energy predominantly to those particles which spend the most time in the resonant layer where the wave frequency is equal to the cyclotron frequency. If the resonant layer is near the magnetic axis, these particles are trapped with bounce angle $\pi/2$, and the repeated heating produces a Maxwellian distribution in energy,

$$F(E, \mu, r) = \delta(E/\mu - \alpha)e^{-E/T}n(r). \tag{8.184}$$

The second case, neutral beam injection, is of interest when the injection is nearly perpendicular, so as to produce a population of deeply trapped particles. To first approximation the particles are all introduced with energy E_m, and slow down with little pitch angle scattering until the energy is reduced substantially. After this they experience significant pitch angle scattering and become isotropized in pitch, and the fraction of particles which are trapped becomes much smaller.

Consider a high energy particle with energy loss rate, primarily from drag by scattering with electrons, of the form

$$\frac{dE}{dt} = -\nu E^p. \tag{8.185}$$

Integrating in time we find the characteristics to be given by

$$E = [E_m^{1-p} - \nu(1-p)(t-t_0)]^{1/(1-p)} \tag{8.186}$$

where $E(t_0) = E_m$. The distribution function then satisfies

$$(\partial_t - \nu E^p \partial_E)F = s(t)\delta(E - E_m) \tag{8.187}$$

where $s(t)$ is a source of particles at the injection energy E_m. The solution for F by the method of characteristics is

$$F(E, t) = \int_0^t dt_0 s(t_0)\delta(E - [E_m^{1-p} - \nu(1-p)(t-t_0)]^{1/1-p}). \tag{8.188}$$

Integrating, we find, for a constant source function s,

$$F(E, t) = \frac{c}{\nu E^p} \tag{8.189}$$

for $E < E_m$. For high energy particles the energy loss is proportional to $E^{3/2}$, $p = 3/2$, so a slowing-down distribution is of the form $F \sim E^{-3/2}$, $E < E_m$.

The dispersion function thus takes the form

$$D(\xi) = \delta W_{MHD} + \delta W_k + \delta I, \tag{8.190}$$

with the MHD, kinetic, and inertial contributions as given in Sec. 8.13. Minimization is carried out order by order in ϵ, as was done with the kink mode. The ratio of the density of the hot component to the background is taken to be of order ϵ^3, and the ratio of temperatures, T_c/T_h of order ϵ^2. The background beta is taken of order ϵ^2, giving a hot trapped particle beta, β_h of order ϵ^3. Thus δW_k is of order ϵ^3 and the lowest order minimization involves only δW_{MHD}. Now consider δW_{MHD}. The integration volume gives a factor ϵ^2. Expand $\vec{\xi}$ in orders of ϵ. Terms in δW_{MHD} of order unity are made zero by taking $\vec{\xi}_0$ to be the lowest order cylindrical form $\vec{\xi}_0 = \xi(r)(\hat{r} + i\hat{\theta})e^{i(\theta-\phi)}$, since this form also makes $\vec{Q}_\perp$ of order ϵ. Terms of order ϵ^2 are made zero by choosing $\vec{\xi}_1$ such that $\nabla \cdot \vec{\xi}_\perp^1 = -2\xi_\perp^0 \cdot \kappa$ and $\nabla \cdot \vec{\xi}^1 = 0$. The minimization of δW_{MHD} then leads to an expression of order ϵ^4. It is a function of the cold plasma β and becomes negative (destabilizing) for β_p greater than a threshold value.

Because of the ordering, $\beta_h \sim \epsilon^3$ the lowest order $\vec{\xi}$, obtained from the MHD minimization, along with the condition $\nabla \cdot \vec{\xi}_\perp^1 = -2\xi_\perp^0 \cdot \kappa$, is sufficient to evaluate δW_k This is in error in one significant respect. The lowest order $\vec{\xi}$ is nonzero only within $q = 1$, and thus in this approximation only particles within the $q = 1$ surface destabilize the mode. Actually, due to the Shafranov shift, higher m components are present, and they extend to the plasma edge. Thus to higher order also particles outside the $q = 1$ surface have a destabilizing influence on the mode, and should be included in the dispersion relation.

To evaluate J, use for trapped particles $\alpha \simeq 1$ and $\nabla \cdot \vec{\xi}_\perp = -2\vec{\xi}_\perp \cdot \vec{\kappa}$, to find

$$J = \frac{1}{2}(\xi_r cos\theta - \xi_\theta sin\theta). \tag{8.191}$$

But we also have from the lowest order cylindrical minimization, $\xi_r = \xi_0 e^{i(\theta-\phi)}$ $\xi_\theta = i\xi_0 e^{i(\theta-\phi)}$, with ξ_0 a step function, giving

$$J = \frac{1}{2}e^{-i\phi}. \tag{8.192}$$

Bounce average with $\phi = \theta/q \simeq \theta$, giving

$$\bar{J} = \frac{\xi_0}{2} \frac{\oint \cos\theta d\theta/\sqrt{1-\alpha B}}{\oint d\theta/\sqrt{1-\alpha B}} = \frac{\xi_0}{2}\left(\frac{\omega_d r R \omega_0}{Eq}\right). \tag{8.193}$$

We then find using $q \simeq 1$

$$\delta W_k = 2^{5/2}\pi^3 \int r dr d\alpha dE E^{1/2} K_b r^2 \omega_d^2 \xi_0^2 \frac{\omega \partial_E F - (1/r)\partial_r F}{\omega_d - \omega}. \tag{8.194}$$

The hot trapped particles also contribute to δW_{MHD}, through the term

$$\delta W_{MHD,h} = -\int d^3x (\vec{\xi}_\perp \cdot \nabla p)(\vec{\xi}_\perp \cdot \kappa). \tag{8.195}$$

Substituting $p = \int d^3v EF$ and using the expression for $\vec{\xi}_\perp$ we find

$$\delta W_{MHD,h} = \xi_0^2 \pi^3 2^{5/2} \int dr dE d\alpha r^2 \omega_d K_b E^{1/2} \partial_r F. \tag{8.196}$$

Combining this term with δW_k we find the total contribution from the hot trapped particles to be

$$\delta W_k + \delta W_{MHD,h}$$
$$= \xi_0^2 \pi^3 2^{5/2} \omega \int dr dE d\alpha \frac{r^3 \omega_d^2 K_b E^{1/2}}{\omega_d - \omega}\left[\partial_E F - \frac{1}{r\omega_d}\partial_r F\right]. \tag{8.197}$$

The cancellation of terms not proportional to ω reflects the fact that the trapped particles do not contribute to the energy of the system at zero frequency. Now estimate the two contributions to this expression, which are of the form given by Eq. 8.6. Using $\omega_d \simeq Eq/r$ the ratio of the term in ∂_E to that in ∂_r is $\sim r/R$ and we thus neglect the ∂_E term. Now integrate by parts in r. We assume that F vanishes at some radius, so that there is no term left from the integration by parts. Also normalize δW through

$$\delta \hat{W} = \frac{2\delta W}{\pi^2 r_s^2 \xi_0^2}, \tag{8.198}$$

giving

$$\delta \hat{W}_k + \delta \hat{W}_{MHD,h} = C_\beta \int dr dE d\alpha E^{1/2} F \partial_r \left(\frac{K_b r^2 \omega_d}{\omega_d - \omega}\right). \tag{8.199}$$

Combine these with the contributions from the core plasma and the inertial layer to find

$$0 = \frac{-i\omega}{\omega_A} + \delta\hat{W}_c + C_\beta\omega \int drdEd\alpha E^{1/2} F\left[\frac{A}{\omega_d - \omega} - \frac{B}{(\omega_d - \omega)^2}\right] \quad (8.200)$$

with $A = \partial_r(K_b r^2 \omega_d)$, $B = K_b r^2 \omega_d \partial_r \omega_d$, and $w_A = V_A/(\sqrt{3}rq')$ is the effective Alfvén frequency.

For the remainder of this section we will use the model particle distributions given in Sec. 8.13, keeping only the leading order in ϵ, as is consistent with the derivation of the dispersion relation, and neglect the dependence of ω_d on r. This last approximation may be regarded as replacing $\omega_d(r)$ with $\omega_d(r^*)$, with r^* an appropriate mean value in the range of integration. The dispersion relation then reduces to

$$0 = \frac{-i\omega}{\omega_A} + \delta\hat{W}_c + C_\beta\omega \int drdEd\alpha E^{1/2} \frac{F\omega_d\partial_r(r^2 K_b)}{\omega_d - \omega} \quad (8.201)$$

and the trapped particle beta is given by Eq. 8.177.

The properties of the solution of the dispersion relation can be examined using a Nyquist analysis. We wish to examine the existence of unstable mode solutions. The dispersion relation has the approximate form

$$0 = D(\omega) = \frac{-i\omega}{\omega_A} + \delta\hat{W}_c + D_k, \quad (8.202)$$

with

$$D_k = C_\beta\omega \int dE \frac{FG}{\omega_d - \omega} \quad (8.203)$$

the contribution from the trapped particles, with G a weighting function, and we have approximated the integrations over α, r by using some mean values. Consider a contour in the ω plane encircling the upper half plane, as shown in Fig. 8.31. Along the real axis the behavior of D_k is shown. It is dominated by the resonance at $\omega = \omega_d$ where ω_d is the average precession frequency for the distribution. Depending on the sign of $\delta\hat{W}_c$ and the magnitude of β_h, the image of the contour C under the map $D(\omega)$ is shown in Fig. 8.32. Clearly zero, one, or two growing modes can appear. If C_β is small the trapped particles have little effect and there is an unstable mode with $\omega_r = 0$ if $\delta\hat{W}_c < 0$. If C_β is large enough the excursions of D_k produce another mode with $\omega_r \simeq \omega_d$.

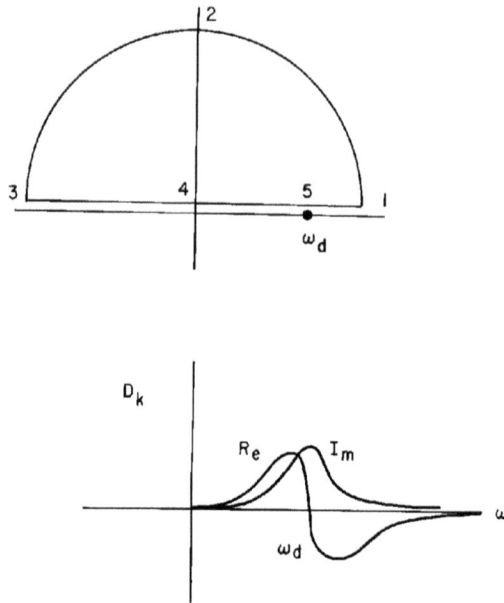

Fig. 8.31   Nyquist contour and D_k along real axis.

Consider the model slowing-down particle distribution for $E < E_m$, $F = n(r)E^{-3/2}\delta(\alpha - \alpha_0)$. Write $\omega_d = \omega_{dm}E/E_m$ and $\Omega = \omega/\omega_{dm}$. We then find $\beta_h = C_\beta E_m \int r\, dr\, K_b(\alpha_0)n(r)$. The dispersion relation then reduces to

$$0 = \frac{-i\Omega\omega_{dm}}{\omega_A} + \delta\hat{W}_c + C\beta_h\Omega ln\left(1 - \frac{1}{\Omega}\right) \qquad (8.204)$$

with $C = \int drn(r)d/dr(r^2 K_b)/\int drn(r)rK_b$ of order unity. At threshold, for $\Omega = \Omega_r$ real, the imaginary part determines the critical β_h,

$$\beta_{hc} \simeq \frac{\omega_{dm}}{\pi\omega_A}. \qquad (8.205)$$

The real part of the frequency is given by the solution to

$$\delta\hat{W}_c = \frac{\omega_{dm}}{\pi\omega_A}\Omega ln\left|1 - \frac{1}{\Omega}\right|. \qquad (8.206)$$

Thus for $\delta\hat{W}_c = 0$ the real frequency is $\Omega = 1/2$, or $\omega = \omega_{dm}/2$.

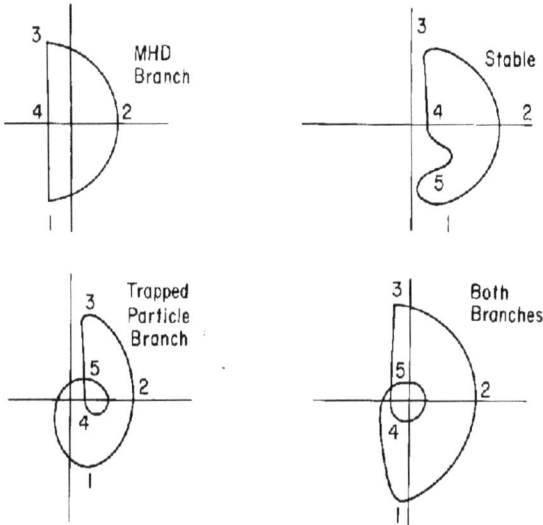

Fig. 8.32   Map of Nyquist contour.

Linearizing the dispersion relation about the threshold using $\beta_h = \beta_{hc} + d\beta_h$ and $\omega = \omega_{dm}/2 + d\omega_r + i\gamma$ we find $d\omega_r = 0$ and $\gamma = \pi^2 \omega_A d\beta_h/4$. The mode thus passes vertically in the complex plane through the real axis at $\omega = \omega_{dm}/2$ when $\beta_h = \beta_{hc}$, and since typically $\omega_A \gg \omega_d$, the growth rate is the same order as the real frequency for fairly small β_h.

This trapped particle branch of the dispersion relation is quite distinct from the ideal branch, which has $\omega_r = \gamma = 0$ for $\delta \hat{W}_c = 0$, and is purely growing for $\delta \hat{W}_c < 0$ and $\beta_h = 0$. Thus we refer to this dispersion relation as describing the kinetic kink mode, possessing two distinct branches.

More realistic particle distributions do not greatly modify this result. The critical trapped particle β and the real frequency are weak functions of the distribution function, since they are determined by integrations over the distribution. If we are concerned only with the fishbone branch, diamagnetic and resistive effects are also not very important.

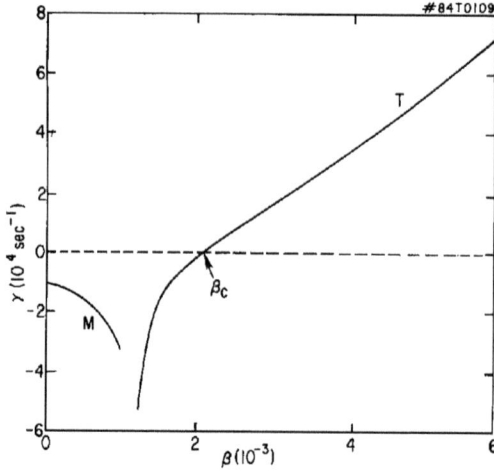

Fig. 8.33   Growth for trapped particle and MHD branch.

To estimate the saturation due to the profile flattening, simply combine Eqs. 8.166 and 8.167 to give again the pendulum equation

$$\frac{d^2Q}{dt^2} = \omega_T^2 cosQ, \qquad (8.207)$$

with $\omega_T^2 = n\partial_E\omega_d\omega\Phi_{mn}J_0(m\theta_b)$. Inserting threshold values we find that the saturation occurs for $\Phi_{mn} \sim E$. This is a very large perturbation level, and the mode does not normally saturate by profile flattening without also causing extensive particle loss through convective motion in a large phase space island.

8.14   THE FISHBONE CYCLE

The resonant beam loss process, along with the mode destabilization mechanism, allow a description of the full fishbone cycle. If we assume that the core plasma component has been heated until it is near marginal stability for the kink mode, and neglect variations in the temperature and current profiles, then the mode cycle is governed by the neutral beam deposition.

Hot trapped particles are added until $\beta_{h,crit}$ is exceeded, at which point the mode rapidly grows, with a real frequency and growth rate given by a weighted average trapped particle precession frequency. For PDX parameters the threshold β_h is typically 2×10^{-3}. In Fig. 8.33 is shown the growth rate of the trapped particle and MHD branches of the dispersion relation versus β_h. $\delta \hat{W}_c$ is positive and the stable MHD branch is further stabilized by the trapped particles. This result is similar to the stabilizing effect of trapped particles on ballooning modes, which was known before the discovery of the fishbone mode (Rosenbluth *et al.*, 1983). When the amplitude is sufficiently large, rapid resonant particle loss depletes the trapped particle population. The mode stops growing when β_h is again reduced to $\beta_{h,crit}$, but resonant particle loss continues as the mode decays until β_h has been driven well below its critical value. The neutral beam injection rate is generally much slower than the loss rate at peak amplitude, so the MHD activity and the losses occur in sharp bursts, with the separation determined by the beam injection rate.

A Monte Carlo guiding center simulation, using PDX parameters with the mode growth rate and frequency calculated using the instantaneous hot particle distribution, is shown in Fig. 8.34. There are no free parameters in the simulation, and comparison with experiment is reasonable. The particle loss is observed in the phase and location predicted by the simulations. Note that particle loss from the machine as opposed to simple expulsion to the region outside the $q = 1$ surface depends on the modes with $m \geq 2$, produced by the 1/1 mode coupling with the Shafranov shift of the high β plasma core. As β_h decreases, and the hot particle distribution is modified by the losses, the simulations show a whistling downward of the mode frequency by as much as 30 percent. This effect is also observed in the experiments. Precise comparison of the mode frequency with experiment is difficult because of imprecise knowledge of the q profile, the neglected destabilizing effect of trapped particles outside the $q = 1$ surface in the theory, plasma rotation, and diamagnetic effects, which also contribute to the mode rotation. Diamagnetic effects are of little importance as long as the frequencies involved are large compared to diamagnetic frequencies. If instead the ion diamagnetic frequency is large, the dispersion relation is significantly altered. This limit was first considered by Coppi and Porcelli (1986). In addition to modification of the fishbone branch, diamagnetic effects play an important role in the behavior of the kink branch, as shown in Sec. 8.9.2.

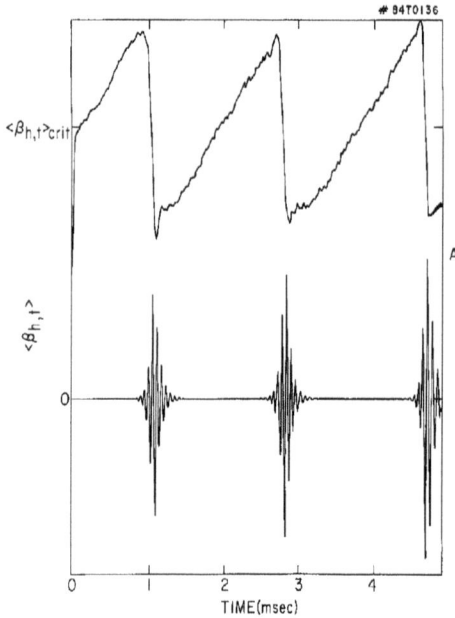

Fig. 8.34   Fishbone cycle.

8.15   RESISTIVE KINETIC INTERNAL KINK

The modification of the internal kink modes by the effects of resistivity was calculated some time ago by Coppi *et al.* (1976) and is well known. The only change in the theory occurs in the resistive layer, *i.e.* the resistivity and diamagnetic frequencies modify the inertial term and the interior solution must be calculated as in Sec. 7.9. The modification of the fishbone mode was considered by Biglari and Chen (1986). The final expression follows from Eq. 7.93 and is

$$\delta W_c + \delta W_k - \frac{8i[\omega(\omega - \omega_{*i})]^{1/2}\Gamma((\Lambda^{3/2} + 5)/4)}{\Lambda^{9/4}\Gamma((\Lambda^{3/2} - 1)/4))\omega_A} \qquad (8.208)$$

where $\Lambda = -i[\omega(\omega - \hat{\omega}_{*e})(\omega - \hat{\omega}_{*i})]^{1/3}/\gamma_R$, $\gamma_R = S^{-1/3}\omega_A$ is the resistive growth rate, S is the magnetic Reynolds number, $\omega_A = v_A/(\sqrt{3}Rrq')$ is the shear Alfvén frequency, with v_A the Alfvén velocity, and R and

r the major and minor radii, respectively, and $q' = dq/dr$ with q the safety factor. The ω_* terms are diamagnetic frequencies of the background plasma with $\omega_{*i} = -(c/neBr)(dp_i/dr)$, $\omega_{*e} = (c/neBr)(dp_e/dr)$, $\hat{\omega}_{*e} = \omega_{*e} + 0.71(c/eBr)(dT_e/dr)$. The term in Eq. 8.208 involving the Γ functions arises from the inertial layer, so all expressions are evaluated at the $q = 1$ surface. The inclusion of diamagnetic terms was carried out by Bussac *et al.* (1977) and Ara *et al.* (1978) generalizing the work of Coppi *et al.* (1976), and considered in Sec. 7.10. The expression δW_c is the minimized ideal variational energy for the internal kink, discussed in Sec. 6.10, and δW_k is the kinetic contribution coming from the trapped particle distribution F.

For drift frequencies less than or comparable to the resistive grow rate, $\omega_d < \gamma_R$ the resistivity enhances the dissipation in the inertial layer and substantially increases the threshold for trapped particle destabilization. Using $\Lambda \ll 1$ in Eq. 8.208 we easily obtain for the threshold

$$\beta_{crit,res} = \left(\frac{\gamma_R}{\omega_d}\right)^{9/4} \beta_{crit,ideal}. \tag{8.209}$$

This modification does not appear to be too relevant, even for devices which are very large and hence have relatively small values for ω_d, because of the decrease of γ_R with increasing temperature.

8.16 STABILIZATION OF THE SAWTOOTH

High energy ions in banana orbits exert a stabilizing influence by adding their inertia to the field line to which they are attached. If they precess toroidally at a rate which is greater than the mode growth rate, they effectively inhibit helical displacement of the field lines. The dispersion relation governing this coupled system is given by Eq. 8.208. The possibility of complete stabilization of the resistive internal kink was first shown by White *et al.* (1989) and we follow this presentation.

It is instructive to consider the ideal limit, $\Lambda \to \infty$ but to keep effects of ω_{*i}, using a model slowing-down distribution for the hot trapped particles

$$F(E, \mu, r) = n(r)\delta\left(\frac{\mu}{E} - \alpha\right) E^{-3/2} \quad for \quad E < E_m \tag{8.210}$$

with μ the magnetic moment, and E the energy. Equation 8.208 then becomes

$$0 = \frac{-i[\omega(\omega - \omega_{*i})]^{1/2}}{\omega_A} + \delta\hat{W}_c + \frac{\beta_h\omega}{\epsilon\omega_{dm}}ln\left(1 - \frac{\omega_{dm}}{\omega}\right) \qquad (8.211)$$

where ω_{dm} is the precession rate of a particle with $E = E_m$. The condition for threshold (ω = real) gives for $\omega_{*i} < \omega < \omega_{dm}$, an equation for the threshold frequency

$$\gamma_I = \frac{[\omega(\omega - \omega_{*i})]^{1/2}}{\pi}ln\left(\frac{\omega_{dm}}{\omega} - 1\right) \qquad (8.212)$$

with $\gamma_I = -\delta W_c\omega_A$ the ideal kink mode growth rate, and the associated value of beta is $\beta_h = \epsilon\omega_{dm}(1 - \omega_{*i}/\omega)^{1/2}/(\pi\omega_A)$, a monotone increasing function of ω. The ideal growth rate is related to the background plasma. For example, for a quadratic q profile and circular flux surfaces γ_I was found by Bussac *et al.* (1975) to have the form

$$\gamma_I = \frac{\omega_A r_s^2 3\pi}{R^2}(1 - q(0))\left(\beta_p^2 - \frac{13}{144}\right) \qquad (8.213)$$

where β_p is the poloidal beta of the background plasma. In general, γ_I depends on the full q profile and plasma shape.

Behavior of the solution is significantly different according to whether γ_i is larger than $\omega_{*i}/2$. We consider here the case $\gamma_I > \omega_{*i}/2$, which is of particular interest because of the possibility of complete stabilization of the resistive internal kink mode without excitation of the fishbone branch. For a discussion of the solution in all domains see White *et al.* (1989).

The right-hand side of Eq. 8.212 is zero for $\omega = \omega_{*i}$ and $\omega = \omega_{dm}/2$, and is positive and real for $\omega_{*i} < \omega < \omega_{dm}/2$. For γ_I small and positive there are two solutions, the lower frequency one corresponding to the stabilization of the kink mode branch and the higher frequency one to the destabilization of the fishbone. We are interested in the case in which these threshold frequencies, and associated values of β_h, are widely separated, so we consider $\omega_{*i} \ll \omega_{dm}/2$. This condition puts a lower bound on the trapped particle energy. If also $\gamma_I \ll \omega_{dm}/(2\pi)$, the threshold frequencies are given approximately by $\omega_1 = \omega_{*i} + \pi^2\gamma_I^2/[\omega_{*i}ln^2(\omega_{dm}/\omega_{*i})]$ and $\omega_2 = \omega_{dm}/2 - \pi\gamma_I/2$, and the corresponding values of β_h indicate that a stable gap can exist between the two modes. The stabilization of the ideal internal kink, which has also been pointed out by Pegoraro *et al.* (1988)

occurs for $\beta_h \simeq \epsilon\omega_{dm}\gamma_I/[\omega_A\omega_{*i}ln(\omega_{dm}/\omega_{*i})]$. For $\omega_{*i} = 0$ the stabilization occurs at $\beta_h = \epsilon\omega_{dm}/(\pi\omega_A)$, as pointed out by White *et al.* (1985).

Now consider resistive modification of these results. For the gap to continue to exist it is necessary and sufficient to require that the arguments of the Γ functions be large (ideal limit) at the threshold locations ω_1, ω_2. If the arguments of the Γ functions are small, it is easy to show that the growth rate tends asymptotically to zero for $\beta_h \to \infty$ so that complete stabilization does not occur. Using ω_1 the condition $\Lambda \gg 1$ gives

$$S \gg \frac{16}{\pi^2} \frac{ln^2(\omega_{dm}/\omega_{*i})\omega_A^3}{\gamma_I^2(\omega_{*i} - \hat{\omega}_{*e})}. \qquad (8.214)$$

Since $\omega_2 \gg \omega_1$, this condition automatically insures that $\Lambda \gg 1$ at $\omega = \omega_2$. It is desirable to achieve a stable gap for S not too large. From Eq. 8.214 we see that this requires taking γ_I large, consistent with there being two solutions to Eq. 8.212, *i.e.* γ_I must remain small compared to $\omega_{dm}/(4\pi)$.

We can now write necessary and sufficient conditions for the existence of a range of hot particle density between the sawtooth stabilization value and that for fishbone destabilization. The limits on the ideal growth rate γ_I are given by Eq. 8.214 and by the maximum value permitting a solution to Eq. 8.212. This gives

$$\frac{4\omega_A}{\pi}\left(\frac{\omega_A}{S(\omega_{*i} - \hat{\omega}_{*e})}\right)^{1/2}ln\left(\frac{\omega_{dm}}{\omega_{*i}}\right) < \gamma_I \leq 0.09\omega_{dm}\left(1 - \frac{2.4\omega_{*i}}{\omega_{dm}}\right).(8.215)$$

It seems perhaps strange that to achieve stabilization using trapped particles the kink mode must be above its ideal threshold ($\gamma_I > 0$), but this is understandable in that it is precisely this instability which preserves the ideal character of the mode. In fact it is common knowledge in the theory of feedback stabilization that modes near marginal stability are very difficult to control, more so than those which are strongly unstable. In the resistive limit the mode cannot be stabilized, and for small values of S the gap described by Eq. 8.215 vanishes. The upper bound is the point at which the kink branch and the fishbone branch coalesce in the complex ω plane. For γ_I larger than this there are no thresholds for any value of β_h. The kink and fishbone branches exchange roles; the kink branch is destabilized by the trapped particles and takes on a large real frequency, and the fishbone branch is stable.

The range of trapped particle beta in which both modes are stable is given approximately by

$$\frac{\omega_{dm}\gamma_I}{\omega_A\omega_{*i}ln(\omega_{dm}/\omega_{*i})} < \frac{\beta_h}{\epsilon} < \frac{\omega_{dm}}{\pi\omega_A}\left(1 - \frac{\omega_{*i}}{\omega_{dm}} - \frac{\pi\gamma_I\omega_{*i}}{\omega_{dm}^2}\right). \quad (8.216)$$

Since ω_{dm} is proportional to the particle energy, Eq. 8.216 is seen to be a condition on the density of the hot trapped species. The upper limit is the fishbone threshold, and the lower limit is the internal kink stabilization point. The stability domain in the β_h, γ_i plane is approximately triangular. Outside this triangle one or both of the branches are unstable. If γ_I is below the lower limit in Eq. 8.215 and β_h/ϵ above the upper limit of Eq. 8.216, sawtoothing and fishbones can occur simultaneously.

Results of a numerical solution are shown in Figs. 8.35 and 8.36 for a hydrogen minority species in JET. A root finding procedure using a Stokes plot was used to follow unstable roots (see Fig. 16.1 and the discussion referring to it). Shown in Fig. 8.35 are the complex frequencies, and the growth rates, as a function of trapped particle density for an approximate JET equilibrium with $R = 296$ cm, for two different values of the magnetic Reynolds number, $S = 10^6$ and 10^7. Both the resistive internal kink branch and the fishbone branch are shown. The trapped particle density n ranges from zero to $2.3 \times 10^{11}/cm^3$. The particles were taken to be a slowing-down distribution with an average energy of 700 keV. The toroidal field was $B = 24$ kG and the averaged trapped particle precession rate was $< \omega_d >= 1.3 \times 10^5/sec$. The shear Alfvén frequency was $\omega_A = 2 \times 10^6/sec$, and the diamagnetic frequencies were $\hat{\omega}_{*e} = -3 \times 10^4/sec$ and $\omega_{*i} = 2 \times 10^4/sec$.

The ideal growth rate was chosen to be $\gamma_I = 1.4 \times 10^4/sec$. As seen, the $S = 10^7$ case is almost ideal, with $\gamma \simeq \gamma_I$ and $\omega_r \simeq \omega_{*i}/2$ at $n = 0$. There is a range of trapped particle density in which both the kink mode branch and the fishbone branch are stable. For smaller values of S the resistive modification of the kink branch makes it more unstable and eliminates this stable gap. For $S = 10^6$, the decrease of the growth rate with increasing hot particle density results in values of $\Lambda \simeq 1$. The behavior for large β_h is then dominated by the small argument of the Γ functions, and the growth rate only asymptotically approaches zero for $\beta_h \to \infty$ and complete stabilization does not occur. The fishbone branch is not noticeably modified by resistive effects, because although the growth rates of the two branches are comparable, the real frequency of the fishbone branch is much larger, giving $\Lambda \gg 1$. In Fig 8.36 is shown the stability domain for these same parameters. Outside of the stable triangle one would expect sawteeth (S),

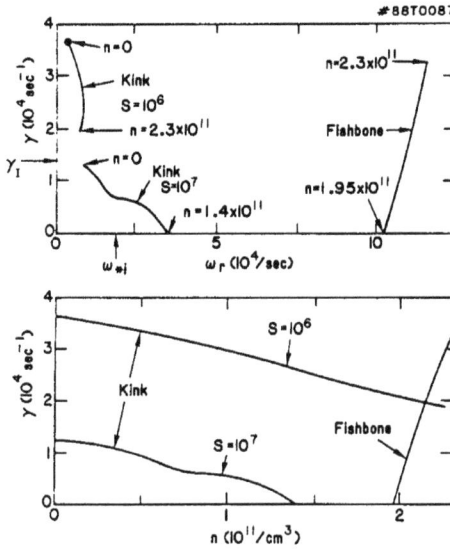

Fig. 8.35 Complex frequencies and growth rate, as a function of trapped particle density, for minority heating in JET.

fishbones (F) or both (S+F). Only the lower limit on γ_I depends on the Reynolds number, and two values are shown.

A Maxwellian particle distribution gives qualitatively the same results as the slowing-down distribution. Using the model distribution given by Eq. 8.184 and neglecting the dependence of ω_d on r we find

$$\delta W_k = \sqrt{\pi}\beta_h\Omega[1/2 + \Omega + \Omega^{3/2}Z(\Omega^{1/2})] \qquad (8.217)$$

where $\Omega = \omega/\langle\omega_d\rangle$ and $\langle\omega_d\rangle$ is the precession frequency of a particle with energy T, and Z is the plasma dispersion function.

Consider again the ideal MHD limit, $\gamma_R \to 0$. Substituting into Eq. 8.208 we find

$$0 = -\frac{\gamma_I}{\omega_d} - i[\Omega(\Omega - \Omega_*)]^{1/2} + \beta_h\sqrt{\pi}[1/2 + \Omega + \Omega^{3/2}Z(\Omega^{1/2})] \quad (8.218)$$

with $\Omega_* = \omega_{*i}/\langle\omega_d\rangle$. The threshold condition is

$$\gamma_I = \frac{[\Omega(\Omega - \Omega_*)]^{1/2}}{\sqrt{\pi}\Omega^{3/2}}[1/2 + \Omega + \Omega^{3/2}ReZ(\Omega^{1/2})] \qquad (8.219)$$

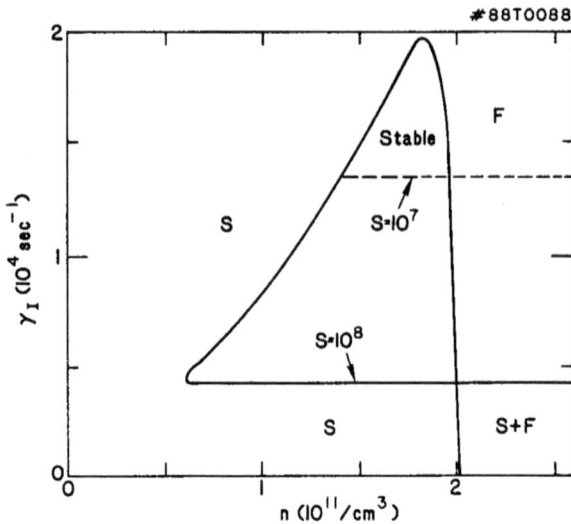

Fig. 8.36   Stability domain for sawtooth and fishbone modes.

and the value of β_h at threshold is given by

$$\beta_h = \frac{[\Omega(\Omega - \Omega_*)]^{1/2}e^{\Omega}}{\pi\Omega^{5/2}}. \tag{8.220}$$

The behavior of the roots is similar to that found for the slowing-down distribution, provided $\Omega_* > 1/2$. If instead $\Omega_* < 1/2$ the function β_h is not a monotone function of Ω, and the fishbone branch can destabilize at a smaller value of β_h than that at which the kink mode branch becomes stable. In this case there is no stable window of operation between the two branches. The deliberate use of a high energy trapped particle component as a means to stabilize MHD activity in a tokamak has only begun to be explored, both experimentally and theoretically, and the practical value of such schemes is not yet clear.

As can be seen from Fig. 8.36 parameter changes, such as a modification of the q profile or a change in the hot particle beta, can cause an abrupt loss of stabilization, thus allowing the plasma to suddenly resume sawtoothing, with what is referred to as a giant sawtooth. As a further complication involving the interaction of high energy particles and MHD, it has been found that sawtooth stabilization can be terminated by the appearance of

EPM and/or TAE modes which expel the stabilizing particles from the plasma core, producing a giant sawtooth (Bernabei, 2000).

8.17   BALLOONING DESTABILIZATION

Weiland and Chen (1985) considered the destabilization of ballooning modes as a candidate for high frequency precursor oscillations often seen with the fishbone mode. The analysis is similar to that performed for the internal kink mode. They analyze the case of a shifted circular magnetic equilibrium and arrive at a dispersion relation similar to that derived for the internal kink mode. For a model slowing-down particle distribution there is a threshold associated with the hot particle pressure which is an order of magnitude lower than the corresponding one for the core plasma. This mode can be expected to occur before the onset of the fishbone mode, for n values typically equal to five or six, so that it has a correspondingly higher frequency than the fishbone. Resistive modifications of this mode were calculated by Biglari and Chen (1986). They also considered finite banana width effects, and find a modification of the threshold if the banana width exceeds the resistive layer width.

8.18   ALPHA PARTICLE EFFECTS

The properties of confinement and the stability of a tokamak may be greatly affected by the presence of a significant alpha particle population. The alphas, by modifying the heat deposition profile, will change both temperature and current profiles, affecting the stability of short wave length modes contributing to global transport and of large scale MHD modes. In addition, destabilization of Alfvén waves may occur, with phase velocity comparable to the particle transit time, introducing a set of alpha driven instabilities (Nazikian *et al.*, 1997; Rewoldt *et al.*, 1987; Spong *et al.*, 1986; Cheng *et al.*, 1992; Fu *et al.*, 1996). In addition, alpha particles are subject to possible rapid loss due to ripple because of their high energy (see Sec. 12.7). This loss is fairly localized and can produce thermal loads on the outer wall which are not acceptable. On the other hand, reducing this loss requires an increase in magnetic field strength, machine size, or number of toroidal field coils, all of which are expensive. Significant theoretical work and experiments will be required to determine the exact nature of the limits placed on machine design by the presence of the alphas.

Instabilities induced by hot trapped particles play an important role in limiting the effectiveness of plasma heating with neutral beam injection and radio-frequency heating. High energy trapped particles can also have a stabilizing effect, and in particular eliminate the sawtooth oscillations. The same mechanisms may also play a role in ignited plasmas, with the fusion alpha particles constituting the high energy population. Because of the larger size and stronger magnetic fields in ignition devices, the precession frequency of 3.5 MeV alpha particles in them is not very different from the precession frequency of beam injected particles in PDX size devices. Details of the q profile and the existence of a magnetic well can greatly modify the effect of the high energy population (Cheng *et al.*, 1991).

The large MHD signals and rapid particle loss associated with the fishbone mode provide excellent diagnostic means for its investigation. The dependence of the loss mechanism on the plasma q profile and on the particle distribution may provide means of control and of deliberate excitation. It is possible that the mode could serve as an efficient means of burn control or ash removal.

Study of fishbone excitation and sawtooth suppression promises to lead to a better quantitative understanding of basic MHD properties of high temperature plasmas. The code NOVA-K has been developed (Cheng and Chance, 1987; Cheng, 1992) which allows analysis of the mode dispersion relation for shaped equilibria and without use of an inverse aspect ratio expansion.

8.19 Problems

1. Consider a simple pendulum, $E = \dot{\theta}^2/2 - cos(\theta)$. Assume a solution of the form $\theta = asin(\omega_b t) + bsin(3\omega_b t)$, giving a maximum excursion of $\theta_b = a - b$. Find ω_b to fourth order in θ_b, and a and b to third order in θ_b.

2. Calculate the threshold β_h for the fishbone in the highly resistive limit using the model slowing-down distribution function. Find expressions for the mode growth rate and frequency in the vicinity of the threshold.

3. Show that the threshold β_h is a monotone function of Ω provided $\Omega_* > 1/2$ for the model Maxwellian dispersion relation.

4. A simple model for the fishbone cycle can be made. Assume the trapped particles are deposited at rate S until the threshold beta is reached. Model the losses as a rigid displacement of the trapped particles toward the wall. Equations giving the trapped particle beta and mode amplitude A are then

$$\frac{d\beta}{dt} = S - A\beta_c$$

$$\frac{dA}{dt} = \gamma_0 \left(\frac{\beta}{\beta_c} - 1 \right) A.$$

Show that the solution to these equations is cyclic, *i.e.* find $F(A, \beta)$ with $dF/dt = 0$. Show that F has a maximum at the fixed point $\beta = \beta_c$, $A = S/\beta_c$. Show that if the losses are diffusive, rather than given by a rigid displacement, e.g. $d\beta/dt = S - A\beta$, then $dF/dt \geq 0$ with $dF/dt = 0$ only at $\beta = \beta_c$ and that the solution thus spirals in to the fixed point.

5 Use the equations of motion to verify explicitly that for a mode with a single n value $\omega \dot{P}_\zeta = n\dot{E}$.

8.20   References

Resonance

- Kolmogorov, A. N. in Proc. Int. Congr. Mathematicians, Amsterdam, Vol 1, 315 (1957), Arnold, V. I., Russ. Math. Surv. 18(5):9, (1963), J. Moser, Math. Phys. Kl. II 1,1 Kl(1):1, (1962).
- White, R. B., High Energy Particles in Tokamaks, in Proc. of the first ITER Summer School, Edited by S. Benkadda (Am. Inst. of Phys. New York, 2008).
- White, R. B. and M.S. Chance, Phys. Fluids 27, 2455 (1984).
- Rosenbluth, M. N., R.Z Sagdeev, J.B. Taylor, G M Zaslavsky Nuclear Fusion 6, 297 (1966).
- Chirikov, B. V. Phys. Rep. 52 263 (1979).
- Lichtenberg A. J. and M. A. Lieberman, Regular and Chaotic Dynamics (Springer-Verlag, New York, 1992) p. 386.
- White, R. B., N. N. Gorelenkov, W. W. Heidbrink, M. A. Van Zeeland, Phys. of Plasmas 17, 056107 (2010a).
- White, R. B., N. N. Gorelenkov, W. W. Heidbrink, M. A. Van Zeeland, Plasmas Physics Controlled Fusion 52, 045012 (2010b).
- Heidbrink W. W. Phys. of Plasmas 15 055501 (2008).
- Todo Y. and T. Sato, Phys. of Plasmas 5, 1321 (1998).
- R. B. White, Plasma Phys. Control. Fusion 53, 085018 (2011).
- White, R. B., Commun. Nonlinear Sci. Numer. Simulat. 17, 2200 (2012).
- Gobbin, M., R. B. White, L. Marrelli, P. Martin, Nucl Fusion 48, 075002 (2008).
- Lundquist S., Phys. Rev. 83, 307 (1951).
- Roberts, P. H., Introduction to Magnetohydrodynamics, p. 46 (Longmans Green, London 1967).

Stabilization by hot trapped particles

- Antonson, T. M., Y. C. Lee, H. L. Berk, M. N. Rosenbluth *et al.*, Phys. Fluids 26, 3580 (1983).
- Berk, H. L., Phys. Fluids 19, 1255 (1976).
- Conner, J. W., R. J. Hastie, T. J. Martin, and M. F. Turner, in Proc. of the 3rd Joint Varenna–Grenoble International Sympo-

sium on Heating in Toroidal Systems (Comm. of European Comm., Brussels, 1982) Vol. 1, p. 65.

- Fleischmann, H. H., Ann. N.Y. Acad. Sci. 251, 472 (1975).
- Lovelace, R. V., Phys. Rev. Lett. 41, 1801 (1976).
- Rosenbluth, M. N., S. T. Tsai, J. W. Van Dam, and M. G. Engquist, Phys. Rev. Lett. 51, 1967 (1983).
- Sudan, R. N., and M. N. Rosenbluth, Phys. Rev. Lett. 36, 972 (1976).

TAE, KTAE, EPM, EAE

- Berk H. L., and B. N. Breizman, Phys Fluids B 2, 2246 (1990).
- Berk H. L., and B. N. Breizman, Phys Fluids B 5, 1506 (1993).
- Betti, R., and J. P. Freidberg, Phys. Fluids B 3, 1865, (1991).
- Breizman, B. N. H. V. Wong, H. L. Berk, M. S. Pekker *et al.*, Bull. Am. Phys. Soc. 39, 1705 (1994).
- Cheng, C. Z., Physics Reports, 211 (1992).
- Cheng, C. Z., L. Chen and M. S. Chance, Ann. Phys. NY 161, 21 (1984).
- Cheng, C. Z., Phys. Plasmas 2, 1427 (1990).
- Fu G. Y. and W. Park, T., Bull. Am. Phys. Soc. 38, 1945 (1993).
- Gorelenkov, N. N., C. Z. Cheng, and W. M. Tang, Phys. Plasmas 5, 3389 (1998).
- O'Neil, T., Phys. Fluids 12, 2255, (1965).
- Wu, Y. and R. B. White, Phys. Plasmas 1, 2733 (1994).
- Zonca, F., F. Romanelli, G. Vlad, and C. Kar, Phys. Rev. Lett 74, 698 (1995).
- Zonca, F., L. Chen, Phys. Plasmas 3, 323 (1996).

Fishbone experiments

- DIII Group, GA Technologies, in Proc. of the 4th International Symposium on Heating in Toroidal Plasmas, edited by H. Knoepfel and E. Sindoni, Rome (1984).
- ISX-B Group, Oak Ridge National Laboratory, Phys. Rev. Lett. 4, 538 (1982).
- JFT Group, Tokai Ibaraki Japan, Nuel. Fusion 21, 993 (1981).
- PDX Group, Princeton Plasma Physics Laboratory, Phys. Rev. Lett. 50, 891 (1983).

- Strachan, J., B. Grek, W. Heidbrink, D. Johnson *et al.*, Nucl. Fusion 25, 863 (1985).

Fishbones-theory

- Biglari, H., and L. Chen, Phys. Fluids 29, 2960 (1986).
- H. Biglari, L. Chen, and R. B. White, Proc. 11th Intl. Conf. on Plasma Physics and Controlled Nuclear Fusion Research, Kyoto, 1986, (IAEA, Vienna, 1987) Vol. 2, p. 119.
- Chen, L., R. B. White, C. Z. Cheng, F. Romanelli *et al.*, in Proc. of 10th International Conference on Plasma Physics and Controlled Nuclear Fusion Research, London, 1984, (IAEA, Vienna, 1985) Vol. 2, p. 59.
- Chen, L., R. B. White, and M. N. Rosenbluth, Phys. Rev. Lett. 52, 1122 (1984).
- Chen, L., R. B. White., G. Rewoldt, Y. P. Chen, *et al.*, in Proc. of 12th International Conference on Plasma Physics and Controlled Nuclear Fusion Research, Nice, 1988 (IAEA, Vienna, 1989).
- Coppi, B., and F. Porcelli, Phys. Rev. Lett. 57, 2272 (1986).
- Pegoraro, F., F. Porcelli, J. Hastie, B. Coppi *et al.*, in Proc. of 12th International Conference on Plasma Physics and Controlled Nuclear Fusion Research, Nice, 1988 (IAEA, Vienna, 1989).
- Weiland, J., and L. Chen, Phys. Fluids 28, 1359 (1985).
- White, R. B., R. J. Goldston, K. McGuire, A. H. Boozer, *et al.*, Phys. Fluids 26, 2958 (1983).
- White, R. B., L. Chen, F. Romanelli, and R. Hay, Phys. Fluids 28, 278 (1985).

Numerical codes for non self adjoint operators

- C. Z. Cheng and M. S. Chance, J. of Comput. Physics 71, 124 (1987).
- Kerner, W., K. Lerbinger, and J. Steuerwald, Computer Physics Communications 38, 27 (1985).
- Kerner, W., K. Lerbinger, R. Gruber, and T. Tsunamatsu, Computer Physics Communication L6, 225 (1985).

Gyrokinetics

- Catto, P.J., W. M. Tang, and D. E. Baldwin, Plasma Phys. 23, 639 (1981).
- Hahm, T.S., W.W. Lee, A. Brizard, Phys Fluids 31, 1940 (1988).
- Brizard, A., J., Plasma Phys. 41, 541 (1989).
- Brizard, A., J., Gyrokinetic Theory, (World Scientific, Singapore, 2012).

Sawtooth

- Ara, G., B. Basu, B. Coppi, G. Laval *et al.*, Ann Phys. 112, 443 (1978).
- Bussac, M. N., R. Pellat, D. Edery, and J. L. Soule, Phys. Rev. Lett. 35, 1638 (1975).
- Bussac, M. N., D. Edery, R. Pellat, and J. L. Soule, Proc. of the 6th International Conference on Plasma Physics and Controlled Nuclear Fusion Research, Berchtesgaden, 1976 (IAEA, Vienna, 1977) Vol. 1, p. 607.
- Coppi, B., R. Galvao, R. Pellat, M.N. Rosenbluth *et al.*, Fiz. Plazmy 2, 961 (1976) [Sov. J. Plasma Phys. 2, 533 (1976)].

Sawtooth stabilization

- Bernabei, S, M. G. Bell, R. V. Budny, E. D. Fredrickson *et al.*, Phys. Rev. Lett. 6, 1212 (2000).
- Pegoraro, F., F. Porcelli, and J. Hastie, Sherwood Theory Conference, Gatlinburg, TN, (1988) paper 2B1.
- White, R. B. , P. H. Rutherford, P. Colestock, and M. N. Bussac, Phys. Rev. Lett. 60, 2038 (1988).
- White, R. B. J. Comput. Phys. 31, 409 (1979).
- White, R. B., M. N. Bussac, F. Romanelli, Phys. Rev. Lett. 62, 539 (1989).
- White, R. B. , F. Romanelli, M. N. Bussac, Phys Fluids B 4, 745 (1990).

Alpha particles

- Fu G, Y., Phys. Plasmas 3, 4036 (1996).
- Nazikian, R. M., Phys Rev. Lett 78, 2976 (1997).
- Rewoldt, G., W. M. Tang, and R. J. Hastie, Phys. Fluids 30, 807 (1987).
- Spong, D. A., D. J. Sigmar, K. T. Tsang, J. J. Ramos *et al.*, ORNL/TM-10200 (1986).

Chapter 9

Cyclotron Motion

9.1 INTRODUCTION

For many purposes guiding center analysis is sufficient, but there are some situations when the full particle orbit including the cyclotron motion is essential. Whenever mode frequencies approach the cyclotron frequency, or mode structures approach the cyclotron radius it may be necessary to use a full cyclotron formalism. Resonant heating of particles in a magnetic field by ion cyclotron waves has been examined by many authors and is of importance in the heating of magnetically confined laboratory as well as extraterrestrial plasmas. For a review see Lieberman and Lichtenberg (1983). Stochastic heating by a lower hybrid wave in a tokamak was investigated by Karney (1979) for wave frequency much larger than the cyclotron frequency. It has also been noted that heating with ion Bernstein waves can be obtained at frequencies above the cyclotron frequency Ω but below the second harmonic, at frequencies of $\omega/\Omega = 3/2$, $4/3$, etc. If mode amplitudes are sufficiently large significant heating can occur even at frequencies only a fraction of the cyclotron frequency.

9.2 SUB CYCLOTRON HEATING

Consider resonant heating of particles by an electrostatic wave propagating perpendicular to a confining uniform magnetic field. With a sufficiently large wave amplitude, significant perpendicular stochastic heating can be obtained with wave frequency at a fraction of the cyclotron frequency.

At sufficiently large wave amplitude, low-frequency wave heating is indeed possible. To this end we consider the simplest problem possible; that of a particle gyrating in a constant magnetic field acted upon by an electrostatic plane wave propagating perpendicular to the field.

The Hamiltonian for this system is

$$H = \frac{(\vec{p} - \vec{A})^2}{2} + \Phi(x, t) \tag{9.1}$$

with the magnetic field given by the vector potential $\vec{A} = -By\hat{x}$. Take the units of time to be given by Ω, the cyclotron frequency, let the electrostatic wave be given by a single harmonic, $\Phi = \Phi_0 cos(kx - \omega t)$, and assume zero velocity parallel to the field, $v_z = 0$. There are then three dimensionless parameters characterizing the heating problem. Define $\rho = v/\Omega$ to be the instantaneous cyclotron radius. Then $k\rho$ characterizes the ratio of cyclotron radius to wave length, and $k^2\Phi_0$, the nonlinearity parameter, characterizes the ratio of particle displacement caused by the wave to wave length, and ω is the ratio of the wave frequency to the cyclotron frequency. The initial particle distribution is also characterized by $k\rho_0$ with ρ_0 a mean cyclotron radius for the distribution.

The equations of motion become $\dot{v}_x = v_y + k\Phi_0 sin(kx - \omega t), v_y = -x + x_0$, giving

$$\frac{d^2x}{dt^2} + x = x_0 + k\Phi_0 sin(kx - \omega t). \tag{9.2}$$

For small wave amplitude near the cyclotron frequency, it is possible to describe the particle response to the wave in terms of oscillation at the cyclotron frequency with a slowly varying cyclotron radius, or energy. In the case of interest here, wave amplitudes are large and wave frequencies different from, but comparable to, the cyclotron frequency, so response of the particle at additional frequencies must be retained. To treat the full problem it is necessary to include particle motion at fractions of the cyclotron frequency, sidebands, harmonics, etc. The particle motion must be written $x = x_0 + \lambda cos(t) - \mu sin(t) + \sum_m [\alpha_m cos(\nu_m t) + \beta_m sin(\nu_m t)]$ with λ, μ, α_m, β_m slowly varying in time compared to one and ν_m, with ν_m giving the set of frequencies necessary to describe the motion. A full analytic treatment is not possible, but analytic approximations can give insight into the nature of the solutions.

First consider Eq. 9.2 for $s \equiv k(x - x_0) \ll 1$. Letting $2T = kx_0 - \omega t$ and keeping only lowest order in s we have

$$\frac{d^2 s}{dT^2} + \left[\frac{4}{\omega^2} - \frac{4k^2 \Phi_0}{\omega^2} cos(2T) \right] s = \frac{4k^2 \Phi_0}{\omega^2} sin(2T), \qquad (9.3)$$

i.e. a driven Mathieu equation with unstable solutions for $\omega \simeq 2/q$ with q integer. Of course this equation is valid only for small s, but it indicates the existence of large amplitude solutions for these values of ω. This response is simply interpretable as due to resonance consisting of an integer number of cyclotron oscillations within one wave oscillation. The fact that there are many such resonances implies that at large amplitude stochastic threshold should be attained, permitting heating.

Now consider a Poincaré section of $k\rho$, $\psi = kx - \omega t$, by taking points when $v_y = 0, \dot{v}_y > 0$. This gives $\psi = \psi_0 - \omega t_j$, with $\psi_0 = kx_0$, and t_j given by the times at which $x = x_0$ and $\dot{x} < 0$. Given $\lambda(t)$, $\mu(t)$, $\alpha_m(t)$, $\beta_m(t)$ one can solve for the Poincaré times t_j. Without loss of generality at $t = 0$ we take x random, v_x random negative and $v_y = 0$, giving $x = x_0$, $\psi(0) = \psi_0$. The values at $t = 0$ then determine one Poincaré point. Others are given by $k\rho(t_j)$, $\psi(t_j) = \psi_0 - \omega t_j$. Fixed points are given by $dv/dt = 0$ and constant phase, or $\dot{\lambda} = \dot{\mu} = \dot{\alpha}_m = \dot{\beta}_m = 0$.

In general these equations are very complicated and the Poincaré section must be examined numerically. For significant heating there must exist resonances. A complete analysis would consist of a determination of all fixed points and then the calculation of the widths of the islands occuring around the elliptic points, followed by an estimate of stochastic threshold due to island overlap. Unfortunately this approach is not feasible, and to make any progress analytically one must be guided by numerical results. A numerical Poincaré plot using Eq. 9.2 is shown in Fig. 9.1 for $k^2 \Phi_0 = 0.1$, $\omega = 1/2$, showing period two fixed points occuring at small wave amplitude.

Guided by numerical results, including a Fourier analysis of the fixed point trajectories, we illustrate the nature of the solutions for this case by considering only the cyclotron motion and the particle response at the wave frequency of $\omega = 1/2$. Employing multiple time scales, express the solution to the equations of motion as $x = x_0 + \lambda cos(t) - \mu sin(t) + \alpha cos(\omega t) - \beta sin(\omega t)$ with $\lambda, \mu, \alpha, \beta$ slowly varying with respect to $1, \omega$. We then find, keeping only leading order in the slow time scale and using $e^{\pm i a sin(b)} = \sum_m J_m(a) e^{\pm i m b}$,

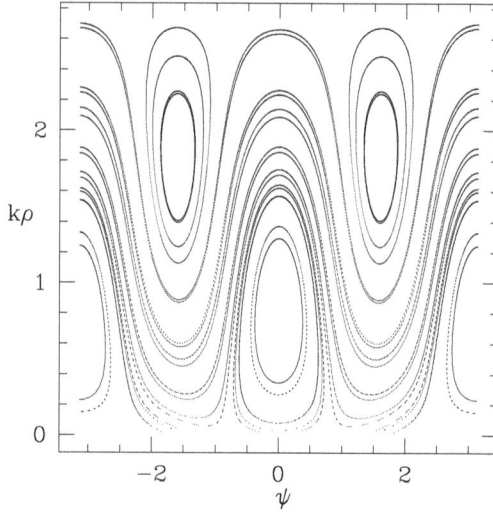

Fig. 9.1 Poincaré section for $C_0 = 1$, $S_0 = 0$, $\omega = 1/2$.

$$-2\frac{d\mu}{dt}cos(t) - 2\frac{d\lambda}{dt}sin(t) + (1-\omega^2)\alpha cos(\omega t) - (1-\omega^2)\beta sin(\omega t) =$$

$$k\Phi_0 \sum_{jklm} J_j(k\lambda)J_k(k\mu)J_l(k\alpha)J_m(k\beta)$$

$$\{sin[(j-k+l\omega-m\omega-\omega)t]cos[\psi_0+(j+l)\pi/2]$$

$$+cos[(j-k+l\omega-m\omega-\omega)t]sin[\psi_0+(j+l)\pi/2]\} \quad (9.4)$$

Taking the frequency to be a fraction of the cyclotron frequency, $\omega = 2/q$ with q an integer and integrating over the short time scales, we have

$$(1-\omega^2)\alpha = k\Phi_0 \sum_{jklm} J_j(k\lambda)J_k(k\mu)J_l(k\alpha)J_m(k\beta)sin(W)\Delta_{\omega+} \quad (9.5)$$

$$(1-\omega^2)\beta = k\Phi_0 \sum_{jklm} J_j(k\lambda)J_k(k\mu)J_l(k\alpha)J_m(k\beta)cos(W)\Delta_{\omega-} \quad (9.6)$$

$$2\frac{d\mu}{dt} = -k\Phi_0 \sum_{jklm} J_j(k\lambda)J_k(k\mu)J_l(k\alpha)J_m(k\beta)sin(W)\Delta_{1+} \quad (9.7)$$

$$2\frac{d\lambda}{dt} = -k\Phi_0 \sum_{jklm} J_j(k\lambda) J_k(k\mu) J_l(k\alpha) J_m(k\beta) cos(W)\Delta_{1-} \qquad (9.8)$$

with $\Delta_{\zeta\pm} = \delta_{j-k+(l-m-1)\omega,\zeta} \pm \delta_{j-k+(l-m-1)\omega,-\zeta}$ and $W = \psi_0 + (j+l)\pi/2$.

To gain an intuitive understanding of the occurences of the nonlinear resonances which permit heating at frequencies well below the cyclotron frequency, we can examine the limit of small wave amplitude, $k^2\Phi_0 \ll 1$ analytically. Then we have $k\alpha \ll 1$, $k\beta \ll 1$.

Now set $\omega = 1/2$. To the leading orders in $k\alpha$ and $k\beta$ we then find the first delta function of $\Delta_{\omega\pm}$ is limited to the values $(j,k,l,m) = (s, s-1, 0, 0)$ and the second delta function to $(s, s, 0, 0)$, with s integer. Similarly, for $\Delta_{1\pm}$, the first delta function is limited to the values $(j,k,l,m) = (s, s-1, 1, 0)$, $(s, s-1, 0, -1)$, $(s, s-2, -1, 0)$, $(s, s-2, 0, 1)$, with s integer, and the second delta function to the values $(j,k,l,m) = (s, s+1, 1, 0)$, $(s, s+1, 0, -1)$, $(s, s, -1, 0)$, $(s, s, 0, 1)$, with s integer. Note that these values are peculiar to the case of $\omega = 1/2$, which is a special degenerate case since $\omega = 1/2$ is the only solution to $1 - \omega = \omega$. There are fewer but different lowest order terms for other fractions. Now in each term replacing the sum $s = -\infty, \infty$ with $s = 2n, 2n+1$, $n = -\infty, \infty$, denoting $C_0 = cos(\psi_0)$, $S_0 = sin(\psi_0)$, and then defining $F_{a,b}(\lambda, \mu) = \sum_{n=-\infty}^{\infty}(-1)^n J_{2n+a}(k\lambda) J_{2n+b}(k\mu)$, we have

$$(1-\omega^2)\alpha = k\Phi_0[C_0(F_{1,0} + F_{1,1}) + S_0(F_{0,0} + F_{0,-1})], \qquad (9.9)$$

$$(1-\omega^2)\beta = k\Phi_0[C_0(F_{0,0} - F_{0,-1}) + S_0(F_{1,0} - F_{1,1})], \qquad (9.10)$$

$$\begin{aligned}
-2\frac{d\mu}{dt} = {} &k\Phi_0 J_1(k\alpha)[C_0(F_{0,-1} + F_{0,-2} + F_{0,1} + F_{0,0}) \\
&- S_0(F_{1,0} + F_{1,-1} + F_{1,2} + F_{1,1})] \\
&+ k\Phi_0 J_1(k\beta)[C_0(-F_{1,0} + F_{1,-1} - F_{1,2} + F_{1,1}) \\
&+ S_0(-F_{0,-1} + F_{0,-2} - F_{0,1} + F_{0,0})], \qquad (9.11)
\end{aligned}$$

$$\begin{aligned}
2\frac{d\lambda}{dt} = {} &k\Phi_0 J_1(k\alpha)[S_0(F_{0,-1} + F_{0,-2} + F_{0,1} - F_{0,0}) \\
&+ C_0(F_{1,0} + F_{1,-1} + F_{1,2} + F_{1,1})] \\
&+ k\Phi_0 J_1(k\beta)[S_0(-F_{1,0} + F_{1,-1} + F_{1,2} - F_{1,1}) \\
&+ C_0(F_{0,-1} - F_{0,-2} - F_{0,1} + F_{0,0})]. \qquad (9.12)
\end{aligned}$$

These equations determine the motion of a Poincaré point in the $k\rho$, ψ plane for small $k^2\Phi_0$. To determine the existence of resonances first look for fixed points of the Poincaré map, with $k^2\Phi_0 \ll 1$. In this case α and β are small, and since $\lambda(0) = -\alpha(0)$ and for a fixed point λ must be constant, it remains small. Keeping only up to first order in λ we find for the existence of a fixed point in the case $\omega = 1/2$ either $C_0 = 0$ or $S_0 = 0$.

For $C_0 = 1, S_0 = 0$ we have

$$(1 - \omega^2)\alpha = k\Phi_0 J_1(k\lambda)[J_2(k\mu) + J_0(k\mu)], \tag{9.13}$$

$$(1 - \omega^2)\beta = -k\Phi_0[J_1(k\mu) + J_0(k\mu)], \tag{9.14}$$

$$0 = J_1(k\lambda)[J_0(k\mu) + J_2(k\mu)][J_2(k\mu) + J_0(k\mu)]$$
$$-J_1(k\lambda)[J_1(k\mu) + J_0(k\mu)][2J_0(k\mu) + 2J_2(k\mu) + J_1(k\mu) + J_3(k\mu)], \tag{9.15}$$

$$0 = J_1^2(k\lambda)[J_0(k\mu) + J_2(k\mu)][2J_0(k\mu) + 2J_2(k\mu) - J_1(k\mu) - J_3(k\mu)]$$
$$-[J_1(k\mu) + J_0(k\mu)][2J_1(k\mu) + J_2(k\mu) - J_0(k\mu)]. \tag{9.16}$$

To lowest order in λ, μ is given by $0 = [J_1(k\mu) + J_0(k\mu)][-2J_1(k\mu) - J_2(k\mu) + J_0(k\mu)]$. The first root is from $J_0 = 2J_1 + J_2$ giving $k\mu = 0.825$. Equation 9.15 then gives $\lambda = 0$, and we find $\alpha = 0$, $\beta = -1.62k\Phi_0$. Since $\lambda = \alpha = 0$, the Poincaré points are given by $t_j = 2j\pi$. The fixed points are then $k\rho = k(\mu - \omega\beta)$, $\psi = 0$ and $k\rho = k(\mu + \omega\beta)$, $\psi = \pi$, agreeing with the values found in the Poincaré section, Fig. 9.1.

No solution is found for $C_0 = 0, S_0 = 1$. The second pair of fixed points in Fig. 9.1 at $\psi = \pm\pi/2$ and $k\rho = 1.84$ is more complex, due to a combination of motion at ω and 3ω. It should be obvious from the above that by including particle response at more frequencies, and allowing larger values of $k^2\Phi_0$, the number of fixed points in the map will increase enormously.

For $\omega \neq 1/2$, but less than one the situation is qualitatively different. To leading order there do not exist any fixed points; rather the fixed points of the map emerge from $\rho = 0$ through the coalesence of complex conjugate pairs as Φ_0 is increased. Nevertheless, such fixed points exist for all integer q, associated with the unstable domains of the associated Mathieu equation. A numerical Poincaré plot is shown in Fig. 9.2 for $k^2\Phi_0 = 0.1$, $\omega = 1/3$, showing period three fixed points which move upward as $k^2\Phi_0$ increases.

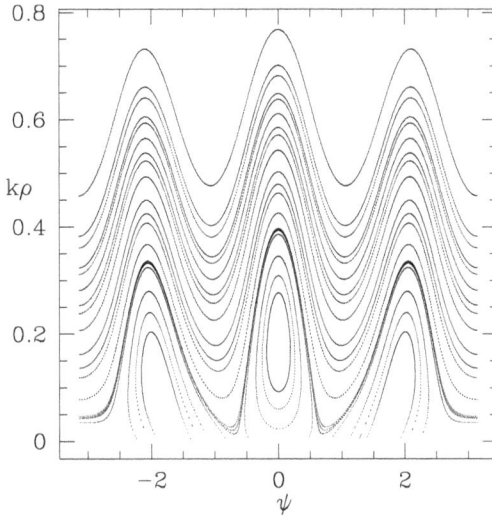

Fig. 9.2   Poincaré section for $k^2\Phi_0 = 0.1$, $\omega = 1/3$.

9.3   STOCHASTIC DOMAIN

Now investigate the approach to chaos and the extent of the chaotic domain, which limits the possible heating obtained. Figure 9.3 shows an example of the extent of the stochastic domain for $\omega = 1/4$, with $k^2\Phi_0 = 0.77$, bounded by good KAM surfaces at large $k\rho$. The initial particle distribution was random with $k\rho < 0.1$. In the domain of the good KAM surfaces the perpendicular energy is only oscillatory, described by the magnetic moment, and becomes an adiabatic invariant for large energy and relatively weak wave amplitude. Heating of an initially cold distribution proceeds to the maximum limit given by the good KAM surfaces in a rather short time; on the order of one to two hundred cyclotron periods. Even at a wave frequency of $1/10$ of the cyclotron frequency a Poincaré plot is quite stochastic for $k^2\Phi_0 = 1$. Note that this is a collisionless result.

Figure 9.4 shows the variation of the extent of the heating domain in $k\rho$ versus wave frequency for $k^2\Phi_0 = 0.36$, 0.8, and 2.6. For small wave amplitude some peaking can indeed be seen at low-order (small) integer fractions, as predicted by the Mathieu equation approximation. As the amplitude increases, however, nonlinear generation of many fixed points

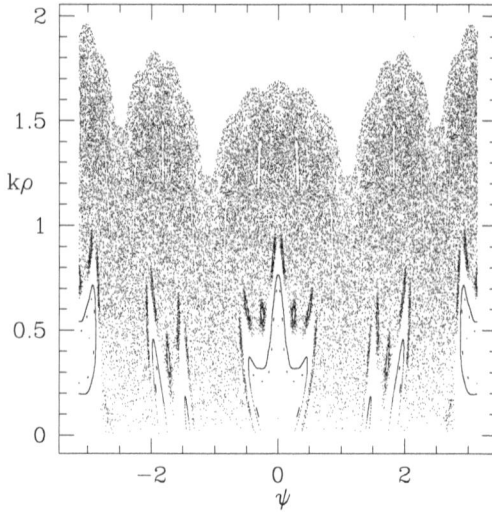

Fig. 9.3   Stochastic domain for $\omega = 1/4$, with $k^2\Phi_0 = 0.77$.

produces chaos which smooths out the resonance structures and makes the extent of the domain almost linear in ω. Of course in the limit of $\omega \to 0$ the motion is not stochastic, and there is no real heating, only large amplitude excursions in the potential. But for large amplitude the motion becomes stochastic, producing true heating for very low frequencies. For the two larger amplitude plots an X indicates the frequency for the onset of chaos. For $k^2\Phi_0 = 0.36$, curve a, there is no chaos, only large scale convective motion, even at $\omega = 1$. The onset of chaos at large wave amplitude as a function of ω is shown in Fig. 9.5. This plot was obtained by visual examination of the Poincaré plot produced by advancing a distribution with initially $k\rho < .01$. Although we have not calculated a precise stochastic threshold, values of $k^2\Phi_0$ above the line give a significantly stochastic plot, with the area of the domain of stochasticity increasing rapidly as one moves away from the line. The onset of chaos is very irregular, it can occur through period doubling, the overlap of islands of various period, or through the stochastic broadening of a separatrix. Of particular interest is the limit $\omega \to 0$, where the value of $k^2\Phi_0$ giving chaos approaches infinity as it must, since in this limit the system is described by a time independent Hamiltonian.

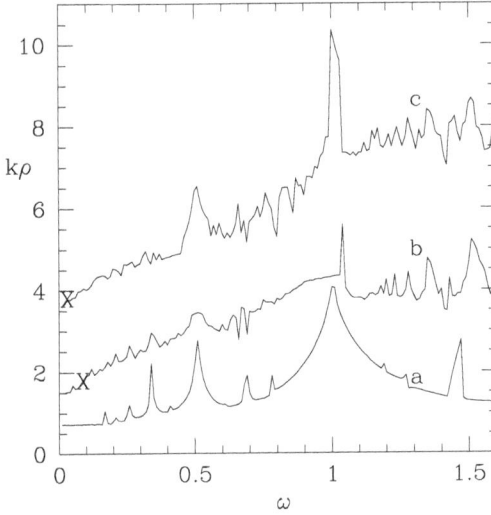

Fig. 9.4 Variation of the extent of the heating domain in $k\rho$ vs wave frequency for $k^2\Phi_0 = 0.36$ (a), 0.8 (b), and 2.6 (c).

Significant perpendicular heating can be obtained at a fraction of the cyclotron frequency. The above analysis treated only the case of a longitudinal wave propagating across a constant magnetic field, but the nonlinear resonance phenomenon should be more fundamental, and may have application in high power radio-frequency heating schemes in laboratory as well as in astrophysical plasmas.

9.4 CYCLOTRON MOTION IN A TOROIDAL SYSTEM

Use a cylindrical coordinate system with $\hat{x}$, $\hat{z}$, $\hat{\phi}$ forming a right-handed system. The particle motion is given by

$$\frac{d\vec{v}}{dt} = \vec{v} \times \vec{B} + \vec{E} \tag{9.17}$$

with $\vec{v} = \dot{x}\hat{x} + \dot{z}\hat{z} + x\dot{\phi}\hat{\phi} \equiv v_x\hat{x} + v_z\hat{z} + v_\phi\hat{\phi}$. Also we have $d\hat{x}/dt = \dot{\phi}\hat{\phi}$, $d\hat{\phi}/dt = -\dot{\phi}\hat{x}$, and find

$$\frac{d\vec{v}}{dt} = \frac{dv_x}{dt}\hat{x} + v_x\dot{\phi}\hat{\phi} + \frac{dv_z}{dt}\hat{z} + v_x\dot{\phi}\hat{\phi} + x\frac{d\phi}{dt}\hat{\phi} - x\dot{\phi}^2\hat{x}, \tag{9.18}$$

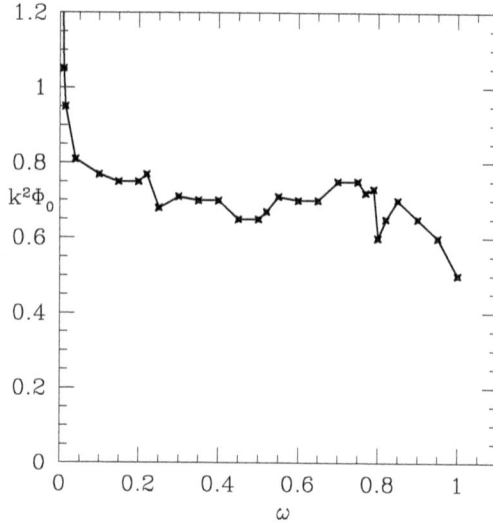

Fig. 9.5   The onset of chaos at large wave amplitude as a function of ω.

giving equations of motion in the field $\vec{B}$

$$\frac{d\vec{v_x}}{dt} = v_z B_\phi - x\dot{\phi}B_z + x\dot{\phi}^2, \tag{9.19}$$

$$\frac{d\vec{v_z}}{dt} = x\dot{\phi}B_x - v_x B_\phi, \tag{9.20}$$

$$\frac{d\dot{\phi}}{dt} = \frac{v_x B_z - v_z B_x}{x} - 2\frac{v_x\dot{\phi}}{x}. \tag{9.21}$$

The canonical momentum is $P_\phi = x(v_\phi + A_\phi) = xv_\phi - \psi_p$.

A magnetic perturbation is introduced by perturbing the poloidal flux, $\delta\psi_p = \alpha_{mn}(\psi)sin(\Omega_{mn})$ with $\Omega_{mn} = n\phi - m\theta - \omega t$, giving

$$\delta\vec{B} = -\nabla\alpha_{mn}(\psi)sin(\Omega_{mn}) \times \nabla\phi, \tag{9.22}$$

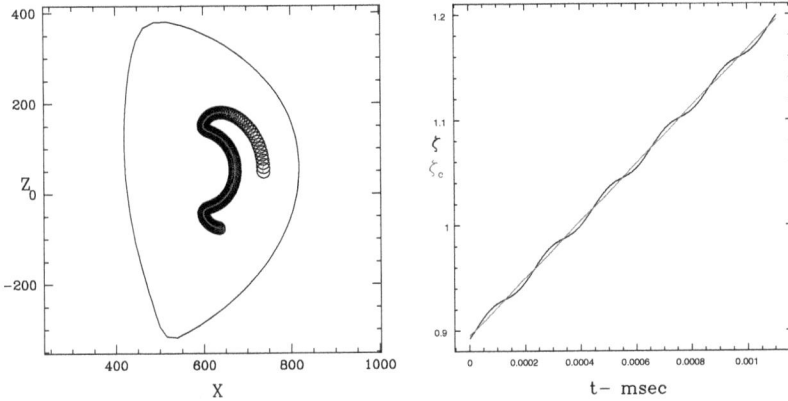

Fig. 9.6 A banana trapped particle and the guiding center trajectory.

and we have

$$\nabla \alpha_{mn}(\psi)sin(\Omega_{mn}) = \alpha'_{mn,\psi}(\psi'_a \hat{a})sin(\Omega_{mn}) - m\alpha_{mn}(\theta'_a \hat{a})cos(\Omega_{mn}) \quad (9.23)$$

with $a = x, z$, plus a term in $\hat{\phi}$, where primes indicate derivatives with respect to the subscript. From $cos\theta = (x - x_c)/r$, and $tan\theta = (z - z_c)/(x - x_c)$, with x_c, z_c the magnetic axis location, we find

$$\delta B_x = \frac{\alpha'_{mn,\psi}\psi'_z sin(\Omega_{mn})}{x} - \frac{m\alpha_{mn}(x - x_c)cos(\Omega_{mn})}{xr^2}, \quad (9.24)$$

$$\delta B_z = \frac{-\alpha'_{mn,\psi}\psi'_x sin(\Omega_{mn})}{x} - \frac{m\alpha_{mn}(z - z_c)cos(\Omega_{mn})}{xr^2}. \quad (9.25)$$

Note also that $\vec{B} = g\nabla\phi + I\nabla\theta + \delta\nabla\psi_p$ gives $g = xB_\phi$ and $I = rB_\theta$ with $B_\theta^2 = B_x^2 + B_z^2$. These equations are realized in a numerical code GYROXY using a fourth order Runge–Kutta stepping algorithm.

The instantaneous gyro center position is given by

$$\vec{X} = \vec{x} + \frac{\vec{v} \times \vec{B}}{B^2}. \quad (9.26)$$

Shown in Fig. 9.6 is an example of the full trajectory and the guiding center trajectory given by Eq. 9.26 for a poloidally trapped particle. We have used a shaped equilibrium with a major radius of 630cm and a field strength on axis of 3.6 kG.

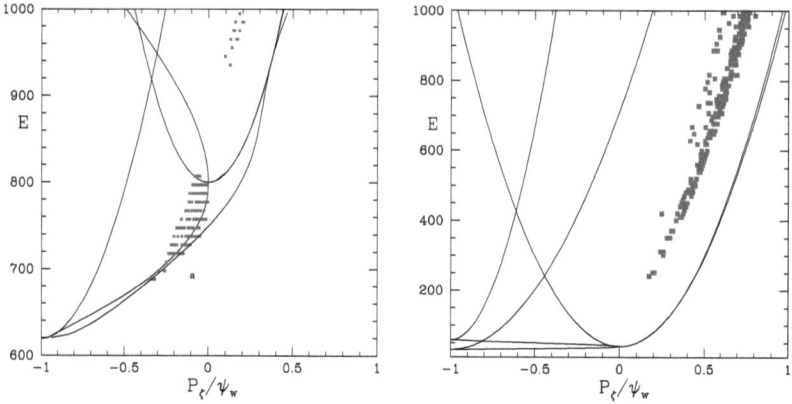

Fig. 9.7   Resonance location for deeply trapped particles, $\mu B = 0.8E$, left, and deeply passing particles, $\mu B = 0.04E$, right.

9.5   CYCLOTRON AND GUIDING CENTER ANALYSES OF RESONANCE

We previously developed methods for finding mode particle resonances capable of modifying particle distrubutions in toroidal plasma confinement devices using guiding center analysis. Since for high energy particles the cyclotron motion of the particle is often larger than the size of the resonances, it is necessary to ascertain whether in fact the guiding center approximation is valid for the examination of these resonances. We now compare the full cyclotron behavior of particle motion due to mode particle resonances with guiding center analysis.

Choose cases where the resonance width is small compared to the cyclotron motion. To this purpose we introduce equivalent perturbations in the two formulations and examine Poincaré plots. The ad hoc perturbation chosen was a single harmonic mode with $m = 2$ and $n = 1$, a broad radial profile, and zero frequency.

In Fig. 9.7 is shown the extent of the particle resonance for two values of μB, giving deeply passing orbits in one case, with very small cyclotron motion, and passing orbits with large cyclotron motion in the second. These plots were obtained using the methods described in Chap. 8. It is seen that the resonance exists over a large range of particle energies and values of μ,

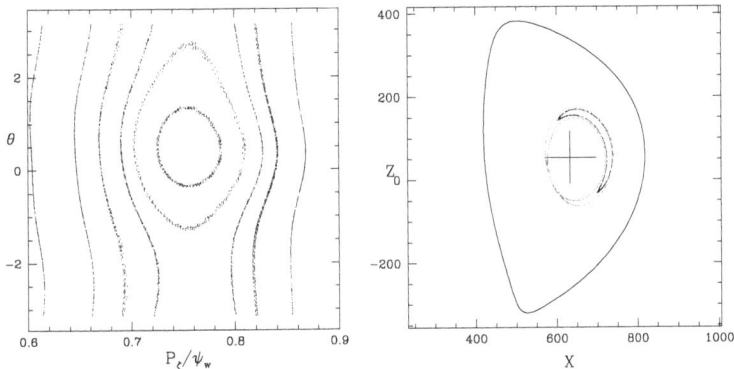

Fig. 9.8   Island produced in guiding center coordinates, $\rho = 36$ cm $\mu B = .04E$.

but is strongest at higher energies. At energies of 1 MeV the particles have gyro radii of 36 cm. These numbers are of course not relevant, only the ratio of the cyclotron motion to the size of the resonance is of importance for this work.

In Fig. 9.8 are plots of the resonance, showing the island produced by a Poincaré sections made with a particle trajectory with a gyro radius of $\rho = 36$ cm and a pitch chosen to be deeply passing, $\mu B = 0.04E$, using the guiding center code, both in the plane of P_ϕ, θ and in the poloidal cross section. In the poloidal cross section we have shown only those Poincaré points bordering and inside the resonance, to display its size. We have chosen a small perturbation so that the resonance width can easily be made smaller than the cyclotron radius. The island width is about 8 cm and the minor radius at the midplane is about 370 cm. It is much easier to see island width in the space of canonical momentum, and we see that the resonance has a width of $\Delta P_\zeta / \psi_w \simeq 0.5$.

In Fig. 9.9 is shown a Poincaré plot made using the full cyclotron motion, again with a particle with a 36 cm cyclotron radius, but with nearly parallel velocity, $v_\parallel = \vec{v} \cdot \vec{B}/B = 0.99$. Shown also is the trajectory in the poloidal plane, and it is evident that the cyclotron excursions about the guiding center are not large compared to the resonance width.

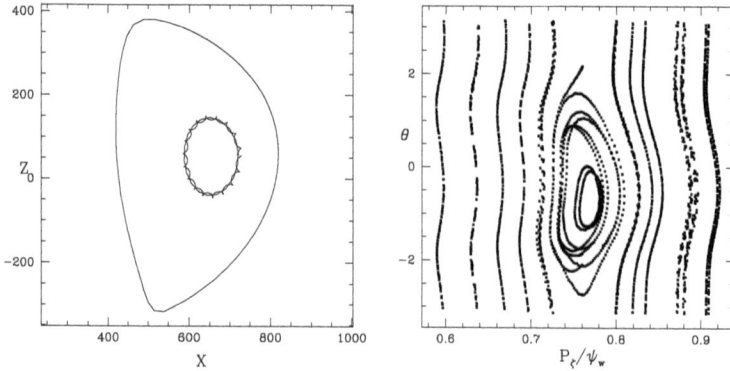

Fig. 9.9   Cyclotron orbit, $\rho = 36$ cm, $\mu B = 0.04E$. $\lambda = 0.99$, and Poincaré plot.

However, even in this case, it is noticed that the Poincaré plot is not as clean as the guiding center case shown in Fig. 9.8. The orbit positions of the resonantly trapped particles are not precisely concentric, they are shifted such that they do not share a common center. This distortion is due to the cyclotron motion. It becomes much worse as the pitch is decreased and the cyclotron excursions become larger. The reason for this is obvious. When the cyclotron motion is large the particle spends only a part of the cyclotron orbit in the resonance. When the particle is in resonance, the canonical momentum is modified, and this leads to local modifications of the orbit, producing an incoherent Poincaré plot. When the cyclotron excursions are small a particle can remain trapped in a resonance for a full cyclotron orbit and its motion traces out the resonance structure.

In Fig. 9.10 is shown the result of a Poincaré section of orbits with smaller pitch and hence larger cyclotron excursions from the guiding center trajectory. Coherent Poincaré plots are not obtained. Even in the domains where good KAM surfaces should exist the lines are broadened and blurred because of the fact that the Poincaré sections do not occur at a fixed value of cyclotron phase. However, the mixing of values of canonical momentum by the resonance still occurs over a range of ΔP_ζ of about 0.05, as in the case of deeply passing orbits. This means that profile flattening should occur over this range of canonical momentum, just as in the case when coherent trapping motion is present.

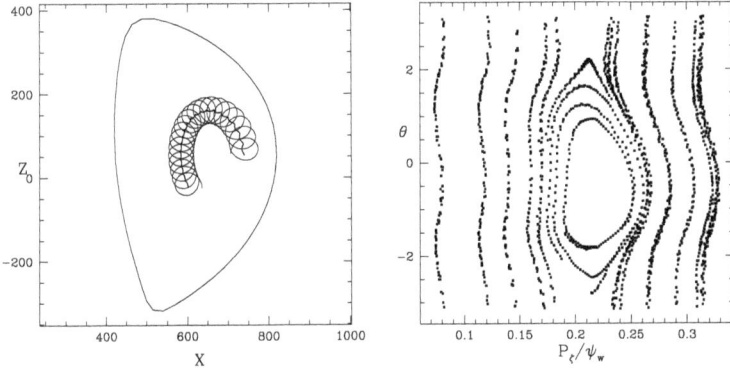

Fig. 9.10 Cyclotron orbit, $\rho = 36$ cm, $\mu B = 0.8E$. $\lambda = 0.6$, and Poincaré plot.

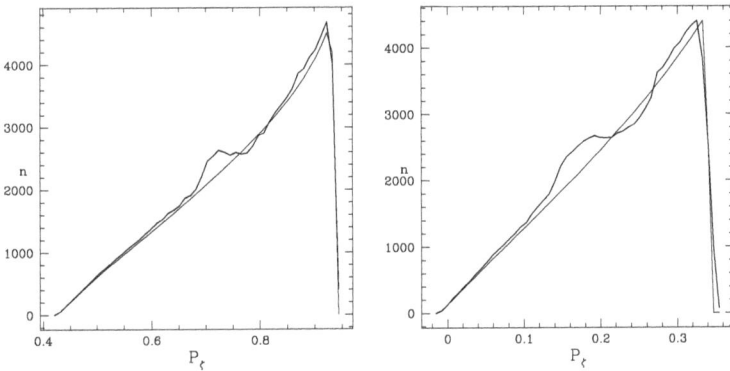

Fig. 9.11 Profile using guiding center code, $\mu B = 0.04E$, $\mu B = 0.8E$.

9.6 PROFILE FLATTENING

To verify the process of profile flattening in cases where the cyclotron motion is larger than the extent of the resonance we carry out simulations of the effect of a resonance on a distribution with an initially steep density profile. In Fig. 9.11 is shown the effect of the resonance on an initially steep density profile for distributions with large pitch, $\mu B = .04E$, and small pitch and consequently large cyclotron orbit $\mu B = 0.8E$ using a guiding center anal-

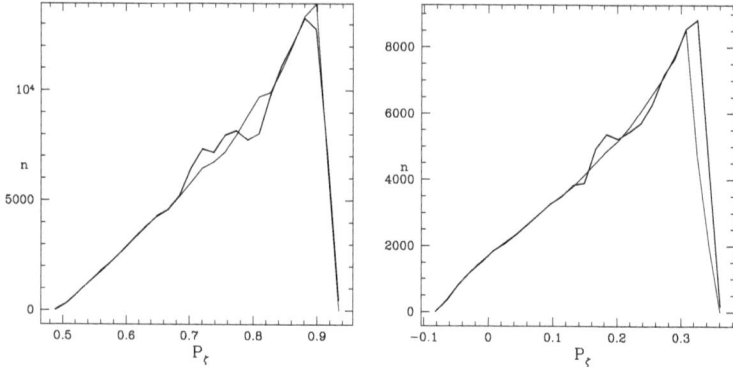

Fig. 9.12   Profile using cyclotron code, $\mu B = 0.04E$, $\mu B = 0.8E$.

ysis. In Fig. 9.12 is shown the effect of the same resonance using the full cyclotron orbit simulation. We conclude that even in the case where the cyclotron radius is large compared to the resonance width the resonances are still capable of producing flattening of the particle distribution.

Thus it is still possible to restrict an analysis to a guiding center treatment even in the case that the cyclotron radius is larger than the resonance islands.

9.7 References

- White, R. B., Commun. Nonlinear Sci. Numer. Simulat. 17, 2200 (2012).
- White, R. B., Plasma Phys. Control. Fusion, 53 085018 (2011)
- White, R. B., M. S. Chance, Phys. Fluids 27, 2455 (1984).
- Lieberman, M. A. and A. J. Lichtenberg, Regular and Chaotic Dynamics, Springer Verlag, New York (1983) p. 87.
- Smirnov, Yu. N. and D. A. Frank-Kamenetskii, Sov. Phys. JETP, 26, 627 (1968).
- Terasawa, T. and M. Nambu, Geo. Res. Lett. 16, 357 (1989).
- McChesney, J. M., P. M. Bellan, and R. A. Stern, Phys. Rev. Lett. 59, 1436 (1987).
- Bailey, A. D., R. A. Stern, and P. M. Bellan, Phys. Plasmas 2, 2963 (1995).
- Johnson J. R. and C. Z. Cheng, private communication (2001).
- Chen, L., Z. Lin and R. B. White, Phys. Plasmas 8, 4713 (2001).
- Karney, C. F. F. , Phys. Fluids 22, 2188 (1979).
- Cranmer, S. R., G. B. Field and J. L. Kohl, Astrophysical J. 518, 937 (1999).
- Gates, D., N. Gorelenkov and R. B. White, Phys. Rev. Lett. 87, 205003-1 (2001).

Chapter 10

Nonlinear Behavior

10.1 INTRODUCTION

To discuss the role of ideal MHD modes or resistive modes in tokamak discharges, their nonlinear behavior must be understood. Simple estimates show that the linear analysis is valid only for very small amplitudes; for ideal modes the displacements are typically small compared to global scale sizes, and for tearing modes the linear analysis fails when the island widths are comparable to the tearing layer thickness, typically on the order of millimeters in tokamaks. In high β plasmas the nonlinear behavior of the plasma often involves a combination of ideal and resistive effects.

Clearly the practical consequences of these modes thus depends on their nonlinear behavior. Nonlinear saturation of all modes at small amplitude could in principle lead to a perfectly acceptable nonaxisymmetric equilibrium state, possessing a sufficient volume of nested topologically toroidal magnetic surfaces to provide adequate confinement. The fact that a tokamak can remain in steady operation for extended periods with reasonable confinement assures us that in some sense this is true. The fact that the energy loss rate is one or two orders of magnitude larger than what would be expected from a quiescent axisymmetric equilibrium by neoclassical processes, and the occurrence of sudden, violent termination of discharges if certain operating parameters are exceeded, indicates that there are at least two types of nonlinear phenomena operating in tokamaks which we must try to understand.

Large scale phenomena occurring on time scales intermediate between the ideal MHD time τ_A, and the resistive time τ_R include the internal disruptions, constituting the rapid phase of the sawtooth oscillations, and the major disruptions. These phenomena play a major role in the behavior of tokamak discharges. Although there is a reasonable consensus that the basic qualitative and quantitative features are understood, there remain puzzling features which are not yet explained. As higher temperatures and densities are achieved it is almost certain that these phenomena will exhibit qualitatively new behavior, and that new phenomena will emerge.

The investigation of nonlinear consequences of the equations of magnetohydrodynamics has from the beginning been motivated by the desire to explain rapid, sometimes violent tokamak behavior observed experimentally, with a hope to finding methods to control or avoid it. The use of resistive MHD in explaining the large scale dynamics has been remarkably successful, perhaps more so than should be expected.

In 1971 Rutherford, Furth, and Rosenbluth examined the nonlinear behavior of the kink mode for small amplitudes and showed that in a constant current profile (shear free) equilibrium the nonlinear terms produce further destabilizing forces, and thus the final state might be expected to be highly distorted.

Kadomtsev and Pogutse showed in 1973 that in fact a shear free plasma can possess a second equilibrium state with large helical vacuum bubbles present, with the energy of this bubble state lower than that of the initial circular equilibrium. The presence of a large bubble might cause a significant plasma volume to make contact with the tokamak wall, and such a process was suspected of being involved in tokamak disruption. It immediately became of interest to discover whether these bubble states were accessible from an initial equilibrium, how much shear was sufficient to prohibit the formation of such states, and to examine the temporal evolution associated with bubble formation.

This task involved numerical analysis, since it was hopeless to attempt to describe the evolution through highly distorted plasma shapes analytically. Fortunately, the free energy reservoir for the helical deformation is large compared to toroidal effects, so the problem could be done in cylindrical approximation. Consideration of a single helicity m/n then means that all quantities are functions only of r and $m\theta - n\phi$, even in the full nonlinear evolution, which reduces the problem from one of three dimensions to one of two. The ideal MHD equations were simplified by an inverse aspect ratio expansion, giving a numerical scheme for the examination of the full nonlinear evolution of an equilibrium linearly unstable to kink perturbations.

This work is described in Sec. 10.4. First results were given at the 1974 IAEA meeting, (White *et al.*, 1975) and a more detailed description was published later (Rosenbluth *et al.*, 1976). At the same time a similar effort was under way at the Kurchatov Institute, led by Dnestrovskii (1975).

Results showed that in a shear free equilibrium the plasma evolves rapidly to a bubble state. The absense of shear means that the plasma is free to change its shape quickly to achieve overall magnetic energy minimum. Cases of severe bubble formation could not be treated using a Fourier decomposition code, because of the very large number of harmonics involved.

This evolution to a bubble state is very effectively impeded by the presence of magnetic shear, and for a parabolic current profile no real bubble states exist, the plasma only assuming a relatively minor helical deformation. For most current profiles an important role of bubble states in the evolution of the plasma is ruled out, but the possibility exists for such deformations to occur in states of very low shear, and we will return to this eventuality in the discussion of the sawtooth and the major tokamak disruption.

The internal kink was a possible candidate for a description of the internal tokamak disruption, or sawtooth, having many of the correct qualitative features. Nonlinearly however, Rosenbluth, Dagazian, and Rutherford (1973) found the mode to saturate at too small an amplitude to account for the experimentally observed negative voltage spike. This analytic result, which was later confirmed with a nonlinear code by Park *et al.* (1980), effectively ruled out the internal kink from playing an important role in tokamak dynamics. It was relegated to a minor role until it was later discovered that it was capable of strongly interacting with high energy trapped particles, as was discussed in Chap. 8.

Nonlinear theory for the tearing mode began in 1973 when Rutherford showed that the mode ceases exponential growth and enters a domain of algebraic growth as soon as the island is larger than the tearing layer. This slowing down of the mode is due to the formation of an inductive current flowing at the island O-point, parallel to the Ohmic current. This current, which opposes island growth, replaces the effect of inertia in the description of the evolution, and the plasma current becomes a function of the magnetic flux.

In the early 1970s Rosenbluth and collaborators explored the full nonlinear resistive evolution of tokamak profiles, using the inverse aspect ratio expansion first developed for the investigation of the kink mode. The results of Rutherford were duplicated, and it was soon shown that the evolution of the magnetic island in the tearing mode to a saturated state could be predicted by a generalization of Δ' to the case of finite island width, *i.e.* that the island state simply found a width determined by a magnetic energy minimum, and then only slowly changed its width to follow the slower resistive evolution of the current profile. This allows for the possibility of stable configurations which possess magnetic islands and exhibit Mirnov oscillations.

By the late 1970s results were being produced with a variety of numerical resistive codes using the tokamak ordering expansion. Some of the results will be discussed in the Sec. 10.8, on the sawtooth, and Sec. 10.9, on the major disruptions. Waddell and collaborators extended the calculations to include many different helicities (Callen *et al.*, 1979). In a real tokamak, helical symmetry could certainly be assumed to be broken, and it was possible that nonlinear coupling of different helicities could be important at large mode amplitude. Monticello and White (1980) included diamagnetic terms to study nonlinear effects on mode rotation, possible modification of the saturation results, and to explore the possibility of feedback stabilization of the mode.

In addition, new analytic results were extending the range of validity of the nonlinear calulations and adding insight to the numerical results. Strauss (1983) extended the formalism to higher order in the expansion parameter to be able to treat cases of high plasma beta. Drake and Lee (1977) and Cowley *et al.* (1986) examined different domains of collisionality and found modifications of the linear theory, but showed that the results of Rutherford still held when the island size became larger than the tearing layer.

Kadomtsev (1975) developed a qualitative scenario for the experimentally observed sawtooth oscillation involving the $m/n = 1/1$ mode. Again, only numerical analysis could discover whether the time scale for the process was in agreement with the experiments. Nonlinear evolution of the $m = 1$ mode showed that the magnetic island grew far beyond the tearing layer width with no modification of the exponential growth rate given by linear analysis. The conditions necessary for the Rutherford derivation of algebraic growth are not satisfied.

One of the important goals of the investigation of nonlinear MHD phenomena was the understanding of the major disruption. Experimental evidence indicated that the $m/n = 2/1$ mode played an important role in the onset of the disruption, both because of the precursor oscillations observed with Mirnov coils, and the tendency for disruption to occur when the value of q at the tokamak wall approached two. By investigating island saturation widths using a single helicity code, White, Monticello, and Rosenbluth (1977) found conditions necessary for the occurence of a large $m = 2$ island.

Large islands occur only when the current profile is fairly flat, and thus steep in the outer region, and when the value of q on axis rises above one. The single helicity analysis is valid for large island development, but cannot explore the ensuing violent behavior.

Waddell and collaborators at Oak Ridge discovered significant nonlinear coupling of different helicities. Upon insertion of a flat profile with $q(0) > 1$, suggested by the single helicity analysis, a nonlinear destabilization of the $m/n = 3/2$ and $5/3$ modes was observed when the large $m = 2$ island overlapped the $q = 3/2$ and $5/3$ surfaces. The ensuing highly stochastic state provided a candidate for the turbulent final stages of the disruption.

10.2   THE REDUCED EQUATIONS

To study the nonlinear evolution of a plasma the variational methods of Chap. 6 are naturally of no use, it being necessary to follow the temporal evolution of some given initial conditions, perhaps chosen to correspond to a linearly unstable mode. Considerable simplification can be obtained by expanding in the inverse aspect ratio ϵ. If the equilibrium is axisymmetric, consisting of nested circular flux surfaces, this expansion results in the decoupling of different helicities, reducing the problem from a three-dimensional to a two-dimensional one. Even if this is not the case, the reduced equations are simpler to implement numerically. The reduced equations were originally obtained by Rosenbluth et $al.$ (1976) to study nonlinear kink mode evolution, using harmonics of a single helicity, for a system with plasma beta of order ϵ^2. They were subsequently extended by Strauss (1976) to systems with beta of order ϵ, and involving multiple helicities: an important extension since it includes ballooning. Later the ϵ expansion was extended by Izzo et $al.$ (1983) to include higher order terms, allowing the treatment of smaller aspect ratio.

Consider an arbitrary magnetic field in toroidal geometry. Use the general representation Eq. 1.36, and choose ζ to be the toroidal angle ϕ. We then have

$$\vec{B} = g\nabla\phi + \nabla\phi \times \nabla_\perp \psi_p. \tag{10.1}$$

Here $\nabla_\perp = \hat{r}\partial_r + (\hat{\theta}/r)\partial_\theta$, *i.e.* $\perp$ means perpendicular to $\hat{\phi}$. Now use the tokamak ordering introduced in Sec. 2.7. Expand the equations (1.5–1.10), giving the time evolution, in the inverse aspect ratio ϵ. Take the plasma beta, $\beta = 2p/B^2$, to be of order ϵ or smaller. Choosing B_ϕ and q of order unity gives $B_\perp \sim \epsilon$. Take the major radius as the scale of distance. All quantities are assumed to vary on the scale of the minor radius in r, θ so $\nabla_\perp \sim 1/\epsilon$ but on the scale of the major radius in ϕ, so that $\partial_\phi \sim 1$. Thus we have $\psi_p \sim \epsilon^2$. Since we allow β to be of order ϵ we have $g = 1 + g_1$ with $g_1 \sim \epsilon$. The current is then

$$\vec{j} = \nabla g \times \nabla\phi + (\nabla_\perp^2 \psi_p)\nabla\phi - \frac{1}{X^2}\partial_\phi \nabla\psi_p - \frac{1}{X}\nabla\phi(\nabla\psi_p \cdot \nabla X). \tag{10.2}$$

The first two terms are of order one and the last two of order ϵ. We then find to order ϵ

$$\vec{j} \times \vec{B} = -\frac{g}{X^2}\nabla_\perp g - \partial_\phi \nabla_\perp \psi_p \times \nabla\phi - \nabla\psi_p(\nabla_\perp^2 \psi_p) + (\vec{B} \cdot \nabla_\perp)g\nabla\phi \tag{10.3}$$

and only the first term is of order one. Assume the most rapid growth rate to be that of the kink mode, *i.e.* given by $\tau_A^{-1} = B_\phi/(\sqrt{\rho}X)$. Ordering $\rho \sim 1$ we have $\partial_t \sim 1$. The plasma motion is given by $\rho d\vec{v}/dt = \vec{j} \times \vec{B} - \nabla p$. Time variation is restricted to be of order one, so $\rho d\vec{v}/dt \sim \epsilon$. We assume v_ϕ to be of order ϵ or smaller. Later we will find that it decouples completely from the other variables and can be taken to be zero. Since $\vec{j} \times \vec{B}$ and ∇p are of order one, to lowest order we have $\vec{j} \times \vec{B} = \nabla p$, or

$$\nabla_\perp(p + g) = 0. \tag{10.4}$$

Now consider the time dependence of the field

$$\partial_t B = \nabla \times [\vec{v} \times \vec{B} - \eta\vec{j}]$$
$$= -(\vec{v} \cdot \nabla)\vec{B} + (\vec{B} \cdot \nabla)\vec{v} - \vec{B}(\nabla \cdot \vec{v}) - \nabla \times (\eta\vec{j}). \tag{10.5}$$

Taking the ϕ component we find that every term is of order ϵ except the $\nabla \cdot \vec{v}$ term, and thus

$$\nabla \cdot \vec{v} = 0 \tag{10.6}$$

within terms of order ϵ. This is a consequence of our assuming the time evolution to be no faster than the kink mode rate, and is important numerically because it eliminates the magnetosonic wave from the system, the fastest wave being the incompressible Alfvén wave.

The resistivity is assumed small, with the resistive time much greater than one. Now from $dp/dt = -\gamma p \nabla \cdot \vec{v}$ we find that this is of order ϵ^2 and thus negligible. Thus

$$\frac{dp}{dt} = 0. \tag{10.7}$$

In order to maintain pressure balance, Eq. 10.4, we then require

$$\frac{dg}{dt} = 0. \tag{10.8}$$

The ϕ component of Eq. 10.5 then determines $\nabla \cdot \vec{v}$ to order ϵ. Physically $\nabla \cdot \vec{v}$ must be regarded as adjusting instantaneously compared to the time scale of the kink, through the compressional sound waves which have been eliminated by means of the ordering. The incompressibility allows the velocity to be written as

$$\vec{v} = \nabla \phi \times \nabla U + v_\phi \hat{\phi} + O(\epsilon^2) \tag{10.9}$$

with U of order ϵ^2. Substitute this expression into the equation of motion, and apply $\hat{\phi} \cdot \nabla \times$. We find using Eq. 10.3 that to lowest order

$$\rho \frac{d}{dt} \nabla_\perp^2 U = (\vec{B} \cdot \nabla) \nabla_\perp^2 \psi_p + 2 \frac{dp}{dz}, \tag{10.10}$$

which reduces to Eq. 7.24 if the pressure term is neglected.

We wish to treat the general problem of many coupled helicities, so we do not introduce the helical flux, but rather work with the poloidal flux. The results for a single helicity can be readily obtained by using Eq. 2.109. For the poloidal flux, ψ_p, one begins by noting that $\vec{B}_\perp = -\nabla \times (\psi_p \nabla \phi)$ and thus the vector potential is $\vec{A} = -\psi_p \nabla \phi + \vec{A}_\perp$ with $\vec{B} = \nabla \times \vec{A}$. Eqs. 1.5 and 1.6 then give $\partial_t \psi_p = X^2 \nabla \phi \cdot [\eta \vec{j} - \vec{v} \times \vec{B}]$. Substituting ψ_p and U, we then find

$$\partial_t \psi_p = \vec{B} \cdot \nabla U + \eta \nabla_\perp^2 \psi_p, \tag{10.11}$$

i.e., similar to Eq. 7.23.

For the ϕ component of the velocity, again using Eq. 10.3 we find

$$\rho\frac{dv_\phi}{dt} = -\vec{B} \cdot \nabla p. \tag{10.12}$$

We now have a complete set of equations for the time advancement of the functions ψ_p, U, p, g, and v_ϕ, namely Eqs. 10.7–10.12. Note that v_ϕ is completely decoupled from the other quantities. Further, we find that $\vec{B} \cdot \nabla p = (g/X^2)\partial_\phi p + (\nabla\phi \times \nabla\psi_p) \cdot \nabla_\perp p$. But we have found that $dg/dt = dp/dt = d\psi_p/dt = 0$ and thus $d/dt\vec{B} \cdot \nabla p = 0$. Thus if initially there is no toroidal acceleration, it remains zero, and if initially also $v_\phi = 0$, then it remains so.

These reduced equations have been used in three-dimensional numerical codes to investigate nonlinear tearing and ballooning modes (White *et al.*, 1979; Strauss *et al.*, 1980) and in a low-β version (Biskamp *et al.*, 1978; Dnestrovskii *et al.*, 1978; Callen *et al.*, 1979) to examine the nonlinear interaction of tearing modes of different helicities. Various numerical methods have been employed. The most successful uses Fourier decompositions in θ and ϕ, and a radial grid. Results obtained using these codes are discussed in the following sections.

10.3   NONLINEAR EXTERNAL KINK

The kink mode consists of the growth of a helical distortion of the plasma-vacuum interface. It was first suggested by Kadomtsev and Pogutse (1973) that this distortion might grow into a large vacuum bubble, displacing the plasma outward, and that contact of the plasma with the limiter might be involved in the major tokamak disruption observed to occur on operating devices. Using the constant current profile model in cylindrical geometry, it is straightforward to find that a final state consisting of a helical vacuum bubble can indeed have lower magnetic energy than the initial circular equilibrium. This calculation does not provide a description of the nonlinear behavior, because it is not obvious that the final bubble state is dynamically accessible. Numerical simulation of the nonlinear behavior soon confirmed (White *et al.*, 1975) that the plasma could evolve to a vacuum bubble state, but that such violent plasma contortion is limited to rather flat (shear free) q profiles.

The nonlinear analysis is restricted to the evolution of a mode of a fixed helical symmetry. In the cylindrical approximation a mode of a given m/n

couples nonlinearly to form harmonics with mode numbers $2m/2n$, $3m/3n$, etc., always with the same helicity.

We consider cylindrical geometry, with $\beta \sim \epsilon^2$. This means we take $X = 1$, $\nabla X = 0$, and $g = 1 + g_2$ with $g_2 \sim \epsilon^2$. Eq. 10.2 gives

$$\vec{j} = \nabla \phi \nabla_\perp^2 \psi_p - \nabla_\perp \partial_\phi \psi_p + \nabla g \times \nabla \phi \tag{10.13}$$

and Eq. 10.3 gives

$$(\vec{j} \times \vec{B})_\perp = -g\nabla_\perp g - (\nabla_\perp \partial_\phi \psi_p) \times \nabla \phi - \nabla^2 \psi_p \nabla_\perp \psi_p. \tag{10.14}$$

Now use $B^2 = g^2 + (\nabla \psi_p)^2$, and substitute the helical flux, ψ through $\psi_p = \psi + (nr^2/2m)g$. We omit a subscript for the helical flux ψ for the remainder of this chapter. We then find from $\rho d\vec{v}/dt = \vec{j} \times \vec{B} - \nabla p$, by simplifying Eq. 10.3,

$$\rho \frac{d\vec{v}_\perp}{dt} = -\nabla_\perp \left(p + B^2/2 - (\nabla_\perp \psi)^2/2 + 2g\psi n/m + (nrg/m)^2/2 \right)$$
$$- \nabla_\perp^2 \psi \nabla_\perp \psi. \tag{10.15}$$

Eq. 10.10 then gives, substituting and using $\partial_\phi \psi = (-n/m)\partial_\theta \psi$

$$\frac{d}{dt}\rho \nabla_\perp^2 U = [\nabla_\perp \psi \times \nabla_\perp (\nabla_\perp^2 \psi)] \cdot \phi. \tag{10.16}$$

We allow for the possibility of a thin surface current and derive a boundary condition for $\partial U / \partial n$ by using the pressure balance across it,

$$p + B^2/2|_p = B^2/2|_v. \tag{10.17}$$

In the vacuum the condition $\vec{j} = 0$ gives, from the ϕ component of Eq. 10.13, $\nabla_\perp^2 \psi_p = 0$ or

$$\nabla_\perp^2 \psi = -\frac{2n}{m}. \tag{10.18}$$

The r component gives $g = (-nr/m)\partial_r \psi_p = f(r)$, with $f(r)$ arbitrary, and the θ component then gives

$$g = 1 - \frac{nr}{m}\partial_r \psi_p. \tag{10.19}$$

Using $B^2 = g^2 + (\nabla \psi_p)^2$ and substituting we then find in the vacuum

$$B_v^2 = 1 + (\nabla \psi)^2 - \left(\frac{nrg}{m}\right)^2 + O(\epsilon^4). \tag{10.20}$$

Now use Eq. 10.14, and replace $p + B^2/2$ by $B_v^2/2$ by pressure balance. Introduce an orthonormal coordinate system at the plasma boundary, with $\hat{n}$ normal and $\hat{s}$ tangent to the surface, and $\hat{n}, \hat{s}, \hat{\phi}$ right-handed. Then Eq. 10.15 gives

$$-\rho\hat{s} \cdot \frac{d}{dt}\vec{v}_\perp = \frac{d}{ds}\left[B_v^2/2 - (\nabla_\perp\psi)^2/2 + ng\psi/m + (nrg/m)^2/2\right]$$
$$+\nabla_\perp^2\psi\partial_s\psi. \quad (10.21)$$

Now use Eq. 10.20 and the fact that ψ is a flux function ($\partial_s\psi = 0$) to find

$$-\rho\hat{s} \cdot \frac{d}{dt}\vec{v}_\perp = \frac{1}{2}\frac{d}{ds}[(\nabla\psi_{vac})^2 - (\nabla\psi)^2]. \quad (10.22)$$

Since we have an equation for advancing $\nabla_\perp^2 U$, we need, to solve the Poisson equation for U, the boundary condition $\partial U/\partial n$. Thus we must advance $\partial U/\partial n$ in time. We have

$$-\frac{d}{dt}\partial_n U = \frac{d}{dt}(\vec{v}_\perp \cdot \hat{s}) = \hat{s} \cdot \frac{d\vec{v}_\perp}{dt} + \vec{v}_\perp \cdot \frac{d\hat{s}}{dt}. \quad (10.23)$$

But we have

$$\vec{v}_\perp \cdot \frac{d\hat{s}}{dt} = [(\vec{v}_\perp \cdot \hat{n})\hat{n} + (\vec{v}_\perp \cdot \hat{s})\hat{s}] \cdot \frac{d\hat{s}}{dt} = (\vec{v}_\perp \cdot \hat{n})\hat{n} \cdot \frac{d\hat{s}}{dt} \quad (10.24)$$

and further

$$\hat{n} \cdot \frac{d\hat{s}}{dt} = \hat{n} \cdot (\hat{s} \cdot \nabla_\perp)\vec{v}_\perp. \quad (10.25)$$

This is best seen by using a collection of discrete points on the boundary with $\hat{s} = (\vec{r}_k - \vec{r}_{k-1})/|\vec{r}_k - \vec{r}_{k-1}|$. We then find

$$\frac{d}{dt}\partial_n U = \frac{1}{2}\hat{s} \cdot \nabla_\perp[(\nabla\psi_{vac})^2 - (\nabla\psi)^2] - (\vec{v}_\perp \cdot \hat{n})\hat{n} \cdot (\hat{s} \cdot \nabla)\vec{v}_\perp. \quad (10.26)$$

In the vacuum we have $\nabla_\perp^2\psi = 2n/m$. The equilibrium is completely specified by $q(r)$ within the plasma, the plasma radius a, the radius of the conducting wall b, and the skin current j^*. The equilibrium ψ_0 is found by solving the differential equation $\psi_0' = r(1/q - n/m)$. In the vacuum we have $q = q_a r^2/a^2$ and thus we find

$$\psi_0(r) = \frac{a^2}{q_a}\ln(r/b) + \frac{n(b^2 - r^2)}{m} + \psi_w \quad (10.27)$$

with the integration constant chosen so that $\psi_0(a) = 0$, $\psi_w = n(a^2 - b^2)/m - (a^2/q_a)ln(a/b)$. The skin current is related to a discontinuity in q at the plasma surface, with

$$j^* = \frac{ma}{n}\left(\frac{1}{q_-} - \frac{1}{q_+}\right).\tag{10.28}$$

Here $q_+ = q_a$ and q increases at $r = a$ if j^* is positive. The linear growth rate for the kink mode in the case of a constant q profile was derived in Sec. 6.9.

10.4 VACUUM BUBBLES

The kink mode consists in the growth of a magnetic island located in the vacuum region which helically deforms the plasma surface. If, as the island grows, it is enveloped by the plasma a final state can be imagined in which the island has completely entered the plasma, forming a bubble. Such a state can in some circumstances possess lower energy than the initial state. There is a conserved integral of the motion, the energy. To find its form begin with Eq. 10.15, dotted with $\vec{v}_\perp$ to give

$$\frac{d}{dt}\int \rho\frac{v^2}{2}d\tau =$$

$$-\int_{pv} ds(\hat{n}\cdot\vec{v})\left[p + B^2/2 - (\nabla_\perp\psi)^2/2 + 2ng\psi/m + (nrg/m)^2/2\right]$$

$$-\int_p \vec{v}\cdot\nabla_\perp\psi\nabla_\perp^2\psi d\tau.\tag{10.29}$$

In the plasma-vacuum integral, use $p + B_p^2/2 = B_v^2/2$ and $d\psi/dt = 0$, and Eq. 10.20 for the vacuum field to find

$$\frac{d}{dt}\int \rho\frac{v^2}{2}d\tau = -\int_{pv} ds(\hat{n}\cdot\vec{v})\left[(\nabla\psi_{vac})^2/2 - (\nabla\psi)^2/2\right]$$

$$+\int_p \partial_t\psi\nabla^2\psi d\tau.\tag{10.30}$$

However,

$$\frac{d}{dt}\int_p (\nabla\psi)^2 d\tau = \int_p \frac{\partial}{\partial t}(\nabla\psi)^2 d\tau + \int_{pv} (\nabla\psi)^2\vec{v}\cdot\hat{n}ds\tag{10.31}$$

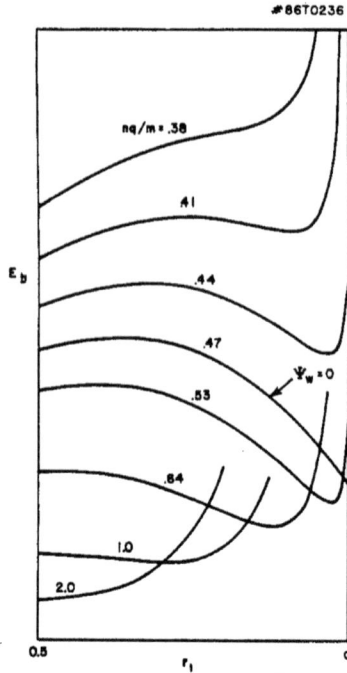

Fig. 10.1　Energy of the bubble state.

and

$$\int_p \frac{\partial}{\partial t} \frac{(\nabla \psi)^2}{2} d\tau = \int_{pv} \partial_t \psi \hat{n} \cdot \nabla \psi ds - \int_p \partial_t \psi \nabla^2 \psi d\tau. \qquad (10.32)$$

On the surface $\partial_t \psi = -\vec{v} \cdot \nabla \psi$ so

$$\oint_{pv} \partial_t \psi \hat{n} \cdot \nabla \psi ds = -\oint_{pv} (\vec{v} \cdot \nabla \psi)(\hat{n} \cdot \nabla \psi) ds - \oint_{pv} \vec{v} \cdot \hat{n} (\nabla \psi)^2 ds \quad (10.33)$$

and thus we find from Eq. 10.30

$$\frac{d}{dt} \int_p [\rho v^2 + (\nabla \psi)^2] d\tau = -\oint_{pv} ds \hat{n} \cdot \vec{v} (\nabla \psi_{vac})^2. \qquad (10.34)$$

To evaluate the right hand side of Eq. 10.34 note that

$$\frac{d}{dt}\int (\nabla\psi_{vac})^2 d\tau = \int \partial_t(\nabla\psi_{vac})^2 d\tau - \oint ds(\hat{n}\cdot\vec{v})(\nabla\psi_{vac})^2 \quad (10.35)$$

and

$$\int \partial_t(\nabla\psi_{vac})^2 d\tau = 2\int (\nabla\psi_{vac}\cdot\nabla)\partial_t\psi_{vac}d\tau, \quad (10.36)$$

which can be expressed as $\oint ds\psi_{vac}(\nabla\partial_t\psi_{vac})\cdot\hat{n}$ since $\nabla^2\psi_{vac} = constant$.

Using a harmonic solution for the vacuum flux function

$$\psi_{vac} = c(t)ln(r/b) + \psi(b,t)$$
$$+\sum\psi_m(t)[(r/b)^m - (r/b)^{-m}]cosm\theta + A(r^2-b^2)/2, \quad (10.37)$$

this integral can be evaluated, and if the value of the flux at the conducting wall is held fixed during the evolution the conserved energy is

$$E = \frac{1}{2}\int_p [\rho v^2 + (\nabla\psi)^2]d\tau - \frac{1}{2}\int_v (\nabla\psi_{vac})^2 d\tau + 2\pi\psi_w c. \quad (10.38)$$

Holding ψ_w fixed is not equivalent to holding q fixed, the total current changes.

If the plasma is shear free, the minimum energy is obtained with a circular cross section bubble. If the bubble radius is r_b the plasma radius is $r_1 = (a^2 + r_b^2)$. Consider the case of a constant current profile, with $q = m/n$ everywhere within the plasma. This is a worst case model, with the field resonant with the perturbation everywhere in the plasma and no shear to impede the motion. Choose $\psi = 0$ in the plasma. The initial equilibrium is given by Eq. 10.27. For the final state, inside the bubble we have $\psi = (r_b^2 - r^2)n/2m$, and outside

$$\psi(r) = n(r_1^2 - r^2)/2m - \frac{ln(r/r_1)n(r_1^2-b^2)}{ln(b/r_1)2m} + \psi_w, \quad (10.39)$$

and the energy of the bubble is

$$E_b = \pi r_1^2 a^2/2 - \pi[(b^2-r_1^2)/2 + \psi_w]^2/ln(r_1/b). \quad (10.40)$$

The energy of the bubble state as a function of r_1 for various values of ψ_w is shown in Fig. 10.1. For a wide range of ψ_w or equivalently q_a for the initial equilibrium, a bubble state is energetically favorable. This range of q values is shown on the plot of the growth rates for the various modes

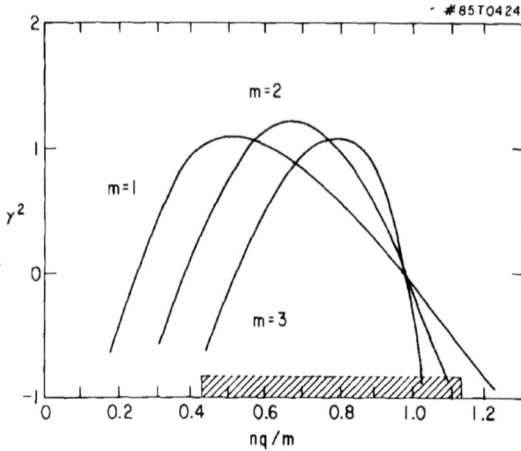

Fig. 10.2   Energetically favorable domain for bubble formation.

of different m values in Fig. 10.2. The bubble region extends beyond the region of linear instability. Thus even some stable equilibria can evolve to bubble states if the initial perturbation is sufficiently large.

Numerical simulations of the plasma evolution showed that in cases in which the bubble state was energetically favorable the plasma in fact evolved to such a state. An example of the time evolution of the plasma-vacuum interface for a $m = 1$ mode is shown in Fig. 10.3. The evolution to state 3 occurs in $\Delta t = 10\tau_A$.

To investigate the extent of bubble formation, and compare with the predictions of the minimum energy bubble states, a damping term was introduced into the equations of motion, allowing the configuration to settle into its minimum energy state. Results are shown in Fig. 10.4 for a mode with $m = 1$, $n = 1$, and initial plasma radius $a = 0.8$. The top two deformed states are not bubble states, as can be seen by the continuity of $\partial_n \psi$ on the surface. This indicates that the vacuum field lines curve gently around the plasma, with no discontinuities in direction. The addition of more computation points in the surface makes the extended tips of the plasma round and smooth. In a true bubble state, such as the bottom two in the figure, $\partial_n \psi$ is discontinuous at the tips, and the inclusion of more

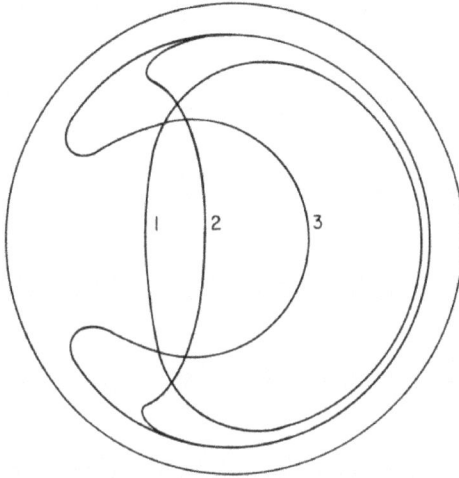

Fig. 10.3   Bubble evolution.

computation points leads to a sharp cusp. The X-point associated with the mode has moved around to the left, and the extended tips of the plasma reach in toward the X-point.

Plasma shear has a strong stabilizing influence on the linear as well as the nonlinear behavior of the kink mode. With a parabolic profile the distortions of the plasma surface assume much milder forms, amounting only to small deformations, and vacuum bubbles are not formed. For this case see Fig. 10.5.

10.5   NONLINEAR INTERNAL KINK

The nonlinear behavior of the internal kink was investigated to see if it could explain the internal plasma disruption (Rosenbluth *et al.*, 1973). They found that the mode saturated at a relatively small amplitude, for a quadratic pressure profile and a step current profile of extent $a < r_s$ given by

$$\frac{\xi}{r_s} = 3.25 \left(\frac{r_s^2}{R^2} \right) \left[\beta_p + \frac{1}{2} \left(1 - \frac{a^2}{r_s^2} \right) - \frac{1}{2} ln \left(\frac{a^2}{r_s^2} \right) \right]. \qquad (10.41)$$

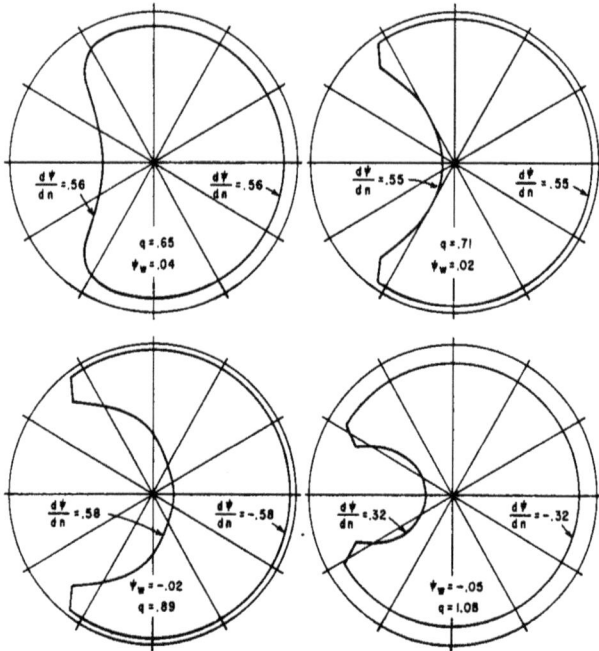

Fig. 10.4   Final bubble states, low shear.

The associated negative voltage spike produced by the plasma shift is qualitatively what is observed experimentally, but is not large enough to provide an acceptable model for the disruption.

The nature of the final state is however interesting. The plasma approaches a state in which there is a true current singularity, and infinite forces. Thus strictly speaking there is no MHD final state equilibrium. Physically of course the MHD description breaks down at some point in the evolution. An arbitrarily small resistivity for example will produce rapid reconnection and island formation in the neighborhood of the current singularity. Thus the nonlinear behavior is dominated by the resistive reconnection, not the ideal effects. The resistive behavior will be treated in Sec. 10.6.

The numerical evolution and saturation of this mode was followed using an ingenious technique (Park *et al.*, 1980). If the MHD equations are advanced in time, approach to the final singular state results in numerical

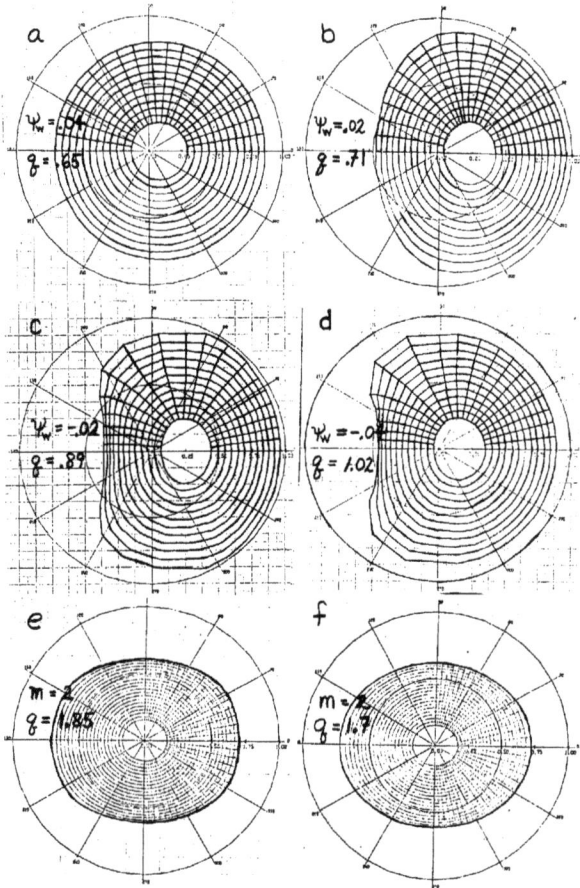

Fig. 10.5   Final bubble states, high shear.

breakdown. An artificial magnetic field, $\vec{B}_a$, was added to the problem. It was taken to behave exactly like the real field, and advanced in time in the same manner, but it was not parallel to the real field, so that as the singular state was approached saturation was achieved because of the trapping of artificial flux. To find the true asymptotic amplitudes numerical runs are made with a sequence of amplitudes of $\vec{B}_a$ and the results extrapolated to $\vec{B}_a = 0$. The analytical expression Eq. 10.41 was confirmed in this manner.

10.6 COMPLETE RESISTIVE RECONNECTION

The $m = 1$ mode is capable of nonlinearly evolving through a sequence
of states which ends with helical flux surfaces forming concentric circles,
just as in the initial state, as shown by Kadomtsev (1975). In this process
the current profile is flattened inside the $q = 1$ surface, resulting in a state
of lower magnetic energy. The reconnection sequence proposed is shown
in Fig. 10.6. The first sketch shows the initial helical flux contours with
the $q = 1$ surface shown as a dotted line. An initial $m = 1$ perturbation
causes the displacement of the central region, reconnection taking place
at the X-point. The two surfaces labeled 1 connect to form one surface
which withdraws from the X-point (b, c). During subsequent evolution the
area inside surface 1 is conserved as resistivity is negligible away from the
rational surface, and also ψ itself is conserved following a fluid element. The
same process occurs with the surfaces labeled 2, 3 until finally the O-point
(4) has been expelled through the X-point, after which the flux contours
relax to an axisymmetric state. The resulting changes in the helical flux
profile, the q profile, and the current profile are shown in Fig. 10.7. In
the final state there is a discontinuity in $d\psi/dr$ at r_4, leading to a negative
current spike. In the final state $q(0) = 1$, and $q(r)$ is appreciably flattened
out to r_4, where it is discontinuous. No changes occur for $r > r_4$. The final
state has lower energy than the initial state, but the dynamical accessibility
of the final state has not been addressed by this "cartoon approximation".

Nonlinear numerical simulations carried out by Sykes *et al.*, (1976), and
Waddell *et al.*, (1976), showed that the process was in fact possible. Further
it was found that the island growth proceeds at approximately the linear
growth rate until the reconnection is mostly complete. The current spike
at the X-point can be very large, leading to some particles being driven to
high energy.

Shown in Fig. 10.8 are the flux surfaces during the evolution of a cylindrical $m/n = 1/1$ mode. In the final state the current profile is quite flat,
and the process confirmed the Kadomtsev reconnection scenario.

10.7 NONLINEAR TEARING MODE ANALYSIS

We are interested in the domain near the rational surface, but outside the
tearing layer, and in magnetic islands which are large compared to the

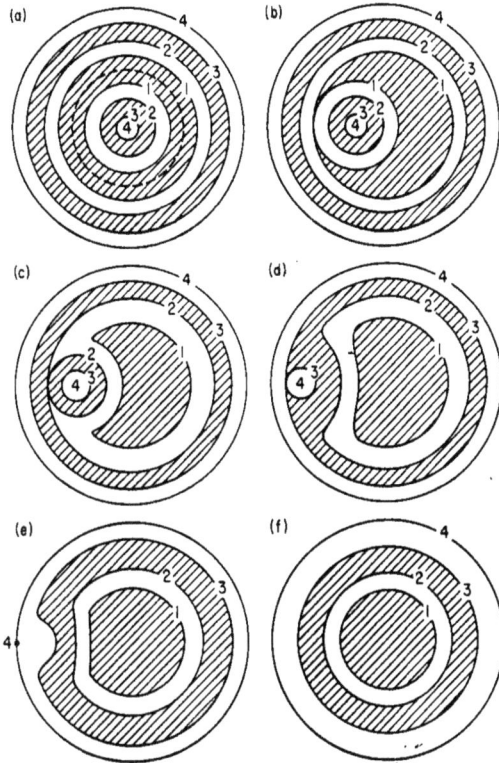

Fig. 10.6   Flux surface evolution during $m = 1$ reconnection.

tearing layer, which is the width at which nonlinear effects begin to play a role. In this domain the helical flux can be approximated as constant for modes with $m \geq 2$ and as linear in the distance from the rational surface for $m = 1$. These approximations suffice to carry out a nonlinear analysis which is valid when the island width is large compared to the tearing layer, but small compared to the system size. We will present here simple heuristic derivations which give good approximate results; for a more rigorous treatment see Thyagaraja (1981). The tearing layer width is given by $w_t = (Sns)^{-1/3} r_s$ for the $m = 1$ mode and $w_t = (Sns)^{-2/5} r_s$ for $m \geq 2$. (See Secs. 7.2, 7.2.2).

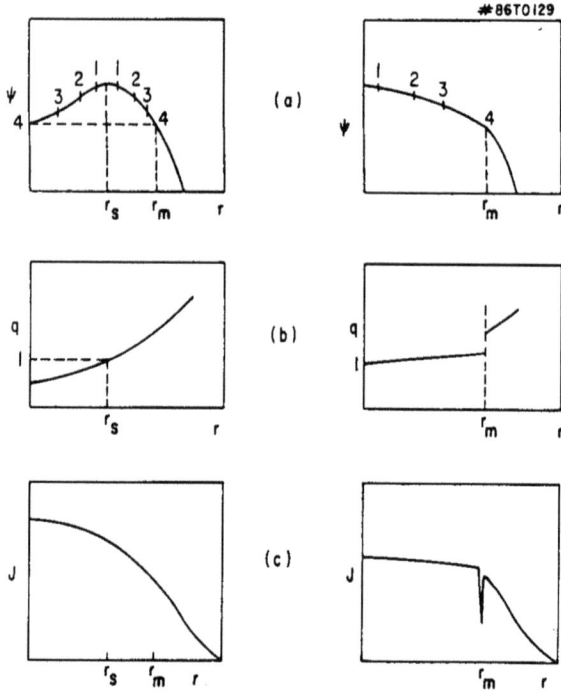

Fig. 10.7 Helical flux, q profile and current profile before and after a Kadomtsev $m = 1$ reconnection.

In this region the linear eigenfunctions have the form

$$\psi_1(x) = \begin{cases} \psi_0'' \xi_0 x \theta(-x) & m = 1 \\ \psi_1 & m \geq 2 \end{cases}, \qquad (10.42)$$

$$\xi(x) = \begin{cases} \xi_0 \theta(-x) & m = 1 \\ \dfrac{\psi_1}{\psi_0'' x} & m \geq 2 \end{cases} \qquad (10.43)$$

where $x = r - r_s$, r_s is the rational surface; $q(r_s) = m/n$. The relation between ψ_1 and ξ is given by the ideal MHD solutions derived in Sec. 7.2. For $m \geq 2$, ψ_1 is approximately constant near $r = r_s$, and $\xi = constant$ near $r = r_s$ follows from the $m = 1$ solution given in Sec. 7.2.2. For $m > 2$ the perturbation produces a magnetic island structure with width

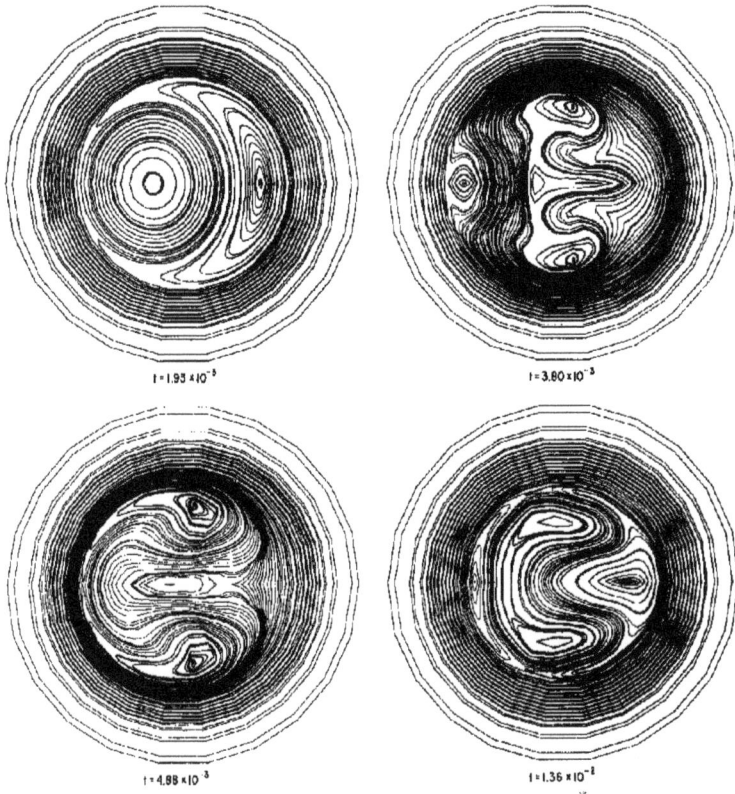

Fig. 10.8 Numerical simulation of sawtooth crash.

$w = 4(-\psi_1/\psi_0'')^{1/2}$ and within the constant ψ approximation the X-point and O-point are located at the tearing layer $r = r_s$. This is not the case for the $m = 1$ mode. In this case surfaces of constant ψ are given by

$$\psi = \psi_0'' x^2/2 + \psi_0'' \xi_0 x(1 - \theta^2/2) \tag{10.44}$$

for small x, θ. The point $\nabla\psi = 0$ is located at $x = -\xi_0$, $\theta = 0$. Expanding ψ about this point we find

$$\psi = const + \psi_0''(x + \xi_0)^2/2 + \psi_0'' \xi_0^2 \theta^2/2 + ... \tag{10.45}$$

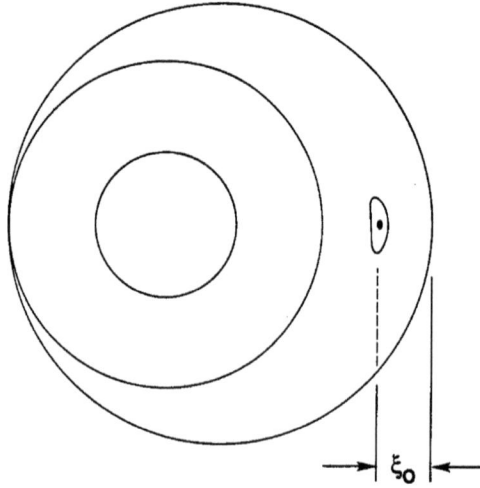

Fig. 10.9   Magnetic surfaces for a large $m = 1$ island.

Thus this is an O-point and the island width is given by $w = 2\xi_0$, which scales directly as the magnitude of ψ_1, not its square root. The flux surface describing the interior separatrix is given by $x - \xi_0(1 + cos\theta)$. Recall that the ideal solution is $\xi = constant$, thus the surfaces inside r_s consist of displaced circles. The X-point occurs at the tangent point of the circles $r = r_s$ and the displaced circle of radius $r_s - \xi_0$, as shown in Fig. 10.9.

To obtain the quasilinear evolution of the mode the convection term is eliminated by averaging over a flux surface. Carry out a flux surface average $\oint dl/\nabla\psi$. The convection term becomes $\oint dl\vec{v}\cdot\hat{n}$ where $\hat{n} = \nabla\psi/|\nabla\psi|$ is normal to the flux surface. By incompressibility this integral is zero. Thus

$$\langle\partial_t\psi\rangle = \langle\eta(j - j_d)\rangle \tag{10.46}$$

where $\langle f\rangle = \oint(dl/\nabla\psi)f/\oint(dl/\nabla\psi)$.

In the case of $m \geq 2$, as the mode grows there is a transition from exponential growth on a hybrid hydrodynamic-resistive timescale to algebraic growth on the resistive timescale (Rutherford, 1973). Consider the equation for the velocity potential, Eq. 7.24, and substitute the linear solutions to find

$$\gamma^2 \tau_A^2 \nabla_\perp^2 r\xi = \frac{m^2}{r} \psi_1^2 \frac{d}{dr} \left(\frac{j_0'}{\psi_0'} \right) + \dots \qquad (10.47)$$

where the nonlinear term is a third order nonlinear force arising through $\delta j_z \times \vec{B}_1$, which impedes the growth of the mode. Using $\xi = \psi_1/\psi_0$ and evaluating both expressions at the edge of the tearing layer, $r - r_s = x_t \ll 1$, we find using the constant ψ approximation that both sides are dominated by the singular $1/\psi_0'$. Using $w^2 = -\psi_1/\psi_0''$ we find that the nonlinear term becomes comparable to the inertia term when the island width w equals the tearing width. The sign of the nonlinear term is such that it replaces the inertia. Thus for island width larger than the tearing width the inertia is negligible. Neglect of the inertia means that

$$\nabla \psi \times \nabla j = 0 \qquad (10.48)$$

or $j = j(\psi)$, this constraint on the form of j replacing the equation for the velocity. In the case of $m = 1$ the inertia cannot be neglected, but the nonlinear behavior can be calculated (Hazeltine *et al.*, 1986) without making any assumptions about the form of j.

The analysis is significantly different for $m = 1$ and $m \geq 2$ so we examine the two cases separately. First consider the case $m \geq 2$.

The use of the linear eigenfunctions reduces the problem to a zero-dimensional determination of a function of time alone. It is common in the literature to perform multiple spatial integrals to determine this single time function. Since the spatial form of the quasilinear solutions is fixed, this is completely equivalent to the use of particular flux surfaces which facilitate this evaluation. In any case the quasilinear treatment can only be regarded as approximate.

Evaluated at the O- and X-points the flux surface averaged equation for $\partial_t \psi$ gives

$$\dot{\psi}_0(r_0) + \dot{\psi}_1(r_0) = \eta(r_0)[j(r_0) - j_d(r_0)], \qquad (10.49)$$

$$\dot{\psi}_0(r_x) + \dot{\psi}_1(r_x) = \eta(r_x)[j(r_x) - j_d(r_x)]. \qquad (10.50)$$

In this case the current is not negligible, and must be evaluated at the X- and O-points, which lie within the tearing layer. By the constant ψ approximation, the exterior solutions can be used for ψ_1 in this region.

Using the narrowness of the island, $w \ll 1$, $\nabla_{\perp}^{2}\psi_1 = j_1$, and the form of ψ_1, we find $j(r_0) = j_0(r_0) + [\Delta'(w)\psi_1/w]cosm\theta$ where $\Delta'(w) = [\psi_1'(w/2) - \psi_1'(-w/2)]/\psi_1(0)$, and similarly for $j(r_x)$. This approximation is in error by a factor of about $\pi/2$ because of the neglect of the flux surface averaging. Allowing perturbations of η and j_d of the form $\eta = \eta_0(r) + \eta_1 cosm\theta$, $j_d = j_{d1}cosm\theta$ we then find, on subtracting X and O-point values and using $r_0 \simeq r_s$,

$$\dot{\psi}_1 = \frac{\eta(r_s)\Delta'(w)\psi_1}{w} - \eta_0(r_0)j_{d1} + j_0(r_0)\eta_1. \qquad (10.51)$$

The first term describes the usual Rutherford algebraic growth (Rutherford, 1973) and saturation (White *et al.*, 1977) of the mode, and the two additional terms describe the effects of local heating and current drive. Heating the island interior causes η_1 to decrease, slowing the growth of the mode, and cooling the island interior causes η_1 to increase, speeding up the growth of the mode. Cooling can be due to increased radiation losses from the island, see Sec. 10.11. Increasing the current in the island interior, $j_{d1} > 0$, stabilizes the island. If $j_{d1} < 0$, which produces more current at the X-point than at the O-point, the island is destabilized. Since the the density is flattened in the interior of the island but not at the X-point, bootstrap current is larger at the X-point, destabilizing the island. This is known as the neoclassical tearing mode. See section 12.5.

In the absence of local heating and current drive, and neglecting bootstrap current, the nonlinear evolution of the mode is thus determined by the quantity $\Delta'(w)$. However, the island width is complicated by the fact that the two island edges are not symmetric with respect to the rational surface r_s, and this asymmetry must be taken into account in calculating $\Delta'(w)$.

This prediction has been compared with numerical simulations using the full nonlinear equations, and gives reasonable agreement. The quantity $\Delta'(w)$ for typical profiles vanishes for some finite w, giving mode saturation. Δ' is related to the energy differential for island formation, and its vanishing means that the island state, with width w, is at a magnetic energy minimum. There is no further energy gain from island growth. In Fig. 10.10 are shown predicted island saturation widths, w, normalized to the minor radius, versus the position of the singular surface. The points are saturated island states obtained in numerical simulations.

Now consider the case of $m = 1$. The $m = 1$ mode analysis differs because of the inward shift of the O-point due to ψ_1 being linear in x.

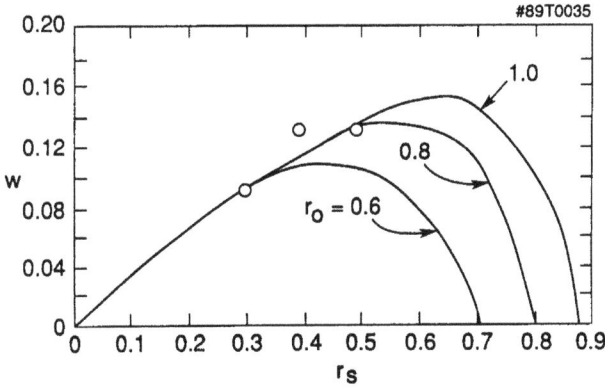

Fig. 10.10   Island saturation width normalized to minor radius, for $m = 2$ mode.

Because of the form of ψ_1, the $m = 0$ part contributes significantly through the inertial $\vec{v} \cdot \nabla \psi$ term. Substitution of the linear eigenfunctions leads to an equation for $\xi_0(t)$,

$$\left\langle \dot{\psi}_0 \right\rangle = -\psi_0'' x \dot{\xi}_0 \left\langle x cos\theta \right\rangle = \left\langle \eta(j - j_d) \right\rangle. \tag{10.52}$$

Now evaluate $\dot{\psi}_0$ quasilinearly, using the linear eigenfunctions. Average the nonlinear equation $\dot{\psi} + \vec{v} \cdot \nabla \psi = \eta(j - j_d)$ in θ, giving

$$\dot{\psi}_0 = -\frac{\gamma \xi_0^2 \psi_0''}{2} + \eta_0 j_0. \tag{10.53}$$

The first term on the right-hand side is an inertial contribution, arising from $\vec{v} \cdot \psi$, and scales as $\eta^{1/3}$. Note that a similar evaluation of the $\vec{v} \cdot \psi$ term for modes with $m \geq 2$ gives zero within the constant ψ approximation. From the linear eigenfunctions the current j is independent of η and the current terms are thus negligible. Then we have, in the exterior region,

$$\frac{\psi_0'' \gamma \xi_0^2}{2} + \psi_o'' \dot{\xi}_0 \left\langle x cos\theta \right\rangle = \left\langle \eta_1 j_0 cos\theta \right\rangle - \left\langle \eta_0 j_{d1} cos\theta \right\rangle. \tag{10.54}$$

For $r < r_s$ but outside the island the flux surfaces are circles, with the origin shifted an amount ξ_0 as shown in Fig. 10.9. Next to the island in this region $x = -\xi_0(1 + cos\theta)$ on a flux surface. We then find $\left\langle x cos\theta \right\rangle = -\xi_0/2$ on this surface, giving in the absence of local heating or current drive $\dot{\xi} = \gamma \xi$, or exponential growth at the linear growth rate. This is what is observed in the simulations, the mode continues at approximately the linear growth rate far into the nonlinear regime. Thus the inertia continues to play a role in the evolution of the $m = 1$ mode in the nonlinear regime, with plasma flow entering into the force balance equation. This means that the plasma does not evolve through a succession of MHD equilibria, as it does for modes with $m \geq 2$.

10.8 SAWTOOTH OSCILLATIONS

Although there do occur sawtooth oscillations which are apparently describable in terms of the nonlinear development of the 1/1 mode, as given in Secs. 10.6 and 10.7, there also exist variations in the sawtooth evolution which are not understood, and which may require serious changes in the theoretical picture. In Fig. 10.11 (a) is shown the evolution in time from the soft X-ray traces in TFTR, operating in conditions of a relatively small radius for the $q = 1$ surface, and large q at the limiter.

These traces, showing the characteristic soft X-ray signals inside and outside the $q = 1$ surface, are what would be expected from a complete reconnection of the Kadomtsev type. The 1/1 precursor oscillations grow in amplitude until the full reconnection leads to a rapid drop in central temperature and a subsequent rise in the temperature outside the $q = 1$ surface. This is referred to as a simple sawtooth (McGuire *et al.*, 1985).

There are, however, at least two other kinds of sawteeth, observed generally under conditions of lower q at the limiter, and a larger $q = 1$ radius. They have been referred to as small and compound sawteeth by McGuire *et al.* In Fig. 10.11 (b) are shown the soft X-ray traces from a small sawtooth event. They generally differ from the traces of a simple sawtooth in that the 1/1 precursor oscillations are significantly smaller, with a more rapid growth, and there generally exist successor oscillations with a fairly long damping time. In Fig. 10.11 (c) are shown soft X-ray traces from a compound sawtooth, consisting of a subordinate followed by a main event. The subordinate event has small precursor oscillations, but large continuous successor oscillations. It is observed to occur at any point during the main sawtooth period. The main sawtooth has a central drop time varying by a factor of 30 within the same discharge. It has a very small precursor oscillation, and medium size successor oscillations.

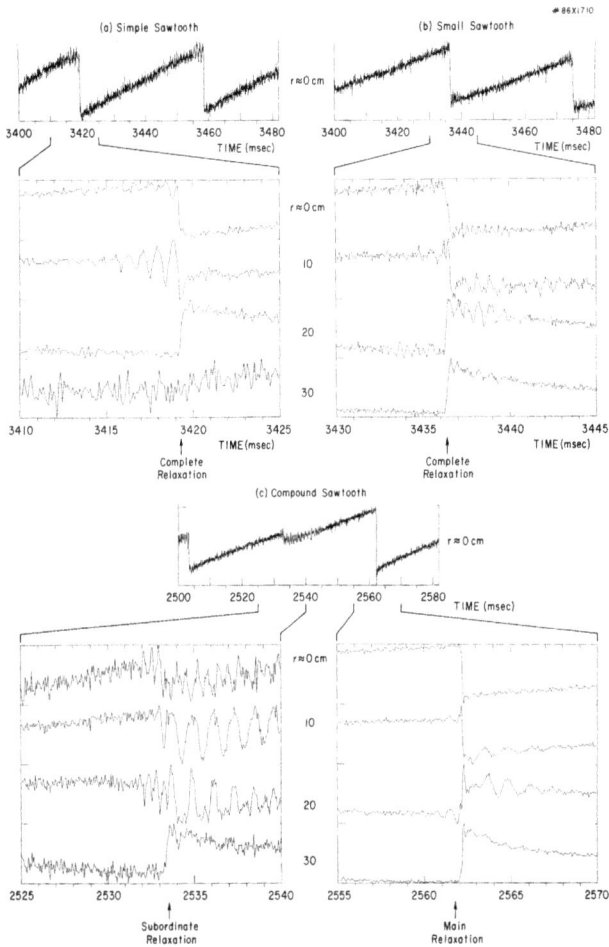

Fig. 10.11   Sawtooth X-ray signals.

The fact that the small and compound sawteeth occur at higher temperatures and with larger values of the $q = 1$ surface suggests that they may be due to slow current penetration rates, and thus fairly flat q profiles. In Fig. 10.12 are shown current density, safety factor, and helical flux for initial states (solid curves) and after reconnection (dotted curves) in the case of a single $q = 1$ surface (a), and partial (b) and full (c) reconnection in the case of two $q = 1$ surfaces. Total current and toroidal field are the

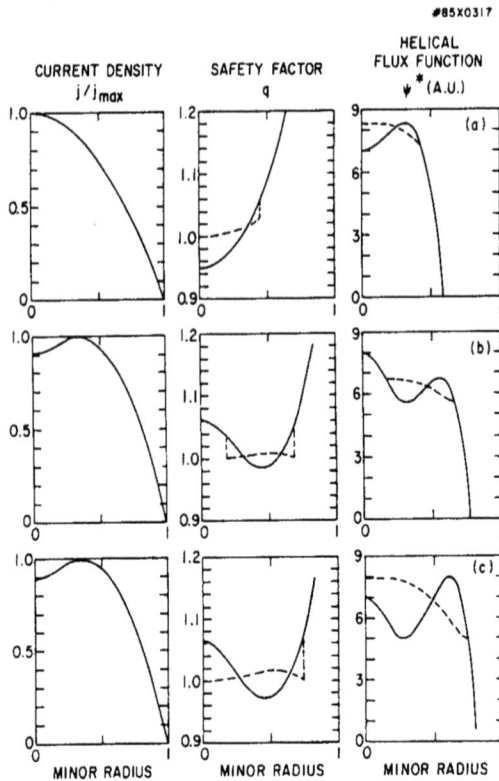

Fig. 10.12   Helical flux, q profile, and current profile before and after $m = 1$ reconnection.

same for all three cases. A slightly hollow current profile occurs readily after a sawtooth oscillation in cases of long current penetration time. The partial reconnection leaves a resonant 1/1 amplitude which could be responsible for successor oscillations, and this reconnection scenario may be responsible for the subordinate relaxation of the compound sawteeth. The full reconnection with two $q = 1$ surfaces also leaves successor oscillations. The main relaxation of the compound sawteeth could be due to complete reconnection involving two $q = 1$ surfaces, and the small sawteeth could involve either one or two $q = 1$ surfaces, as a flat current profile slowly evolved from slightly hollow to slightly peaked. Some of the complexity of structure observed in experiments is understandable in terms of this model.

The wide range in time scale for the drop in central temperature is less easily understood. In particular some events appear to be too fast to explain with simple resistive reconnection models, requiring modification of the reconnection layer through the collisionless skin depth (Wesson, 1990) or kinetic effects (Drake and Lee, 1977; Waelbroeck, 1989; Zakharov and Rogers, 1992; Zakharov *et al.*, 1993). See also the discussion of fast reconnection in Sec. 7.1. In the presence of a very flat current profile small perturbations can lead to very large islands, but the 1/1 growth rate is also lower. From Eq. 7.48 the growth rate is proportional to $s^{2/3}$ with $s = r_s q'/q$ the shear. It is also possible that some ideal instability (Bussac *et al.*, 1984) or the onset of turbulence (Dubois and Samain, 1980) is responsible for the very rapid drop in central temperature observed in some cases. Accurate knowledge of the q profile during a sawtooth is very difficult to obtain and generally not available, so comparison with experiment is difficult. Numerical simulation of the sawtooth cycle, however, also shows a wealth of complex detail and variety. The first nonlinear simulations of the reconnection phase were performed by Waddell *et al.* (1976). Sykes and Wesson (1976) followed repeated oscillations by assuming Spitzer resistivity and introducing an equation for the temperature including Ohmic heating and perpendicular thermal diffusion, but the oscillations were decaying in time.

Park *et al.* (1985) and Denton *et al.* (1986, 1987), found periodic oscillations by introducing a large thermal conductivity along the field lines. Without this large parallel conduction, a nonlinear $m/n = 1/1$ stable convection cell forms in the plasma center and inhibits the temperature from peaking on axis after reconnection. Vlad and Bondeson (1988) found that very rapid sawtooth crash times could be reproduced using plasma parameters close to the experimental values providing an anomalously small value was used for the viscosity. Thus it is still possible that the basic Kadomtsev model is essentially correct, but that additional physics is needed to give correct values for tearing layer width and effective resistivity and viscosity during the evolution of a discharge.

Perhaps the most difficult challenge for this scenario is experimental evidence primarily from TEXTOR (Soltwisch *et al.*, 1986) that in some cases the value of q on axis remains significantly below one during the whole sawtooth cycle, including a complete flattening of the central temperature, a condition definitely at odds with complete reconnection.

One model which has been suggested, but which has not been simulated, involves the development of a small $m = 1$ island, followed by nonlinear

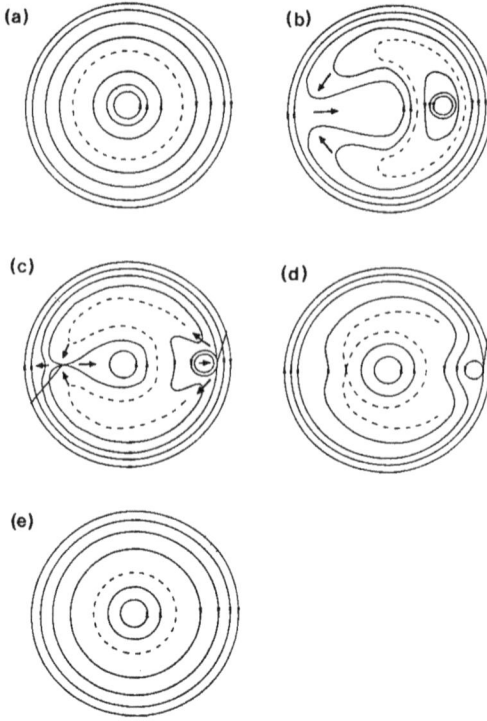

Fig. 10.13   Reconnection leaving $q(0) < 1$.

coupling to modes with $m/n < 1$ with sufficient amplitude to make the domain inside $q = 1$ completely stochastic. This would cause a rapid flattening of the temperature, but only a small modification of the current profile. It has been shown that in toroidal geometry the $m = 1$ mode can be stable to a current profile which has flattening near the $q = 1$ surface (Manickam, 1984). Thus it is possible that a large amount of stochasticity near the separatrix could stabilize the island, making the whole process cyclic. Such a non-Kadomtsev sawtooth cycle is still quite conjectural, since there has been no simulation and the experiments showing $q < 1$ are still not definitive, but it does appear to be a possibility. It has also been shown (Park, 1988) that neoclassical resistivity can account for a rapid increase in the central current density following complete reconnection, resulting in a decrease of q on axis to a value significantly below one. This process naturally produces a current profile with a central peak and strong flattening near the $q = 1$ surface. The possibility of central q values being near one for only a very brief time could explain the TEXTOR results.

A possible model for a reconnection event leaving $q < 1$ on axis has been constructed by Kolesnichenko *et al.* (1992). The sequence of flux surfaces is shown in Fig. 10.13. The essential modification of the Kadomtsev sequence is that after the $m = 1$ island has grown to some size, a large cold bubble invades the interior of the crescent (Fig. 10.13 b) through an interchange instability. This cold center then causes the $m = 1$ O-point to undo itself, being pushed to the outside of the plasma. During the whole process the value of q on axis remains below one, the key to this being that the "Kadomtsev" reconnection of the original island is never completed. In the original Kadomtsev picture (see Fig. 10.6) the island O-point expels the original O-point and becomes the new plasma center. In the Kolesnichenko model the original plasma center remains the center, and the q profile and hence the current profile are practically unchanged by the process, but the plasma is in a lower energy state because hot plasma has been moved outward, *i.e.* the temperature profile is the driving free energy, not the current profile. However, this process has not been simulated by any numerical MHD code, and although it would explain some experiments, it is not certain that it actually occurs.

Another model for incomplete reconnection concerns diamagnetic effects, discussed in Section 7.10. As the sawtooth commences, hot plasma from the interior is flowing toward the X-point from the inside, and cold plasma from the exterior toward the outside. This increases the temperature gradient at the X-point and hence increases the diamagnetic frequency, and this can stabilize the mode, causing the sawtooth reconnection to cease. See Beidler and Cassak (2011) for a simple model and for references to work by Rogers and Zakharov. This model is promising but it also has not been confirmed by large scale fully three-dimensional simulations.

10.9   DISRUPTIONS

Disruption in a plasma discharge refers to rapid loss of confinement, sometimes involving only an internal flattening of the temperature gradient,

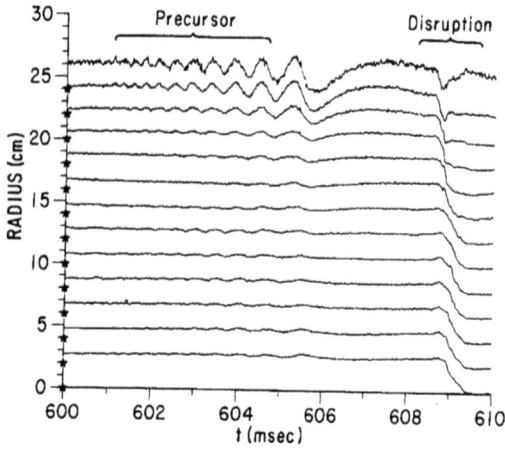

Fig. 10.14   Soft X-ray signals during a major disruption.

which is termed an internal disruption or a thermal quench, and sometimes
involving significant loss of plasma to the vacuum vessel, termed a major
disruption. There are many variations in the form of tokamak disruptions,
because of the great variety of instabilities present as limits are approached
for stable current, pressure, and density profiles. The theoretical analysis
is also rich enough to allow for many variants.

At low plasma β the free energy in the current profile provides the
driving force for disruption, and a canonical sequence is the following: some
time is spent executing sawtooth oscillations, with the current sufficiently
large so that the value of q at the limiter does not greatly exceed two. The
sawtoothing ceases, and a precursor oscillation, an $m = 2$ mode, is observed
to grow at a rate consistent with the Rutherford rate for a magnetic island,
followed immediately by disruption.

Fig. 10.14 (Sauthoff *et al.*, 1978) shows the soft X-ray diagnostic signal observed during such a disruption on PLT. The fact that the sawtooth
oscillations stop a few msec before disruption is consistent with some broadening of $j(r)$; probably $q(0)$ has risen to a value above unity. As shown by
White *et al.* (1977) and shown in Fig. 10.15, a small change in a current
profile, in particular $q(0)$ becoming larger than one, can lead from a case
in which the $m = 2$ mode is stable to one in which the saturation width is

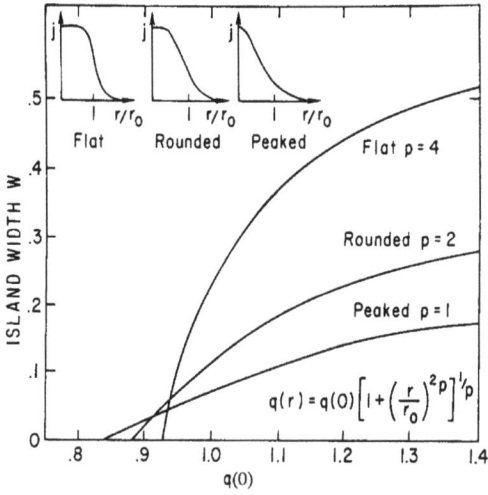

Fig. 10.15   Island saturation width vs $q(0)$.

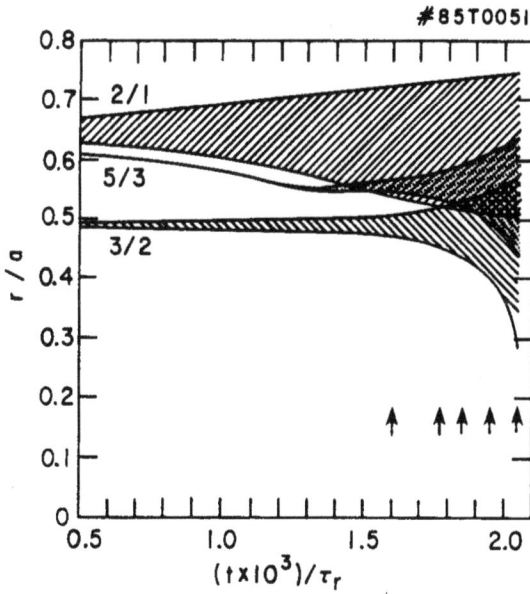

Fig. 10.16   Magnetic island evolution.

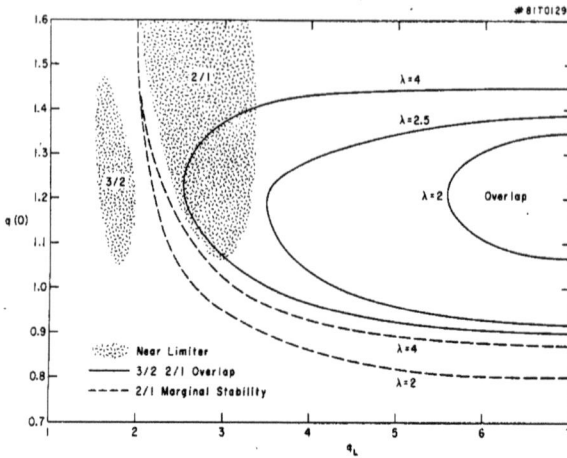

Fig. 10.17   Domains of 2/1, 3/2 overlap.

a large fraction of the minor radius. The precursor oscillation is followed by the disruption itself; a rapid drop in central temperature and often a total loss of plasma. The disruption is thought to be caused by an onset of stochasticity induced by the growing $m = 2$ mode interacting with other modes, and possibly also with the limiter.

In Fig. 10.16 is shown the result of a nonlinear multihelicity simulation (Carreras *et al.*, 1980). It is observed that mode overlap produces destabilization of modes with higher m values which are not unstable in the initial configuration, the 2/1 mode primarily destabilizing the 3/2 mode. Since the nonlinear evolution of a mode is quite well predicted by a single helicity analysis up to the point at which overlap occurs, a single helicity analysis can map out profile parameters leading to mode overlap. Such an analysis was carried out by White and Monticello (1980). The profile parameters leading to island saturation widths large enough to cause 2/1, 3/2 overlap or limiter contact are shown in Fig. 10.17.

The final stages of the disruption are not adequately understood theoretically. There is qualitative understanding of the explosive growth of stochasticity in the final stages, but a detailed quantitative description is not available. This phase is important because the localization and rate of deposition of the plasma on the limiter or wall of a reactor during disruption severely influences design requirements. One obvious recourse would

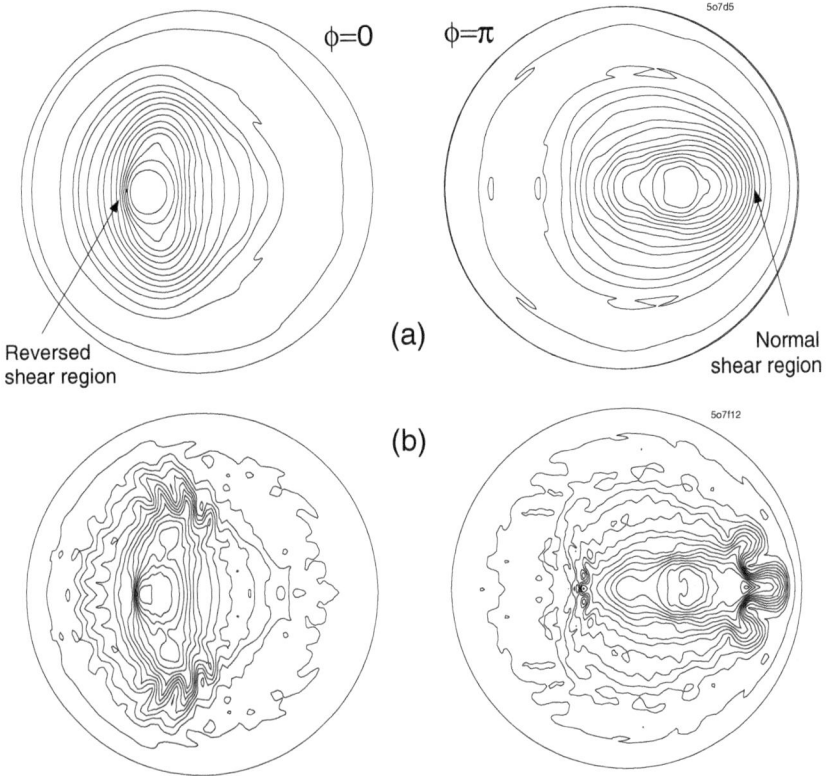

Fig. 10.18   High beta disruption.

be to operate at plasma parameters which avoid disruption. Unfortunately, high current operation is desired in order to reach high β, and the natural limit on current is furnished by the sawtooth oscillation, which prevents q from dropping significantly below one. Repeated sawtooth periods then have a tendency to flatten the current density, leading to a profile which is very disruption prone.

Another form of disruption occurs frequently at high β rather than being associated with particular limits on the current profile. These high β disruptions can effectively limit the parameters of operation. Park *et al.* (1995, 1997) simulated this phenomenon using the resistive MHD code MH3D (Park *et al.*, 1992). At high β the destabilization of a low n mode can lead to a three-dimensional equilibrium which contains toroidally and radially localized steep pressure gradients. In Fig. 10.18 (a) is shown the regions of locally steep pressure caused by the evolution of a $n = 1$ mode, the inner one occuring at $\phi = 0$ and at flux surfaces where the shear is reversed, $q' < 0$, and the outer one at $\phi = \pi$ and in a normal positive shear region. In the normal shear steep pressure region the pressure gradient was responsible for the destabilization of localized high n ballooning modes. In Fig. 10.18 (b) is shown the subsequent evolution of the pressure surfaces, showing very localized deformation due to the development of localized ballooning in the normal shear domain. This flux deformation leads to a rapid thermal quench of the plasma. The toroidally localized steep pressure gradient inside the reversed shear region, although much stronger than the one outside, remains stable, showing the advantage of a reversed shear profile.

10.10   EMPIRICAL LIMITS

As discussed in Sec. 6.19, an optimization over profile shape carried out numerically shows that a limit is provided by ballooning modes in conjuction with the $n = 1$ external kink mode (Troyon *et al.*, 1984). Although the limit was discovered by a numerical optimization procedure, it can be understood by a simple analysis using cylindrical geometry (Wesson, 1987). As pointed out in Sec. 6.13 the ballooning mode threshold scales as $\beta \sim a/(Rq^2)$. A more accurate determination is given from the first stability boundary in the $s - \alpha$ plane, as described in Sec. 6.14, giving $\alpha = 0.8s$. The local ballooning limit is then

$$-p' = \frac{0.4rq'B_\phi^2}{q^3 R}. \qquad (10.55)$$

Using $\beta = 2\bar{p}/B_\phi^2$ and integrating by parts to find $<p> = -(1/a^2)\int_0^a r^2 p' dr$ we find the limiting β to be given by

$$\beta = \frac{0.8}{Ra^2} \int_0^a \frac{r^3 q' dr}{q^3}. \qquad (10.56)$$

For a broad class of current profiles this expression is approximately linear in $1/q(a)$, and thus in the total current. Evaluating the integral for a parabolic current profile, which is weakly unstable to kink modes, gives

$$\beta = 3I/aB \qquad (10.57)$$

with I in Mamps, a in meters, and B in Tesla, about the same as was found numerically. The coefficient in Eq. 10.57 depends to some degree on plasma shaping, *i.e.* ellipticity and triangularity, and so cannot be taken as an exact limit for arbitrary equilibria. Experimental determination of this coefficient for a given tokamak in a particular configuration is a means of quantifying the onset and effectiveness of the ballooning and kink modes. The numerical optimization produces a profile which has $q(0) > 1$ to insure Mercier stability near the axis. From Eq. 10.57, higher beta can be obtained by increasing the current, but this eventually produces a low shear equilibrium with $q \simeq 2$ at the plasma edge, which is highly unstable to the external kink mode, at high β driven by pressure as well as by the current. As seen in Sec. 10.4, in a low shear plasma the external kink can lead to large vacuum bubble formation and severe deformation of the equilibrium surfaces. In the nonlinear phase in a resistive plasma, the effects of this mode would undoubtedly be very similar to the development of a large $m = 2$ magnetic island, and hence lead to disruption.

This possible role of the external kink in a tokamak disruption has been examined by Zakharov (1981). The success of the Troyon result makes it very plausible that the kink mode plays a role in high β disruptions. Simulation of the formation of vacuum bubbles (see Sec. 10.4) in low shear resistive plasmas has been carried out by Kurita *et al.* (1986), and this process is a good candidate for the disruptions occuring near the Troyon limit.

There is an experimentally observed limit on plasma density in Ohmic discharges, as noted by Murakami *et al.* (1976), above which the plasma disrupts. The limit is given by $\bar{n} = B/R$ with the line average density in units of 10^{13}cm^{-3}, B in Tesla, and R in meters. The limiting density has been shown to result from the criterion that for stable operation the Ohmic heating power deposited in the current-carrying channel must exceed the power radiated within the channel by impurity ions (Perkins and Hulse, 1985). This requirement gives a limit which is independent of the model used for the electron thermal conductivity, and also is quite insensitive to the central temperature, but strongly dependent on the impurity content.

The disruptions are associated with current channel shrinkage, caused by increased resistivity due to the radiative cooling, the disruption occuring when the $q = 2$ surface approaches the plasma edge. Thus this limit can be exceeded by providing alternate heating and a particle source which does not excessively cool the plasma edge, such as pellet injection.

10.11   THE GREENWALD DENSITY LIMIT

There is another empirically observed limit to plasma density in tokamaks, known as the Greenwald limit (Greenwald, 1988), given by

$$\bar{n}(10^{20}m^{-3}) < \frac{I(MA)}{\pi a^2}. \tag{10.58}$$

This limit has been recently understood (Gates and Delgado-Aparicio, 2012) as being due to a collapse caused by radiation-driven islands. The physical mechanism is fairly simple. Once a magnetic island forms, typically at the $q = 2$ surface, the interior of the island is shielded from heat coming from the plasma core due to the fact that field lines flow around it. Thus the temperature of the island interior is given by the balance of local Ohmic heating and loss due to radiation, affected by the local plasma density and impurity content. When the radiation dominates over the heating, the cooling of the interior raises the plasma resistivity and the current in the island decreases, causing island growth. See Sec. 10.7. This leads to runaway explosive island growth, stochastization of the magnetic field in the radiation domain, and a limit on the plasma density.

A schematic diagram of these concepts is shown in Fig. 10.19. The power balance inside the island is then given by a simple balance between the Ohmic heating interior to the island and the radiation loss from the same volume.

The equation that describes the evolution of a nonlinear tearing mode can be written

$$\frac{k_0}{\eta}\frac{dw}{dt} = \Delta' r_s - C_1 \left(\frac{w}{w^2 + w_x^2} \right) + \frac{C_2}{w^3} + C_3 w + \frac{C_4}{w} \tag{10.59}$$

where w is the island width, η is the plasma resistivity, k_0 is a constant, Δ' is the classical tearing parameter, r_s is the radius of the rational surface, C_1, C_2, C_3, and C_4 are coefficients that determine the strength of the bootstrap and Pfirsch–Schlüter currents, the polarization current, the Ohmic current perturbation due to temperature changes, and the Ohmic current perturbation due to impurity increases inside the island, respectively. It is assumed that the pressure driven terms (C_1 and C_2) are not important. In addition it is assumed that, for small island widths, C_4, which is due to build up of impurities inside an already formed island, is zero.

The equation then becomes

$$\frac{k_0}{\eta}\frac{dw}{dt} = \Delta' r_s + C_3 w \qquad (10.60)$$

where the coefficient C_3 is defined by

$$C_3 = 3\frac{r_s}{s}\frac{\delta P}{n_e \chi T_e}. \qquad (10.61)$$

The onset criteria for the radiation driven tearing mode is satisfied when the right hand side of Eq. 10.61 becomes positive. If Δ' is taken to be zero the island growth is unstable to arbitrarily small perturbations as long as C_3 is positive. C_3 becomes positive when δP becomes positive, which we

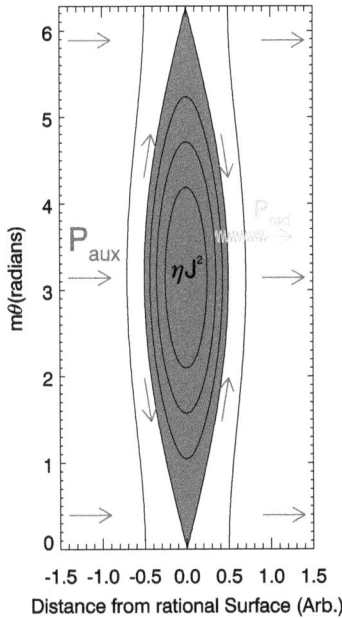

Fig. 10.19   Representation of single lobe of a magnetic island schematically showing the heat flow from the auxiliary heating around the island (red arrows), the resistive heating inside the island (blue area), and the radiation losses from within the island interior (green arrow).

approximate as:

$$P_{rad} > \eta J^2, \qquad n_e \langle n_z L_z \rangle > \frac{m_e \nu_{ei}}{e^2 n_e} J^2 \qquad (10.62)$$

giving

$$P_{rad} > \eta J^2, \qquad n_e > \sqrt{\frac{m_e \nu_{ei}}{e^2 \langle n_z L_z \rangle}} J \qquad (10.63)$$

which is reminiscent of the Greenwald limit except that the quantities are local. Here P_{rad} is the radiated power per unit volume, J is the local Ohmic current, n_e is the local electron density, $n_z L_z$ is the cooling rate for impurity Z, ν_{ei} is the electron ion collision frequency, and m_e is the electron mass. The sign of C_3 varies from negative to positive as the radiated power increases, changing sign when the radiated power exceeds the Ohmic heating power (when δP changes sign). The growth of classical tearing modes has a stabilizing nonlinear term, see Eq. 10.51. For plasma current corresponding to the class of profiles at the Greenwald limit, $\Delta'(w)$ is initially small and positive, and goes to zero for relatively small island width. This implies that there is no threshold effect for the onset of the radiation drive.

10.12   STABILIZATION OF TEARING MODES

The equations for the nonlinear evolution of resistive modes make obvious two means of stabilization, and provide estimates of the necessary amplitudes. The two cases are significantly different so we again discuss them individually. Some attempt has been made to implement feedback stabilization of tearing modes experimentally, but the most interesting application, elimination of the disruptive instability by active stabilization of the $m = 2$ mode, has not been demonstrated. The theoretical analysis indicates that modest amounts of current drive, if well localized and capable of modulation on a time scale of the diamagnetic frequency, should be capable of stabilizing a $m = 2$ mode at a moderate island width. As seen in Sec. 10.9, a fairly large $m = 2$ island is necessary to initiate disruption, and the slow growth of the island, at the Rutherford rate, leaves ample time for a magnetic loop feedback system to respond to its development.

In some devices the occurence of a locked mode, a magnetic perturbation which ceases to rotate and becomes locked to toroidal irregularities in the conducting wall, would make necessary toroidally distributed sensing devices capable of detecting its presence.

Stabilization of the sawtooth mode by this means is, however, a much more difficult, if not impossible, task. The mode grows much more quickly than modes with $m \geq 2$, and also reqires a much larger modulation of the current profile to achieve stabilization. A scheme which appears to be more practical is discussed in Chap. 10.

First consider the case $m \geq 2$. From the quasilinear evolution equation we find that stabilization can be achieved using current drive of magnitude

$$\frac{j_{d1}}{j_0} = \frac{\Delta'\psi_1}{wj_0} = \frac{\Delta'w}{16} \frac{1}{(2q/rq') - 1}. \tag{10.64}$$

Stabilization is thus achieved typically by a local increase of the current density of a few percent at the island O-point. The necessary modification of the resistivity for stabilization is of the same form,

$$\frac{\eta_1}{\eta_0} = -\frac{\Delta'w}{16} \frac{1}{(2q/rq') - 1}, \tag{10.65}$$

but of the opposite sign. Stabilization is thus achieved by a decrease in η at the O-point, or local heating, again of a few percent.

Numerical simulation of $m = 2$ mode stabilization has been carried out, and the results confirm these expressions. The possibility of the use of local heating to control the $m = 2$ mode, and thus major disruptions, rests primarily on the efficacy of local heating in a narrow radial band in phase with a rotating $m = 2$ perturbation. Assuming a resistivity of the form $\eta \sim T^{-3/2}$ we find the required local heating to be given by

$$\frac{\delta T}{T} = \frac{\Delta'w}{24} \frac{1}{(2q/rq') - 1}. \tag{10.66}$$

The power balance for the temperature in the island is

$$-\int \kappa_\perp \nabla T \cdot d\vec{s} = \int P dv \tag{10.67}$$

where P is the net power deposited in the island (Ohmic and auxiliary heating, radiation losses, etc.). For a thin island $\nabla T = -2(\delta T/w)$. Using the fact that the island volume to surface area is $dv/ds = w$, we find

$$P = \frac{\kappa_\perp T \Delta'}{12w} \frac{1}{(2q/rq') - 1}. \tag{10.68}$$

Note also that heat loss which is greater from the island than from the surrounding plasma, due to radiation, for example, provides a destabilizing effect (Rebut *et al.*, 1986).

Now consider the case of $m = 1$. From the quasilinear evolution equation the local heating or current drive necessary for stabilization of the $m = 1$ mode must be localized within the island with amplitude given approximately by

$$\frac{\eta_1}{\eta_0} = \frac{\gamma \psi_0'' \xi_0^2}{j_0 \eta_0} \tag{10.69}$$

or by j_1/j_0 being of the same magnitude, but of the opposite sign. Simulation results have also been carried out in this case, and confirm this result. However, in this case the scaling of the stabilizing term as $\eta^{-1/3}$ means that feedback control is more difficult at high temperatures.

The scaling of the current drive and local heating necessary to stabilize the $m = 1$ resistive kink as $\eta^{-2/3}$ is due to the proximity of this mode to ideal instability. If the ideal internal kink mode is made sufficiently stable then the linear growth rate reduces from $\gamma \sim \eta^{1/3}$ to $\gamma \sim \eta^{3/5}$ (see Eq. 7.66) and the nonlinear behavior of the mode assumes the character of the tearing mode, with $m > 2$. Ideal stabilizing forces would thus make the $m = 1$ mode amenable to feedback stabilization with local heating or current drive. Some stabilization exists due to toroidal curvature (Bussac *et al.*, 1975). The stabilizing contribution to δW is

$$\delta W_T = 3\nu \Delta q \left[\frac{13}{48(\nu + 4)} - \beta_p^2 \int_{r_0}^{r_2} \frac{dr}{r_0} \left(\frac{r}{r_0} \right)^{\nu - 5} \right] \tag{10.70}$$

for a q profile of the form $1 - q(r) = \Delta q[l - (r/r_0)^\nu]$, $q(r_0) = 1$, $q(r_2) = 2$, and

$$\beta_p = \frac{2}{B^2} \int_0^{r_0} \left(\frac{r}{r_0} \right)^2 \left(-\frac{dp}{dr} \right) dr. \tag{10.71}$$

Unfortunately, to have a significant effect δW must be of magnitude $\delta W \gg \eta^{-1/3}$ or $\delta W > 10^{-2}$ for present day tokamaks. This is difficult to achieve since δW_T is of the form $\delta W_T \sim \Delta q (a/R)^2$ and Δq is typically small. The expression for δW due to triangularity of the plasma cross section is similar, so significant ideal stabilization by this means is also difficult.

10.13 Problems

1. Complete the derivation of the expression for the force imbalance on a plasma in cylindrical geometry Eq. 10.15.

2. Calculate the energy of a bubble state in the case of a shear free plasma, Eq. 10.40. Note that the bubble can be taken to be centered at $r = 0$, since moving it inside the plasma does not change the energy.

3. Consider the tearing stability of a slab model with an equilibrium magnetic field $\vec{B}_0 = B_z\hat{z} + b\tanh(x/L)\hat{y}$, where B_z and b are constants with $b \ll B_z$. (from Allan Reiman, see Problems 1–3, Chap. 7.). Calculate $\Delta'(w)$, determining the nonlinear growth of the tearing mode. Calculate the saturated island width for the case where $w/L \ll 1$.

4. For a tearing mode with $m \geq 2$ use the constant ψ approximation to show that if the current profile is locally flat there is no quasilinear saturation, i.e. $\Delta'(w) > \Delta'(0)$.

10.14　　References

Empirical limits

- Greenwald, M., Plasma Phys. Controlled Fusion 44, R27 (2002).
- Greenwald, M., J. Terry, S. Wolfe, S. Ejima *et al.*, Nuclear Fusion 28, 12, 2199 (1988).
- Gates, D. A. and L. Delgado-Aparicio, Phys Rev. Lett. 108, 165004 (2012).
- Murakami, M., J. D. Callen, and L. A. Berry, Nucl. Fusion 16, 347 (1976).
- Troyon, F., R. Gruber, H. Saurenmann, S. Semenzato *et al.*, Plasma Physics and Controlled Fusion 26, 209 (1984).

Nonlinear kink

- Dnestrovski, Y. N., L. E. Zakharov, D. P. Kostomorov, A. C. Kukushikin *et al.*, Pisma Zh. Tekh. Fiz. 1, 45 (1975) [Sov. Tech. Phys. Lett. 1, 18 (1975)].
- Kadomtsev, B. B., and O. P. Pogutse, Zh. Eksp. Teor. Fiz. 65, 575 (1973) (Sov. Phys. JETP 38, 283 (1973)].
- Kadomtsev, B. B., Fiz. Plazmy 1, 710 (1975), Sov. J. Plasma Phys. 1, 389 (1975).
- Rosenbluth, M. N., D. A. Monticello, A. R. Strauss, and R. B. White, Phys. Fluids 19, 1987 (1976).
- Rutherford, P. H., H. P. Furth, and M. N. Rosenbluth in Plasma Physics and Controlled Nuclear Fusion Research (IAEA, Vienna 1971) Vol. 1, p. 533.
- White, R. B., D. Monticello, M. N. Rosenbluth, H. Strauss, in Proc. of the Fifth International Conference on Plasma Physics and Controlled Nuclear Fusion Research (IAEA, Vienna 1975) Vol. 1, p. 495.

Nonlinear internal kink

- Park, W., D. A. Monticello, R. B. White, and S. C. Jardin, Nucl. Fusion 20, 1181 (1980).
- Rosenbluth, M. N., R. Y. Dagazian, and P. H. Rutherford, Phys. Fluids 16, 1894 (1973).

Nonlinear ballooning

- Monticello, D. A., Proc. of the 8th International Conference on Plasma Physics and Controlled Nuclear Fusion Research, (IAEA, Vienna 1981) Vol. 1, p. 227.

Disruption

- Callen, J. D., B. V. Waddell, B. Carreras, M. Azumi *et al.* in the Proc. of the 7th International Conference on Plasma Physics and Controlled Nuclear Fusion Research (IAEA, Vienna, 1979), Vol. 1, p. 415.
- Carreras, B. H., R. Hicks, J. A. Holmes, and B. V. Waddell, Phys. Fluids 23, 1811 (1980).
- Hicks, H. R., J. A. Holmes, B. A. Carreras, D. J. Tetreault *et al.*, in Proc. of the International Conference on Plasma Physics and Controlled Nuclear Fusion Research, Brussels, 1980 (IAEA, Vienna, 1981) Vol. 1, p. 259.
- Ivanov, N. V., A. M. Kakurin, and A. N. Chudnovskii, Sov. J. Plasma Phys. 10, 38 (1984).
- Kurita, G., M. Azumi, T. Takizuka, T. Tuda *et al.*, Nucl. Fusion 26, 449 (1986).
- Park, W., Phys. Fluids B 4, 2033 (1992).
- Park, W., Phys. Rev. Lett. 75, 1763 (1995).
- Park, W., International Conference on Plasma Physics and Controlled Nuclear Fusion Research, Montreal, 1996 (IAEA, Vienna, 1997) Vol. 2, p. 411.
- Perkins, F. W., and R. A. Hulse, Phys. Fluids, 28, 1837 (1985).
- Rebut, P. H., M. Brusati, M. Hugon, P. Lallia, Plasma Physics and Controlled Nuclear Fusion Research, Kyoto, 1986 (IAEA, Vienna, 1987) Vol. II, p. 187.
- Sykes, A., and J. A. Wesson, Phys. Rev. Lett. 44, 1215 (1980).
- Troyon, F., R. Gruber, H. Saurenmann, S. Semenzato, *et al.*, Plasma Physics and Controlled Fusion 26, 209 (1984).
- Wesson, J., Tokamaks (Oxford Science Publications, Oxford, 1987).
- Wesson J. A., and M. F. Turner, Nucl. Fusion 22, 1069 (1982).
- White, R. B., D. A. Monticello, and M. N. Rosenbluth, Phys. Rev. Lett. 39, 1618 (1977).
- White, R. B., and D. A. Monticello, Princeton University Plasma Physics Laboratory, Report PPPL-1674 (1980).

- Zakharov, L. E., Sov. J. Plasma Phys. 7, 8 (1981).
- Zakharov, L. E., Sov. Phys. JETP Lett. 31, 714 (1981).

Avalanche

- S. M. Kaye, M.G. Bell, R.E. Bell, S. Bernabei, J. Bialek, T. Biewer, W. Blanchard, J. Boedo, C. Bush, M.D. Carter et al Nuclear Fusion 45 S168-S180 (2005)
- R. B. White and M. S. Chance Phys. Plasmas [6], 226 (1999)
- N. N. Gorelenkov, C. Z. Cheng, and G Fu, Phys Plasmas 6, 2802 (1999)
- A. Pankin, D. McCune, R. Andre, G. Bateman, and A. Kritz, Comp.Phys.Comm. 159,157 (2004)
- R. B. White, N. Gorelenkov, M. Gorelenkova, M. Podesta, S. Ethier, Y. Chen, Plasma Physics and Controlled Fusion, 58, 2016
- M. Zhou and Roscoe White, Plasma Phys. Control. Fusion xx xx (2016)
- R B White, The Theory of toroidally confined plasmas, third edition, Imperial College Press (2014)
- Chen, Y., R. B. White, Guo-Yong Fu, Raffi Nazikian, Phys. Plasmas [6], 226 (1999)
- R. B. White Commun Nonlinear Sci. Numer. Simulations 17, 2200 (2012)
- R B White Plasma Phys. Control. Fusion 53 (2011) 085018 item A. N. Kolmogorov, Proc. Int. Congr. Mathematicians, Amsterdam, Vol 1 315 (1957), V. I. Arnold, Russ. Math. Surv. 18(5):9, 1963, J. Moser, Math. Phys. Kl. II 1,1 Kl(1):1, 1962.

Nonlinear tearing mode

- Biskamp, D., Phys. Lett. 87, 357 (1982).
- Biskamp, D., Phys. Fluids 29, 1520 (1986).
- Cowley, S., R. Kulsrud, and T. H. Hahm, Phys. Fluids 29, 3230 (1986).
- Dnestrovskii, Yu. N., D. P. Kostomarov, V. G. Pereverzev, and K. N. Tarasyan, Fiz. Plazmy 4, 1001, Sov. J. Plasma Phys. 4, 557 (1978).
- Drake, J. F., and Y. C. Lee, Phys. Rev. Letters 39, 453 (1977).
- Izzo, R., D. A. Monticello, H. R. Strauss, W. Park *et al.*, Phys. Fluids 26, 3066 (1983).

- Monticello, D. A., and R. B. White, Phys. Fluids 23, 366 (1980).
- Park, W., D. A. Monticello, and R. B. White, Phys. Fluids 27, 137 (1984).
- Rutherford, P. H., Phys. Fluids 16, 1903 (1973).
- Strauss, H. R., Phys. Fluids 19, 134 (1976).
- Strauss, H. R., Nucl. Fusion 2–3, 649 (1983).
- Thyagaraja A., Phys. Fluids 24, 1716 (1981).
- Wesson, J., Tokamaks (Oxford Science Publications, 1987).
- White, R. B., W. Park, D. A. Monticello, and H. R. Strauss, in Proc. of the Symposium on Disruptive Instabilities in Toroidal Devices (Garching, 1979).
- White, R. B., D. A. Monticello, M. N. Rosenbluth, and B. V. Waddell, Phys. Fluids 20, 800 (1977).
- White, R. B., D. A. Monticello, M. N. Rosenbluth, and B. V. Waddell, Proc. of the Sixth International Conference on Plasma Physics and Controlled Nuclear Fusion Research (IAEA, Vienna 1977), Vol. 1, p. 569.

Stabilization

- Bussac, M. N., R. Pellat, D. Edery, J. L. Soule, Phys. Rev. Lett. L5, 1638 (1975).
- White, R. B., P. H. Rutherford, H. P. Furth, W. Park, *et al.*, in Magnetic Reconnection and Turbulence, edited by M. A. Dubois, (Les Editons de Physique, Orsay, 1955) p. 299.
- White, R. B., in Proc. of the Workshop on Resistive Reconnection, Cargese, 1985.
- Yoshioka, Y., Nucl. Fusion 24, 565 (1984).

Diagnostics

- Danilov, A. F., Yu. N. Dnestrovskii, D. P. Kostomorov, and A. M. Popov, Fiz. Plazmy 2, 167 [Sov. J. Plasma Phys. 2, 93 (1976)].
- Mirnov, S. V., and I. B. Semenov, Sov. Phys. JETP L3, 1134 (1971).
- Sauthoff, N. R., S. von Goeler, and W. Stodiek, Nucl. Fusion 18, 1445 (1978).
- von Goeler, S., W. Stodiek, and N. Sautoff, Phys. Rev. Lett. 33, 1201 (1974).

- von Goeler, S., in Proc. of the 7th European Conference on Plasma Physics and Controlled Nuclear Fusion (Lausanne, 1975).

Internal disruption, sawteeth

- Biskamp, K., in Magnetic Reconnection and Turbulence, edited by M. A. Dubois, (Les Editons de Physique, Orsay, France, 1985) p. 19.
- Bussac, M. N., D. Edery, R. Pellat, J. L. Soule, and M. Tagger, Phys. Rev. Letters 109, 331 (1985).
- Denton, R., J. F. Drake, and R. G. Kleva, Phys. Rev. Lett. 56, 2477 (1986).
- Denton, R., J. F. Drake, and R. G. Kleva, Phys. Fluids 30, 1448 (1987).
- Drake, J. F., and Y. C. Lee, Phys. Rev. Lett., 39, 453 (1977).
- Dubois, M., and A. Samain, Nucl. Fusion 20, 1101 (1980).
- Hazeltine, R. D., J. D. Meiss, P. J. Morrison, Phys. Fluids 29, 1633 (1986).
- Kadomtsev, B. B., Fiz. Plazmy 1, 710 (1975) [Sov. J. Plasma Phys. 1, 389 (1975)].
- Kolesnichenko, Ya. I., Yu. V. Yakovenko, D. Anderson, M. Lisak, and F. Wising, Phys. Rev. Lett. 68, 3881 (1992).
- Manickam, J., Nucl. Fusion 24, 595 (1984).
- McGuire, K., in Proc. of the 12th European Conference on Controlled Fusion and Plasma Physics (IAEA, Vienna, 1985) p. 134.
- Soltwisch, H., W. Stodiek, J. Manickam, and J. Schlüter in Plasma Physics and Controlled Nuclear Fusion Research, Kyoto, 1986 (IAEA, Vienna, 1987), Vol. 1, p. 263.
- Sykes, A., and J. A. Wesson, Phys. Rev. Lett. 37, 140 (1976).
- Vlad, G., and A. Bondeson, in Joint Varenna-Lausanne Workshop on Theory of Fusion Plasmas (Lausanne, Switzerland, 1988).
- Waddell, B. V., M. N. Rosenbluth, D. A. Monticello, and R. B. White, Nucl. Fusion 16, 528 (1976).
- Waelbroeck, F. L., Phys. Fluids B 1, 2372 (1989).
- Wesson, J. A., P. Kirby, M. F. Nave, in Proc. of the 11th International Conference on Plasma Physics and Controlled Nuclear Fusion Research (IAEA, Vienna 1986), Vol 2, p. 3.

- Wesson, J. A., Nuclear Fusion 30, 2545 (1990).
- Zakharov, L., and B. Rogers, Phys. Fluids B 4, 3285 (1992).
- Zakharov, L., B. Rogers, and S. Migliuolo, Phys. Fluids B 5, 2498 (1993).

Chapter 11

Mode Chirping

11.1 INTRODUCTION

Complex behavior of Alfvén modes is often observed in tokamak discharges, including rapid frequency changes referred to as chirps, occurring at time scales much shorter than the typical time for changes in the equilibrium. Aside from being an interesting test of the capability of numerical simulation, the existence of chirping can significantly modify high energy particle distributions. In general, spherical tokamaks exhibit Alfvénic wave chirping more prominently than conventional tokamaks, which has been explained in terms of the reduced relative microturbulence at the ion scale in spherical machines. Alfvénic avalanches, during which a significant fraction of the fast ions are typically lost are often preceded by a sequence of chirping events. In this chapter we examine unstable Alfvén modes with damping a significant fraction of the drive, using the guiding center code Orbit and a delta f formalism to reproduce the chirping behaviour. The code, which can use a numerically produced NSTX equilibrium, mode eigenfunctions produced by NOVA, and the numerical beam particle distribution produced by TRANSP, has been described in Chapters 3 and 10. The modeling of chirping is an major tool for describing mode-induced fast ion losses in present tokamaks, as well as in ITER, where chirping cannot be ruled out. Because of the strong rotation shear in NSTX, the TAE gap can close, making it easy for a mode to contact the Alfvénic continuum. This provides a sudden damping source for the mode, which can act to produce a chirp. Frequency chirping is typically seen in a simulation when large mode damping produces a collapse in the mode amplitude to a small value, and consists in the production of a phase space particle clump and hole, diverging above and below from the nominal mode frequency.

The equilibrium magnetic field is given by

$$\vec{B} = g\nabla\zeta + I\nabla\theta + \delta\nabla\psi_p, \tag{11.1}$$

where θ and ζ are poloidal and toroidal coordinates and ψ_p is the poloidal flux, and in an axisymmetric equilibrium using Boozer coordinates g and I are functions of ψ_p only. The perturbation has the form $\delta\vec{B} = \nabla \times \alpha\vec{B}$ and α and an electric potential Φ have the Fourier expansions

$$\alpha = \sum_{m,n} A_n\alpha_{m,n}(\psi)sin(\Omega_{mn}), \quad \Phi = \sum_{m,n} A_n\Phi_{m,n}(\psi)sin(\Omega_{mn}), \tag{11.2}$$

where n refers to a single mode with definite toroidal mode number and frequency, and the sum is over poloidal harmonics m with $\Omega_{mn} = n\zeta - m\theta - \omega_n t - \phi_n$, with ϕ_n the mode phase. For ideal modes the electric potential Φ is chosen to cancel the parallel electric field induced by $d\vec{B}/dt$, requiring

$$\sum_{m,n} \omega_n B\alpha_{m,n}cos(\Omega_{mn}) - \vec{B} \cdot \nabla\Phi/B = 0.$$

giving in Boozer coordinates

$$(gq + I)\omega_n\alpha_{mn} = (nq - m)\Phi_{mn}.$$

The perturbation α is related to the ideal displacement $\vec{\xi}$, through

$$\alpha_{mn} = \frac{(m/q - n)}{(mg + nI)}\xi_{mn}^{\psi}.$$

The eigenfunctions produced with the code NOVA are normalized with the largest harmonic $\xi_{mn}^{\psi}(\psi_p)$ having maximum amplitude 1. Thus the amplitude A_n is the magnitude of the ideal displacement caused by this harmonic, normalized to the major radius R. The perturbed radial magnetic field is approximately given by

$$\frac{\delta B_r}{B} \simeq \frac{mR}{r}A_n. \tag{11.3}$$

The modification of the particle distribution by the mode is carried out using a delta f formalism. Write the particle distribution as $f = f_0 + \delta f$ where the distribution in the absense of the modes f_0 is a function of E and

P_ζ and is independent of time. Following particle orbits $df/dt = 0$, and to order α

$$\frac{d}{dt}\delta f = -\partial_E f_0 \dot{E} - \partial_{P_\zeta} f_0 \dot{P}_\zeta. \tag{11.4}$$

The numerically loaded and evolved distribution function is $g(\psi_p, \theta, \zeta, \rho_\|, t)$ with $dg/dt = 0$. The distribution g has the Klimontovich representation

$$g(\psi_p, \theta, \zeta, \rho_\|, t) = \sum_j \delta(\psi_p - \psi_{p,j}(t))\delta(\theta - \theta_j(t))\delta(\zeta - \zeta_j(t))\delta(\rho_\| - \rho_{\|,j}),$$
$$\tag{11.5}$$

with j the particle index and δf is represented by

$$\delta f(\psi_p, \theta, \zeta, \rho_\|, t)$$
$$= \sum_j w_j \delta(\psi_p - \psi_{p,j}(t))\delta(\theta - \theta_j(t))\delta(\zeta - \zeta_j(t))\delta(\rho_\| - \rho_{\|,j}(t)). \tag{11.6}$$

Define $w = \delta f/g$ giving from Eq. 11.4

$$dw/dt = -(1/g)df_0/dt = -(f_0/g)dln(f_0)/dt = -(f/g - w)dln(f_0)/dt,$$
$$\tag{11.7}$$

and f/g is constant in time and given by the value at $t = 0$. Normally simulations assume that the initial perturbation of the distribution δf is zero, so initially $w(0) = 0$.

Stepping equations for the mode amplitude and phase were derived in Chapter 8, Eq. 8.74

$$\frac{dA_n}{dt} = \frac{-\nu_A^2}{D_n \omega_n A_n} \sum_{j,m} w_{n,j} \left[\rho_\| B^2 \alpha_{mn}(\psi_p) - \Phi_{mn}(\psi_p)\right] cos(\Omega_{mn}) - \gamma_d A_n,$$
$$\tag{11.8}$$

$$\frac{d\phi_n}{dt} = \frac{-\nu_A^2}{D_n \omega_n A_n^2} \sum_{j,m} w_{n,j} \left[\rho_\| B^2 \alpha_{mn}(\psi_p) - \Phi_{mn}(\psi_p)\right] sin(\Omega_{mn}), \tag{11.9}$$

with $D_n = 4\pi^2 \sum_m \int \xi_{mn}^2(\psi_p)d\psi_p$, j the particle index and ψ_p, θ, ζ the position of particle j, and $\rho_\|$ the normalized parallel velocity. The linear damping rate γ_d is due to the continuum, electron and thermal ion Landau damping, and radiation, and all terms in the sums are evaluated at the coordinates of particle j, and $w_{n,j}$ is the δf weight of particle j for mode n.

From these equations it is clear that there is a much more significant modulation of the mode frequency compared to the mode growth if both the drive and the damping are large. The modified mode frequency is given by $\omega_n + d\phi_n/dt$. Aside from the difference of sine vs cosine, the drive for ϕ_n and A is the same, but the damping diminishes the effect of the drive on the amplitude but not on the frequency.

Analysis of the mode frequency spectrum is done using the Wigner distribution of quasi-probability (Wigner, 1932)

$$W(t, \omega_k) = \int_{-T}^{T} \Phi^*(t+q)\Phi(t-q)e^{-2i\omega_k q}dq, \qquad (11.10)$$

where we consider a single mode with frequency ω and $\Phi(t) = [cos(\omega t + \phi(t)) + isin(\omega t + \phi(t))]$. The spectrum given by ω_k is chosen to span the mode frequency with a range large enough to include the chirps. This has advantages over a Fourier analysis with a Gaussian window centered on time t. In particular, if $\Phi = constant$ then in the limit of large T, $W(t, \omega_k) = \delta(\omega_k)$. If the mode consists of a single poloidal harmonic $\Phi = e^{i\omega t}$ then $W(t, \omega_k) = \delta(\omega_k - \omega)$, and if Φ is a single chirp, $\Phi = e^{i\omega t^2}$ then $W(t, \omega_k) = \delta(\omega_k - \omega t)$. In addition, W is second order in Φ, so it gives a stronger resolution of relative amplitudes than a linear Fourier analysis.

Introducing units of time given by ω_0^{-1}, where $\omega_0 = eB/(mc)$ is the on axis gyro frequency, and units of distance given by the major radius R, the basic unit of energy becomes $m\omega_0^2 R^2$, which can also be written as $(mv^2/2)(2R^2/\rho^2)$, the gyro radius is $\rho = v/B \ll 1$, and the magnetic moment $\mu = v_\perp^2/(2B)$ is of order ρ^2.

In section 11.2 we discuss the numerical results from simulations, beginning with an axisymmetric equilibrium with a canonical q profile, quadratic in the minor radius, and a mode with a typical Alfvén frequency of 100 kHz, and then discuss simulations of chirping obtained using numerical NSTX equilibria where chirps were experimentally observed, with negative shear in the plasma core. In section 11.3 is the conclusion.

11.2 NUMERICAL RESULTS

We have experimented with the range of the spectrum ω_k, and the magnitude of the time average T in the analysis of, and the search for, chirping. Results differ significantly by varying these parameters, resulting in greater or lesser clarity of the process. The time step used in the code for these simulations was $dt \simeq 10^{-8}sec$ and the mode frequencies were around 100 kHz,

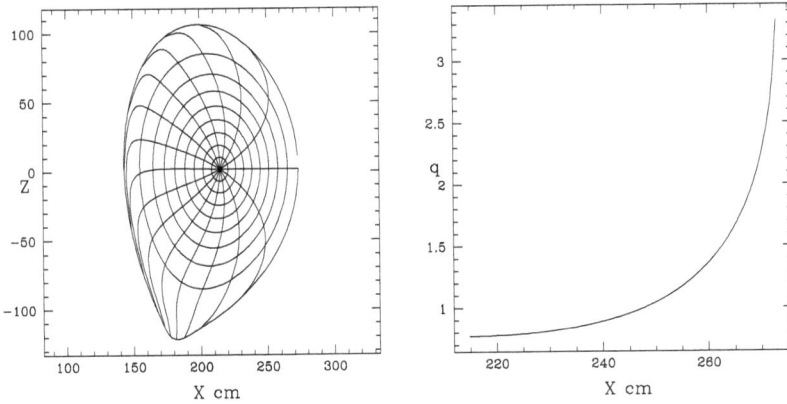

Fig. 11.1 Equilibrium and helicity profile $q(\psi_p)$.

so $\omega dt = .01$, sufficiently small to give good cancellation of incoherent rapid time dependence in the integrals. We assume the time dependence of $A(t)$ and $\phi(t)$ to be much slower than ω_k, and more on the order of the linear growth rate γ. We have chosen at least 51 frequencies ω_k spanning a range of frequencies including the nominal mode frequency ω. Typically the range of the frequencies includes up to a 20 percent change from the mode frequency ω and the time average T was chosen to be a few hundred toroidal transit times.

11.2.1 *A quadratic q profile equilibrium*

Initial investigations were with a conventional equilibrium with a q profile quadratic in the minor radius. In Fig. 11.1 is shown the equilibrium and q profile used. The field on axis was 4.9 kG.

A perturbation consisting of an ideal mode was used with a single harmonic $\xi(\psi_p)$ with $m = 6$ and $n = 5$, with a simple Gaussian radial profile. There is also present an electric potential chosen to cancel the electric field parallel to $\vec{B}$. The use of a single harmonic greatly speeds up the simulation compared to cases with modes of many harmonics. The instability drive was given by a simple local constant gradient in canonical momentum, equivalent to a gradient in minor radius, but a gradient in energy can also be used without changing results.

We find that a simple way to produce a chirp is to initiate a strongly unstable mode with no collisions and strong damping, the damping reducing

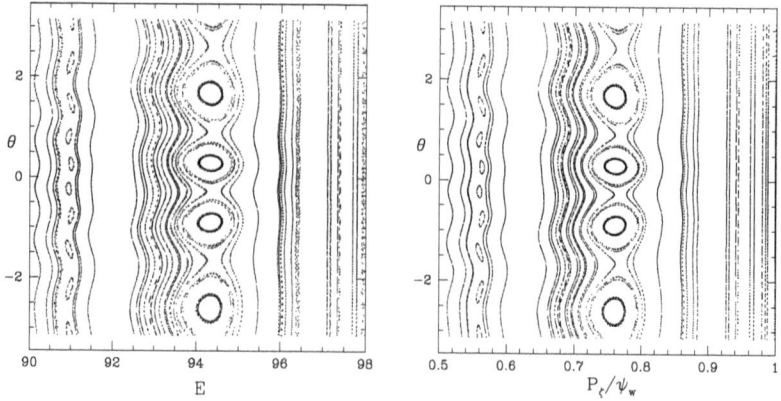

Fig. 11.2   Kinetic Poincaré plots in canonical momentum and in energy, with $\mu B = 2$ keV, mode frequency of 100 kHz, and large mode amplitude of $A = 10^{-3}$. The range of the particle distribution must completely cover the resonance at maximum amplitude. The resonance has a poloidal structure of four elliptic points.

the growth to a small fraction of the linear growth rate. The eigenfunction is associated with the unstable mode. Without collisions the mode grows only until it has flattened the distribution function within the resonance, and then abruptly decays due to the damping. The strongly correlated particles in the resonance are then suddenly left without the structure provided by the perturbed field, and as the mode decays they typically leave by forming a clump and a hole giving Fourier sidebands both above and below the mode frequency. The direct modification of the particle density is not observable with existing techniques because changes in the particle density occur in limited domains of energy and momentum. What can be observed are the Fourier sidebands departing from the principle mode frequency as the chirp occurs.

The code is initiated with the mode having a small amplitude, $A = 10^{-4}$ or smaller, with different instability strengths and mode damping values. To achieve chirping the linear growth rate γ_L and also the damping γ_d must be large. We considered primarily cases of collisionless plasmas, but included in some cases also energy drag and collisions. (See White, Duarte, Gorelenkov, Fredrickson, Podesta and Berk, 2019) The particle distribution was chosen with a single value of magnetic moment μ, with $\mu B = 2keV$, deeply passing, conserved in the absense of collisions. From 200,000 to 500,000 particles were used in the simulations. Note that particles on good KAM

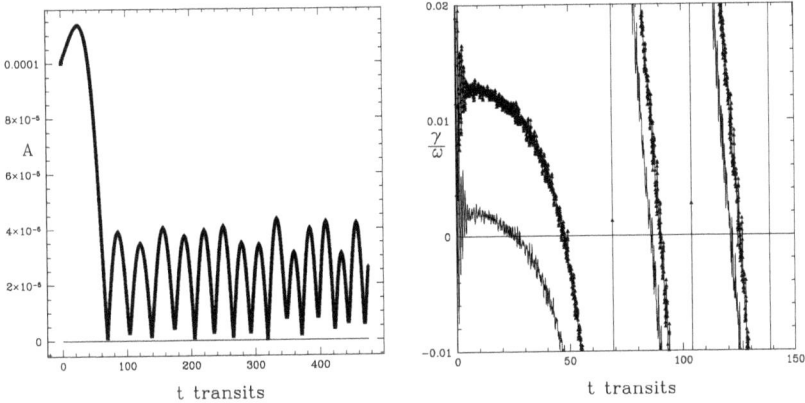

Fig. 11.3   Time evolution of mode amplitude and of growth rate showing strong chirping, 100 kHz. Shown is the growth rate γ/ω as a function of time as well as the value obtained subtracting the damping, giving the actual amplitude evolution. The time scale in this and later plots is in terms of the trorodal transit time of a characteristic high energy particle.

surfaces can not exchange energy with the mode except in an adiabatic manner, and thus cannot participate in mode growth or chirp formation. Only that part of the distribution associated with the resonance need be represented. Changing the value of μ modifies the location of the resonance, and thus including a significant range in μ would require a much larger distribution. The value of μ is chosen to be characteristic of the particle distribution, and results are found to be relatively insensitive to this choice. To demonstrate this we use two very different values of μ for the NSTX cases considered.

The kinetic Poincaré plot of the resonance is shown in Fig. 11.2, for mode amplitude $A = 10^{-3}$ and $\mu B = 2$ keV. The equilibrium and perturbation were selected to give a single well isolated resonance, surrounded by good KAM surfaces in both E and P_ζ. Both the energy distribution and that in canonical momentum were chosen to have several times the width of the largest resonance observed during the simulation. As shown in Fig. 11.2 the energy of the resonance was about 94 keV.

The parameters used for producing this figure were $\gamma_L/\omega = 0.04$, $(\gamma_L - \gamma_d)/\omega = 0.007$ and toroidal transit time $4.7\mu s$. The total time in this plot is about 0.1 msec. In Fig. 11.3 is shown a typical time history of the mode amplitude and growth rate, for a case producing chirping. The mode is

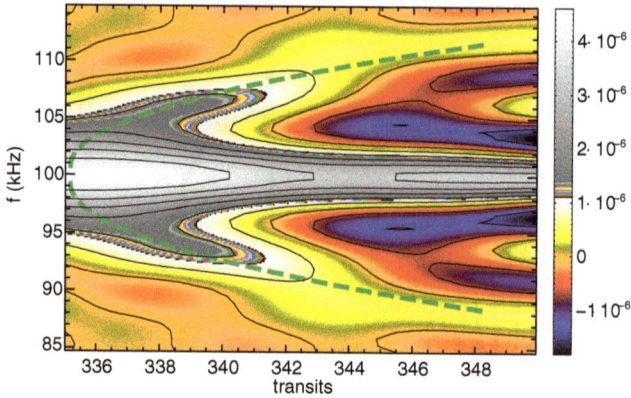

Fig. 11.4 Time evolution of frequency spectrum showing strong chirp, 100 kHz. The green curve represents departures $\delta f = \pm \frac{16\sqrt{2}}{\pi^2 3\sqrt{3}} \gamma_L \sqrt{\gamma_d t}$ from the original eigenmode frequency, as the Berk-Breizman prediction.

initiated at small amplitude with no collisions or energy drag. It grows due to the initial density gradient within the resonance, but when this free energy supply is exhausted the large value of damping produces a rapid decay of the mode amplitude. Shown is the linear growth rate γ_L as well as the actual drive of the mode, $\gamma_L - \gamma_d$. When the mode crashes to a low amplitude the particle distribution in the resonance is restored, so the mode again grows up to the point of flattening the internal density gradient and the process repeats. As long as the linear growth rate is large, and the damping a significant fraction of this, chirping is observed practically at every amplitude crash, although the form of the chirp can vary. Cases with smaller damping fail to chirp, even though the amplitude may decay to a small value. The experimental situation is similar, in that what is typically observed is a stable saturated Alfvén mode suddenly experiencing chirping. We have also simulated this scenario, by adjusting collision rates and mode drive to produce a stable saturated mode, and then suddenly increasing the mode damping. The results are the same, but this scenario requires significant more work and time than simply initiating a mode with strong drive and strong damping.

The wave chirping has been proposed to be associated with self-consistent nonlinear structures that spontaneously move in phase space. The moving resonance allows for a self-organized state in which the extraction of energy occurs at previously unexplored regions of phase space, which can act

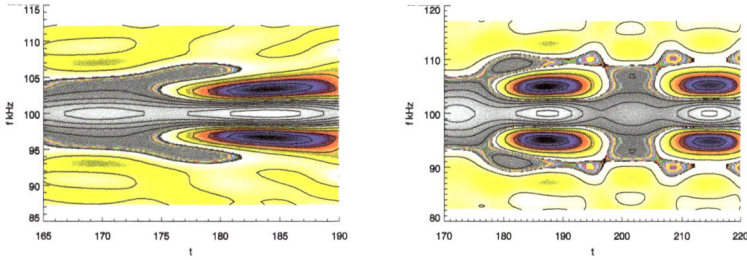

Fig. 11.5   Time evolution of frequency spectrum showing strong chirp, 100 kHz, $\gamma_L/\omega = 0.038$, $\gamma_d/\omega = 0.03$ and in the second case, with the subsequent reabsorption of the clump and hole into the main frequency, followed by repeated chirping. Growth rate and damping were $\gamma_L/\omega = 0.035$, $\gamma_d/\omega = 0.02$.

to overcome background damping. An adiabatic state was analytically explored, in which the mode amplitude is assumed to be roughly preserved during one chirp event. Under this simplification, the frequency excursion has been found to be given by

$$\delta f = \pm \frac{16\sqrt{2}}{\pi^2 3\sqrt{3}} \gamma_L \sqrt{\gamma_d t}. \qquad (11.11)$$

In Fig. 11.4 is shown a 100 kHz chirp with the square root theoretical time dependence added with dashed lines. (Berk, Breizman, and Petviashvili, 1997) Good agreement with the theory is observed, although the amplitude drops by about 30 percent in the period shown.

In Fig. 11.5 are examples of two chirps. For most cases we see the side-band frequencies, after separating from the main frequency as $\sqrt{t}$, tracking the main frequency for a significant amount of time. In some cases as the mode amplitude rebounds from its lowest value the clump and hole are reabsorbed into the main frequency, as shown in the second case. Occasionally the chirping does not follow the simple emission of a clump and a hole, but shows additional period doubling and branching of the ejected frequencies. Examples are shown in Fig. 11.6. The first case had energy drag of $21/sec$, the second had none. In Fig. 11.9 of the next section is shown an experimentally observed example of this compound chirping. Finally in Fig. 11.7 is shown an example of a history of repeated chirping, with significant clump-hole production practically at every amplitude crash.

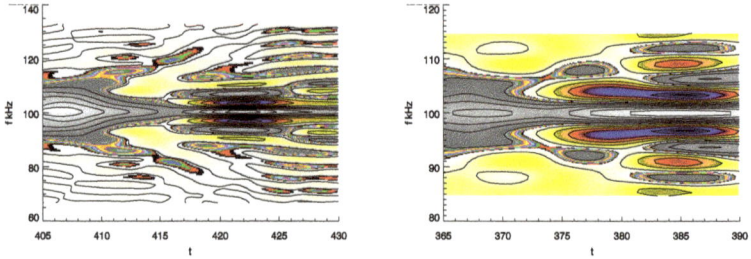

Fig. 11.6   Time evolution of frequency spectrum showing strong chirps with secondary doubling, 100 kHz. Growth rate and damping were $\gamma_L/\omega = .058$, $\gamma_d/\omega = .053$. The first case had energy drag of $21/sec$, the second had none.

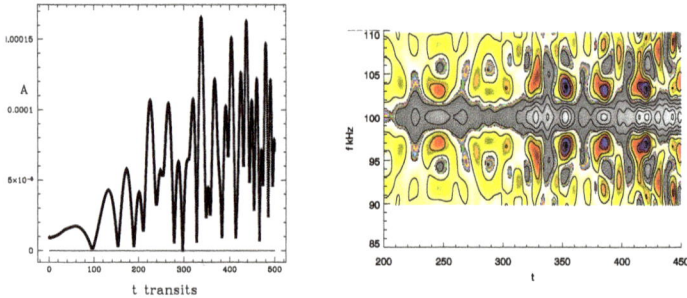

Fig. 11.7   Time evolution of mode amplitude and frequency spectrum with multiple chirps, $f = 100$ kHz, Growth rate and damping were $\gamma_L/\omega = .058$, $\gamma_d/\omega = .053$.

It has been observed that the formation of holes and clumps only happens when the simulation is started with the mode sufficiently close to its linear marginal stability, more specifically when $\gamma_d > 0.4\gamma_L$.

Shown in Fig. 11.8 is the domain in the space of γ_L/ω and γ_d/ω in which chirping is observed in the present simulations, including points obtained from simulations both from this section and also from the next sections, using the reversed shear equilibria of NSTX. We find no significant difference in this plot for simulations using different equilibria and mode frequency, and points are included from each of the three cases studied. Empty triangles are cases which did not chirp, solid triangles are normal chirps, and solid squares are cases with complex chirping including subsequent period doubling and a profusion of clumps and holes. There is a threshold for chirping in γ_L/ω of about 0.01, and above this chirping is

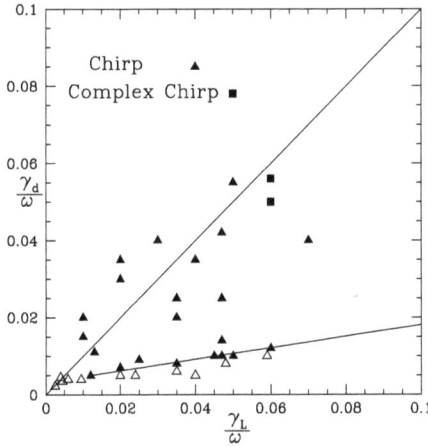

Fig. 11.8 Domain in which chirping is observed. Empty triangles are simulations in which no chirps were observed, solid triangles are examples of chirping, and solid squares are complex chirps. There is a threshold in γ_L/ω of about 0.01, and after that chirping is observed provided approximately that $\gamma_d/\gamma_L > 0.2$.

observed provided approximately that $\gamma_d/\gamma_L > 0.2$. For larger values of both γ_L and γ_d there also occur complex chirps, with subsequent frequency doubling and secondary clump emission after the separation of the initial clump and hole, as seen in Fig. 11.6.

11.2.2 *NSTX Shot 205072*

We now examine equilibria and modes associated with particular discharges in NSTX in which chirping was observed. NSTX shot 205072 exhibited a strong chirp at t = 0.36 sec. The numerical equilibrium was produced by TRANSP and the mode structure by NOVA. Mode structure is not observed to change appreciably in NSTX during a chirp, which justifies the use of a fixed mode structure throughout the mode nonlinear evolution. The magnetic field on axis was 6.2 kG. The mode had a toroidal structure of $n = 1$ and poloidal components from $m = 0, 10$. It was a negative frequency mode with $f = -130kHz$. For this case we choose a distribution with $\mu B = 2$ keV, so it is deeply passing.

In Fig. 11.9 is an experimental plot of the chirp. Note that the mode makes contact with two branches of the continuum at the time of the chirp,

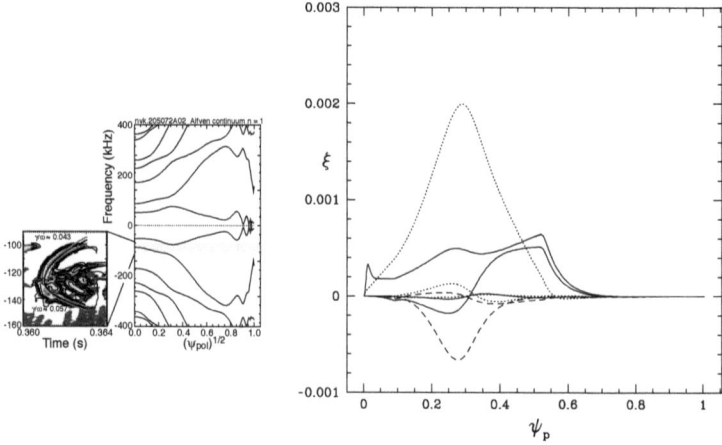

Fig. 11.9   Chirp observed in NSTX, shot 205072. The mode had a frequency of -130 kHz. The lines on the left frequency plot show continuum modes. The eigenmode, shown at the right, is large near the axis, and makes contact with them at two different locations. The clump and hole are fit with Eq. 11.11 using $\gamma_L/\omega = \gamma_d/\omega$ of 0.043 for the upper branch and 0.057 for the lower branch.

indicating possibly a large increase in the value of the mode damping. Because of the small aspect ratio in NSTX the gap in the continuum closes near the magnetic axis, making contact with a continuum mode more probable. The experimentally obtained Fourier plot uses the perturbed frequency signal due to the mode, and hence includes, unlike our analysis, also the the mode amplitude. However, the mode amplitude changes little during the chirp, and in addition the experimental plot uses a log scale, so the resulting plots are not dissimilar.

In Fig. 11.10 is shown the equilibrium and q profile. The equilibrium possessed strong reversed shear in the plasma center. In Fig. 11.11 is shown the determination of the resonance in the plane of E and P_ζ for a value of μB of 2 keV, and a Poincaré plot of the resonance for large mode amplitude of $A = 10^{-3}$. The blue line with large negative slope is the line $\omega P_\zeta - nE = K$, along which the Poincaré plot is made. The mode is produced entirely by the $m = 1$ component of the perturbation, determined by examining Poincaré plots selecting one value of m at a time. It was found that the addition of other harmonics produced no change in the form of the resonance, so further simulations were made with this poloidal harmonic

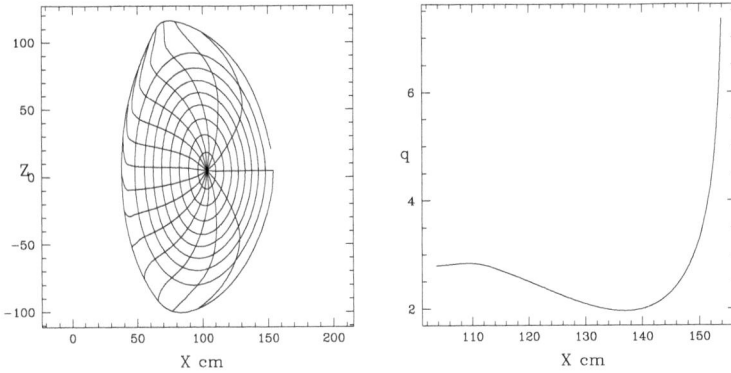

Fig. 11.10   NSTX equilibrium and q profile, shot 205072

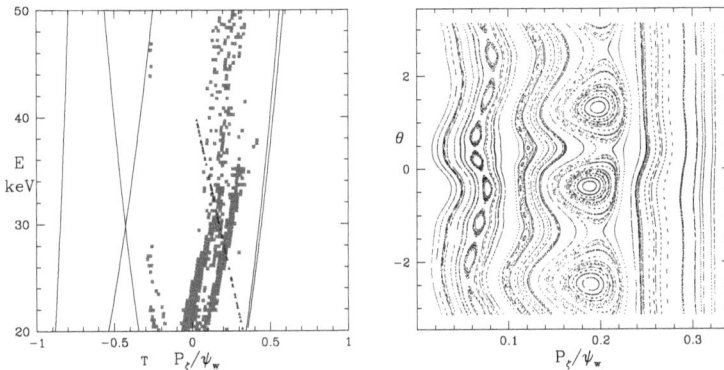

Fig. 11.11   Resonance determination, NSTX shot 205072, showing the domain in the space of energy and canonical momentum where KAM surfaces are broken, and a Poincaré plot obtained along the line $\omega P_\zeta - nE = K$. The resonance has a poloidal structure of three elliptic points.

alone. The radial mode structure is much broader than the resonance width, so there is no effect on mode evolution due to its width. Also note that all chirp simulations are limited to small mode amplitudes.

Now attempt to recreate the situation observed in NSTX, namely the sudden chirping of a stable saturated Alfvén mode. As seen in the previous sections, it is necessary that there be strong damping to produce a significant chirp. We thus hypothesize that the saturated Alfvén mode,

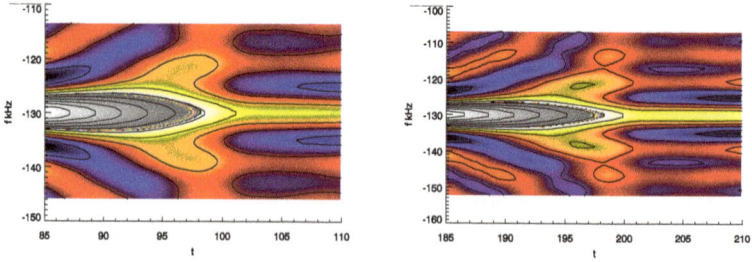

Fig. 11.12   Examples of simulated chirping using the equilibrium associated with NSTX shot 205072. Growth rate and damping were $\gamma_l/\omega = .01$, $\gamma_d/\omega = .02$.

due to small changes in the equilibrium, makes contact with the continuum and thus suddenly encounters strong damping. We simulate this by first allowing a mode in the presence of collisions to saturate, and then suddenly increase the mode damping. This produces the sequence of events observed in the simulations shown above, with rapid collapse of the mode amplitude and multiple chirps. In Fig. 11.12 are shown examples of chirps using this equilibrium.

11.2.3   *NSTX Shot 139048*

NSTX shot 139048 exhibited a strong chirp at t = 0.266 sec. The equilibrium and q profile were very similar to that shown for shot 205072, but the magnetic field was much lower, with $B = 4.9$ kG. The mode had a positive frequency with $f = 100kHz$. For this case we chose a distribution with $\mu B = 25keV$, not as deeply passing as used for shot 205072. The mode is produced entirely by the $m = 1$ component of the perturbation, determined by examining Poincaré plots selecting one value of m at a time. It was found that the addition of other harmonics produced no change in the form of the resonance, so further simulations were made with this harmonic alone. The radial mode structure is much broader than the resonance width, so there is no effect on mode evolution due to its width. Unlike shot 205072, the resonance has a poloidal structure of a single elliptic point.

In Fig. 11.13 is an experimental plot of the chirp. Note that the initial part of the chirp is asymmetric, with the frequency chirping only downward, followed immediately by a more symmetric burst. We have thus far not observed any asymmetric chirping in simulations, possibly due to

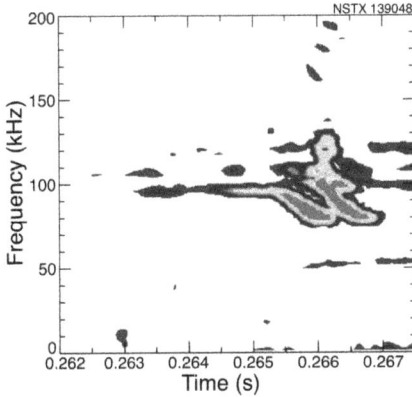

Fig. 11.13   Chirp observed in NSTX, shot 139048

energy drag, decreasing the kinetic energy of particles in the resonance, or to plasma rotation with shear producing the asymmetry.

In Fig. 11.14 is shown a simulation using the equilibrium of NSTX for shot 139048 and the mode structure provided by NOVA, showing the mode amplitude as a function of time for initiation with no collisions and strong damping. Also shown is the linear growth rate of $\gamma_L/\omega = 0.04$ along with the value including the damping, reducing the growth to about half its value. With no collisions the mode initially grows until it has exhausted the density gradient with the resonance, and then collapses to a small value, and in this process produces a chirp, seen in Fig. 11.15.

A significant collision frequency destroys the coherence of the particles trapped in the resonance and makes chirping impossible. Physically, diffusion acting on the resonant particles destroys the phase-space correlations necessary to sustain coherent chirping structures. This effect was studied in NSTX employing nonlinear global gyrokinetic simulations.

From analytical theory, it has been shown that solutions of mode evolution near marginality that blow up occur when $\nu_{eff}/(\gamma_L - \gamma_d) < 2$. This explosive behavior signals that the mode is entering a hard nonlinear scenario, associated with the emergence of chirping. When scattering collisions

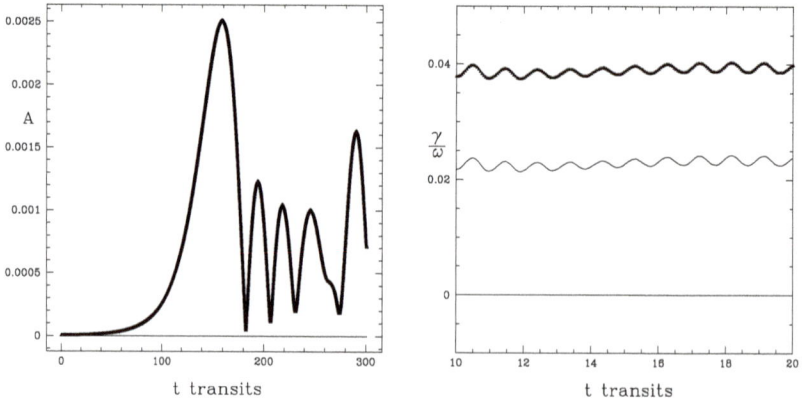

Fig. 11.14   Simulation of chirping in NSTX shot 139048, which had a mode with a frequency of 100 kHz. Shown is the amplitude and γ_L/ω as well as the value including damping as a function of time. The damping reduces the growth almost to half its linear value.

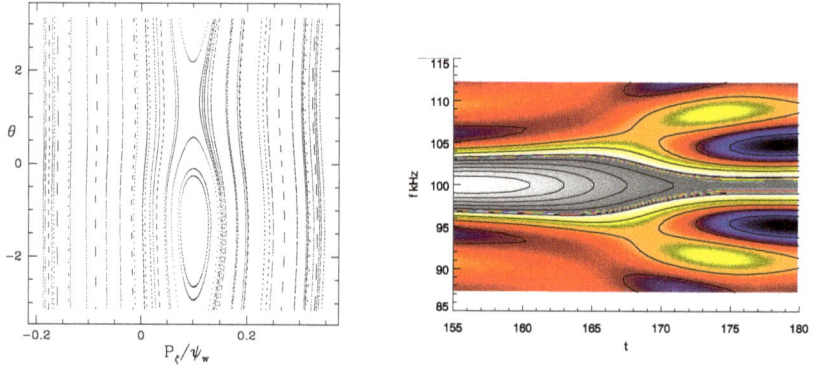

Fig. 11.15   NSTX shot 139048, mode resonance, and simulation of a Chirp with the initial growth rate and damping as shown in Fig. 11.14. Growth rate and damping were $\gamma_L/\omega = .04$, $\gamma_d/\omega = .018$. Unlike shot 205072, the resonance has a poloidal structure of a single elliptic point.

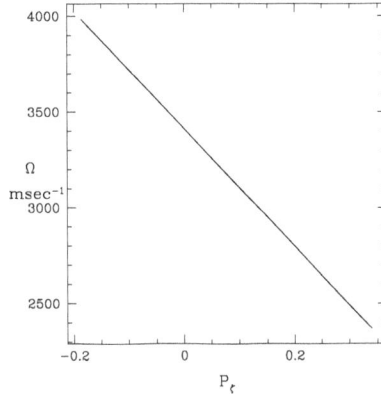

Fig. 11.16   NSTX shot 139048, $\Omega = n\omega_\zeta - p\omega_\theta$ versus P_ζ for fixed K, $n = p = 1$. The equilibrium parameters were B $= 4.9$ kG, $\mu B = 25 keV$.

are frequent enough, the mode evolves quasi-steadily to saturation, in the absence of period bifurcation. In this case, the predicted spectral response is oscillation with a constant frequency.

From Eq. 6 of Berk-Breizman, we have

$$\nu_{eff}^3 \simeq 2\nu_\perp R^2 \left[E \frac{B_{pol}^2}{B^2} + \mu B \right] \left(\left. \frac{\partial \Omega}{\partial P_\zeta} \right|_{K,\mu} \right)^2 . \qquad (11.12)$$

with $\Omega = n\omega_\zeta - p\omega_\theta$ versus P_ζ for fixed K, where ω_ζ is the mean value of $\Delta\zeta/\Delta t$ and ω_θ is the mean value of $\Delta\theta/\Delta t$ for particles launched along the line $K = \omega P_\zeta - nE$, all with $\mu B = 25$ keV, and $p = 1$ is the number of islands in the resonance. In Fig. 11.16 is shown a numerical determination of $d\Omega/dP_\zeta$. The canonical momentum P_ζ is in code units.

Fig. 11.17 illustrates the transition between the chirping and the quasi-steady oscillation types, induced by extrinsic stochasticity. Fig. 11.17 (a) shows the result of collisionless simulation, where the particles are expected to maintain their bounce coherence and, therefore, to lead to wave chirping. In Fig. 11.17 (b) and (c), scattering collisions are included. The effective frequency felt by particles at a given resonance is computed via Eq. 10. Panels (b) and (c) used $\nu_\perp = 8.5/s$ and $\nu_\perp = 26/s$, which correspond to

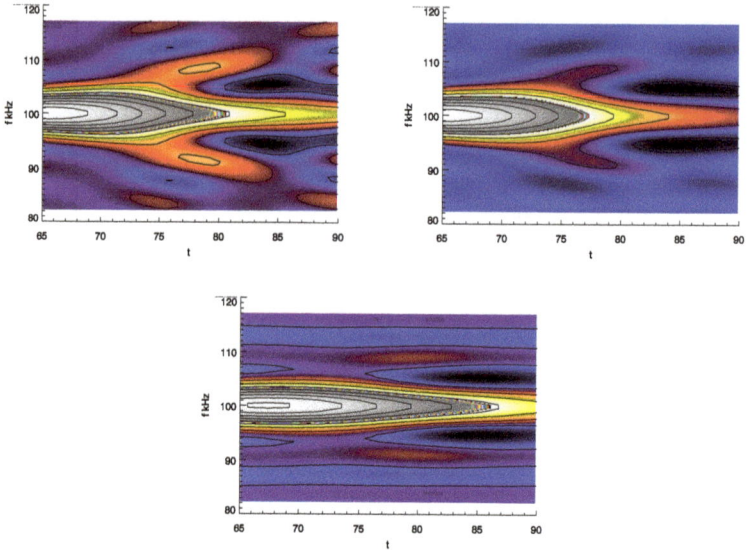

Fig. 11.17   Effect of collisions using energy conserving pitch angle scattering. A significant collision frequency destroys the coherence of the particles trapped in the resonance
and makes chirping impossible. Shown is a chirp with no collisions, with $\nu = 8.5/sec$
and $\nu = 26/sec$. Growth rate and damping were $\gamma_L/\omega = .047$, $\gamma_d/\omega = .015$.

$\nu_{eff}/(\gamma_L - \gamma_d) = 2.17$ and $\nu_{eff}/(\gamma_L - \gamma_d) = 3.15$, respectively. While (b)
represents a situation where the holes and clumps are barely formed and
detached, with a minor degree of chirping, (c) shows a situation with fully
suppressed chirping structures.

11.3   CONCLUSION

Multiple examples of classical mode chirping are observed with the guiding center code ORBIT. The essential ingredient is the presence of strong
mode damping along with significant drive. The damping must be at least
20 percent of the linear growth rate and γ_L/ω must be larger than 0.01. The
simulations performed included a numerical equilibrium with a classical q
profile quadratic in minor radius as well as the use of NSTX discharges with
strongly reversed shear in the core. No significant change in the conditions

necessary for chirping are found using the different equilibria. The shape of the frequency satellites agrees with theory. The necessary strong damping for the production of chirping leads to the conjecture that chirping is caused at least in some cases when small changes in the equilibrium cause a saturated Alfvén mode to come in contact with the continuum and suddenly experience strong damping. Collisions inhibit chirping by destroying the coherence of the particle distribution in the resonance, and the necessary magnitude of the collisions to do this agrees with theory.

11.4   References

- R. B. White, V. N. Duarte, N N Gorelenkov, E D Fredrickson, M Podesta, H L Berk, Phys. Plasmas 26, 092103 (2019)
- H. L. Berk, B. N. Breizman, and Petviashvili, Phys Lett A 234, 213 (1997)
- H. L. Berk, B. N. Breizman, J. Candy, M. Pekker and V. Petviashvili, Phys Plasmas 6, 3102 (1999)
- J. Candy, H. L. Berk, B. N. Breizman and F. Porcelli, Phys Plasmas 6, 1822 (1999)
- S. D. Pinches, H.L. Berk, D.N. Borba, B.N. Breizman, S. Brigulio, A. Fasoli, G. Fogaccia, M.P. Gryaznevich, V. Kiptily, M.J. Mantsinen et al, Plasma Phys. Control. Fusion 46, S47-57 (2004)
- M. K. Lilley, B. N. Breizman, and S. E. Sharapov, Phys. Plasmas 17, 092305 (2010)
- M. K. Lilley and R. M. Nyqvist, Phys. Rev. Lett. 112, 155002 (2014)
- J. Lang, G.Y. Fu, and Yang Chen, Phys. Plasmas 17 042309 (2010)
- X. Wang, S. Brigulio, L. Chen, C. Di Troia, G. Fogaccia, G. Vlad and F. Zonca, Phys. Rev. E 86, 045401(R) (2012)
- H. Zhang, Z. Lin and I Holod, Phys. Rev. Lett. 109 025001 (2012)
- M. Lesur, Phys. Plasmas 20, 055905 (2013)
- J. Zhu, Z.W. Ma and G.Y. Fu, Nucl. Fusion 54 123020 (2014)
- G. Vlad, V. Fusco, S. Brigulio, G. Fogaccia, F. Zonca and X. Wang, New J. Phys. 18 105004 (2016)
- A. Bierwage and K. Shinohara, Phys. Plasmas 23 042512 (2016)
- B. J. Q. Woods, V.N. Duarte, A.J. DeGol, N.N. Gorelenkov and R.G.L. Vann, Nucl. Fusion 58 082015 (2018)
- K. G. McClements and E. D. Fredrickson, Plasma Phys. Control. Fusion 59, 053001 (2017)
- V. N. Duarte, H. L. Berk, N. N. Gorelenkov, W. W. Heidbrink, G. J. Kramer, R. Nazikian, D. C. Pace, M. Podest'a, B. J. Tobias, and M. A. V. Zeeland, Nucl. Fusion 57, 054001 (2017)
- E. D. Fredrickson, N.A. Crocker, R.E. Bell, SD.S. Darrow, N.N. Gorelenkov, G.J. kramer, S. Kubota, F.M. Levinson, D. Liu, S.s. Medley, M. Podesta, K. Tritz, R.B. White and H. Yuh, Phys. Plasmas 16, 122505 (2009)
- R. B. White and M. S. Chance, Phys Fluids 27, 2455 (1984)

- S. M. Kaye, M.G. Bell, R.E. Bell, S. Bernabei, J. Bialek, T. Biewer, W. Blanchard, J. Boedo, C. Bush, M.D. Carter et al Nuclear Fusion 45 S168-S180 (2005)
- C. Z. Cheng and M. S. Chance, Phys. Fluids 29, 3695 (1986). [21] A. Pankin, D. McCune, R. Andre, G. Bateman, and A. Kritz, Comp.Phys.Comm. 159,157 (2004)
- Chen, Y., R. B. White, Guo-Yong Fu, Raffi Nazikian, Phys. Plasmas 6, 226 (1999)
- R. B. White, N. Gorelenkov, M. Gorelenkova, M. Podesta, S. Ethier, Y. Chen, Plasma Physics and Controlled Fusion, 58, (2016)
- M. Zhou and R. B. White, Plasma Phys. Controlled Fusion 58, 125006 15(2016)
- V. N. Duarte, N. N. Gorelenkov, M. Schneller, E. D. Fredrickson, M. Podest'a, and H. L. Berk, Nucl. Fusion 58, 082013 (2018)
- E. Wigner, Phys. Rev. 40, 749 (1932)
- M.J. Bastiaans, J. Opt. Soc. Am, Vol 69, 1710 (1979)
- A. N. Kolmogorov, Proc. Int. Congr. Mathematicians, Amsterdam, Vol 1 315 (1957), V. I. Arnold, Russ. Math. Surv. 18(5):9, 1963, J. Moser, Math. Phys. Kl. II 1,1 Kl(1):1, 1962
- R. B. White, Commun Nonlinear Sci. Numer. Simulations 17, 2200 (2012)
- R B White, Plasma Phys. Control. Fusion 53 (2011) 085018
- M. Podest'a, R. Bell, A. Bortolon, N. Crocker, D. Darrow, A. Diallo, E. Fredrickson, G.-Y. Fu, N. Gorelenkov, W. Heidbrink, G. Kramer, S. Kubota, B. LeBlanc, S. Medley, and H. Yuh, Nucl. Fusion 52, 094001 (2012)
- M. Podesta, E.D. Fredricksin and M. Gorelenkova, Nucl Fusion 58, 082023 (2018)
- B. N. Breizman, H. L. Berk, M. S. Pekker, F. Porcelli, G. V. Stupakov and K. L. Wong, Phys. Plasmas 4, 1559 (1997)
- V.N. Duarte, H.L. Berk, N.N. Gorelenkov, W.W. Heidbrink, G.J. Kramer, R. Nazikian, D.C. Pace, M. Podest'a and M.A. Van Zeeland, Phys. Plasmas 24, 122508, 2017

Chapter 12

Transport

12.1 INTRODUCTION

For fusion-generated power production, a plasma must attain sufficiently effective confinement to achieve ignition. If the heating is supplied primarily by nuclear fusion of the fuel, at ignition the power density produced by the fusion alpha particles, $p_\alpha \simeq (1/4)E_\alpha n^2 \overline{\sigma v}$ watts/m^3 with $E_\alpha = 3.5$ MeV the α-particle energy, and $\overline{\sigma v}$ the velocity averaged fusion cross section, must equal the power density loss rate to the walls. The loss rate is given approximately by $p_L = 3nT/\tau_E$ watts/m^3 with τ_E the energy confinement time, n the density, and T the ion and electron temperatures, assumed equal. Requiring that the alpha produced power be greater than the loss rate gives the Lawson criterion,

$$n\tau_E > 12T/(E_\alpha \overline{\sigma v}) m^{-3} sec \qquad (12.1)$$

which is approximately 3×10^{20} m^{-3}sec for reactor parameters, $T \simeq 15$ keV. The confinement must be sufficiently good to contain a reasonably high density plasma and to restrict the rate of energy loss.

A confined plasma cannot be in thermodynamic equilibrium. Although we have described MHD equilibria in Chap. 2, from the point of view of kinetic theory there are no confined equilibria. Thermal equilibrium for a plasma is described by a distribution function $f \sim e^{-H/T}$ with T the temperature and H the Hamiltonian. But the Hamiltonian has the form

$$H = \sum (p_i - A_i)^2/2m_i + e_i\Phi = \sum[mv_i^2/2 + e_i\Phi] \qquad (12.2)$$

and thus there is no spatial dependence in H arising from the magnetic field. In a true thermal equilibrium the particles collisionally diffuse out to infinity. A MHD equilibrium remains confined, if it is stable, on ideal and resistive time scales but on collisional time scales the particles and the energy leak out, and it is essential to understand the rate of these processes.

While slow relaxation of a MHD equilibrium to a diffuse state may be acceptable, a more rapid relaxation to thermodynamic equilibrium must be avoided. There are three ways the free energy due to the confinement can be lowered: plasma expansion, removal of velocity space anisotropy, and magnetic field rearrangement. The expansion free energy drives interchange, or flute modes, low frequency drift modes, and the ballooning mode. Most such pressure-driven modes in tokamaks are of short scale length and contribute to increased transport rather than to global plasma stability. The ballooning modes, which possibly play a role in limiting plasma beta, are discussed in Chap. 8. The anisotropy free energy is best known for the loss cone modes in a mirror system. Normally in a tokamak such modes do not play an important role, but in special circumstances where deviation from axisymmetry (for example, toroidal field ripple) produces a loss cone consisting of trapped particles, instabilities driven by this energy source could act to feed particles into the loss cone. The magnetic free energy leads to kink modes and tearing modes, and is the most important driving mechanism for global scale instabilities in a stellarator or tokamak. These global modes are discussed in Chaps. 6, 7, and 10.

There are also three constraints working to prevent this energy release. Perfect conductivity, or the MHD constraint, freezes the particles to the magnetic field lines, which are to some degree anchored by external field coils. Conservation of the adiabatic invariants μ and J prevents spreading in velocity space, and the Liouville theorem, demanding that entropy remain constant, limits the class of possible motions. Unfortunately, none of these constraints is perfect. A plasma is not perfectly conducting, and as discussed in Chap. 7, the tearing mode utilizes plasma motion across the magnetic field to release magnetic energy. The adiabatic invariants μ and J can be modified by high frequency instabilities and, in the case of J, by deviation from axisymmetry.

To examine particle confinement it is necessary to delve beneath the formalism of magnetohydrodynamics to the equations governing individual particle motion, *i.e.* the collisional Vlasov equation, or Boltzmann equation,

$$\frac{\partial f}{\partial t} = \vec{v} \cdot \nabla f + (\vec{E} + \vec{v} \times \vec{B}) \cdot \nabla_v f = C(f). \tag{12.3}$$

The collision term $C(f)$, known as the Fokker–Planck operator, causes a continuous flow of particles in velocity space. A heuristic derivation of its form can be found, for example, in Schmidt (1966). In the absence of collisions this equation (the Vlasov equation) simply expresses the constancy of f along an orbit. Thus an arbitrary function of the constants of the individual particle motion E, P_ζ, μ, will be a solution.

The Liouville theorem freezes particles to volumes in phase space in analogy to the way the MHD constraint freezes particles to field lines, preventing the distribution function from becoming too scrambled. Consider a distribution function, f, and a functional $G(f)$. Define a generalized entropy S, a Casimir invariant of the Vlasov equation,

$$S = \int G(f) d^3 x d^3 v. \tag{12.4}$$

It follows that $dS/dt = 0$ for any motion described by the Vlasov equation. If G is chosen to be $ln(f)/f$, S is the usual entropy.

We would like to know what is the lowest energy of a system compatible with the constraint given by S. First we note that if the distribution function does not depend on $\vec{x}$ or the orientation of $\vec{v}$, but on energy only, $f = f(E)$, and $\partial_E f < 0$, then the system is stable (see the problems at the end of this chapter). As a matter of fact we know the state of lowest energy to be the Maxwellian, but we find that Vlasov instabilities will not produce a Maxwellian from another monotonically decreasing f. Particle collisions are required for this.

A plasma is completely described by a distribution function $f(\vec{x}, \vec{v}, t)$ giving the time dependent histories of all the particles of which it is comprised. Particle positions evolve in time under the Lorentz force $(e/m)(\vec{E} + \vec{v} \times \vec{B})$, with $\vec{E}$ and $\vec{B}$ the local electric and magnetic fields. The Hamiltonian nature of the motion implies phase space conservation, or

$$\frac{\partial f}{\partial t} + \vec{v} \cdot \nabla f + (e/m)(\vec{E} + \vec{v} \times \vec{B}) \cdot \nabla_v f = 0 \tag{12.5}$$

with ∇_v the velocity space gradient operator. Solution of this Vlasov equation in general is equivalent to solving the classical electromagnetic many-body problem, an impossible task. However, the distribution function contains vastly more information than is needed or can be possible measured, so it is essential to simplify the description of the plasma. To smooth the discreteness of the distribution function f, an ensemble average is carried out, resulting in an additional term from the correlations due to close encounters, the collision term. The long range effect of particles on each other is accounted for by the self-consistent electric and magnetic fields of the plasma. Only very close interactions of particles, at distances smaller than a Debye length, are not treated by the Vlasov equation.

Approximations are made in two directions. First the equations of motion are simplified by expanding in the available small parameters describing the system, the ratio of the cyclotron radius to the scale length of the fields, the ratio of the collision frequency to typical time scales of plasma response, and the ratio of electron to ion mass. Neglecting collisions and keeping only the lowest order in the expansion of the equations of motion in cyclotron radius the particles simply follow the field lines. The guiding center approximation includes the particle drifts, one order higher in this expansion. Gyrokinetic equations are obtained by carrying out the analysis to still one higher order in the gyro radius expansion. Provided that the resulting equations are Hamiltonian in nature, if collisions are neglected phase space conservation results in a Vlasov equation at each level.

Secondly, the magnetic fields are simplified by making use of the fact that they describe, at least very closely, a toroidal equilibrium, allowing long time plasma confinement. Perturbations of the equilibrium are assumed to consist of small amplitude waves with scale lengths much larger than the cyclotron radius and frequencies small compared to the cyclotron frequency.

The Vlasov equation allows for the stable coexistence of a plasma with very different ion and electron temperatures, since each species separately obeys its own Vlasov equation and conserves its own entropy.

It is not possible, even numerically, to examine confinement in realistic geometries using the Boltzmann equation. The approach has been to identify the principal mechanisms for cross field transport, calculate an associated diffusion coefficient, and incorporate these expressions into simplified transport equations. Often the Vlasov equation is too difficult to use even for this purpose, because it includes the very rapid cyclotron motion, and the beginning point for the derivation of diffusion coefficients is the drift kinetic equation, to be discussed in the next section.

Finally, using the distribution function for each species, $f(\vec{x}, \vec{v})$ one can construct fluid theory functions depending only on position and time by tak-

ing velocity moments, such as density, $n(\vec{x}) = \int d^3v f$, particle flux density
$n\vec{V}(\vec{x}) = \int d^3v f\vec{v}$, stress tensor $P_{\alpha,\beta}(\vec{x}) = \int d^3v f m v_\alpha v_\beta$, and energy flux
density, $\vec{Q}(\vec{x}) = \int d^3v f m v^2 \vec{v}/2$. Taking moments of the ensemble averaged
kinetic equation then results in fluid equations relating these quantities,
but each quantity is coupled to higher order moments and one must make
approximations to give closure to these equations. The first few moments
give, for a single species, and evaluated in the rest frame, the Braginskii
equations for density, momentum, and pressure

$$\frac{dn}{dt} + n\nabla \cdot \vec{V} = 0,$$

$$mn\frac{d\vec{V}}{dt} + \nabla p + \nabla \cdot \vec{\pi} - en(\vec{E} + \vec{V} \times \vec{B}) = \vec{F},$$

$$\frac{3}{2}\frac{dp}{dt} + \frac{5}{2}p\nabla \cdot \vec{V} + \pi_{\alpha,\beta}\partial_\alpha V_\beta + \nabla \cdot Q = W, \qquad (12.6)$$

where $P_{\alpha,\beta} = p\delta_{\alpha,\beta} + \pi_{\alpha,\beta}$, with p the pressure and π the viscosity tensor,
and $\vec{F}$ and W are sources. The coefficients of the terms describing cross-
field particle and energy flux, momentum diffusion, bootstrap current and
resistivity, are dependent on the nature of the particle orbits in the system
considered. For toroidal confinement systems these are called neoclassical
transport coefficients. If there exist fluctuating magnetic or electric fields,
due to MHD or kinetic instabilities causing turbulence, this will produce
additional "anomalous" transport.

The basic processes of energy transport in toroidal devices include electron and ion energy sources, transfer between species, and losses. Energy
sources include Ohmic heating, which heats mainly the electrons in the hot
core of the tokamak, negative or neutral ion beam heating, which heats
primarily ions, and wave heating of either species using propagating waves
launched by an antenna near the plasma edge.

In an Ohmically heated plasma, the central core, with $T_e > T_i$, loses
energy by collisional equilibration with the ions, transport by electron conduction to the colder edge, and electron energy loss by high Z impurity
radiation. Energy is lost from the colder edge region, with $T_e \simeq T_i$, by
convection (electron and ion) and conduction (electron) to the limiter, by
low Z impurity radiation, by charge exchange with cold neutrals with the
energy going to the wall, and by ionization losses associated with ionizing
incoming neutral gas.

Model particle and energy transport equations, in a one-dimensional (cylindrical) form, are

$$\frac{\partial n}{\partial t} = \frac{1}{r}\frac{\partial}{\partial r}(rD\frac{\partial n}{\partial r}) + S_{ionization}, \tag{12.7}$$

$$\frac{3}{2}\frac{\partial}{\partial t}(nkT_e) = \frac{1}{r}\frac{\partial}{\partial r}r(n\chi_e\frac{\partial kT_e}{\partial r} + \frac{3}{2}kT_e D\frac{\partial n}{\partial r})$$
$$-\frac{3}{2}\frac{nk(T_e - T_i)}{\tau_{ei}} - P_{rad} + \eta j^2 + P_{aux}^e, \tag{12.8}$$

$$\frac{3}{2}\frac{\partial}{\partial t}(nkT_i) = \frac{1}{r}\frac{\partial}{\partial r}r(n\chi_i\frac{\partial kT_i}{\partial r} + \frac{3}{2}kT_i D\frac{\partial n}{\partial r})$$
$$+\frac{3}{2}\frac{nk(T_e - T_i)}{\tau_{ei}} - P_{cx} + P_{aux}^i, \tag{12.9}$$

where D is the particle diffusion coefficient, $\chi_e, (\chi_i)$ the electron (ion) thermal conductivity, $\tau_{ei} \sim (m_i/m_e)\tau_{ei}^{coll}$ the equilibration time, P_{cx} the charge exchange loss, P_{aux} the auxiliary heating sources, and P_{rad} the radiation losses. All the sources and losses appearing here must be understood to be flux surface averaged quantities. A full understanding of toroidal transport thus involves all the associated radiative and atomic processes contributing to energy transport, as well as a treatment of all plasma waves leading to microturbulence, which also contributes to heat and particle transport.

Transport in a fixed toroidal magnetic field possessing nested flux surfaces, with no magnetic or electric fluctuations, due to collisions alone, is referred to as neoclassical. Classical diffusion across a constant magnetic field by collisional change of the gyro center is modified by toroidal geometry, helical deformation of equilibria, toroidal ripple, etc. Full neoclassical theory is very complicated and expressions for particle and heat transport have not been obtained in all regimes. For a review see Hinton (1983).

A detailed treatment of neoclassical diffusion is beyond the scope of this book. We restrict ourselves to a discussion of the drift kinetic equation and estimates of a few neoclassical effects, and then discuss the effect of magnetic field fluctuations which partially or wholly destroy magnetic surfaces.

12.2 THE DRIFT KINETIC EQUATION

Particle trajectories, described by classical Hamiltonian mechanics, obey the Liouville theorem. As shown in Sec. 3.2, guiding center drift can also be put in Hamiltonian form and thus in the absence of collisions the particle density in phase space, as given by the guiding center drift, is conserved following a trajectory. The mathematical expression of this is

$$\frac{\partial f}{\partial t} + \vec{v} \cdot \nabla f = C(f) \qquad (12.10)$$

where f, the distribution function, is a function of ψ_p, θ, ζ, E, μ and $\sigma = sign(v_\parallel/v)$, $C(f)$ is the modification of f due to collisions, and $\vec{v}$ is the guiding center drift velocity.

In the absence of collisions this equation is exactly analogous to the Vlasov equation, and is solved for f immediately if the equations for the single particle orbits have been solved. Any distribution f which is a function only of the constants of the motion is a solution. For a symmetric equilibrium any function of E, μ, P_ζ is a solution of Eq. 12.10 with $C(f) = 0$.

The long range effect of particles on each other is accounted for by the self-consistent electric and magnetic fields of the plasma. Only very close interactions of particles, at distances smaller than a Debye length, are not treated by the guiding center drift equations. Thus the collisions can be approximated as being local and instantaneous. That is, the individual particle orbits are given by the drift motion neglecting collisions, and $C(f)$ describes the continuous modification of the distribution due to collisions. This modification consists of a flow of particles in velocity space only, since the collisions are local in physical space. Further, for a Maxwellian distribution, f_M, $C(f_M) = 0$, since the Maxwellian is the state of highest entropy.

In general, for the particle species described by f, the collision term $C(f)$ is a sum of contributions due to each species present. For ions and electrons the collision term for the ion distribution f has the form $C_{ii}(f_i) + C_{ie}(f_i, f_e)$.

Rather than derive the form of the collision operator for the drift kinetic equation we restrict our analysis to two general properties of this operator, responsible for two of the important characteristics of collisional diffusion. First, the collisions conserve the number of particles of each species, and second, in a quasi-symmetric system the sum over all particles of the canonical toroidal momentum P_ϕ is conserved. These conditions are local, *i.e.* they hold on each flux surface. The canonical momentum is $P_\phi = g(\psi_p)\rho_\| - e\psi_p$ with e the sign of the charge. The time derivative of the total canonical momentum is

$$\frac{d}{dt}\langle P_\phi\rangle = \sum_s \int d^3v P_\phi \frac{df}{dt} = \sum_s f d^3 v P_\phi C(f) \qquad (12.11)$$

where $\sum_s$ indicates a sum over species. The $e\psi_p$ terms in P_ϕ do not contribute to this expression, because for each species

$$\int d^3v C(f) = 0. \qquad (12.12)$$

This is simply the statement that collisions conserve the total number of particles of each species. Thus the conservation of total canonical momentum, following from symmetry, gives at each flux surface ψ_p

$$0 = \sum_s \int d^3v \rho_\| C(f). \qquad (12.13)$$

This result has important consequences for particle diffusion, as we will see in the next section.

Transport predicted by the drift kinetic equation can be determined from the steady state distribution function f, by forming appropriate moments to give heat transport, particle transport, etc. The steady state distribution is assumed to be close to a local Maxwellian, and the drift kinetic equation is solved iteratively. That is, thermal equilibration is assumed to be practically achieved on each flux surface, but across flux surfaces the slow rate of equilibration allows nonzero density and temperature gradients to exist. We illustrate this method for a high aspect ratio tokamak. Using the expression for the magnetic field

$$\vec{B} = \nabla\phi + \frac{r^2}{q}\nabla\theta \qquad (12.14)$$

we find the drift velocity, from Eq. 3.10,

$$\vec{v} = \rho_\|[\hat{\phi}(1 + \rho_\| j_\phi + \nabla\rho_\| \cdot \nabla\psi_p) + \hat{\phi} \times \nabla P_\phi]. \qquad (12.15)$$

The time-independent drift kinetic equation reduces to

$$\vec{v} \cdot \nabla f = (\rho_\| \hat{\phi} \times \nabla P_\phi) \cdot \nabla f = C(f) \qquad (12.16)$$

and P_ϕ is a constant of the motion in a symmetric system (see Sec. 3.2). Now treat the effect of the collisions iteratively (Rutherford, 1970)

$$f = f_v + f_c, \tag{12.17}$$

$$\vec{v} \cdot \nabla f_v = 0, \tag{12.18}$$

$$\vec{v} \cdot \nabla f_c = C(f_v) \tag{12.19}$$

with $f_c \ll f_v$ and the subscript indicates that f is a solution of the Vlasov equation. The collisionless equation is satisfied by taking f_v an arbitrary function of P_ϕ. Further it is assumed that f is close to a local Maxwellian

$$f_v = f_M(E, P_\phi) + h(E, \mu, \sigma, \psi_p) \tag{12.20}$$

and h is assumed to be first order in the gyro radius. Thus we have

$$f_v = f_M(E, \psi_p) - \frac{\partial f_M}{\partial \psi_p} + h(E, \mu, \sigma, \psi_p). \tag{12.21}$$

To lowest order in the gyro radius we find

$$\frac{\partial f_c}{\partial \theta} = \frac{C(f_v)B}{\rho_\| \vec{B} \cdot \nabla \theta}. \tag{12.22}$$

Integrating this equation in θ at constant ψ_p gives an equation for h,

$$\int d\theta \frac{C(f_v)B}{\rho_\| \vec{B} \cdot \nabla \theta} = 0. \tag{12.23}$$

Since $C(f_v)$ is normally proportional to the collision frequency, the resulting equation for h is independent of its value. It is only necessary that the collision rate be small enough so that the drift orbits are not strongly modified over one orbit in θ, e.g. that banana orbits exist. We can thus expect two kinds of effects, those arising from h, such as the bootstrap current, independent of the collision frequency, and those such as cross-field diffusion, given by f_c and proportional to the collision frequency.

12.3   CROSS FIELD DIFFUSION

In this section we describe the mechanisms leading to the diffusion of particles across equilibrium magnetic flux surfaces, or neoclassical diffusion. If there exist fluctuating magnetic and electric fields, due to microinstabilities, which partially destroy these surfaces, this will produce additional transport. These "anomalous" transport effects will be discussed in Secs. 12.10 and 12.12.

In a uniform magnetic field, cross-field particle diffusion occurs because of a particle's gyro radius. Collisions causing a 90 degree deflection in particle velocity result in a displacement of the particle gyro center by one gyro radius. A random walk approximation then gives for the classical diffusion coefficients

$$D_e = \frac{1}{2}\nu_{ei}\rho_e^2, \qquad D_i = \frac{1}{2}\nu_{ie}\rho_i^2 \qquad (12.24)$$

where ν_{ei} (ν_{ie}) is the electron-ion (ion-electron) collision frequency for a 90 collision and ρ_e (ρ_i) the electron (ion) gyro radius. These values are equal since the gyro radii are proportional to the square root of the mass and the collision frequencies inversely proportional to mass. This expression is also valid in general geometry if the collision frequency is high enough to prevent significant cross-field drift between collisions.

The first modification of classical diffusion to be found is due to Pfirsch and Schlüter, and bears their names. From Sec. 3.3 the class of trapped particles, having $v_\parallel < (r/R)^{1/2}v_\perp$, have a bounce frequency less than $\omega_b = (r/R)^{1/2}(v_T/Rq)$, with v_T the thermal velocity. For collision frequencies larger than ω_b so that banana orbits do not exist, there still exist passing particle orbits with excursions from the flux surface of $\Delta r = q\rho$. Collisions produce a random walk with a step of this magnitude, and we obtain in this domain

$$D_{PS} = \frac{1}{2}\nu q^2 \rho^2 \qquad (12.25)$$

for the Pfirsch–Schlüter diffusion coefficient, which in a tokamak can be significantly larger than the classical value.

If the collision time is long enough so that banana orbits can be completed, *i.e.* $\nu < \omega_b$, the drift kinetic formalism of Sec. 12.2 is applicable. This collisionality domain is called the banana regime. Assuming toroidal symmetry with nested magnetic surfaces the particle flux across a surface

ψ_p is given by

$$\Gamma = \int d^3 v \frac{f \vec{v} \cdot \nabla \psi_p}{|\nabla \psi_p|}. \tag{12.26}$$

Substituting the expression for $\vec{v}$ from Sec. 12.2 we find

$$\vec{v} \cdot \nabla \psi_p = \rho_{\parallel} \frac{\partial \rho_{\parallel}}{\partial \theta} \frac{1}{q} \tag{12.27}$$

where we have used the lowest order expression $\nabla \psi_p = (r/q)\hat{r}$. This drift term has the opposite sign for electrons. Substituting we find for the average flux over a magnetic surface

$$\overline{\Gamma} = \pm \int d^3 v \frac{d\theta}{2\pi} f \frac{\partial}{\partial \theta} \left(\frac{\rho_{\parallel}^2}{2} \right) \tag{12.28}$$

with the $\pm$ corresponding to the sign of the charge. Integrating by parts and using Eqs. 12.13 and 12.16 we find for the average flux for a single species

$$\overline{\Gamma} = \pm \frac{1}{4\pi} \int d\theta d^3 v \rho_{\parallel} q(r) C(f_v). \tag{12.29}$$

Now from Eq. 12.13 we can draw two important conclusions. First, the flux due to collisions among particles of a single species is zero. Thus, even in the presence of a strong density gradient, single species collisions produce equal flux of particles in each direction and the density gradient will be maintained, there is no net diffusion. Note that the flux is zero at every point, not simply in the average sense, and thus a localized density profile cannot expand. Secondly, in a two component plasma, the ion and electron particle fluxes are equal, *i.e.* collisional diffusion is ambipolar.

Changing integration variables through $d^3 v = \sum_\sigma 2\pi dE d\mu / |\rho_{\parallel}|$ where the summation is over the two signs of $\rho_{\parallel}$, and substituting a Lorentz collision term,

$$C(f) = \nu \rho_{\parallel} \frac{\partial}{\partial \mu} \left(\rho_{\parallel} \mu \frac{\partial f}{\partial \mu} \right), \tag{12.30}$$

the modification of the distribution function can be found explicitly (Rutherford, 1970). It is given by

$$h = \frac{\partial f_v}{\partial \psi_p} \int_\mu^{\mu_T} \frac{d\mu}{< \rho_{\parallel} >} \tag{12.31}$$

for $\mu < \mu_T$ and zero otherwise, as can be verified by substitution into the equation determining h from Sec. 12.2, where μ_T is the trapped-passing boundary value. Substitution of this expression for h then gives for the average particle flux

$$\bar{\Gamma} = -1.46 \left(\frac{r}{R}\right)^{1/2} \frac{\nu}{B_\theta^2} \frac{dp}{dr}, \tag{12.32}$$

as originally found by Galeev and Sagdeev (1968).

These results can also be obtained in a heuristic fashion. We demonstrate this by showing absence of single species diffusion and ambipolarity for classical ion-electron diffusion. Consider the effect of a binary collision on the positions of the guiding centers of the two particles. Take the collision point to be the origin of the coordinate system. Before colliding, the guiding center of a particle is located at

$$\vec{r} = \frac{\vec{p} \times \vec{B}}{eB^2} \tag{12.33}$$

with e the signed charge. During the collision the particle position does not change, but the change in momentum Δp changes the guiding center position by

$$\Delta \vec{r} = \frac{\Delta \vec{p} \times \vec{B}}{eB^2}. \tag{12.34}$$

Since the other particle changes its momentum by $-\Delta\vec{p}$ we can immediately draw two conclusions; the same as found in general from Eq. 12.29. First, particles of like charge are displaced oppositely by a collision. If the charges are equal the displacements are opposite and equal.

Second, particles of opposite charge move in the same direction upon colliding, and if the magnitude of the charges are equal, they move the same distance. Thus collisions between protons and electrons produce ambipolar diffusion.

The diffusion rate in the banana regime can also be obtained heuristically. The diffusion is dominated by the trapped particles, because of the large excursion of the banana orbits. In this case during a collision the step size becomes the banana width, $\Delta r = 2q\rho(R/r)^{1/2}$. In addition, since trapped particles must have $v_\parallel/v_\perp < (r/R)^{1/2}$, a corresponding smaller collisional modification of the velocity direction is necessary to change the orbit. For multiple small collisions $(\Delta\theta)^2 = \nu t$, so the effective collision

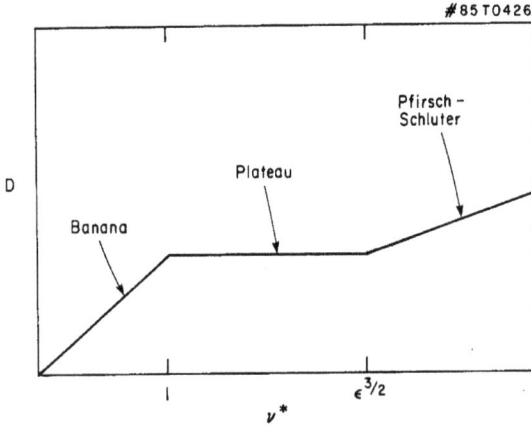

Fig. 12.1   Neoclassical diffusion.

frequency for a trapped particle is $\nu_{eff} = (R/r)\nu$. In addition, only a fraction $(r/R)^{1/2}$ of the particle population is trapped, giving for the diffusion coefficient in the banana regime

$$D = 4\nu q^2 \rho^2 \left(\frac{R}{r}\right)^{3/2}, \qquad (12.35)$$

in agreement with Eq. 12.32 except for the numerical constant. Plotted versus the effective collision frequency normalized to bounce frequency, $\nu^* = \nu_{eff}/\omega_b$, the neoclassical diffusion coefficient then has the approximate form shown in Fig. 12.1, as first found by Galeev and Sagdeev (1967). The transition region between banana and Pfirsch–Schlüter domains is called the plateau domain. This plot shows the three domains clearly, because it is for a large aspect ratio system. For lower aspect ratios the transitions are blurred, and the plateau domain is practically invisible.

Note that the diffusion scales as $\rho^2\nu$, with ν the effective collision frequency for 90 degree scattering. This quantity thus scales with mass as $m^{1/2}$. In a symmetric configuration the ion and electron rates must be equal, as required by Eq. 12.29. In fact they are experimentally found to be a few times the neoclassical value.

Experiments indicate that in tokamaks ion heat flow rates are typically a few times neoclassical, and electron heat flow rates are almost two orders of magnitude anomalously large. The dominant mechanisms for particle and

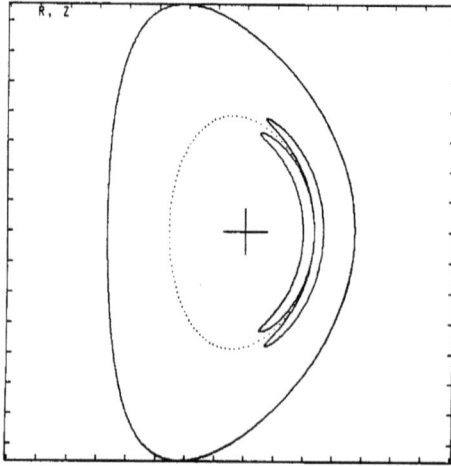

Fig. 12.2   Banana orbits originating at $\theta = 0$ with opposite pitches.

heat transport in a tokamak have yet to be identified, although certainly neoclassical processes play a role. This will be discussed in Sec. 12.10.

12.4   BOOTSTRAP CURRENT

The collisional trapping and detrapping of particles in the banana regime gives rise to a net toroidal current which is called the bootstrap current, named after the reported ability of Baron von Münchausen to lift himself by his bootstraps (Raspe, 1785, 2005). In a purely collisionless plasma a net current can be due only to asymmetric initial conditions and is thus not of interest. The bootstrap current is a net toroidal current in a collisional plasma. It results from a small perturbation of the particle distribution from that of uniform pitch, and is due to toroidal curvature and the existence of banana orbits. It can be calculated by using the perturbed distribution function (f_v, f_c) of Sec. 12.2, but it can also be arrived at directly.

Consider a flux surface ψ, and the two associated ion banana orbits with opposite velocities at $\theta = 0$ on the flux surface of their common origin, as shown in Fig. 12.2. To be trapped they must have pitch $-\sqrt{r/R} <$

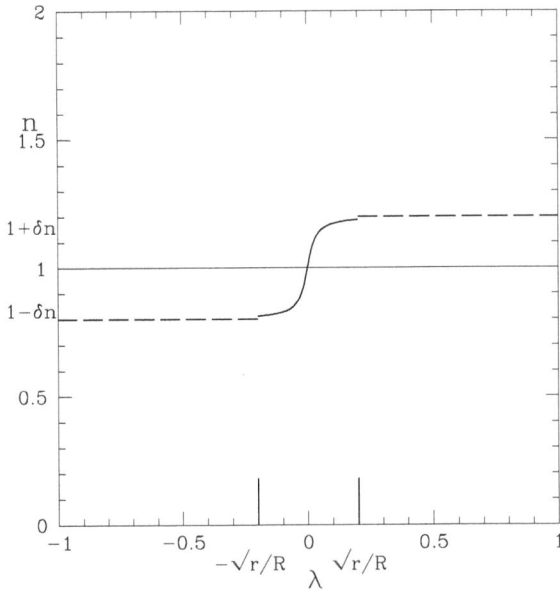

Fig. 12.3   Particle density vs pitch at the midplane. Particles with pitch $-\sqrt{r/R} < \lambda < \sqrt{r/R}$ are trapped. There is a greater density of co-moving trapped particles than counter-moving. Pitch angle scattering then produces a uniform passing distribution in λ, as shown by the dashed lines.

$\lambda < \sqrt{r/R}$ at the midplane. The inside banana is co-moving at its initial point, and the outside banana is counter-moving. Because of the density gradient, on this flux surface the density of co-moving trapped particles will be larger than the counter-moving by a factor $\delta n = (dn/dr)W_b/2$, with W_b the banana width, as shown in Fig. 12.3. The bootstrap current is due to the associated asymmetry in the co- and counter-moving passing particles necessary to maintain steady state through small pitch angle scattering. Pitch angle scattering causes a diffusion in λ, and the constant source of trapped particles leads to a uniform pitch distribution of passing particles, with a density of $1 + \delta n$ for co-moving, and $1 - \delta n$ for counter-moving, as shown in Fig. 12.3. Substituting $W_b = (R/r)^{1/2}q\rho$, multiplying by the average velocity of the passing particles, $\rho/2$, and using the expression for the pressure, $p = n\rho^2/2$, we find for the bootstrap current $(r/R \ll 1)$

$$j_{BS} = - \left(\frac{r}{R}\right)^{1/2} \frac{1}{B_\theta} \frac{\partial p}{\partial r}. \tag{12.36}$$

This expression is for a given pressure, independent of mass. Electrons produce a flow in the opposite direction, and thus a current is also given by Eq. 12.36. For an evaluation of the exact numerical coefficient for this expression, see Rosenbluth *et al.* (1972). Note that this current, although dependent on the collisions for the equilibration of trapped and passing populations, is independent of the collision frequency. The canonical toroidal momentum is $P = g(\psi)\rho_\| - \psi_p$. The trapped particle in the collision is moved down the density gradient, increasing ψ_p, and the passing particle increases $\rho_\|$, with the total momentum conserved.

Collisional relaxation of the nonisotropic flow of passing particles produces a net transfer of particles from the more populated inner banana orbits of Fig. 12.2 to the outer one. Thus the bootstrap current is driven at the expense of net outward particle diffusion, and is not completely without a price. With no outward flow there can be no bootstrap current.

Numerical simulation of neoclassical processes has been greatly improved by the introduction of what is termed the δf method, which consists of representing the equilibrium (Maxwellian or monoenergetic) part of the plasma analytically, and using particle simulation to represent only the deviations from the equilibrium. Thus the particle distribution is written as $f = f_0 + \delta f$ and the particles, or markers, have weights associated with them and are evolved in time in a manner which correctly produces the modification of the equilibrium distribution. See Kotschenruether (1993), Parker and Lee (1993), Hu and Krommes (1994), Chen and White (1997). For example, to simulate bootstrap current the equilibrium distribution f_0 is taken to be a monoenergetic collisional particle distribution in the vicinity of a particular equilibrium flux surface with no density gradient, and no net toroidal current. Then δf is used to simulate the modification of this distribution produced by a density gradient, collisional equilibration in the presence of the density gradient producing a net bootstrap current through the process discussed above. For a monoenergetic distribution the collision operator is taken to be pitch angle scattering only, so that energy is conserved for each particle.

In Fig. 12.4 is shown a δf simulation of bootstrap current, normalized to the theoretical value, Eq. 12.36, as a function of νT, with ν the collision

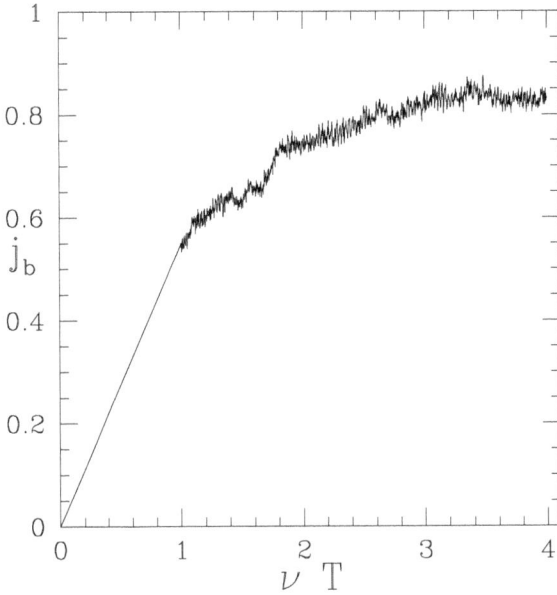

Fig. 12.4   Bootstrap current simulation.

frequency and T the time. The simulation results are normally somewhat smaller than the theoretical value since it is obtained in the collisionless limit, and numerically feasible values of collisionality give slightly smaller values. Results for $\nu T < 1$ were not kept, and as can be seen the bootstrap current takes a few collision times to reach steady state. See Wu and White (1993) for a description of simulations of the bootstrap current and its modification by the presence of magnetic field perturbations.

In a stellarator the bootstrap current is more complex because of the possibility of resonances, and important because it may be the only source of current in the equilibrium, and thus can significantly modify the equilibrium produced by the field coils. Particle motion in any equilibrium depends only on the magnitude of the magnetic field, which can be expanded in Fourier harmonics with respect to the poloidal and toroidal variables θ and ϕ

$$B = \sum_{m,n} B_{m,n} cos(m\theta - n\phi). \qquad (12.37)$$

The $B_{m,n}$ are perturbations of the magnitude of $\vec{B}$ only, and do not produce magnetic islands in the equilibrium. The analytic expression for the bootstrap current for a monoenergetic distribution is $j_b = -G_{bs}f_t E dn/d\psi$ with G_{bs} a geometric factor given by

$$G_{bs} = <H_1> + \frac{H_2 <B^2> (gq + I)}{2B_0^2}$$
$$- \frac{3q(gq + I) <B^2>}{4f_t B_0^2} \int_0^1 d\lambda \frac{W(\lambda)}{<v_\parallel/v>} \qquad (12.38)$$

where the definitions of the various terms and derivations can be found in Shaing *et al.* (1989), and Watananbe *et al.* (1992). Both $W(\lambda)$ and H_2 are given by sums over harmonic terms and may contain resonant denominators. In an equilibrium with a single magnetic perturbation $B = 1 + \epsilon_h cos(m\theta - n\phi)$, $W(\lambda) = 0$ and this expression reduces to (see Boozer and Gardner, 1990)

$$j_b = -f_t E q \frac{m}{m - nq} \frac{dN}{d\psi} \qquad (12.39)$$

with f_t the fraction of trapped particles in the helical well, given by a constant times $\sqrt{\epsilon_h}$ in large aspect ratio approximation, ψ the toroidal flux, E the particle thermal energy, and N the particle density. Although of theoretical interest and great simplicity, such an equilibrium is of no practical interest because of very poor particle confinement, except for the axisymmetric case $n = 0$, but this expression displays a difficulty with the present theory, namely the existence of resonances in the vicinity of which the assumptions of the model have broken down.

Orbits which take longer to complete than the time for them to collisionally scatter from near the X-point (see Fig. 3.2) to the separatrix do not contribute to the bootstrap current. In the axisymmetric case (a tokamak) the correction has been analytically calculated by Hinton and Rosenbluth (1973) by finding the collisionally modified distribution function in the boundary layer between the trapped and untrapped particle regions of phase space. The result is that in the banana regime the bootstrap current is decreased by a function of the dimensionless collision frequency ν^* with $\nu^* = \nu q(R/r)^{3/2}R/v$. Here ν is the pitch angle scattering frequency and v is the thermal velocity.

A similar modification should collisionally limit the magnitude of the resonance in the stellarator expression, again decreasing the bootstrap current, analysis which has not been completed.

Bootstrap current is especially important because of the possibility of producing almost all of the current profile for a tokamak in steady state. This is much easier to do if the desired current profile has very little current on axis, or equivalently very high q on axis. This is called an advanced tokamak configuration, and is of special interest because of the very low levels of transport discovered to exist in configurations with very small or even negative shear.

12.5 NEOCLASSICAL TEARING

A straightforward result following from the nonlinear analysis of the tearing mode, Sec. 10.7, and the existence of the bootstrap current, Sec. 12.4, is the existence of bootstrap-driven magnetic islands, which have become known as neoclassical tearing modes. For a review, see Hegna (1998). From Eq. 10.51 (see also Fig. 7.5) we note that a current perturbation producing positive current at the O-point and negative current at the X-point is stabilizing, and the opposite is destabilizing. Now consider an island large enough so that the plasma pressure has approximately equilibrated on the perturbed flux surfaces; roughly this implies an island large compared to particle gyro orbits. From Eq. 12.36 the bootstrap current is driven by the local pressure gradient. In the equilibrated island there can be no pressure gradient in the vicinity of the island O-point, but there will be a pressure gradient in the vicinity of the X-point, and thus the local bootstrap current will act to destabilize the island if the pressure gradient is large enough. This effect acts to diminish the quality of plasma confinement at high β, has been observed in high β discharges (see Chang *et al.*, 1994) and is an important consideration in reactor-grade plasmas.

12.6 WARE PINCH

Normally in a tokamak there is a toroidal electric field E_ϕ, driving the toroidal current, and all particles experience an inward $\vec{E} \times \vec{B}$ drift, of the magnitude $v_d = E_\phi B_\theta / B_\phi^2$, independent of both charge and mass. For typical tokamaks this velocity is less than 10 cm/sec, and is overwhelmed by the outward particle transport. However the inward pinch effect is much larger for trapped the class of particles, as shown by Ware (1970). Consider the equation for the toroidal canonical momentum, as given in Sec. 3.2,

$$\frac{d}{dt}P_\phi = E_\phi. \tag{12.40}$$

Substituting the expression for P_ϕ, and neglecting variation in g, we find $g\dot\rho_{\parallel} - \dot\psi_p = E_\phi$. For a trapped particle we can evaluate this expression at two successive bounce points, where $\rho_{\parallel} = 0$, giving

$$-\frac{d\psi_p}{dr}\Delta r = E_\phi \Delta t. \tag{12.41}$$

But $d\psi_p/dr = rB_\phi/q = B_\theta$, so we find

$$\frac{\Delta r}{\Delta t} = -\frac{E_\phi}{B_\theta}, \tag{12.42}$$

a net inward pinch, $1/\epsilon^2$ stronger than that experienced by passing particles.

12.7 MAGNETIC FIELD RIPPLE TRANSPORT

All perturbations of the field leading to destruction of toroidal symmetry produce classes of trapped particles, and new loss mechanisms. Most of them cover small volumes in phase space and do not affect appreciable amounts of plasma. In this section we treat one of the most common perturbations in a tokamak, that due to the discrete field coils used to produce the toroidal field. This analysis is also applicable to a quasi-symmetric stellarator, additional helical components of the field destroying the symmetry being equivalent to the toroidal field ripple in a tokamak.

The magnetic field in a tokamak has periodic toroidal variations in strength, or ripple, due to the finite number of toroidal field coils. A simple model of the field including the lowest order toroidal correction and the effect of N toroidal field coils is given by writing the field strength as

$$B = B_0(1 - r cos\theta + \delta cos N\phi). \tag{12.43}$$

Along a field line $\phi = q\theta$, $r = constant$ so ripple wells exist if $r|sin\theta| < Nq\delta$. The magnetic field strength following a field line is shown in Fig. 12.5. To be trapped in one of these wells a particle must have $v_{\parallel}/v, < \sqrt\delta$.

Because it has widely spaced superconducting coils, Tore Supra has large ripple, reaching 5 percent at the plasma edge. The ripple well domain

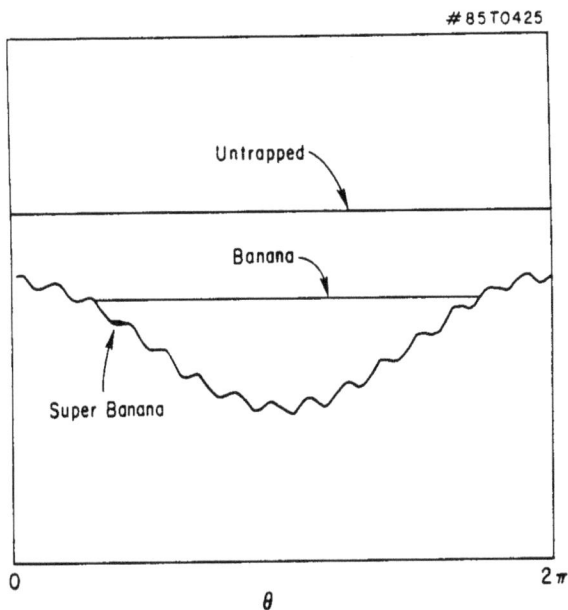

Fig. 12.5   Magnetic field strength variation with ripple.

in Tore Supra for a typical equilibrium is shown in Fig. 12.7. All banana orbits with bounce tips within this domain are ultimatly ripple trapped, due to the fact that near the lower banana tip, where the parallel velocity is very small, the particle is drifting outwards toward larger ripple. Thus the well depth seen by the particle is increasing as the particle passes over it, and if the toroidal phase is such that the well is attractive, the particle is trapped. As the particle precesses around the device, the phase of the ripple is different at each bounce point, and eventually the particle will just barely make it over a peak into a ripple well, and then become trapped. Note that a particle confined exactly to a flux surface can never become trapped, the trapping probability depends on how much the well depth changes as the particle moves over it, and thus increases with the drift rate. A ripple-trapped particle drifts downward at a rate given by $v_d = (\rho/R)^2 R\omega_0$ with ρ the gyro radius, ω_0 the gyro frequency, and R the major radius. For a one keV proton in Tore Supra $\rho/R \sim 5 \times 10^{-4}$ and the gyro frequency is $\omega_0 = 3 \times 10^8/\text{sec}$, giving a velocity of about 3×10^4 cm/sec, so that once a

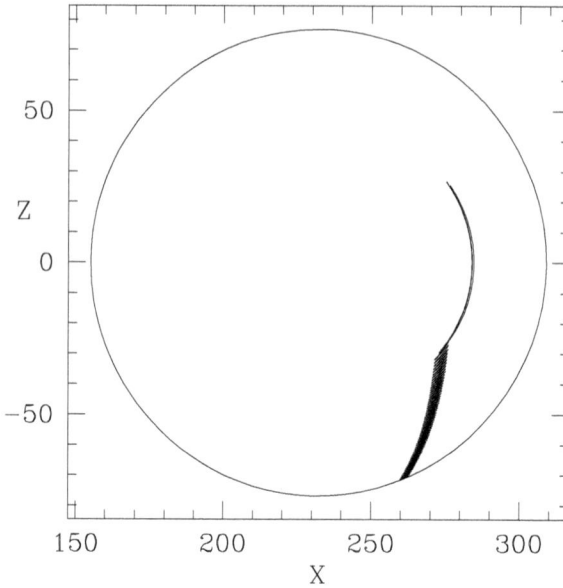

Fig. 12.6   A typical ripple-trapped loss orbit.

particle is ripple trapped in a loss orbit leading to the wall, the loss process typically takes only a fraction of a millisecond. A typical loss orbit is shown in Fig. 12.6.

Particles deeper inside the plasma which are ripple trapped drift downward a distance z_0 until they leave the ripple well domain, with z_0 determined by the geometry of the domain, shown in Fig. 12.7. Due to this drift across flux surfaces, the particles are shifted outwards by a distance $\Delta r \simeq z_0^2/r$ (Yushmanov, 1982) while they are ripple trapped. If a particle exits the ripple trap domain before being lost, it then continues to execute normal banana orbits until again becoming ripple trapped. Particles thus execute a series of superbanana orbits, with each orbit shifted outwards approximately by a distance z_0^2/r, leading ultimately to an orbit which reaches the wall.

During collisional motion in superbanana orbits the diffusive motion is dominated by the radial drift at velocity v_d, which lasts for one collision time, with the collision frequency necessary to free the particle from a ripple

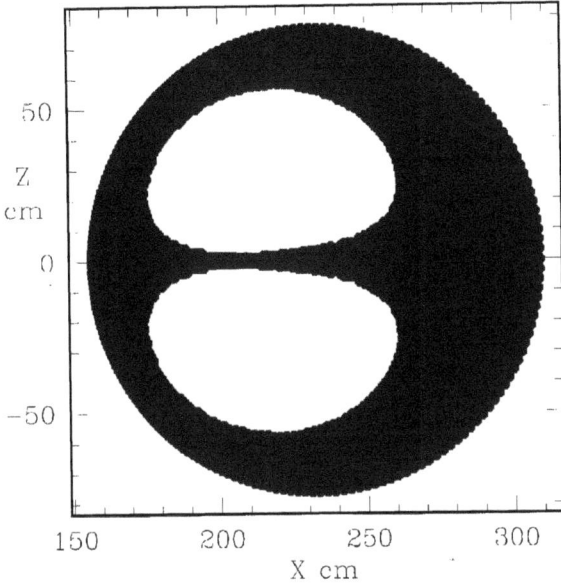

Fig. 12.7 Ripple well domain in Tore Supra (shaded).

well. The radial excursion is thus v_d/ν_{eff}. The diffusion produced is

$$D_{SB} = f \frac{v_d^2}{\nu_{eff}}, \qquad (12.44)$$

where f is the fraction of particles in superbanana orbits, v_d the drift speed, and ν_{eff} the effective collision frequency, which inhibits the drift. This is because without collisions the particles are surely lost; collisions can only detrap them and prevent loss. The fraction is $f = (\delta B/B)^{1/2} = \sqrt{\delta}$ and $\nu_{eff} = \nu/\delta$; giving

$$D_{SB} = \frac{\rho^2}{R^2} \frac{v_T^2}{\nu} \delta^{3/2} \qquad (12.45)$$

where ν is the collision frequency for a 90 degree scattering.

This diffusion is superimposed, but not to scale, on the plot of the neo-classical diffusion in Fig. 12.8. Note that for very small ν the superbanana orbits drift directly to the wall and form a loss cone, and the diffusion again

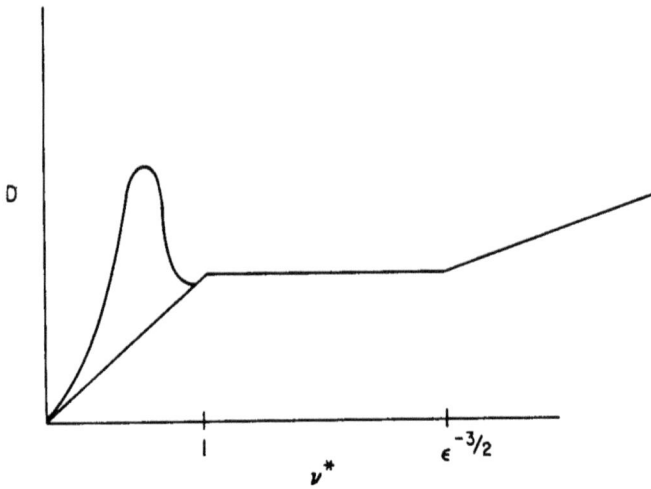

Fig. 12.8   Modification of neoclassical diffusion due to superbanana orbits.

becomes proportional to ν, determined by the rate at which the particles scatter into the loss cone.

Even if superbanana orbits do not exist, the existence of ripple can have a dramatic effect on the transport of trapped particles, because the ripple changes the location of the bounce point, and modifies the orbit. Untrapped particles are not strongly affected.

To calculate the shift across magnetic surfaces of the bounce point caused by ripple, we integrate the cross surface drift with and without ripple and take the difference. Consider a deeply trapped particle, with energy $E = v_\parallel/2 + \mu B$ and $v\,\|^2\,/2 \ll E$ throughout the orbit. The position is given by

$$\psi_p = \int d\theta \frac{\dot{\psi}_p}{\dot{\theta}} = \int d\theta \frac{g\mu\partial_\theta B}{v_\parallel B} \tag{12.46}$$

plus terms of higher order in ρ, with $v_\parallel = (2E)^{1/2}(1 - \mu B(\theta) + \delta cos N\phi)/E)$. The presence of δ moves the bounce point and changes the drift. To separate these effects write

$$B(\theta) + \delta cos N\phi = B(\theta) + \delta cos N\phi_b + \delta(cos N\phi - cos N\phi_b) \tag{12.47}$$

with the new bounce angle given by $E/\mu = B(\theta_b) + \delta cos N\phi_b$. Expanding $v_\parallel$ we find

$$
\begin{aligned}
v_\parallel = \sqrt{2E} &\left(1 - \frac{\mu}{E}(B(\theta) + \delta cos N\phi_b)\right)^{1/2} \\
&\times \left(1 - \frac{\mu\delta[cos N\phi - cos N\phi_b]}{2E[1 - (\mu/E)(B(\theta) + \delta cos N\phi_b)]}\right)
\end{aligned}
\tag{12.48}
$$

and substituting into Eq. 12.46 we find

$$
\Delta\psi_p = \psi_p(\delta) - \psi_p(0) = \int_0^{\theta_b} d\theta \frac{\partial_\theta B}{B} \frac{\rho}{2R} \frac{\delta(cos N\phi - cos N\phi_b)}{[1 - B(\theta)/B(\theta_b)]^{3/2}}, \tag{12.49}
$$

where a factor of two is included because the drift has the same sign approaching and receding from the bounce point. Writing $\phi = q\theta + \phi_0(t)$ and expanding near θ_b we find the expression

$$
\Delta\psi_p = \frac{\delta\rho}{2R} \int_0^{\theta_b} d\theta \frac{\partial_\theta B}{B} \frac{e^{i(Nq\theta+N\phi_0)} - e^{i(Nq\theta_b+N\phi_0)}}{[1 - B(\theta)/B(\theta_b)]^{3/2}}, \tag{12.50}
$$

with the real part understood.

Now assume the entire contribution to the integral comes from $\theta \simeq \theta_b$, and Taylor expand the functions in the integral

$$
\Delta\psi_p = \frac{\delta\rho}{2R} \left(\frac{\partial_\theta B}{B}\right)^{-1/2} e^{iN\phi_b} I \tag{12.51}
$$

where, upon letting $z = Nq(\theta_b - \theta)$,

$$
I = (Nq)^{1/2} \int_0^\infty dz \frac{e^{-iz} - 1}{z^{3/2}}. \tag{12.52}
$$

But $\int_\epsilon^\infty dz z^{-3/2}[cos z - 1] = (1/2)[e^{i\pi/4}\Gamma(-1/2, i\epsilon) + e^{-i\pi/4}\Gamma(-1/2, -i\epsilon)] - 2/\sqrt{\epsilon}$, giving

$$
I = 2\sqrt{\pi Nq} e^{-i3\pi/4}, \tag{12.53}
$$

$$
\Delta\psi_p = \frac{g\delta\rho}{R} \frac{(\pi Nq)^{1/2} sin(N\phi_b - \pi/4)}{(\partial_\theta B/B)^{1/2}}. \tag{12.54}
$$

In cylindrical geometry this reduces to

$$
\Delta r = \left(\frac{q}{r}\right)^{3/2} \left(\frac{\pi N}{sin\theta_b}\right)^{1/2} \rho\delta sin(N\phi_b - \pi/4). \tag{12.55}
$$

Notice that this expression is not valid near $\theta_b = 0$, because of the approximations used to evaluate the integral in Eq. 12.50. A general evaluation of orbit modification due to field perturbations (Mynick, 1986), leads to closed form expressions involving generalized Bessel functions.

There are three processes that can change the phase ϕ of the banana tip from one bounce to the next. They are toroidal precession, Coulomb collisions and radial drift of the banana orbit. Corresponding to the rate of change of ϕ there are four different regimes of ripple transport of banana particles.

If the change in ϕ per bounce is small, the motion of the banana tip radially is periodic. Collisions can then induce a random walk, giving a diffusion rate

$$D = \Delta^2 \nu. \tag{12.56}$$

This is the banana drift regime. The change in ϕ due to toroidal precession and radial drift are proportional to the square root of the energy, that due to Coulomb collisions scales as the inverse square. Thus the banana drift regime occurs for a range of energy, $W_1 < W < W_2$. Below the lower limit W_1 the Coulomb modification of the toroidal phase dominates. The phase then changes randomly at each reflection and the banana orbit diffuses randomly with step size Δ given by Eq. 12.54 and time given by the bounce frequency

$$D = \Delta^2 \omega_b. \tag{12.57}$$

This is the ripple plateau regime (Boozer, 1980; Linsker and Boozer, 1982) with D independent of collision frequency. At energies greater than W_2, first the toroidal precession and then the radial drift become significant. If the toroidal precession dominates, the change in the phase is the same each bounce, since the precession is constant. Thus only those particles which precess through an integral number of ripple periods during each bounce experience significant radial motion, which is not diffusive. This is called the resonance regime (Yushmanov, 1983). It does not contribute significantly to the global diffusion because the resonances are well localized in minor radius, and particles quickly become nonresonant.

Finally, when the radial drifts dominate, the phase space of banana tip trajectories can become stochastic, leading to diffusion at the plateau rate, $D = \Delta^2 \omega_b$, and rapid loss of all trapped particles. The change of the bounce point due to the radial drift can be calculated using $\phi_b = 2q\theta_b$ and the constancy of B at the bounce points, $B(\psi_b, \theta_b) = E/\mu$. This gives

$$\frac{d\phi_b}{d\psi_b} = 2\theta_b \frac{dq}{d\psi_p} - 2q \frac{\partial_{\psi_p} B}{\partial_{\theta_b} B}. \tag{12.58}$$

The motion becomes stochastic approximately when the modification of the toroidal bounce point is as large as a ripple phase (Goldston *et al.*, 1981), or

$$N \frac{d\phi_b}{dr} \Delta > 1. \tag{12.59}$$

Substituting the expression for Δ, we find this limit on the ripple magnitude to be

$$\delta < \frac{(\partial_{\theta_b} B/B)^{1/2}}{\theta_b dq/d\psi_p - q \partial_{\psi_p} B/\partial_{\theta_b} B}, \tag{12.60}$$

which in a low β, cylindrical equilibrium, for an average bounce point $\theta_b \simeq \pi/2$, gives, for the stochastic ripple transport threshold

$$\delta \leq \frac{1}{\rho q'} \left[\frac{r}{\pi R N q} \right]^{3/2}. \tag{12.61}$$

This is an important restriction on the ripple, normally only for energies above 1 MeV, and thus this stochastic regime is relevant only for trapped fusion alpha particles, high energy beam, or ICRF heated particles, unless q or the ripple magnitude is very large.

For a more precise analysis of the stochastic threshold for ripple transport, the expressions for the changes in radial position and phase can be used to construct an area-preserving map which describes banana motion at high energies where the Coulomb collisions are negligible. It has the form

$$r_{j+1} = r_j + \Delta \cos N \phi_j, \tag{12.62}$$

$$\phi_{j+1} = \phi_j + \phi_b(r_{j+1}) + \phi_p(r_{j+1}), \tag{12.63}$$

$$r_{j+2} = r_{j+1} + \Delta \cos N \phi_{j+1}, \tag{12.64}$$

$$\phi_{j+2} = \phi_{j+1} - \phi_b(r_{j+2}) + \phi_p(r_{j+2}), \tag{12.65}$$

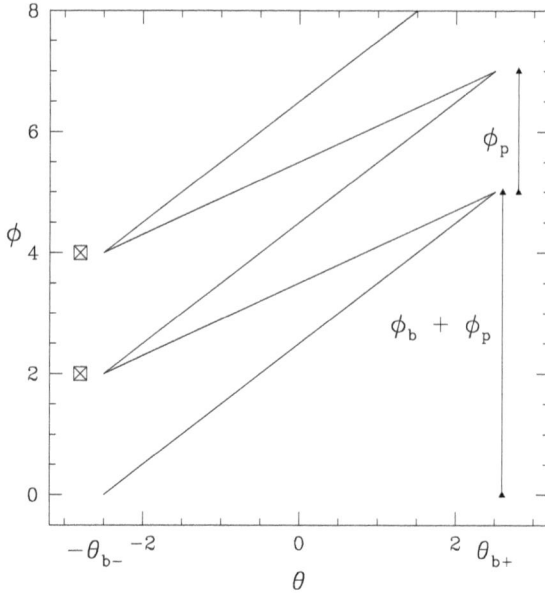

Fig. 12.9 Trapped particle trajectory showing precession resonance. The locations of the field coils are denoted by the boxes with X inside.

with ϕ_b due to the bounce motion and ϕ_p due to precession. This banana tip map has been used to study the onset of stochastic trapped particle ripple loss. It is a composite of two mappings similar to the standard map of Sec. 1.8, but the calculation of the stochastic threshold is much more involved. Here we sketch this analysis, which is given for general up-down asymmetric configurations in White (1998). In Fig. 12.9 is shown the trajectory of a toroidally drifting trapped particle, in precession resonance, with the field coils indicated on the left.

The total precession during one half bounce, given by Eq. 3.59, has the form $\phi_p = \rho P(\Psi_p)$ with P a geometry-dependent integral independent of gyro radius. The map thus depends on the functions of position ϕ_p, ϕ_b, and Δ, which contain all the essential information about the particle precession and bounce motion, the equilibrium, and the field ripple.

Now we search the banana tip map for fixed points, which are the X- and O-points of resonances. Approximate ϕ_b and ϕ_p as linear functions in the variable r, a simplification certainly valid over the scale of the resonance spacing. Conservation of energy E and magnetic moment μ makes the bounce angle in the unperturbed orbit a function of r through $E = \mu B(r, \theta_b)$, and thus $\phi'_b = q'\theta_{b+} + q'\theta_{b-} - q\partial_r B/\partial_\theta B_+ - q\partial_r B/\partial_\theta B_-$. The derivative of ϕ_p must be evaluated using Eq. 3.59. We wish to identify all fixed points, calculate island widths at the resonant flux surfaces, and identify conditions for stochastic threshold. The total change in the position of the upper banana tip in one bounce is given by

$$dr = \Delta sin(N\phi) + \Delta sin(z) \qquad (12.66)$$

and

$$d\phi = 2\phi_p + 2\phi'_p\Delta sin(N\phi) + (\phi'_p - \phi'_b)\Delta sin(z) \qquad (12.67)$$

where $z = N\phi + 2w + 2w'\Delta sin(N\phi)$, $w = N\phi_b/2 + N\phi_p/2$.

The fixed points of the map are qualitatively different for different values of ϕ'_p, ϕ'_b. Refer to the flux surfaces $\phi_p = k\pi/N$ as precession resonance surfaces. These are surfaces at which the trapped particles considered (fixed values of r, E, θ_b) precess an integer number of coil spaces in one bounce. Let $r = r_{p,k} + dr$, with $r_{p,k}$ a precession resonance surface, the neighboring surfaces being located at $dr = \pm\pi/(N\phi'_p)$. This spacing is typically a small fraction of the minor radius; there can be as many as one hundred resonances across the plasma. In Fig. 12.10 is shown a pair of precession resonances for small ripple magnitude. Period one fixed points are given by

$$sin(N\phi) + sin(z) = 0, \qquad (12.68)$$
$$N\phi'_p dr - w'\Delta sin(z) = 0 \qquad (12.69)$$

with $w = w_k + w'dr$ and the subscript k indicating evaluation at the precession resonance surface $dr = 0$. If the term in Δ is negligible in Eq. 12.69 and in z, only $dr \simeq 0$ is a solution, *i.e.* the only resonances are the precession resonances, and $N\phi$ and $N\phi - \pi$ are both solutions to Eq. 12.68; the X- and O-points are separated by π. Physically the particle precesses through an integer number of field coils in one bounce, with the forward and backward movement due to the bounce motion cancelling out.

Increasing the term in Δ initially simply modifies the position of the precession resonances, but there are additional solutions to Eqs. 12.68, 12.69 if this term is sufficiently large. These driven resonances can only be obtained numerically. The envelope of the domain of additional solutions is given by $|dr| \leq |1 - \phi'_b/\phi'_p|\Delta/2$.

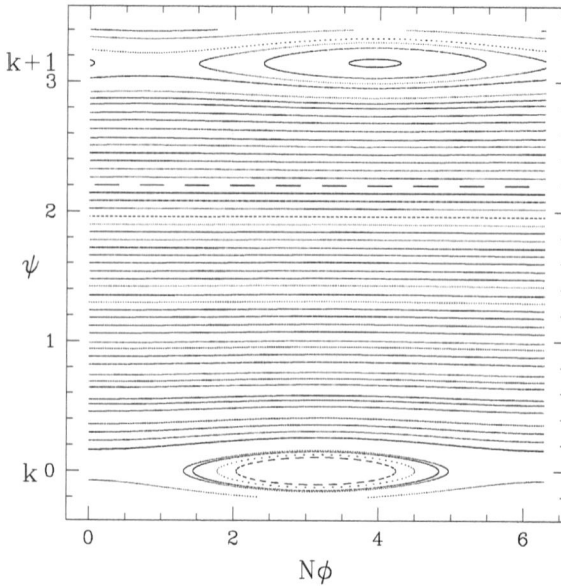

Fig. 12.10   Poincaré plot showing ripple resonances.

To estimate stochastic threshold it is necessary to calculate island widths and compare them to separation. We find for threshold due to precession motion

$$\Delta_p = \frac{2}{N\phi_p'} \left[\frac{1}{D_k^{1/4} + D_{k+1}^{1/4}} \right]^2 \qquad (12.70)$$

with $D_k = 2 + 2cos(2w_k)$, $w_{k+1} = w_k + \pi(1 + \phi_b'/\phi_p')/2$.

However, if the extent of the bounce resonances is comparable to the precession resonance spacing, the results are very different, and the approach to stochastic threshold is dominated by the generation of bounce resonance islands. In the limit $|\phi_b'/\phi_p'| \gg 1$ the map can again be analytically integrated, giving a web extending outward from each precession resonance. We then find from web overlap the stochastic threshold due to bounce motion

$$\Delta_b = \frac{2}{|N\phi_b' - N\phi_p'|}. \qquad (12.71)$$

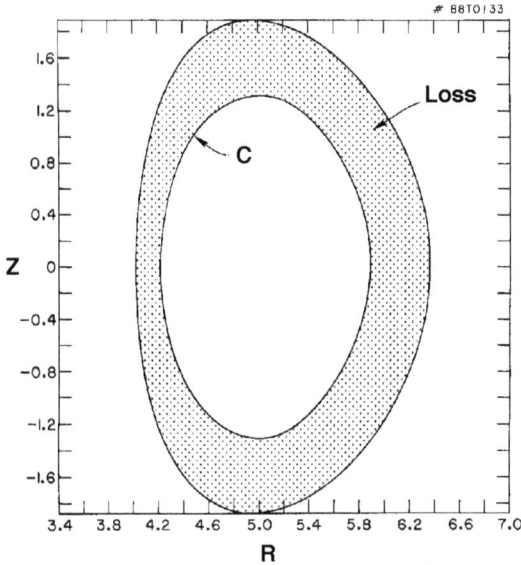

Fig. 12.11   Stochastic loss domain in INTOR.

Examination of the fixed points and islands in the banana tip map results in complete understanding of the approach to chaos. Chaotic orbits are produced either by the overlap of precession islands, or by the growth of a bounce resonance web. All resonances are due to particle precession through an integer number of field coils in one bounce. An expression for stochastic threshold is then given by the smaller of Δ_p and Δ_b, provided $N\phi'_p\Delta|1 - (\phi'_b/\phi'_p)^2| > 3\pi$, otherwise by Δ_p. The transition from one expression to the other occurs when the extent of the bounce resonance web equals the precession island width. This threshold expression gives an improvement to the estimate made by Goldson *et al.* Eq. 12.60.

Examination of 3.5 MeV fusion product alpha particle loss using guiding center codes indicates that ripple induced stochastic loss can be the dominant mechanism for loss of high energy alpha particles. In addition, since the ripple magnitude and the value of q increase with r, the condition given by Eq. 12.60 describes a loss domain. All trapped alpha particles with banana tips within this domain are lost in a time which is very short compared to the alpha particle slowing-down time. Shown in Fig. 12.11 is the stochastic loss domain for the INTOR design, with $q = 2.5$ at the

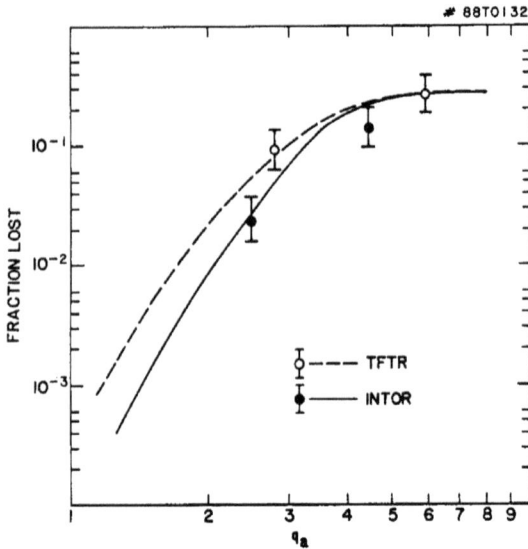

Fig. 12.12   Fraction of alpha particles lost vs q at the limiter.

plasma edge. The ripple magnitude $\delta B/B$ is 1 percent at the edge. Because of the strong q dependence of the stochastic threshold, as given by Eq. 12.60, the loss boundary C is very sensitive to q values.

A simple Monte Carlo analysis can then be used to generate an alpha particle distribution and determine the loss fraction according to the threshold criterion (White and Mynick, 1989). This allows a rapid examination of the dependence of alpha particle loss on equilibrium parameters. In Fig. 12.12 is shown for TFTR and INTOR the obtained dependence on the q profile. The strong q dependence arises from the sensitivity of the threshold condition to q and the steepness of the alpha distribution. For very large q the flattening of the loss curve reflects the fact that all trapped particles are lost.

Because trapped alpha particles are lost through small modification of the banana tip position, they make impact with the wall near the midplane, where the banana width is largest. They are fairly localized, and this is of concern because of the increased wall loading. The limitation on ripple caused by first wall loading is a more serious concern for reactor design than that given by simple energy loss considerations.

For a stellarator, ripple induced loss restricts the possible choice of reactor designs to be very close to quasi-symmetric. Otherwise excessive alpha particle loss makes wall loading too excessive and energy confinement too poor for a successful design.

12.8 DIFFUSION IN A STOCHASTIC FIELD

Consider time independent magnetic field perturbations of the form $\delta B = \nabla \times \alpha \vec{B}$, with a simple cylindrical equilibrium given by Eq. 2.97. The structure of the perturbed magnetic surfaces is dominated by the component $\delta \vec{B} \cdot \nabla \psi$, which is described in general by this perturbation.

Representing α in a Fourier series

$$\alpha = \sum \alpha_{mn}(\psi) sin(n\zeta - m\theta - \delta_{mn}), \tag{12.72}$$

we find, following the calculation of Sec. 1.7, magnetic islands produced at the rational surfaces $q = m/n$ with width

$$\delta \psi = 4q \left[\frac{\alpha_{mn}}{q'} \right]^{1/2}. \tag{12.73}$$

with $q' = dq/d\psi_p$.

First, consider diffusion of the magnetic field lines produced by these islands. Of course, if the islands do not overlap, there is no diffusion. If there is overlap there are field lines which pass from one rational surface to another. The quasi-linear result for the diffusion constant for motion in ζ is (Rosenbluth *et al.*, 1966)

$$D_Q = \sum_{n=m/q} \pi q |b_{mn}|^2 \tag{12.74}$$

with $b_{mn} = m\alpha_{mn}/r$ equal to the m, n component of $\delta B_r/B$. The sum is restricted to modes which are resonant locally, because only resonant perturbations contribute to diffusion.

It is nontrivial to numerically verify this result, and there are published papers which claim that it is wrong (Duchs *et al.*, 1991; Reidel, 1992). A correct simulation, which verified this expression, was carried out by White and Wu (1993). Care must be taken to produce a sufficiently large domain of stochastic field so that the measured evolving field lines have time to form a Gaussian envelope before reaching the bounding limits of the domain. A spectrum consisting of 24 harmonics with amplitudes chosen so that the overlap criterion, S, was independent of position, was necessary to allow sufficient time for the distribution to develop beyond the period in which short time correlations play a strong role.

In the absense of collisions and in the small gyro radius limit, particles simply stream along the field lines, so the particle diffusion is simply given by the field diffusion, using the particle velocity to transform from motion in toroidal angle ζ to time.

Even below stochastic threshold, island structure has a particularly large effect on electron transport because the effect of islands dominates over the diffusion due to the relatively small banana width of electron orbits. Each island chain effectively decreases the size of the confinement domain by allowing rapid transport across it.

Although ion and electron particle loss rates must be equal, heat loss rates are not, and magnetic field perturbations provide a possible explanation of anomalously large electron heat transport. Determinations of low-collisionality particle confinement using Monte Carlo techniques are prohibitive for large-scale systems involving very many modes. Because the diffusion rate of the electrons, for typical fusion parameters, can vary by three or four orders of magnitude with small changes in the field amplitudes, it is useful to have a means of numerically estimating the diffusion rate. Modifications of particle diffusion rates due to stochasticity and magnetic islands can be calculated to lowest order in gyro radius.

Now approximate the effect of the stochastic field by expressing the equations of motion in terms of a discrete map. Examine the drift orbit equations of Sec. 3.2 to zero order in ρ, ignoring corrections of order ϵ and the ψ dependence of α. Write $\Delta\theta = \theta - n\phi/m$ and expand q in ψ around ψ_r with $q(\psi_r) = m/n$, giving

$$\frac{d(\Delta\theta)}{d\phi} = \frac{n^2}{m^2}q'(\Delta\psi), \tag{12.75}$$

$$\frac{d(\Delta\psi)}{d\phi} = \frac{m}{\epsilon}q'\alpha_{mn}cos(m\Delta\theta). \tag{12.76}$$

Now replace these differential equations with discrete time-step equations, stepping through $\Delta\phi = 2\pi/n$, that is, $\Delta t = 2\pi/(n\epsilon\rho_\parallel)$. Substituting $x =$

$-2\pi(n/m)q\Delta\psi$, $y = m\Delta\theta$ gives the equivalent area-preserving discrete map

$$y' = y + x$$
$$x' = x + kcosy' \tag{12.77}$$

with $k = (2\pi)^2\alpha_{mn}q/\epsilon$. This has the form of the standard map of Sec. 1.8. In the limit of large k there is stochastic motion in x. The magnitude of the diffusion can be calculated by using the random phase approximation, that is, $(\delta x)^2 = k^2cos^2y \simeq k^2/2$, giving for the stochastic diffusion rate in ψ, $D_s = (\delta\psi)^2/2t$

$$D_s = \pi\left(\frac{m}{n}\right)^2\frac{n\alpha_{mn}^2|\rho_{\parallel}|}{\epsilon}. \tag{12.78}$$

For several modes this expression must be summed over resonant m and n as in Eq. 12.74, and for a distribution of particles it must be averaged over the directions of $\vec{v}$. For an isotropic velocity distribution the average is $< |\rho_{\parallel}| >= (2E)^{1/2}/\pi$.

This diffusion rate is valid only for particles streaming along the field line in one direction, *i.e.* for times short compared to the collision time. It is generally valid for example for runaway electrons, which are very collisionless.

For times long compared to the collision time, the particles move along the field line diffusively, *i.e.* $(\Delta\phi)^2 = D_{\parallel}t$. Thus $D_s \sim t^{-1/2}$ giving zero for large time; there is no cross field diffusion due to diffusive motion along the field line. However, another effect enters, as found by Rechester and Rosenbluth (1978) and Stix (1978), namely the shift of the particle to a different field line during a collision, due to the gyro radius. If the field is stochastic, nearby field lines on the average diverge exponentially, with the distance between lines given by $d = d_0e^{hz}$, z the distance along the line and h the Kolmogorov entropy, as discussed in Sec. 1.7. Then in a time $\tau_h = L_h^2/D_{\parallel}$ the particle will be carried a distance $\delta_{\perp}$ from the initial line, with $L_h = (1/h)ln(\delta_{\perp})$ and $\delta_{\perp}$ the wave length across the unperturbed field, defining the decorrelation distance of the initial field.

The particle diffusion is then

$$D = D_m v\left(\frac{\lambda_{\nu}}{L_h}\right) = \frac{D_m D_{\parallel}}{L_h}. \tag{12.79}$$

This diffusion rate has been numerically simulated using a collision operator, which simply reverses the direction of the map and displaces the trajectory in x by a distance ρ, simulating a collisional displacement by a gyro radius (Rax and White, 1992). High numerical accuracy is required, because to correctly simulate the effect, the trajectory, with $\rho = 0$, must retrace its path in the stochastic field. A numerical coefficient of $k = 0.5$ was found for the expression given by Eq. 12.79.

12.9 ISLAND INDUCED DIFFUSION

Structure in particle orbits produces diffusion through the particle collisions. A collision changes the detailed nature of the particle orbit in a time which is very short compared to the time required to traverse magnetic field scale lengths, thus producing a step across flux surfaces the size of the orbit drift excursion away from the flux surface. Even with perfect flux surfaces, the particles diffuse at a rate determined by some width Δ given, for example, by the banana width or the gyro radius, depending on the collision rate. Magnetic field structure on a scale smaller than Δ is thus irrelevant, not because the particles do not sense such structure, but because diffusion due to it is small compared to that given by Δ. Ignoring structure smaller than a critical size Δ means that the exact field consists of a sequence of topologically toroidal shells, where within each shell the field can be described as either toroidal, island, or stochastic.

The effect of magnetic islands is significant only if the island size is large compared to the magnitude of the orbit drift excursion from the magnetic surface. To estimate the effect of the islands, the particle can be considered to be following the field line. Provided the collision time is long compared to the time it takes for the particle to circumnavigate the island elliptic point, the diffusion rate due to the island can be estimated as

$$D = \frac{W^2 \nu}{2} f \qquad (12.80)$$

with W the island width and f the fraction of particles with sufficiently long collision time. The time required to move around an island is $t_I = \pi q_I / \epsilon \rho_\|$ and thus the fraction of particles f is given by all those with $|\rho_\|| > \pi \nu q_I / \epsilon$.

These results can be used to estimate the diffusion rate over the whole domain. In Fig. 12.13 is shown the local diffusion rate as a function of ψ, normalized to the neoclassical rate, for 1 keV electrons, in the field shown in Sec. 1.7. The electrons have banana width $\Delta \psi = 3 \times 10^{-4}$, small com-

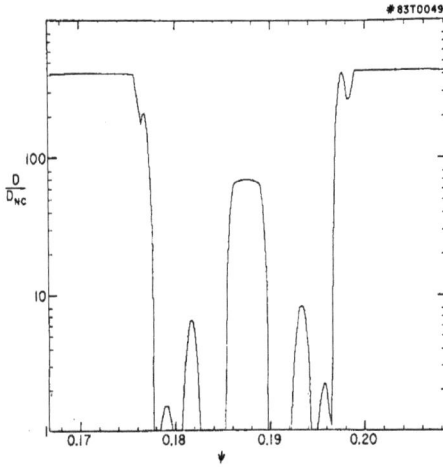

Fig. 12.13   Neoclassical diffusion with field perturbations.

pared to the islands of the first three Fibonacci orders. For this energy and collisionality the major island chains and stochastic domains have effectively infinite diffusion rates, and the higher-order Fibonacci islands also contribute significantly. The situation is considerably different for 1 keV deuterium ions, with banana width $\Delta\psi = 2 \times 10^{-2}$. Even the large islands have small effect on the diffusion, and higher-order islands and stochasticity are irrelevant.

Assuming steady-state diffusion, the local density gradient is inversely proportional to the local diffusion rate, and an average diffusion rate for the whole domain is given by

$$\frac{1}{D^*} = \int d\psi \frac{1}{D(\psi)}. \tag{12.81}$$

This gives for 1 keV electrons in this field, $D^* = 4D_{NC}$. Values of D^* obtained in this manner are shown by the solid lines in Fig. 12.14. The modification of the diffusion over neoclassical is shown as a function of perturbation amplitudes. The field perturbations taken were those of Fig. 1.5. Good agreement with Monte Carlo determinations (points) is obtained below stochastic threshold, $S = S_c$, where the main assumption involved in the calculation, that is, that the different domains form distinct toroidal bands, is well satisfied. Above stochastic threshold, however, the Poincaré

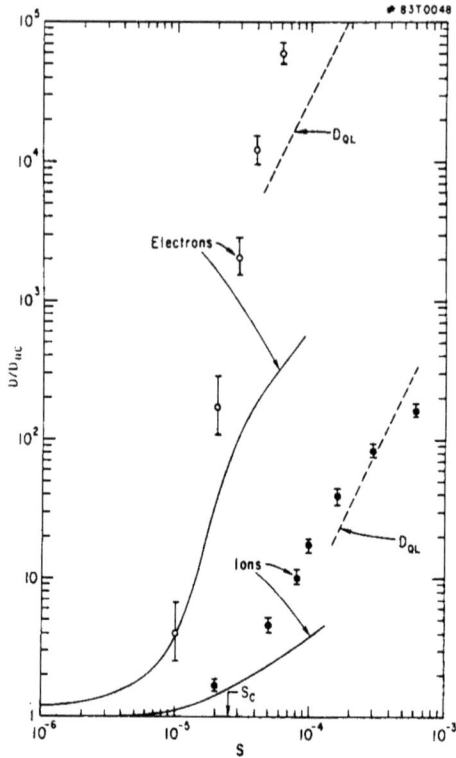

Fig. 12.14   Diffusion vs field perturbation amplitude S.

plot takes on the form of a stochastic sea containing isolated islands. In this configuration the diffusion rate across the islands is slower than that in the stochastic sea, but particles can easily diffuse around them, and thus their effect is over-estimated by this analysis. The behavior of the diffusion near stochastic threshold in the absence of collisions has also been examined numerically (White, 1987) and found to be of the form

$$D \sim a(s - s_c)^3, \tag{12.82}$$

i.e. to exhibit the same threshold behavior as the standard map (Sec. 1.8).

A confined plasma will quickly develop an electric potential, which makes the diffusion ambipolar, and the correct particle diffusion rate must be determined by the solution of this self-consistent problem. However, the potential will have little effect on the high-energy particles responsible for heat conduction. Thus, the results obtained here, neglecting electric fields, are of interest in the analysis of the anomalously large electron thermal conductivity.

The modification of the drift orbits due to electric perturbations is similar to that due to magnetic perturbations, as is obvious from the guiding center equations derived in Sec. 3.2. Perturbations $\delta\Phi_{mn}$ produce islands in the drift orbit trajectories at $q = m/n$, even though the magnetic surfaces are unchanged. Thus the preceeding analysis of the diffusion applies equally well to electric fluctuations.

Whether the anomalous energy confinement times observed in tokamaks is partly described by the effects of magnetic or electric fluctuations cannot be answered without a detailed knowledge of the field fluctuation spectrum. As seen in Fig. 12.14 the electron diffusion rate is very sensitive to the proximity to stochastic threhold.

12.10   ANOMALOUS TRANSPORT

The observed transport in toroidal devices is much larger than values given by neoclassical theory, and is associated with the existence of short wavelength turbulence. Electron heat transport is typically almost two orders of magnitude above neoclassical, and at high densities ion heat transport is also anomalous. Particle transport is typically a few times the ion neoclassical rate. Different instabilities are undoubtedly responsible for these effects in different parts of a discharge and in different regimes of operation. Presently no first principles theory exists which is capable of predicting the observed transport, although the linear theory of the microinstabilities involved has been obtained (Tang, 1978).

Particle transport is described by an averaged continuity equation of the form

$$\frac{\partial}{\partial t} \langle n \rangle = -\frac{\partial}{\partial x} \Gamma \tag{12.83}$$

where the average is over a flux surface. A complete theory would allow the calculation of the flux Γ. Consider the continuity equation, Eq. 1.8, for a single species

$$\frac{\partial}{\partial t} n = -\nabla \cdot (n\vec{v}).$$ (12.84)

Assume there are no magnetic fluctuations, the effect of magnetic perturbations having been treated in Sec. 12.8. Averaging Eq. 12.84 over the
magnetic flux surface we find

$$\Gamma = \langle \delta n \delta v_x \rangle$$ (12.85)

where x is the direction across the flux surfaces. Thus to predict the transport due to fluctuations in density and velocity, the amplitudes of these
fluctuations as well as their correlation must be known, which depends on
delicate nonlinear balance in the turbulent state. If the transport is local
the flux can be written in terms of a local diffusion coefficient through

$$\Gamma = -D \frac{\partial \langle n \rangle}{\partial x}.$$ (12.86)

The prevalent theoretical opinion is that for low beta discharges the
dominant contribution to the electron heat transport χ_e is due to electrostatic drift waves, and in particular to the ion temperature gradient (ITG)
instability (Tang and Rewoldt, 1994; Romanelli *et al.*, 1991). These modes
are the dissipative and collisionless drift instabilities and are driven by temperature gradients and by trapped particle effects. In the presence of small
electric field perturbations the particles oscillate at the diamagnetic frequency. To lowest order the ions and electrons $\vec{E} \times \vec{B}$ drift together, and
there is no instability. If the two species are driven out of phase by dissipative processes, such as collisions and wave particle resonances, the particles
can give energy to the waves and destabilize them. The transport is caused
by the associated $\vec{E} \times \vec{B}$ motion. Simple estimates using mixing length formulae, $\chi_e \sim \gamma/k^2$ with the mixing length $1/k$ given approximately by the
gyro radius and γ the linear growth rate, which typically has a magnitude
of a fraction of ω_{*e}, are not very successful in predicting the turbulence
level. What is missing is a complete knowledge of the nonlinear saturation
amplitudes. These instabilities produce naturally two of the important effects observed in Ohmic discharges, neo-Alcator scaling, and saturation of
the energy confinement time at high density. Models giving these results
will be discussed in Sec. 12.11.

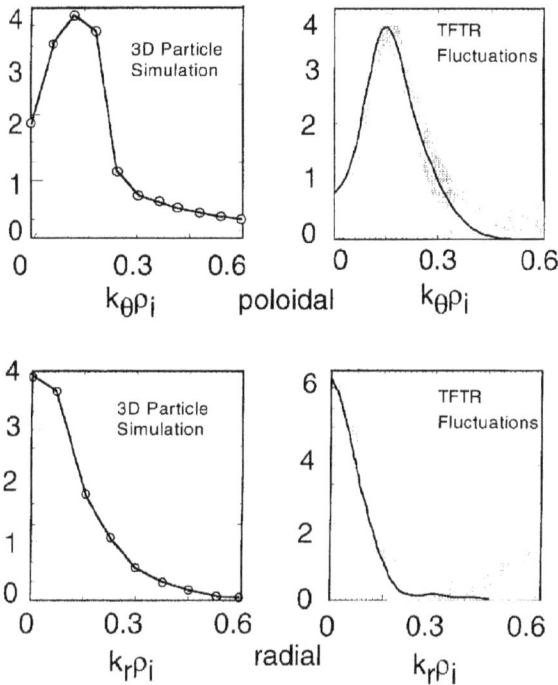

Fig. 12.15   Fluctuation spectrum.

Theoretical understanding of this turbulent transport is being obtained through the use of large scale gyrokinetic and gyrofluid simulations by several groups. See Lee (1983); Hammett *et al.* (1993); Parker *et al.* (1993); Waltz *et al.* (1994), Dimits *et al.* (1996); Lin *et al.* (1998). In Fig. 12.15 are shown the experimental and numerically obtained spectrum for drift wave turbulence. The fully three dimensional toroidal electrostatic gyrokinetic simulations (Parker *et al.*, 1993) of the ion temperature gradient instability reproduce reasonably well the experimental fluctuation spectrum both in the poloidal and the radial directions. The experimental results were obtained using beam emission spectroscopy on TFTR (Fonck *et al.*, 1993). In the simulations the ions are fully gyrokinetic, including trapped particles, and the electrons are treated adiabatically.

These instabilities, however, fail to describe the poor confinement observed near the plasma edge. No good model exists yet for the edge region, although a few attempts have been made. Microtearing modes, perhaps nonlinearly destabilized, are a candidate for this region. Rippling modes also possibly play a role in the plasma periphery. Models thus far simply use ad hoc large transport in the domain with $q > 2$ and are reasonably successful. It is very possible that the edge region has sufficient stochastic field perturbations that all orbits are lost; it is simply a region of no confinement.

At high beta, resistive ballooning modes are thought to play a role. Destabilization requires a critical pressure gradient, and thus these modes could determine the form of optimized profiles. A calculation of the renormalized nonlinear response of a plasma to an unstable spectrum of resistive ballooning modes was calculated by Carreras *et al.* (1983). They found that the dominant nonlinear effect is due to the pressure convective nonlinearity. The saturated mode level was calculated and used to estimate the anomalous transport. It was shown that the dominant loss channel is electron heat conduction. The electron thermal conductivity has the form

$$\chi_e \simeq \frac{3}{2}(v_T a)\frac{q}{S}\left(\frac{\beta R q^2}{a s L_p}\right)^{3/2},\qquad(12.87)$$

with L_p the pressure gradient scale length, $S = T_R/T_A$, and s the shear. This expression was found to give a reasonable fit to the experimental χ_e profiles in high power neutral beam injection plasmas where values of β_p are near the ballooning threshold.

Some progress has been made by making use of the empirically observed fact that the form of plasma temperature profiles is rather insensitive to discharge parameters, the so-called principle of profile consistency (Coppi, 1980). Models of anomalous transport for ohmically heated plasmas which give reasonable agreement have been constructed by Perkins (1984), Tang (1986), Romanelli *et al.* (1986), and Dominguez and and Waltz (1987). The discharge is divided into three domains spatially. Heat transport inside the $q = 1$ surface is assumed dominated by the sawtooth mode, which keeps the temperature essentially flat. In the second domain, between the $q = 1$ and the $q = 2$ surface, the electron heat transport is taken to be dominated by the dissipative trapped electron modes, and profile consistency is used to determine the radial dependence. The local magnitude of the thermal transport coefficient is approximated by $\chi_e = \gamma/k^2$ with γ the growth rate and k the perpendicular wave number of the mode. The obtained expression for the electron heat transport produces neo-Alcator scaling. The ion thermal transport in this second domain is taken to be neoclassical (Chang and Hinton, 1982) plus an anomalous conductivity given by the ITG, unstable when $\eta_i = dln(T_i)/dln(n)$ is greater than one. The ion thermal conductivity becomes important either at large density, where the neoclassical transport is large, or for relatively flat density profiles, when the η_i mode becomes important.

Finally, a semi-empirical model for transport in the domain $q > 2$ is assumed. Either an ad hoc deterioration of confinement is assumed, or without specifying the mechanism, the experimental observation of Gaussian temperature profiles is used to apply the model for the electron heat transport also outside the $q = 2$ surface. The anomalously large transport in the edge region is relevant only in describing the high q discharges. The reasonable success of these models gives some confidence that the dominant contributions to anomalous transport come from these electron and ion microinstabilities, but such models are still far from giving a derivation of the heat loss rate believable enough to extrapolate to fusion reactor parameters.

Experimentally there have been observed enhanced confinement regimes in tokamaks (Mazzucato, 1996; Burrell, 1997; Synakowski, 1997; Shirai, 1998; Wang, 1998), which are understood to be due to turbulence suppression by poloidal $\vec{E} \times \vec{B}$ induced plasma flow. The poloidal flow has small radial scale structure, and is responsible for breaking up the large scale structures in the fluctuation potential and thus regulating the radial transport. In DIII-D, discharges achieved the theoretical minimum neoclassical level of transport (Stambaugh, 2000). It has been shown analytically that an asymptotic residual flow develops in the presence of linear flow damping (Rosenbluth and Hinton, 1998). The development of this flow has been reproduced in gyrokinetic simulations (Lee, 1983) of microturbulence in toroidal geometry, and it is observed that the turbulence driven zonal flows can substantially reduce turbulence (Lin, 1998). These simulations are producing insight into the nonlinear state of turbulent transport in tokamaks, and may lead to means of regularly producing enhanced confinement states. The stabilizing influence of very small or even negative shear has led to the exploration of configurations with large q on axis, referred to as an advanced tokamak configuration. However, there are associated restrictions due to the strong q dependence of stochastic ripple loss, see Sec. 12.7, and difficulty of thermal equilibrium, see Sec. 12.13, associated with these configurations.

12.11   CONFINEMENT SCALING

It is useful to find empirical scaling laws for energy confinement in tokamaks, for the purpose of predicting capabilities of new devices, as well as a means of testing predictions of models for the mechanisms producing losses exceeding the neoclassical rates. The first scaling law, obtained by analyzing low density Ohmic discharges, was neo-Alcator scaling (Hugill, 1983)

$$\tau_E \simeq \bar{n} a R^2 \qquad (12.88)$$

where $\bar{n}$ is the line average density, a the minor radius, and R the major radius. This result was, however, not found to hold at higher densities, where τ_E saturated and even decreased with increasing density. Initially the saturation was ascribed to neoclassical ion thermal losses, which become important at high density. However, further experiments showed that often the density at saturation was too low for the losses to be explained in this manner, and other mechanisms were sought. This effort is hampered on the one hand by the fact that little is known experimentally concerning the nature of the fine scale fluctuations present in the plasma, and presumably responsible for the anomalous transport, and on the other hand by the fact that, while many microinstabilities exist as candidates to explain anomalous transport (see Tang, 1978), their nonlinear behavior is poorly understood theoretically.

Connor and Taylor (1977) have investigated the restrictions placed on the form of possible confinement scaling laws by the theoretical framework assumed. They note that if the basic equations of plasma behavior are invariant under a group of transformations, then any scaling law derived from them must have the same invariance. They consider a number of basic plasma models and for each model determine the constraints which are placed on the scaling laws by the invariance properties of the model. As the simplest example, they consider the collisionless Vlasov model. The plasma distribution is described by Eq. 12.3 with $C(f) = 0$, and the electric field is determined by charge neutrality. In the electrostatic limit the magnetic field is fixed. Write the heat flux per unit area per unit time as

$$\vec{Q} = \int \vec{v} v^2 f(x,v) d^3 v \tag{12.89}$$

and express its magnitude in terms of a scaling law,

$$Q = \sum c_{pqrs} n^p T^q B^r a^s. \tag{12.90}$$

Three linear transformations on the independent and dependent variables f, v, x, B, t, E are found which leave the basic equations invariant. Requiring that Q be also invariant under these transformations imposes three restrictions on the exponents of Eq. 12.90, namely $p = 1$, $3p + 2q + r = 6$, and $r - s = 0$. The energy confinement time τ is proportional to naT/Q and is thus restricted to the form

$$B\tau = \left(\frac{T}{a^2 B^2} \right)^q, \tag{12.91}$$

i.e. there is only one free exponent. If in addition it is assumed that a local transport coefficient $\kappa(n, T, B)$ exists with $Q = -\kappa \partial_x T$, the scaling is completely determined to be

$$\tau \sim a^2 B/T \sim a^2/(\omega_c \rho^2) \tag{12.92}$$

with ω_c the cyclotron frequency and ρ the gyro radius, which is Bohm diffusion, thus determined to be the only local transport model consistent with the collisionless Vlasov equation in the electrostatic limit.

If instead the transport is assumed to be dominated by fine scale turbulence on the scale of the gyro radius, other scalings exist, leading to

$$\tau \sim (a^2 B/T)(a/\rho), \tag{12.93}$$

which is known as gyro-Bohm scaling.

Similar constraints on the powers appearing in Eq. 12.90 were found for collisional and collisionless plasmas at low and high beta, and for ideal and resistive MHD models. If the plasma is only ohmically heated the plasma temperature is no longer an independent variable, but is determined by the other parameters and the confinement time.

This analysis should provide useful clues, as the data base increases in size, in identifying what physical mechanisms are responsible for the

502

Transport

anomalous transport. Note that dependence of τ_E on dimensionles parameters such as the aspect ratio or the safety factor q cannot be determined in this way.

Several studies of confinement scaling have been carried out using the ever growing tokamak confinement data base. See Pfeiffer and Waltz (1979); Goldston (1984); Kaye (1985). That given by Goldston (1984) has the form, for low Z_{eff}, low radiation loss, Ohmic and sawtoothing discharges

$$\tau_E(sec) = 7.1 \times 10^{-22} n^w (cm^{-3}) a^x (cm) R^y (cm) q^z \qquad (12.94)$$

with $w = 1$, $x = 1.04$, $y = 2.04$, and $z = 0.5$. Connor and Taylor constraints for collisional or collisionless low β models are satisfied within the errors, but those given by ideal or resistive MHD models are not.

While providing some guidance for design of future devices, the scaling laws obtained still allow too much freedom to give complete confidence either in extrapolation to ignition parameters or in selecting appropriate physical models for anomalous confinement.

12.12   NONLOCAL TRANSPORT

The topic of particle transport in a magnetized plasma is still an open issue: on the one hand, comprehending transport and devising a model to analyze particle and impurity losses is a practical necessity for confinement in a fusion reactor; on the other hand, the problem of linking microscopic particle motion and macroscopic transport models is one of the oldest problems of modern physics.

The topic is of general interest for fusion devices. Transport is usually modelled as diffusive, $\Gamma = -D\nabla n$, where diffusion arises from neoclassical effects and turbulence (drift wave or magnetic). Experiments have shown the need for corrections to the diffusive model, and such corrections are often in the form of a plasma flow, $\Gamma = v \cdot n$ ("pinch" velocity). The explanations proposed for a correction term are numerous: thermodiffusion (hot particles in the core diffuse faster than cold particles in the edge, and so $v \approx \nabla T/T$); "curvature" pinch (diffusion proportional to $\nabla(n/B)$, so the ∇B introduces a term proportional to the toroidal field curvature); microscopic phenomena, such as particle trapping in a more or less rapidly varying electrostatic potential; spatial inhomogeneities of the underlying turbulence felt by particles (this reduces to thermodiffusion if the magnetic field is believed to be chaotic); electric fields due to impurity transport; and finally, collisional effects due to temperature gradients (which are also linked to thermodiffusion, given the dependence on temperature of the collision frequency).

The simplest model for nonlocal transport is given by a sandpile, with a continuous source of grains of sand. When locally the density gradient exceeds a threshold an avalanche occurs, causing a stream of sand extending a long distance below this point. Thus the transport at points below the initiation point does not depend only on the local gradient, but also on values located above it and for earlier times. Long distance, non-diffusive particle displacements are referred to as Lévy flights, and it is the statistical distributions of the time of occurance, and of the distance covered, that determine the transport. The combination of these two distributions can give rise to transport with $< r^2 > \sim t^\beta$, with different values of β. Diffusive transport has $\beta = 1$, $\beta < 1$ is subdiffusive, and $\beta > 1$ is superdiffusive.

While in almost all of the interpretations the "pinch" term can be seen as an actual flow of the plasma (i.e. a translational particle velocity averaged over a statistical ensemble), a complementary interpretation sees in the "pinch" term the correction to the first order of a transport which is inherently non-diffusive. One way to treat such behavior is to obtain new equations (Fractional Kinetic Equations, FKE) which embed the non-diffusive particle motion directly in the differential equations governing macroscopic transport.

The relation between microscopic particle trajectories and macroscopic transport coefficients (such as D, v) is still obscure, and a variety of experimental results still lack a self-consistent interpretative scheme. While corrections to a pure diffusion are almost ubiquitously used, the state-of-the-art of transport theory relates the convective part of the particle flux to spatial inhomogeneities in temperature or field curvature, to some radial dependence of the diffusion coefficient, or to particle trapping in electric potentials or chaotic magnetic fields. In particular, in the reversed field pinch (RFP) case, transport has been described by the Rechester–Rosenbluth formalism, but this approach seems to be inadequate to fit the density and impurity profiles. In particular, the magnetic field in the RFP is only slightly above stochastic threshold, not nearly chaotic enough to warrant the use of the random phase approximation needed for the derivation of quasilinear diffusion or Rechester–Rosenbluth diffusion.

In the field of RFPs, examination of ion transport in the RFX, using a guiding center code, has shown that the pinch effect is actually a manifestation of the nonlocal subdiffusive motion of particles in the chaotic field produced by saturated tearing modes (Spizzo *et al.*, 2007). Particles do not move in a diffusive way, but follow the chaotic field lines across the original equilibrium flux surfaces; the resulting flight statistics are of the Lévy type.

The RFP is unique among confinement devices in that it possesses a chaotic field which is well known and relatively stable, thus providing an excellent test of theoretical models. But in addition, the tearing mode spectrum produces a magnetic field structure that is only slightly above stochastic threshold, so a random phase approximation is not valid; there still exist long scale correlations.

A quite general nonlocal transport equation is given by Montroll (1965)

$$
\begin{aligned}
\partial_t n(r,t) = & S(r,t) + \int_0^t dt' \int_{-\infty}^{\infty} dr' p(r', r - r', t - t') n(r', t') \\
& - \int_0^t dt' n(r, t') \int_{\infty}^{\infty} dr' p(r, r - r', t - t')
\end{aligned}
\tag{12.95}
$$

where $n(r,t)$ is the density, normalized by $\int n(r,t) dr = 1$, and $S(r,t)$ is a source, $p(r', r - r', t - t')$ the probability of motion from r', t' to r, t, with $\int d\tau \int dr p(r', r - r', \tau) = 1$. The first integral consists of particles arriving at r and the second integral those that leave r. This equation captures the possibility of a long distance Lévy flight distribution in the continuous-time random walk (CTRW) probabilistic approach. The final term guarantees the conservation of the total number of particles in the absence of a source. A simple diffusion equation results by taking $p(\Delta, t) = e^{-\Delta^2/4\sigma^2} \delta(t)/(2\sigma\sqrt{\pi}t_w)$.

Model the transport by considering particle motion in the stochastic field and defining a flight to be a trajectory with one sign of the pitch, describing the motion as a sum over flights. This differs from the usual CTRW formalism, first in that there is a discrete smallest flight time, the trapped particle bounce time. At the end of each flight, in the presence of stochasticity, the particle is found to have moved some distance across equilibrium flux surfaces. Particles have limited velocity, so the distance they travel along a field line and hence the amount of stochastic wandering across equilibrium surfaces is a function of the duration of the flight. During one poloidal transit, or bounce in the case of a trapped particle, the excursion along the field is insufficient to sample the field stochasticity, so the contribution to the transport is negligible compared to neoclassical diffusion. Thus there is a minimum flight time, Δt. In addition, the field in the RFP is not at all uniform, so the probability of making such a displacement is a function of the initial position. Thus we conclude that the probability must have the general form $p(r', dr, dt)$ with dr the distance moved across the equilibrium surfaces and dt the flight time.

The Montroll equation is generally given as a simplified version of Eq. 12.95 in which the kernel $p(r', dr, dt)$ is factored into functions $p(dr)$ and $\phi(dt)$. In this case the memory function $\phi(dt)$ can be calculated in terms of a waiting time distribution $\psi(dt)$ by a recursive sum over all flights leading from t' to t. If the flight time distribution given above is taken as a waiting time distribution, the resulting memory function $\phi(dt)$ is highly oscillatory on a small scale. But this approach is incorrect for our model, as this recursive sum ignores causality, *i.e.* there is no limit in this model to the length of a flight in an infinitesimal time interval. In addition, in this case there is also a minimum time, that of a poloidal transit, below which no flight occurs. Zaslavsky (2002) has also demonstrated that factoring $p(dr, dt)$ is not correct for the standard map, web map, and other more complicated models of chaos.

Furthermore, it is necessary in our case to construct a two fluid model, consisting of the trapped and passing particle densities. This is clearly shown from the guiding center simulations. Trapped particles only diffuse, passing particles follow the stochastic field for long distances, and are the only ones to experience Lévy flights. Thus the minimum flight time is the poloidal transit time.

There is no way of discovering the distribution of the flights analytically. The magnetic field in the RFP, while known very accurately, is only slightly above stochastic threshold and very complex. The distribution of flights must be discovered using a guiding center code to follow particles initiated at a particular flux surface until the pitch changes sign. The RFP is unique in that it is a system in which the stochastic field is known and stationary, allowing a detailed analysis of nonlocal transport.

One concern regarding the construction of a flux surface averaged, one-dimensional model is the treatment of the origin. A flight starting at surface ψ and passing through the origin arriving again at surface ψ is counted as a flight of length zero. This identification is what is required for the Montroll equation, which is a one-dimensional representation of the transport.

A description of the transport in the RFP using volume averaged values for the flight distributions, consisting only of a local analysis showed (Spizzo *et al.*, 2007) that the transport is subdiffusive and due to Lévy flights, and that this produces a phenomenological pinch effect. This analysis was extended to permit global simulations in the full geometry of the RFP. The very different behavior of the trapped and passing particles leads to a two fluid model which includes effects of inhomogeneity, finite boundaries and causality, difficult if not impossible to include in a formalism based on fractional derivatives, but inherently capable of description by a generalized Montroll equation.

12.12.1   *Determination of the Lévy flight distribution*

In the following, consider only a monoenergetic single species of ions, and use units of time given by the on-axis toroidal transit time of these particles. Distances are given in units of the device minor radius.

The flight time distribution is given by the pitch angle scattering operator, given by

$$\lambda' = \lambda(1 - \nu dt) \pm \sqrt{(1 - \lambda^2)\nu dt}, \qquad (12.96)$$

where ν is the collision frequency, $\lambda = v_\parallel/v$, with $v_\parallel$ the velocity parallel to $\vec{B}$ and dt the numerical time step. This is not a simple diffusion operator, it acts in the limited space $-1 < \lambda < 1$, and reproduces the Lorentz collision operator. A numerical simulation shows that this operator determines a flight time distribution, defined as the time between changes of the sign of the pitch, of

$$\psi(t) = \frac{a}{(t_0 + t)^{1.4}}, \qquad (12.97)$$

normalized to 1 in $(0, \infty)$. Changing the collision frequency simply changes the time scale, and thus the values of t_0 and a, but not the large t behavior.

This function is truncated at small t by the trapped particle bounce frequency, assuming that this is larger than the collision frequency, *i.e.* $t_0 \ll 1/\nu \ll 1$. This is verified in the RFP, where a typical value of the bounce frequency is given by $0.66 \div 1.2 \times \tau_{\text{tor}}$ (with τ_{tor} the on-axis toroidal transit time), and the collision frequency for an ion of energy 250 eV is about $\nu_{ii}\tau_{\text{tor}} = 0.4$, corresponding to a collision time of 2.5 toroidal transits. A

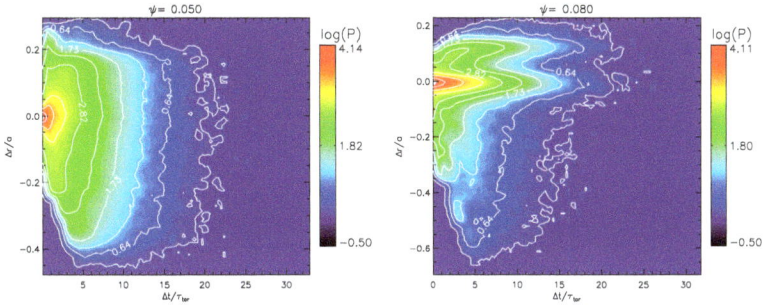

Fig. 12.16   Probability matrices: left, at $\psi_p = 0.05$ ($r/a = 0.53$), right $\psi_p = 0.08$ ($r/a = 0.74$). The vertical axis is the flight distance, and the horizontal axis is the flight time.

trapped particle changes sign of pitch during each bounce, independent of the collisions.

Exact behavior near small t is not important, since there are no flights of very short duration, it is the asymptotic large t behavior that matters. Note that $<t>$ is infinite.

The flight distance probability $p(r', r - r', t - t')$, depending on initial position and flight direction and duration to describe the Lévy flights of the passing particles, is determined using the guiding center code. Particles are launched at a particular flux surface r' and followed until the pitch changes sign, at which point the new position r and the flight time $t - t'$ are recorded. Examples are shown in Fig. 12.16 for two values of r'. Note that this method of recording the flights takes account of the local variation of the level of stochasticity, as well as the finite boundaries of the plasma. The asymmetry of these plots in the space variable and the dependence on the initial flight location indicate the variation of the degree of stochasticity of the field. Note also that the probability vanishes for large distance and small time, reflecting the causal nature of the propagation.

The probability of flight distance must be normalized through

$$\int_0^\infty d\tau \int_0^1 dr p(r', r - r'; \tau) = 1. \qquad (12.98)$$

The second term in Eq. 12.95 guarantees the conservation of the total number of particles in the absence of sources or sinks.

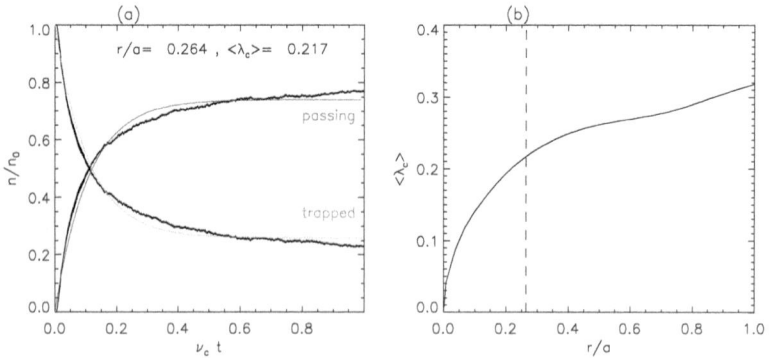

Fig. 12.17   Collisional transfer simulation: (a) an initial load of 10,000 trapped particles is evolved through the pitch angle scattering operator at $r/a \simeq 0.25$ to determine the
transfer coefficients A and B; solid lines represent the analytical fit with the solutions
of Eq. 12.100; (b) the radial profile of the (average) critical pitch $\langle \lambda_c \rangle$, which represents
the trapped particle fraction, $B/(A + B)$.

12.12.2   Two fluid Montroll equation

Take the domain of plasma to be $0 < r < 1$. Introduce passing and trapped
particle densities n_p and n_t coupled together through pitch angle scattering
and Lévy flights,

$$\partial_t n_t(r,t) = S_t(r,t) - A(r)\nu n_t + B(r)\nu n_p + D(r)\partial_r^2 n_t$$
$$+ \int_0^t dt' \int_0^1 dr' p(r', r - r'; t - t')n_p(r', t'),$$
$$\partial_t n_p(r,t) = S_p(r,t) + A(r)\nu n_t - B(r)\nu n_p + D(r)\partial_r^2 n_p$$
$$- \int_0^t dt' n_p(r,t')[\int_0^1 dr' p(r, r' - r; t - t')].$$

$$(12.99)$$

where $D(r)$ is the local particle diffusion rate due to neoclassical and classical scattering and S_p and S_t are local sources. Since a Lévy flight of a
passing particle ends because the pitch reaches zero (or nearly), at the end
of a flight the particle is trapped, hence the first integral in Eq. 12.99. The
integral in the second equation is the loss of passing particles on flights,
and gives conservation of total particle number in the absense of sources.

    To determine the coefficients $A(r)$ and $B(r)$, which are functions of
radius, we study Eq. 12.99 in the absence of the diffusion terms, flights and

sources:

$$\partial_t n_t(r,t) = -A(r)\nu n_t + B(r)\nu n_p,$$
$$\partial_t n_p(r,t) = A(r)\nu n_t - B(r)\nu n_p.$$
(12.100)

This equation can be easily solved analytically with initial conditions $n_t = n_0$, $n_p = 0$, giving $n_t/n_0 = 1/(A+B) \cdot (B + Ae^{-(A+B)\nu t})$ and $n_p/n_0 = A/(A+B) \cdot (1 - e^{-(A+B)\nu t})$, $\forall r$. The analytical solution is then used to fit the result of the scattering of 10,000 particles through the pitch angle scattering operator (12.96), defining trapped particles to be those with pitch sub-critical $\lambda < \langle \lambda_c(r) \rangle$, where $\langle \lambda_c(r) \rangle$ is the critical pitch averaged over the poloidal angle θ, at a given radius r. It is easily shown that $\langle \lambda_c(r) \rangle = \frac{B(r)}{A(r)+B(r)}$. Averages over the flux surface θ are performed to reduce the problem to a one dimensional transport simulation.

A sample result is shown in Fig. 12.17(a): the initial load of trapped particles is evolved using as threshold the value $\langle \lambda_c(r) \rangle = 0.2$, $r/a \simeq 0.25$. The fit with the analytical solutions is represented by solid lines. The final values of A and B are consistent with the trapped fraction profile in a RFP, represented by the profile of $\langle \lambda_c(r) \rangle$ shown in Fig. 12.17(b).

This two component Montroll equation was found to reproduce the sub-diffusive character of the transport in the RFP, allowing one to model the effect of the stochastic field on the particle motion. However, there is no way to derive the nature of the transport using the harmonic content of the magnetic field; it is necessary to follow particles in the field to obtain the Lévy flight distributions. Fig. 12.18 shows the outcome of a simulation integrating the Montroll equation. These simultions describe the hollow density profiles common in the RFP, due to nonlocal Lévy type transport, without corrective "pinch" velocity terms.

12.13   BURN CONTROL

As we have seen, a plasma equilibrium can be determined by the choice of the pressure and current density profiles, provided they are chosen within bounds allowing a solution to the Grad– Shafranov equation. Stability of an equilibrium to ideal and resistive MHD perturbations additionally restricts the parameter space of operation. In experiments without significant alpha particle production, where the plasma heating is supplied externally, there is some degree of control over the temperature profile and, through the

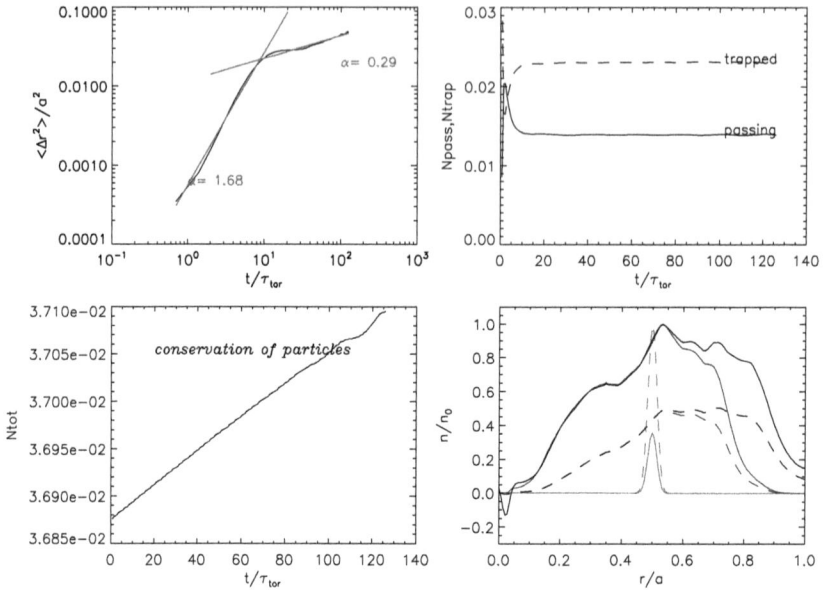

Fig. 12.18   Montroll equation results. Top left $< r^2 >$ as a function of time; top right equilibration between trapped and passing particles; botttom left particle conservation; botom right the evolution of the density profile from an initial Dirac delta, solid=passing, dashed=trapped.

conductivity, the current profile. In an ignited, burning plasma, with the heat source coming from the fusion alpha particles, there will be much less control. It is then necessary to ascertain whether the pressure and current profiles, desired from the point of view of MHD stability, will be maintained during the burn. If the reactor is pulsed, which may be the case with a tokamak since there is no efficient way to maintain the toroidal current, the burn is of relative short duration and this may not be a great problem, but if steady state is desired it must be consistent with the fusion-derived heat source. In the limit of no external control, the alpha heating power is proportional to the square of the pressure profile

$$H_\alpha(r) = cp^2(r) \qquad (12.101)$$

and radial power balance in steady state implies

$$\frac{1}{r}\frac{d}{dr}r\chi(r)\frac{dp}{dr} = cp^2(r) \tag{12.102}$$

with $\chi(r)$ the thermal conductivity. Thus the pressure profile determines the form of the conductivity profile for steady state. If the conductivity is in fact different from this the profiles will change, and the plasma may move towards a MHD unstable state. Burning plasma dynamics has been examined for the ITER design by Wang (1997) and for an advanced tokamak scenario (large q on axis with most of the toroidal current supplied by bootstrap current, see Sec. 12.4) by Moreau and Voitsekhovich (1999). The advanced tokamak configuration presents real problems for burn control in steady state. Note that this problem is completely eliminated if alpha particles are not confined and the plasma heating is supplied externally through the use of neutral beams, see Chap. 15.

12.14   Problems

1. Prove that $dS/dt = constant$, with the entropy S given by Eq. 12.4. Start with

$$\frac{\partial S}{\partial t} = \int \frac{\partial G}{\partial f}\frac{\partial f}{\partial t} d^3 x d^3 v,$$

substitute using the Vlasov equation, and integrate by parts to find the possible time evolution of $G(f)$, showing that the volume occupied by any $G(f)$ is fixed. (Rosenbluth lecture notes)

2. Using the defined entropy function and perturbation theory show that if $f = f_0 + f_1$ where $f_0 = f_0(E)$, $\partial_E f_0 < 0$ and f_1 is consistent with $dS/dt = 0$, then the perturbation f_1 raises the energy of the system.

3. Derive the diffusion coefficient in the Pfirsch–Schlüter regime using the drift equation. Consider the drift equation

$$q\frac{B_\theta}{Br}\frac{\partial f^1}{\partial \theta} - [\frac{mq}{eRBr}\partial_\theta(Rq)]\partial_r f^0 = \frac{mvq}{B}\frac{\partial}{\partial \mu}(q\mu\frac{\partial f^1}{\partial \mu})$$

with

$$f^0 \sim \frac{n(r)}{T^{3/2}}e^{-E/T}.$$

Since this equation satisfies all collision regimes one could get Pfirsch–Schlüter diffusion. In this limit

$$\frac{\partial f}{\partial t}|_{coll} \gg q\frac{B_\theta}{Br}\frac{\partial f^1}{\partial \theta}.$$

The steps one should follow to solve this problem are:

a. To lowest order find something which makes the right-hand side of the drift equation vanish, i.e. $f_0^1 = f_0^1(\theta)$ only.

b. To first order

$$\frac{mvq}{B}\frac{\partial}{\partial \mu}(q\mu)\frac{\partial f_1^1}{\partial \mu} = \frac{qB_\theta}{Br}\frac{\partial f_0^1}{\partial \theta} - \frac{mq}{eRBr}\frac{\partial}{\partial \theta}(Rq)\frac{\partial f^0}{\partial r}.$$

Solve this explicitly for f_1^1.

c. To next order one needs a solubility condition from:

$$\frac{mvq}{B}\frac{\partial}{\partial \mu}(q\mu)\frac{\partial f_2^1}{\partial \mu} = \frac{qB_\theta}{Br}\frac{\partial f_0^1}{\partial \theta} - \frac{mq}{eRBr}\frac{\partial}{\partial \theta}(Rq\frac{\partial f^0}{\partial r}),$$

where we have the expansion in inverse powers of ν

$$f_1 = f_0^1 + f_1^1(\nu^{-1}) + f_2^1(\nu^{-2}).$$

d. The solubility condition you should apply is conservation of number of particles. This is done in step c by setting

$$\int \left.\frac{\partial f}{\partial t}\right|_{coll} d^3v = 0,$$

using the proper volume in phase space. This should give a differential equation for $f_0^1(\theta)$.

e. One can then calculate the flux

$$\frac{F}{Area} = \frac{\int Rd\theta d^3v f_0^1 v_{dr}}{Area}.$$

(Rosenbluth lecture notes).

4. Consider the derivation of the bootstrap current. Write expressions for all co-moving and counter-moving banana particles passing through interval dr. Show that the difference is $-(r/R)^{1/2}w_b(dn/dr)dr$, *i.e.* it is proportional to the total density gradient, not the trapped particle density gradient.

5. Consider a hot plasma in slab geometry with an $\vec{x}$-directed gravitational field which acts on the ions, but is negligible for the electrons (due to their much smaller mass). The ions and electrons obey the respective Vlasov equations

$$\frac{\partial f_i}{\partial t} + \vec{v}_i \cdot \nabla f_i + (\frac{e}{M}\vec{E} - g\hat{x}) \cdot \frac{\partial f_i}{\partial \vec{v}} = 0,$$

$$\frac{\partial f_e}{\partial t} + \vec{v}_e \cdot \nabla f_e + (\frac{e}{M}\vec{E}) \cdot \frac{\partial f_e}{\partial \vec{v}} = 0.$$

a. Find equilibrium solutions to this pair of equations which have the property that

$$n_e = \int f_e d^3\vec{v} = n_i = \int f_i d^3\vec{v}.$$

b. Develop a formalism for computing the perturbed electron and ion distribution functions δf_e, δf_i resulting from a perturbed electric potential Φ of the form $\delta\Phi = \delta\Phi(x)cos(ky - \omega t)$.

6. The Lorentz scattering operator is given by

$$\frac{\partial f}{\partial t} = \frac{\nu}{2} \frac{\partial}{\partial \lambda} (1 - \lambda^2) \frac{\partial f}{\partial \lambda}$$

where $\lambda = v_{\parallel}/v$ is the pitch. Use this collision operator to evaluate the diffusion in the banana regime.

7. If a trapped particle is bouncing in a domain of the poloidal cross section in which ripple wells exist, it will eventually be trapped because of the outward drift, and hence increase in ripple well depth occurring during the bounce. Find in terms of particle energy and equilibrium and ripple parameters the probability of trapping, and hence the number of bounces expected before trapping occurs.

12.15 References

Classical transport

- Balescu, R., Transport Processes in Plasmas (North Holland, Amsterdam, 1988) Vol. I.
- Braginskii, S. T. in Reviews of Plasma Physics, edited by M. N. Leontovich, (Consultants Bureau, New York, 1965).
- Montgomery, D. C., and D. A. Tidman, Plasma Kinetic Theory (McGraw-Hill, New York, 1964).
- Rosenbluth, M. N., W. M. MacDonald, and D. L. Judd, Phys. Rev. 107, 1 (1957).
- Schmidt, G., Physics of High Temperature Plasmas (Academic Press, New York, 1966).

Neoclassical transport

- Artsimovich, L. A., A. V. Glukhov and M. P. Petrov JETP Letters 11, 304 (1970).
- Balescu, R., Transport Processes in Plasmas, (North Holland, 1988) Vol. II.
- Boozer A. H. and H. J. Gardner, Phys. Fluids B 10, 2408 (1990).
- Chang, Z., J. D. Callen, C. C. Hegna, E. D. Fredrickson *et al.*, Phys. Rev. Lett. 74, 4663 (1994).
- Chang, C. S., and F. L. Hinton, Phys. Fluids 259, 1493 (1982).
- Chang, C. S., and F. L. Hinton, Phys. Fluids 29, 3314 (1986).
- Galeev, A. A., and R. Z. Sagdeev, JETP 26, 223 (1968).
- Galeev, A. A., and R. Z. Sagdeev, Reviews of Plasma Physics, edited by M. N. Leontovich, (Consultants Bureau, New York 1979).
- Hinton, F. L., and M. N. Rosenbluth, Phys. Fluids 16, 836 (1973).
- Hinton, F. L., and R. D. Hazeltine, Rev. Mod. Phys. 48, 239 (1976).
- Hinton, F. L., Handbook of Plasma Physics, edited by M. N. Rosenbluth and R. Z. Sagdeev (North Holland, Amsterdam, 1983), Vol. I, p. 147.
- Hegna, C. C., Phys. Plasmas 5, 1767 (1998).
- Hirshman, S. P. and D. J. Sigmar, Nucl. Fusion 21, 1079 (1981).
- Hu, G. and J. Krommes, Phys. Plasmas 1, 863 (1994).

- Kotschenruether, M., in Proc. of the 14th International Conference on the Numerical Simulation of Plasmas Paper PT20 (Office of Naval Research, Arlington, VA, 1991).
- Parker, S. and W. W. Lee, Phys. Fluids B5, 77 (1993).
- Pytte, A., and A.H. Boozer, Phys. Fluids 24, 88 (1981).
- Raspe, R., Adventures of Baron Münchausen (London, 1785; Dover, New York, 2005).
- Rosenbluth, M. N., R. D. Hazeltine, and F. L. Hinton, Phys. Fluids 15, 116 (1972).
- Rutherford, P. H., Phys. Fluids 13, 482 (1970).
- Shaing K. C., B. A. Carreras, N. Dominguez, V. E. Lynch *et al.*, Phys. Fluids B 1, 1663 (1989).
- Ware, A. A., Phys. Rev. Lett. 25, 15 (1970).
- Watanabe I., N. Nakajima, M. Okamoto, Y. Nakamura *et al.*, Nucl. Fusion 32, 1499 (1992).

Ripple transport

- Anderson, O. A., and H. P. Furth, Nucl. Fusion 12, 207 (1972).
- Boivin, R. L., S. J. Zweben, and R. B. White, Nucl. Fusion 33, 449 (1993).
- Boozer, A.H., Phys. Fluids 23, 2283 (1980).
- Connor, J. W., and R. J. Hastie, Nucl. Fusion 13, 221 (1973).
- Goldston, R. J., R. B. White, and A. H. Boozer, Phys. Rev. Lett. 47, 647 (1981).
- Kolesnichenko, Ya. I., R. B. White, and Yu. V. Yakovenko, Phys. Plasmas 9, 2639 (2002).
- Linsker, R., and A. H. Boozer, Phys. Fluids 25, 143 (1982).
- Mynick, H. E., Nucl. Fusion 26, 491 (1986).
- Redi, M. H., M. C. Zarnstorff, R. B. White, R. V. Budny *et al.*, Nucl. Fusion 35, 1191 (1995).
- Redi M. H., R. V. Budny, D. S. Darrow, H. H. Duong *et al.*, Nucl. Fusion 35, 1509 (1995)
- Stringer, T. E., Nucl. Fusion 12, 689 (1972).
- White, R. B., A. H. Boozer, R. J. Goldston, R. Hay *et al.*, in Proc. of the Ninth International Conference on Plasma Physics and Controlled Nuclear Fusion Research, Baltimore, 1982 (IAEA, Vienna, 1983).

- White, R. B., M. S. Chance, J. L. Johnson, H. E. Mynick, *et al.*, Proc. of the 12th. International Conference on Plasma Physics and Controlled Nuclear Fusion Research, Nice, 1988 (IAEA, Vienna, 1988).
- White, R. B., R. J. Goldston, M. H. Redi and R. V. Budny Physics Plasmas 3, 3043 (1996).
- White, R. B., Phys. Rev E 58, 1774 (1998).
- White, R. B., Transport in Systems with Destroyed Magnetic Flux Surfaces, in Proc. of the Cargese Workshop, edited by D. Gresillon and M. A. Dubois (Editions de Physique, Orsay, France, 1987).
- White, R. B., Phys. Rev E 58, 1774 (1998).
- White, R. B. and H. E. Mynick, Phys. Fluids B 1, 980 (1989).
- Wu, Y. and R. B. White, Phys. Fluids B5, 3291 (1993).
- Yushmanov, P. N., Nucl. Fusion 22, 315 (1982).
- Yushmanov, P. N., Nucl. Fusion 23, 1599 (1983).

Anomalous transport

- Burrell, K. H., Phys. Plasmas 4, 1499 (1997).
- Carreras, B. A., P. H. Diamond, M. Murakami, J. L. Dunlap *et al.*, Phys. Rev. Lett. 50, 503 (1983).
- Dimits, A. M., T. J. Williams, J. A. Byers, B. I. Cohen, Phys. Rev. Lett. 77, 71 (1996).
- Dominguez, R., and R. E. Waltz, Nucl. Fusion 27, 65 (1987).
- Duchs, D., Nucl. Fusion 17, 565 (1977).
- Fonck, R. J., Phys. Rev. Lett. 70, 3736 (1993).
- Hammett, G. W., Plasma Phys. Control. Fusion 35, 973 (1993).
- Krommes, J. A., in Basic Plasma Physics II, edited by A. A. Galeev and R. N. Sudan (North Holland, Amsterdam, 1984) Chap. 5.5.
- Lee W. W., Phys. Fluids 26, 556 (1983).
- Leonov, V. M., in the Proc. of the 8th International Conference on Plasma Physics and Controlled Nuclear Fusion Research, Brussels, 1980 (IAEA, Vienna, 1981).
- Liewer, P. C., Nucl. Fusion 25, 543 (1985).
- Lin, Z., T. S. Hahm, W. W. Lee, W. W. Tang *et al.*, Science 281, 1835 (1998).
- Mazzucato E., S. H. Batha, M. Beer, M. Bell *et al.*, Phys. Rev. Lett. 77, 3145 (1996).
- Molvig, K., S. Hirschman, Phys. Rev. Lett. 43, 582 (1978).

- Ohkawa, T., Phys. Letters 67A, 35 (1978).
- Parker, S. E., W. W. Lee, and R. A. Santoro, Phys. Rev. Lett. 71, 2042 (1993).
- Perkins, F. W., Proc. of the 4th International Symposium on Heating in Toroidal Plasmas (Rome, 1984), 2, 977.
- Romanelli, F., W. M. Tang, and R. B. White, Nucl. Fusion 26, 1515 (1986).
- Romanelli, F., L. Chen, and S. Briguglio, Phys. Fluids B 3, 2496 (1991).
- Rosenbluth, M. N., and F. L. Hinton, Phys. Rev. Lett. 80, 724 (1998).
- Shirai, H., Phys. Plasmas 5, 1712 (1998).
- Stambaugh, R. D., Phys. Plasmas, submitted (2000).
- Synakowski, Phys. Rev. Lett. 78, 2972 (1997).
- Tang, W. M., Nucl. Fusion 26, 1605 (1986).
- Tang, W. M., and G. Rewoldt, Phys. Fluids B 5, 2451 (1993).
- Waltz, G. D., G. D. Kerbel, J. Milovich, Phys. Plasmas 1, 2229 (1994).
- Wang, G., Phys. Plasmas 5, 1328 (1998).

Transport in stochastic fields

- Duchs, D. F., A. Montvai, and C. Sack, Plasma Phys. Control. Fusion 33, 919 (1991).
- Krommes, J. A., C. Oberman, and R. G. Kleva, J. Plasma Phys. 30, 11 (1983).
- Rax, J. M., and R. B. White, Phys. Rev. Lett. 68, 1523 (1992).
- Rechester, A. B., and M. N. Rosenbluth, Phys. Rev. Lett. 40, 38 (1978).
- Reidel, K., Phys. Fluids B 4, 299 (1992).
- Stix, T. H., Nucl. Fusion 18, 353 (1978).
- White, R. B., in Statistical Physics and Chaos in Fusion Plasma, edited by C. W. Horton and L. E. Reichl, (John Wiley, New York, 1984).
- White, R. B., in Magnetic Reconnection and Turbulence, edited by M. A. Dubois, D. Gresillon and M. N. Bussac (Les Editions de Physique, Orsay, France, 1985).
- White, R. B. and Yanlin Wu., Plasma Phys. Control. Fusion 35, 595 (1993).

Microinstabilities

- Liewer, P. C., Nucl. Fusion 25, 543 (1985).
- Tang, W. M., Nucl. Fusion 18, 1089 (1978).

Confinement scaling

- Connor, J. W., and J. B. Taylor, Nucl. Fusion 17, 1047 (1977).
- Goldston, R. J., Plasma Physics 26, 87 (1984).
- Hugill, J., Nucl. Fusion 23, 331 (1983).
- Kaye, S. M., Phys. Fluids 28, 2327 (1985).
- Pfeiffer, W., and R. E. Waltz, Nucl. Fusion 19, 51 (1979).

Profile consistency

- Biskamp, D., Comments on Plasma Physics and Controlled Fusion, Vol. X, 165 (1986).
- Coppi, B., Comments on Plasma Physics and Controlled Fusion, 5, 261 (1980).

Nonlocal Transport

- Del Castillo-Negrete, D., B. A. Carreras and V. E. Linch, PRL 91, 018302 (2003).
- Montroll, E. W., J. Math. Phys. 6, 167 (1965).
- Spizzo, G., R. B. White and S. Cappello, Phys. of Plasmas, 14, 102310 (2007).
- Spizzo, G., R. B. White, S. Cappello, and L. Marrelli, Plasma Phys. Control. Fusion 51, 124026 (2009).
- Zaslavsky, G. M., Phys. Rep. 371, 461 (2002).

Burn Control

- Moreau, D. and I. Voitsekhovich, Nuclear Fusion, 5, 685 (1999).
- Wang, J. F., T. Amano, Y. Ogawa, *et al.* Fusion Technology 32, 590 (1997).

Chapter 13

Collisions and Resonances

13.1 INTRODUCTION

Particle resonances, as analyzed in Chapter 4 through Poincaré sections and discussed in Chapter 5 regarding their location are modified in the presence of particle collisions. There are two main effects, discussed in this chapter. First, collisions significantly broaden the extent of a resonance. Background turbulent fields act to diffuse particle orbits in such a way as to increase the effective resonance width. In principle, any mechanism that leads to resonant particle phase randomization such as pitch angle scattering should contribute to the broadening.

Secondly, collisions replenish the gradients driving an instability, so mode saturation is affected by the collision rate. Scattering leads to non-resonant particles being kicked into resonance while kicking resonant particles out of resonance. Therefore, intuitively, scattering increases the effective resonance extension and also replenishes the local gradients, extending the mode growth.

13.2 PARTICLE DYNAMICS

The unperturbed orbits in an axisymmetric tokamak are uniquely determined by values of toroidal canonical momentum P_ζ, energy E and magnetic moment μ. The guiding center dynamics formalism we employ exploits the particle actions, which are invariants of the unperturbed motion, as the relevant variables to describe the effects of the perturbation. For the interaction of energetic particles with low-frequency Alfvénic modes, μ

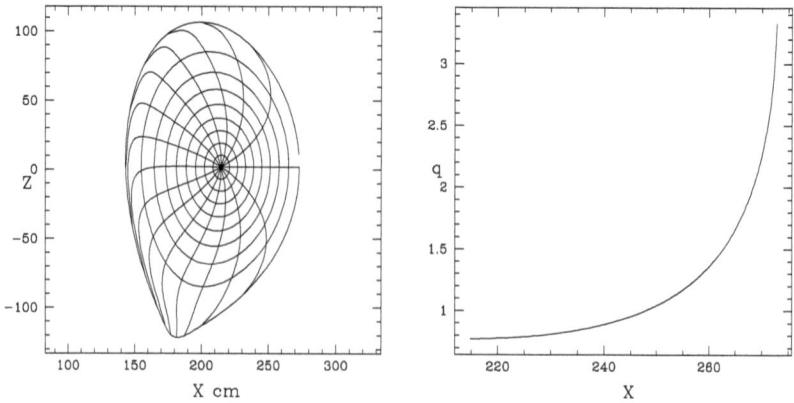

Fig. 13.1   Equilibrium and q profile using Boozer coordinates.

is approximately conserved and the other two actions are the toroidal and poloidal canonical momenta, defined as $P_\zeta = g\rho_\parallel - \psi_p$ and $P_\theta = \psi + \rho_\parallel I$, respectively, where ψ and ψ_p are the toroidal and poloidal fluxes, $d\psi/d\psi_p = q(\psi_p)$ is the field line helicity. $g(\psi_p)$ and $I(\psi_p)$ are the toroidal and poloidal covariant components of the equilibrium magnetic field and $\rho_\parallel$ is defined as the parallel velocity divided by the strength of the magnetic field. Hamilton's equations are

$$\dot{\theta} = \frac{\partial H}{\partial P_\theta} \qquad \dot{P_\theta} = -\frac{\partial H}{\partial \theta}$$
$$\dot{\zeta} = \frac{\partial H}{\partial P_\zeta} \qquad \dot{P_\zeta} = -\frac{\partial H}{\partial \zeta}, \tag{13.1}$$

where the poloidal and toroidal angles are θ and ζ. The Hamiltonian is $H = \rho_\parallel^2 B^2/2 + \mu B$.

13.2.1   *EQUILIBRIUM CHOICE*

We use a tokamak equilibrium with a field on axis of 4.9 kG. The major radius was 215 cm and the minor radius 62 cm. The equilibrium and q profile are shown in Fig. 13.1. The perturbation was an Alfvén mode with a simple broad Gaussian radial form. The particle distribution consisted of 85 keV to 98 keV deuterium ions.

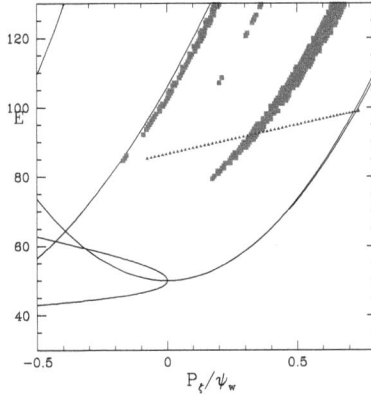

Fig. 13.2 The plane of E, P_ζ for $\mu B_0 = 50 keV$, showing an isolated resonance in the co-passing domain, extending from 80 keV upwards, for amplitude $A = 4 \times 10^{-5}$, with a frequency 130 kHz, and poloidal and toroidal mode numbers $m/n = 6/5$. The red squares in a nearly vertical strip in the center is the resonance, and the particle distribution is along the near horizontal line cutting across it near the plasma center. The right edge of this line is at the magnetic axis, and the left end is near the plasma edge. ψ_w is the poloidal flux at the plasma edge.

The Alfvén mode is chosen with a given toroidal mode number n and with an eigenfrequency ω. The Alfvén velocity is a function of the plasma density and our choice for the toroidal and poloidal mode numbers is to have the mode located near the $q = 1$ surface. We have chosen the equilibrium to possess a strong well isolated resonance near the plasma center.

Assuming the Hamiltonian to be a function of ψ_p, θ and $n\zeta - \omega t$, we have from $\dot{P}_\zeta = -\partial_\zeta H$, and $\dot{E} = \partial_t H$ that there is a constant of the motion given by

$$K = \omega P_\zeta - nE, \qquad (13.2)$$

simply related to the particle energy in the frame rotating with the mode. The kinetic Poincaré recurrence plots are produced at constant values of μ and K. Those plots involve either P_ζ or E as a function of θ on the orbit. Each point in the Poincaré plot is obtained when the recurrence condition $n\zeta - \omega t = 2\pi k$, k integer is satisfied.

13.2.2   *PARTICLE POPULATION*

A passing population was chosen with all particles having the same value of magnetic moment μ and K. The equations of motion are easily generalized to include flute-like perturbations of the form $\delta\vec{B} = \nabla \times \alpha\vec{B}$ with $\vec{B}$ the equilibrium field and $\alpha = \sum_{m,n} \alpha_{m,n}(\psi_p)sin(n\zeta - m\theta - \omega_n t)$. The perturbation α has units of a length, simply related to the cross field ideal displacement produced by the mode. A well isolated resonance was found for $\mu B_0 = 50$ keV with B_0 the on-axis field strength, with a perturbation with a single harmonic with poloidal and toroidal mode numbers of $m/n = 6/5$, and a frequency of 130 kHz. The plane of E, P_ζ in Fig. 13.2 shows a radially isolated resonance extending from 80 keV upwards in the co-passing domain, and the particle distribution is shown as a line at fixed K intersecting the resonance approximately at mid radius, for $P_\zeta = 0.31\psi_w$ and $E = 93$ keV. The mode can move particles only along this line, not across it. The amplitude A is the maximum value of the ideal eigenfunction of the perturbation, normalized to the major radius of 210 cm.

13.3   COLLISIONAL BROADENING

The width of an isolated linear resonance, as observed with a kinetic Poincaré plot, is proportional to the square root of the perturbation amplitude. If the mode structure does not change considerably within the resonance and the perturbation amplitude is small enough to prevent significant higher-order Fibonacci resonances, the dynamics of a resonant particle can be described by a nonlinear pendulum framework. See Chapter 4. In this case, the maximum resonance width in terms of frequency $\Delta\Omega$ can be shown to be $\Delta\Omega = 4\omega_b$, where ω_b is the bounce (or trapping) frequency.

The line broadened quasilinear (QL) model, relies on tuning the constant coefficients a and c of the resonance broadening

$$\Delta\Omega = a\omega_b + c\nu_{eff} \tag{13.3}$$

in such a way that the evolution of single modes meet the expected saturation levels resulting from analytic theory in both limiting cases of very close to and very far from linear marginal stability. In the presence of collisions Poincaré plots become blurred and therefore cannot be used as a diagnostic tool for the resonance width. For that purpose, instead, we perform the

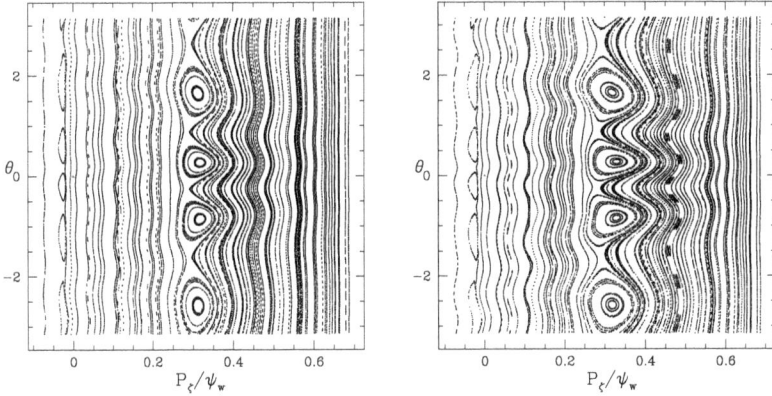

Fig. 13.3 Poincare plots for A $= 10^{-4}$ and A $= 2 \times 10^{-4}$ without collisions. The mode frequency is 130 kHz, with m/n $= 6/5$. The mode is chosen to have a large region of relatively undisturbed KAM surfaces on each side of the resonance.

analysis by means of density plots, using the flattening as an indication of the resonant extension.

To examine the effect of collisions on the resonance associated with a fixed amplitude perturbation, a particle distribution is initiated with a range of ψ_p, energy and P_ζ, all with the same values of μ. Particles are launched along a value of constant K, with a strong density gradient in P_ζ, corresponding to a gradient in minor radius, and in energy E. The profile is produced using standard Monte Carlo methods. We examine the modification of the resonance width in P_ζ.

In Fig. 13.3 are shown the Poincaré plots for the two amplitudes used for our examination of profile flattening. The mode was chosen to produce an isolated resonance near the center of the distribution with wide ranges of unperturbed KAM surfaces on each side of the resonance.

The collisions are energy conserving pitch angle scattering, and directly change the value of P_ζ through $\rho_\parallel$. The gradient must be maintained in time by re-injecting lost particles using the Monte Carlo procedure. The action of the resonance produces a modification of the particle density $n(P_\zeta, t)$, consisting of a local flattening in the vicinity of the resonance at $P_\zeta = P_{res}$ as shown in Fig. 13.6.

The modifications of the density profile are shown in Fig. 13.4. Six hundred thousand particles are launched using standard Monte Carlo methods

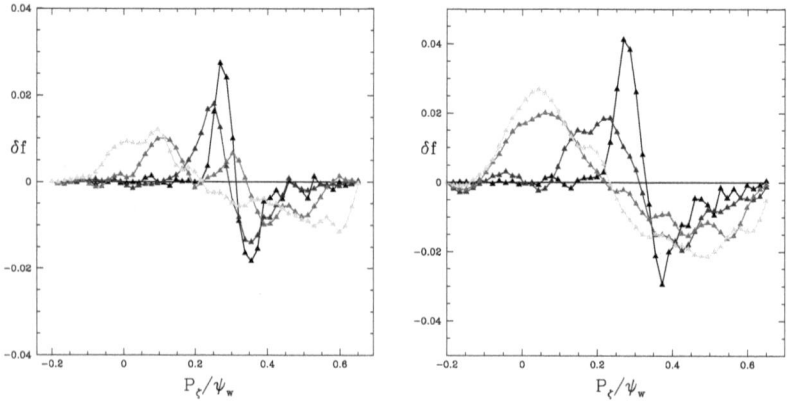

Fig. 13.4 Left: Density profile modification with perturbation $A = 10^{-4}$. The collisionality values shown are $\nu_\perp = 0$ (black), 0.6 Hz (blue), 6 Hz (red), and 22 Hz (green). Right: For $A = 2 \times 10^{-4}$ shown are $\nu_\perp = 0$ (black), 2Hz (blue), 15 Hz (red) and 22 Hz (green).

to produce the initial density profile. After a delay to allow phase mixing, a time average of the density using five hundred toroidal transits is taken in each of the 50 bins in P_ζ, to reduce statistical noise.

We choose to show δf, the difference between the distribution with and without the presence of the mode, giving a much clearer picture of the effect of the resonance. However, for each value of the collision rate, the density profile is modified in time also in the absence of the mode so a determination of the collisional mode free density profile must be made using the same time interval as used in the presence of the mode, and the unperturbed density profile subtracted from the perturbed one to give δf. Note that in these plots, density flattening appears as a negative slope. Thus with no collisions, the negative slope occurs only between the edges of the Poincaré resonance. Outside, on the two sides of the resonance, the slope of δf is in fact positive. Collisionless flattening of the density profile occurs only within the island structure, due to the bounce frequency mixing. Outside the resonance island, the KAM surfaces are distorted outward, as seen in Fig. 13.3, leading to a local increase in the density, visible as a positive slope in the plot of δf in Fig. 13.4. This increase is adiabatic, since there is no mixing occuring in these distorted surfaces, and the density and particle energy will return to their initial values upon decay of the mode.

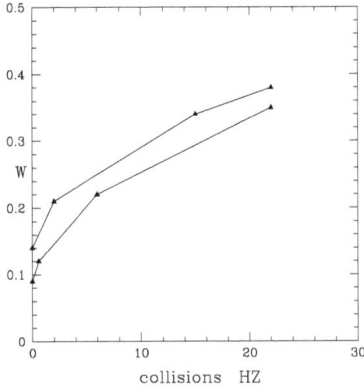

Fig. 13.5 A plot of resonance vs collision frequency using the data from the plots of Fig. 13.4. The upper points are for $A = 2 \times 10^{-4}$ and the lower ones for $A = 10^{-4}$. The accuracy of the points is very poor, because of the difficulty of reading a precise value.

The collisionality values shown are, for $A = 10^{-4}$ the values $\nu_\perp = 0$ (black), 0.6 Hz (blue), 6 Hz (red), and 22 Hz (green). For $A = 2 \times 10^{-4}$ shown are $\nu_\perp = 0$ (black), 2 Hz (blue), 15 Hz (red) and 22 Hz (green). The collisions are seen to initially decrease the depth of the density modification but to extend the width.

Note that these collision rates are all below the values needed for mode saturation at these amplitudes, as can be seen in Fig. 13.9, but the final value of 22 Hz is very near the saturation value for these amplitudes. Note also that the mean bounce time in the resonance is about 10 transit times, but the collision time is more than 1000 times this.

In Fig. 13.5 is a plot of resonance width versus collision frequency for the two cases of Fig. 13.4. It is very difficult to obtain accurate values, but a rough linear dependance on collision frequency can be seen except for very low frequency values, and a rough square root dependance on mode amplitude, as given by Eq. 13.3.

13.4   SATURATION DYNAMICS

A density gradient in energy E or P_ζ drives the mode amplitude. Particle rotation within the resonance at the bounce frequency partially flattens the distribution within the island, a process which delivers energy to the mode, causing growth. An example of this flattening is shown in Fig. 13.6.

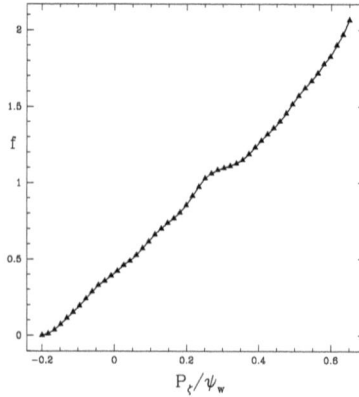

Fig. 13.6   The flattening of the distribution due to a resonance without collisions. The positive gradient in P_ζ, corresponding to a negative gradient in the minor radius, is destabilizing. The resonance is located at $P_\zeta = 0.31\psi_w$ and the mode amplitude is $A = 2 \times 10^{-4}$.

The resonance is at $P_\zeta = 0.31\psi_w$, where ψ_w is the poloidal flux at the last closed flux surface. Complete flattening of the distribution is not attained because the island is at full width only near the elliptic points, with zero width near the hyperbolic points. Averaging over the distribution gives the partial flattening shown. Without particle collisions to replenish the density gradient within the resonance, mode growth ceases within roughly one average particle bounce time within the resonance. Mode saturation with collisions occurs when the rate at which the resonance is flattened, represented by the internal rotation (mixing) frequency ω_b, becomes linearly proportional to the effective frequency of collisional replenishment of the density gradient, represented by the effective scattering frequency ν_{eff}, which was introduced theoretically to combine the effect of Coulomb collisions and the rate of particle decorrelation from the resonance location coming from other stochastic mechanisms such as turbulent scattering and RF diffusion. The exact proportionality constant between the two is determined by the initial linear drive and background damping rates.

The dynamics of this process can be observed by launching particles within the resonance, both with and without collisions, at single values of energy E, μ, and canonical momentum P_ζ, but with different values of θ and ζ. In the unperturbed equilibrium all these particles describe the same orbit - they are distinguished only by their phase relations with respect to

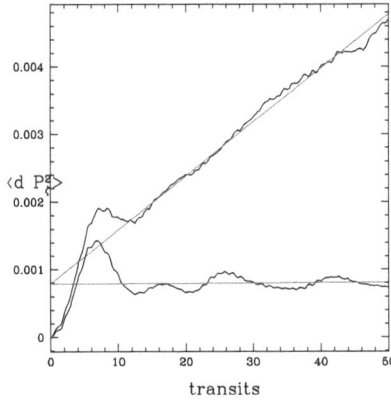

Fig. 13.7 Collisionless evolution and collisional diffusion, $A = 2 \times 10^{-4}$, $\nu_\perp = 10^{-4}/T$, $dP_\zeta^2 = 7.8 \times 10^{-4}$, $S = 8 \times 10^{-5}$ with T the on axis particle transit time. S is the slope of the diffusive plot. The mixing time T_m is given by the first crossing of the collisionless curve with the asymptotic value, $T_m = 10$, in units of toroidal transits. T_d, the time to diffuse across the resonance, is given by $< dP_\zeta^2 > /S$.

the mode. The position of the initial particle launch to determine the mean internal rotation and the time for local diffusion to replenish the density gradient in the resonance is at E $= 93$ keV and $P_\zeta/\psi_w = 0.31$.

Without collisional scattering one observes a simple damped oscillation, as the initial distribution is mixed by the internal rotation, and the energy and P_ζ spread out to the bounding KAM surfaces. The oscillation can be observed either through the mean square energy or the mean modification of P_ζ^2 of the particle distribution. We choose to examine dP_ζ^2, but these variables are simply related through the conservation of K. The magnitude of the final distribution width gives the mean width of the resonance. Since the rotation rate is a function of the distance between the O-point and the separatrix, dropping to zero at the separatrix, as time goes on the mixing occurs at smaller and smaller scale lengths, and becomes an irreversible process, increasing the entropy of the system due to dissipation.

The collisionless simulation determines the mean bounce time of particles in the resonance and the width of the resonance. The collisional simulation determines the time it takes for particles to diffuse across the resonance, giving the time for replenishment of the local distribution. Note that the diffusion rate is given by motion away from the isolated resonance, it is collisional diffusion across the surrounding good KAM surfaces.

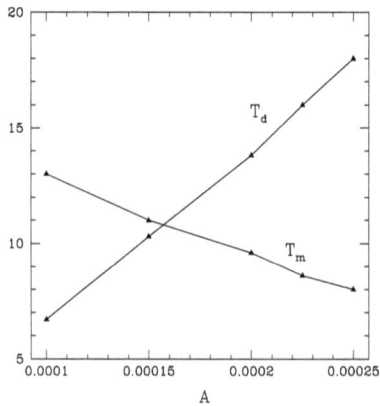

Fig. 13.8   An example of the estimation of the saturation amplitude A.   A plot of the mixing period T_m and the diffusion period T_d vs A showing intersection giving an estimation of the saturation amplitude.   Collisions are fixed at $\nu_\perp = 10^{-4}/T$ for each value of A.

Collisions in the ORBIT code are given by energy conserving pitch angle scattering, in a Monte Carlo representation.

Fig 13.7 presents a determination of effective mixing time and the local diffusion time. The unit of time T in these plots is the unperturbed toroidal transit time of a particle near the major axis, equal to 4.7 μ s in this case. Replenishment of the density gradient within the island is provided by collisions, with a characteristic time given by the time for nearby particles to diffuse across the width of the resonance T_d, and is given by the square of the resonance width divided by the slope of the collisional plot S, which in this case is 10 transits or 47 μ s. The period of rotation in the resonance, approximately given by the first crossing of the collisionless plot with its asymptotic value, is also 10 transits, indicating that this mode amplitude corresponds approximately to a saturation amplitude for this value of the collision frequency.

For a fixed value of collision frequency $\nu_\perp$ a plot of the mixing time T_m and the diffusion time T_d versus mode amplitude A shows a crossing of these values. This occurs because for increased amplitude the mixing time decreases, whereas because of the increased width of the resonance the time to diffuse across the resonance increases. In Fig. 13.8 is shown an estimate of the saturation value using this method.

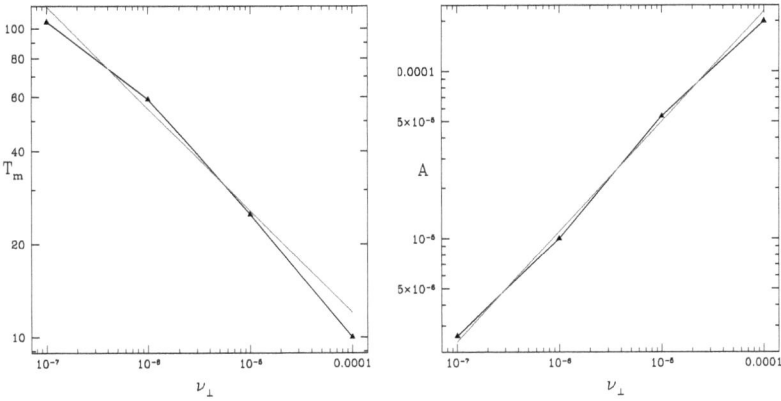

Fig. 13.9 Values at saturation. T_m vs $\nu_\perp$, showing $T_m \sim \nu_\perp^{-1/3}$, and A vs $\nu_\perp$, showing $A \sim \nu_\perp^{2/3}$. The red lines are $T_m = 0.57\nu_\perp^{-1/3}$, and $A = 0.36\nu_\perp^{2/3}$.

The scaling of the saturation level with collisions can be determined by equating T_m and T_d. Since $T_m \sim 1/\omega_b$ and $\omega_b \sim \sqrt{A}$ we have $T_m \sim 1/\sqrt{A}$. The time to diffuse across the resonance is given by $T_d = dP_\zeta^2/S$ with S the slope of the diffusion curve and the resonance width $dP_\zeta \sim \sqrt{A}$. The diffusion across the nearby unperturbed KAM surfaces is given by $dP_\zeta^2 \sim \nu_\perp t$ giving $S \sim \nu_\perp$. Thus setting $T_m = T_d$ gives $A_{sat} \sim \nu_\perp^{2/3}$. The linear dependence on A of T_d and the $1/\sqrt{A}$ dependence of T_m can be seen in Fig. 13.8.

For each value of the collision frequency a number of determinations of T_m and T_d are performed in order to find an estimate for the saturation amplitude. In Fig. 13.9 are shown the values of T_m and A vs collision frequency $\nu_\perp$ for the values at saturation given by this method. We see that $T_m \sim \nu_\perp^{-1/3}$, giving $\omega_b \sim \nu_\perp^{1/3}$ and $A \sim \nu_\perp^{2/3}$. Thus, dynamically, the effect of collisions is to increase the saturation level by supplying the drive given by the density gradient.

13.5 References

- W.Heidbrink and R. White, Phys. Plasmas 27, 2802 (2020)
- V. N. White, R. B. Gorelenkov, N. N. Duarte, and H. L. Berk, Phys. of Plasmas 25, 1 (2018)
- R. B. White, Plasma Phys. Control. Fusion 57, 115008 (2015)
- R. B. White, Physics of Fluids B: Plasma Physics 2 4, 845 (1990)
- M. Podesta, M. Gorelenkova, N. N. Gorelenkov, and R. B. White, Phys. Plasmas Control Fusion 59, 095008 (2017)
- S. P. Hirschman and J. C. Whitson, Phys. Fluids 26, 3553 (1983)
- N.N. Gorelenkov, S.D. Pinches and K. Toi, Nucl. Fusion 54, 125001 (2014)
- C.S. Collins, W.W. Heidbrink, M.E. Austin, G.E. Kramer, D.C. Pace, C.C. Petty, L. Stagner, M.A. Van Zeeland, R.B. White, Y. B. Zhu, Phys. Rev. Lett 116, 095001 (2016)
- R. B. White, N. Gorelenkov, W. W. Heidbrink, M. A. Van Zeeland, Phys. Plasmas 17, no. 5 (2010)
- N. Gorelenkov et al Nucl. Fusion 58, 082016 (2018)
- Berk H.L., Breizman B.N. and Pekker M., Plasma Phys. Rep. 23, 778 (1997). [11] G. Meng, N.N. Gorelenkov, V.N. Duarte, H.L. Berk, R.B. White and X. Wang, Nucl. Fusion 58, 082017 (2018)
- Drummond W. and Pines D., Nucl. Fusion 2 1049 (1962)
- T. H. Dupree, Phys. Fluids 9, 1773 (1966)
- A. N. Kaufman, Phys. Fluids 15(6), 1063-1069 (1972)
- M. Zhou and R.B. White, Plasma Phys. Control. Fusion, 58, 125006, (2016)
- N.N. Gorelenkov, Y. Chen, R.B. White, H.L. Berk, Phys. Plasmas 6, 629 (1999)
- K. Ghantous, H. L. Berk, and N. N. Gorelenkov, Phys. Plasmas 21, 032119 (2014)

Chapter 14

Avalanche

14.1 INTRODUCTION

When several Alfvén modes are present in a discharge, overlap of mode resonances can lead to large modification of the high energy population and a growth of the modes to large amplitude, a phenomenon refered to as an avalanche. The processs can stop with a local redistribution of high energy particles, or it can lead to large scale particle loss. We examine weakly unstable damped Alfvén modes present in NSTX preceding an avalanche, using the guiding center code ORBIT and a delta f formalism to reproduce the avalanche behaviour. The code uses a numerically produced NSTX equilibrium, mode eigenfunctions produced by NOVA, and the numerical beam particle distribution produced by TRANSP.

14.2 EQUILIBRIUM

The equilibrium magnetic field is given by

$$\vec{B} = g\nabla\zeta + I\nabla\theta + \delta\nabla\psi_p, \tag{14.1}$$

where θ and ζ are poloidal and toroidal coordinates and ψ_p is the poloidal flux, and in an axisymmetric equilibrium using Boozer coordinates g and I are functions of ψ_p only. The perturbation has the form $\delta\vec{B} = \nabla \times \alpha\vec{B}$ and α and an electric potential Φ have the Fourier expansions

$$\alpha = \sum_{m,n} A_n \alpha_{m,n}(\psi)sin(\Omega_{mn}), \quad \Phi = \sum_{m,n} A_n \Phi_{m,n}(\psi)sin(\Omega_{mn}), \tag{14.2}$$

where n refers to a single mode with definite toroidal mode number n and frequency ω_n, and the sum is over toroidal and poloidal harmonics m with

$\Omega_{mn} = n\zeta - m\theta - \omega_n t - \phi_n$, with ϕ_n the mode phase. For ideal modes the electric potential Φ is chosen to cancel the parallel electric field induced by $d\vec{B}/dt$, requiring

$$\sum_{m,n} \omega_n B \alpha_{m,n} cos(\Omega_{mn}) - \vec{B} \cdot \nabla\Phi/B = 0.$$

giving in Boozer coordinates

$$(gq + I)\omega_n \alpha_{mn} = (nq - m)\Phi_{mn}.$$

The perturbation α is related to the ideal displacement $\vec{\xi}$, through

$$\alpha_{mn} = \frac{(m/q - n)}{(mg + nI)}\xi_{mn}^\psi.$$

The eigenfunctions produced with the code NOVA are normalized with the largest harmonic $\xi_{mn}^\psi(\psi_p)$ having maximum amplitude 1. Thus the amplitude A_n is the magnitude of the ideal displacement caused by this harmonic, normalized to the major radius R.

Stepping equations for the mode amplitude and phase were previously derived, see Eq. 8.74

$$\frac{dA_n}{dt} = \frac{-\nu_A^2}{D_n\omega_n A_n} \sum_{j,m} w_{n,j} \left[\rho_\| B^2 \alpha_{mn}(\psi_p) - \Phi_{mn}(\psi_p)\right] cos(\Omega_{mn}) - \gamma_d A_n,$$

(14.3)

$$\frac{d\phi_n}{dt} = \frac{-\nu_A^2}{D_n\omega_n A_n^2} \sum_{j,m} w_{n,j} \left[\rho_\| B^2 \alpha_{mn}(\psi_p) - \Phi_{mn}(\psi_p)\right] sin(\Omega_{mn}), \quad (14.4)$$

with $D_n = 4\pi^2 \sum_m \int \xi_{mn}^2(\psi_p)d\psi_p$, j the particle index and ψ_p, θ, ζ the position and $\rho_\|$ the normalized parallel velocity of particle j. The modes are resonant with and destabilized by the high energy injected beam, so the particles refer to beam ions. The linear damping rate γ_d is due to the continuum, electron and thermal ion Landau damping and radiation, all terms in the sums are evaluated at the coordinates of particle j, and $w_{n,j}$ is the δf weight of particle j for mode n.

The perturbed distribution δf is represented by

$$\delta f(\psi_p, \theta, \zeta, \rho_\|, t)$$
$$= \sum_j w_{n,j}\delta(\psi_p - \psi_{p,j}(t))\delta(\theta - \theta_j(t))\delta(\zeta - \zeta_j(t))\delta(\rho_\| - \rho_{\|,j}(t)).$$

(14.5)

and the particle weights are stepped by

$$\frac{dw_{n,j}}{dt} = \frac{w_{n,j} - f/g}{f_0}[\partial_E f_0 \dot{E} + \partial_{P_\zeta} f_0 \dot{P}_\zeta] \qquad (14.6)$$

with g the marker distribution given by $\psi_{p,j}$, θ_j, ζ_j, and $\rho_{\|,j}$, and f_0 the unperturbed initial beam distribution. The particle energy is E and $P_\zeta = g\rho_\| - \psi_p$ is the toroidal canonical momentum, both conserved in time in the absense of the modes. The change in time of the particle energy and canonical momentum due to the modes are given by

$$\frac{dE}{dt} = -\rho_\| B^2 \partial_t \alpha + \partial_t \Phi,$$
$$\frac{dP_\zeta}{dt} = \rho_\| B^2 \partial_\zeta \alpha - \partial_\zeta \Phi. \qquad (14.7)$$

14.3 NSTX SHOT 141711

To simulate a specific example we consider NSTX shot 141711 at t = 470 msec, shortly before the observation of a major avalanche. The equilibrium and q profile are shown in Fig. 14.1. The beam particle distribution is shown in Fig. 14.2, and the distribution in pitch was strongly peaked near $\mu B = 20 keV$. The modes are destabilized by the gradient in the distribution in P_ζ, equivalent to a gradient in the minor radius. The gradient in energy E is negative and mildly stabilizing.

It is not necessary to use the full numerical beam particle distribution in these simulations. Doing so is in fact difficult, because in order to smooth the distributions sufficiently enough to allow the partial derivatives required for Eq. 14.6 one finds that the magnitude of the resulting growth rates do not match the values given by NOVA or the KICK model. But all energy exchange between modes and particles occurs at the resonances. Particles on good KAM surfaces only exhibit adiabatic oscillations of energy and canonical momentum, and cannot contribute to mode growth. Thus simulating modes agreeing with theoretical or experimental growth rates requires only the numerical imposition of the correct energy and momentum gradients at the locations of the resonances, not the use of the full particle distribution. In addition, what is known accurately, either from experiment or theory, are the linear growth rates for the modes. Thus the gradients at the resonances can be adjusted to produce the correct growth rates.

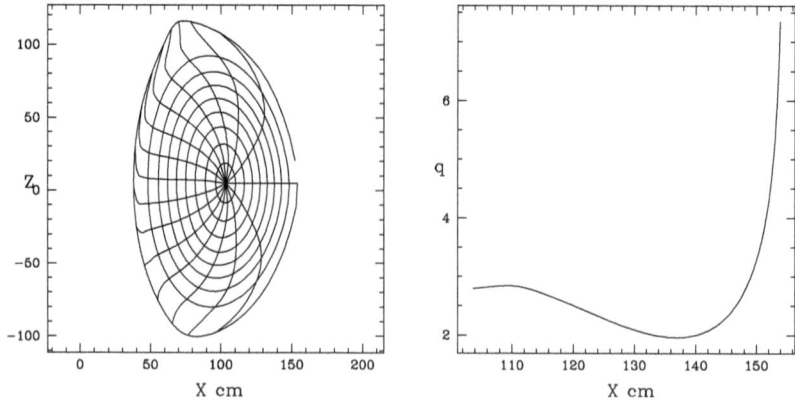

Fig. 14.1   NSTX 141711 at t = 470 ms, showing the equilibrium flux surfaces and the *q* profile

14.3.1   *Alfvén Modes*

There were four Alfvén modes present in the discharge before the observed avalanche. Each mode consists of ten or more poloidal harmonics, but only a few of them are large enough to significantly affect the particle distribution through resonance formation. A Poincaré plot showing the location of resonances can only be performed for a fixed value of mode frequency and toroidal mode number n. Thus it is impossible to show the resonances of all modes together using this method. However, one can show the location of the destroyed KAM surfaces for all modes together, giving the location of all resonances. In Fig 14.3 is shown the domain of destroyed KAM surfaces for all modes, with amplitudes fixed at $A = 3 \times 10^{-4}$. The domain in energy E and canonical momentum P_ζ is that of co-passing ions. The right boundary is the magnetic axis and the left boundary is the plasma edge.

All resonances are in the outer half of the plasma, and extend from 40 keV up to the maximum energy in the beam, about 90 keV. The resonances are far from overlapping at this amplitude, and only produce local flattening of the particle distribution. The method of destroyed KAM surfaces shows the location of the resonances, but we wish also to associate each individual resonances with the mode responsible for it. For this we perform Poincaré plots for each mode separately, showing also the location and number of elliptic points in the poloidal plane.

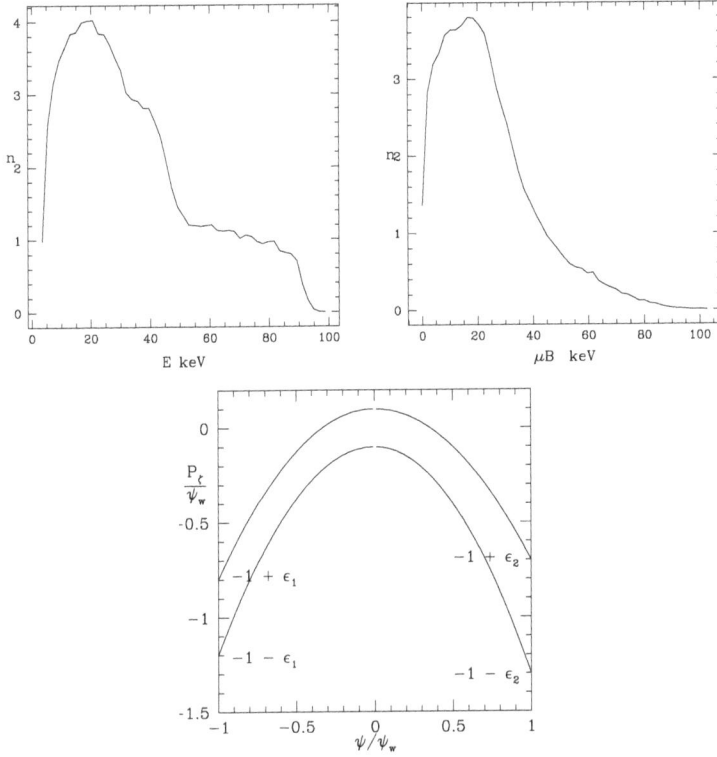

Fig. 14.2   NSTX 141711 beam particle distribution at t = 470 ms, energy, magnetic moment, and canonical momentum distributions

Kinetic Poincaré plots can be made only with fixed ω and μB. In Figs 14.4 and 14.5 are shown the poloidal harmonics and the resonances for each of the four modes. The resonances are shown for a value of $\mu B = 20 keV$, characteristic of the high energy particle distribution. The location and nature of the resonances is only weakly dependent on this value. For clarity, the resonances are shown in the variable P_ζ, the toroidal canonical momentum, for a large mode amplitude, of $A = 1.5 \times 10^{-3}$. It is interesting that the number of islands poloidally for the major resonance in each case is equal to the n value. Modes with $n = 2, 3, 4$ display mostly good KAM surfaces except for the one significant resonance near $P_\zeta/\psi_w = -0.1$. The $n = 2$ mode has a smaller resonance at $P_\zeta/\psi_w = -0.37$. The $n = 3$ mode has two small resonances at $P_\zeta/\psi_w = -0.3$ and -0.4. The $n = 4$ mode has

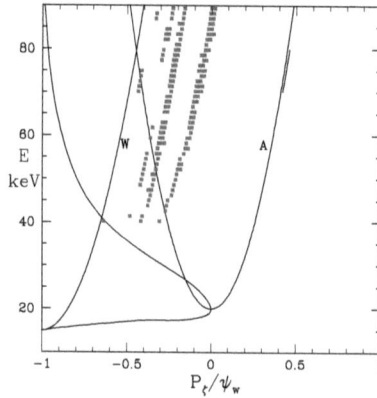

Fig. 14.3   Resonance location for all modes, $\mu B = 20 keV$, $A = 3 \times 10^{-4}$. The large domain in the upper right of the diagram is that of co-passing particles. The right edge of this domain is the magnetic axis, and the left edge is the plasma boundary.

two small resonances at $P_\zeta/\psi_w = -0.24$ and -0.35. The mode with $n = 5$, in addition to the major resonance at $P_\zeta/\psi_w = -0.1$ has other significant resonances at lower values of P_ζ/ψ_w. The primary resonance has 5 elliptic points in the poloidal plane, but there are also resonances with 6, 7,and 8 elliptic points, and one can also see higher order Fibonacci resonances, one in between the 6 and 7 with 13 elliptic points, and one in between the 7 and 8 with 15 elliptic points. In addition to these higher order resonances, the coupling of the modes with different n values will produce more. Hence with larger mode amplitude these resonances can possibly provide a loss channel to the plasma edge. The plasma edge is to the left in these plots, at about $P_\zeta/\psi_w = -0.5$, and the line at the right is the magnetic axis.

For a mode of given toroidal mode number n and frequency ω, the particle energy E and momentum P_ζ are related in the presence of the mode through

$$\omega P_\zeta - nE = K, \tag{14.8}$$

with K a constant depending on the initial particle conditions. The energy range for the data in these plots is from 70 to 80 keV. Both E and P_ζ are constant in the absense of the modes.

The numerical particle distribution in the simulations is chosen to well include all resonances at large mode amplitude, as well as a full radial

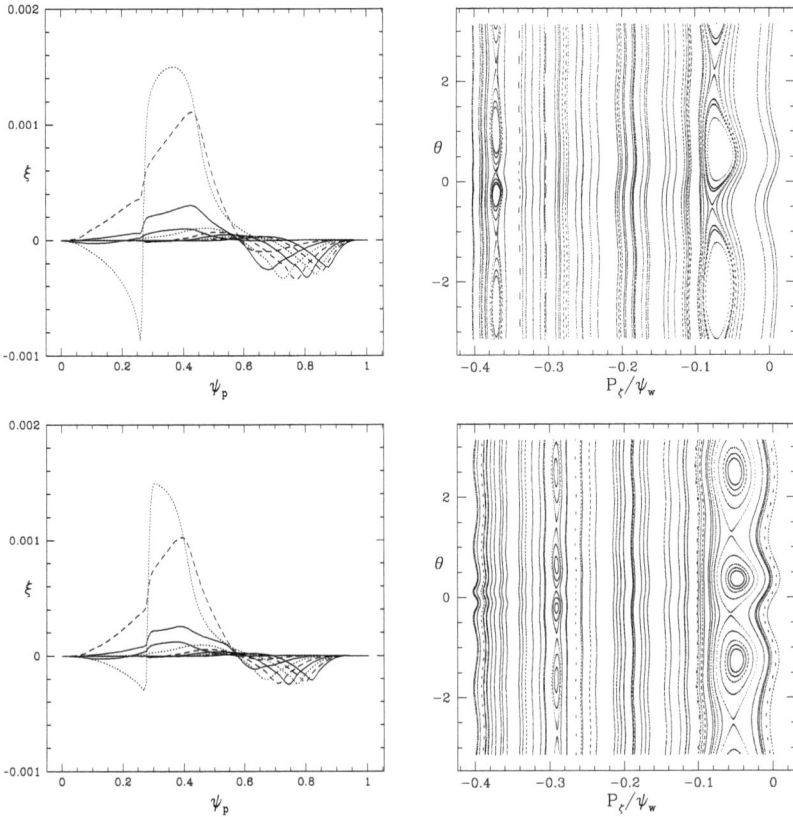

Fig. 14.4   Above: Mode $n = 2, m = 1 - 10$, $\mu B = 20$ keV, f $= 88.362$ kHz. Below: Mode n=3, m = 1-10, $\mu B = 20$ keV, f $= 109.507$ kHz. The amplitude is $A = 1.5 \times 10^{-3}$ to clearly show the resonance.

distribution, in order to be able to describe the full extent of an avalanche.

$$
\begin{pmatrix}
\multicolumn{5}{c}{TABLE \qquad I} \\
n & f\ kHz & \gamma_N/\omega & \gamma_K/\omega & \gamma_d/\omega \\
2 & 88.4 & 2.4 \times 10^{-4} & 1.8 \times 10^{-3} & 2.85 \times 10^{-3} \\
3 & 109.5 & 5.0 \times 10^{-4} & 8.9 \times 10^{-3} & 3.36 \times 10^{-3} \\
4 & 126.8 & 6.0 \times 10^{-4} & 14.5 \times 10^{-3} & 1.91 \times 10^{-3} \\
5 & 161.6 & 6.7 \times 10^{-4} & 9.4 \times 10^{-3} & 7.5 \times 10^{-4}
\end{pmatrix}
$$

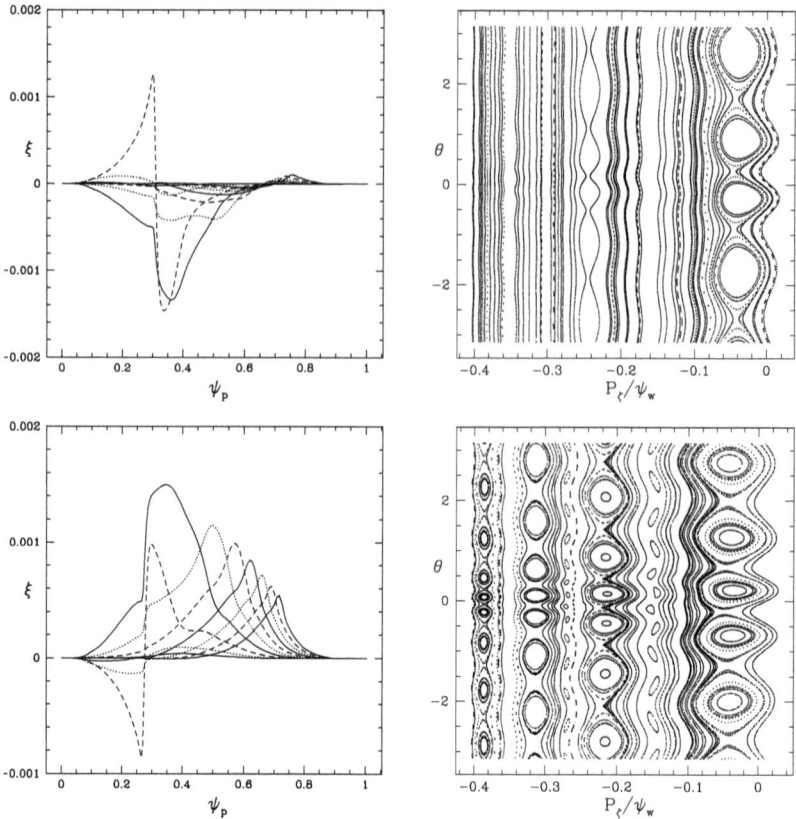

Fig. 14.5   Above: Mode $n = 4$, $m = 1 - 10$, $\mu B = 20$ keV, $f = 126.758$ kHz. Below: Mode $n = 5$, $m = 1-10$, $\mu B = 20$ keV, $f = 161.635$ kHz. The amplitude is $A = 1.5 \times 10^{-3}$ to clearly show the resonance.

In table I are shown the growth rates obtained for the mode from NOVA-K (γ_N) and from the KICK model (γ_K). The two methods give values which differ by as much as an order of magnitude. Note that, unlike the other modes, the $n = 2$ mode has a growth rate much smaller than the damping, and is thus stable.

Also shown are the theoretically calculated values of the damping for each mode, (γ_d). Although the $n = 2$ mode is stable, the modes with $n = 3, 4, 5$ are located directly above it in P_ζ, and thus they naturally steepen the density gradient for the $n = 2$, destabilizing it provided they become large enough.

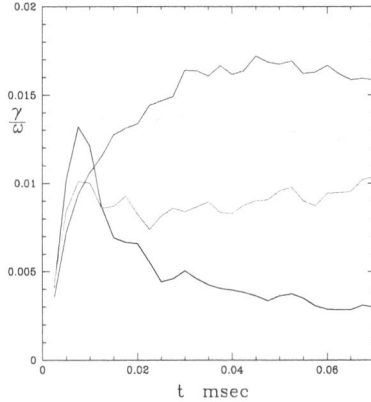

Fig. 14.6 Numerically determined growth rates, modes with $n = 2$ (black), $n = 3$ (red), $n = 4$ (blue), $n = 5$ (green).

14.3.2 *Mode time history*

The results of a simulation are shown in Fig. 14.6 giving the growth rates for the modes. This plot was obtained by keeping the mode amplitude fixed at $A = 10^{-5}$, but allowing the particle weights to generate the growth rates. The plot shows the modes with $n = 2$ (black), $n = 3$ (red), $n = 4$ (blue), $n = 5$ (green), with γ/ω less than 0.02 for all modes. The pitch angle scattering rate for the simulations was fixed at the theoretical value of 2/sec for the observed plasma density. Growth rates are approximately the values given by the KICK model.

In Fig. 14.7 is shown the time evolution of these modes for a period of 20 msec, with a characteristic collision frequency of 2/sec, at the normal drive magnitude and at a drive of 10 percent higher. Both simulations included the theoretical damping values. The plots show the modes with $n = 2$ (black), $n = 3$ (red), $n = 4$ (blue), $n = 5$ (green). For the normal drive the modes exhibit significant fluctuations but amplitudes remain mostly below $A = 10^{-3}$. The n=2 mode, although initially stable, is slowly destabilized by the other modes, but the amplitude remains below $A = 10^{-3}$. Note that the $n = 5$ mode, the only one with resonances extending toward the plasma edge, remains at a low amplitude.

The simplest example of an avalanche is given by the sandpile model. Addition of sand to one part of the sandpile can lead to a cascade of particles in that region, causing an increase in the density gradient in nearby

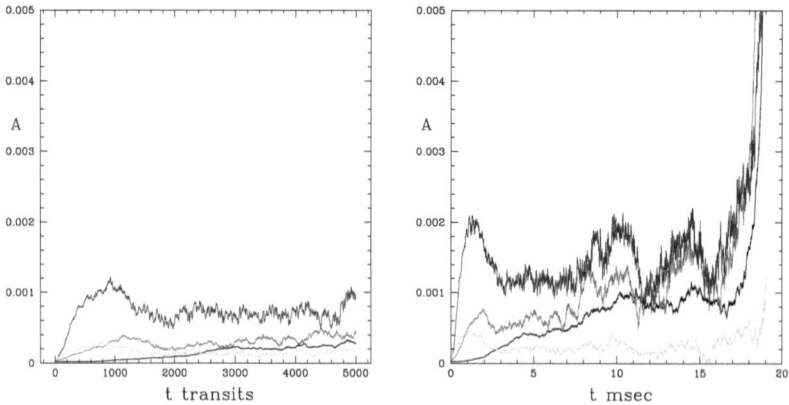

Fig. 14.7   Mode evolution, modes with $n = 2$ (black), $n = 3$ (red), $n = 4$ (blue), $n = 5$ (green), with a collision frequency of $\nu = 2/\text{sec}$, for 20 msec. Evolution with normal drive (left) and 10 percent increase in drive (right). With the theoretical growth rates the modes fluctuate with amplitudes mostly below 10^{-3}, but with a ten percent increase in the drive there is an explosive growth of all modes.

domains, both above and below this point, and stimulating additional particle flow down the gradient, the effect being thus nonlocal. A similar phenomonon is observed in the present case. Growth of the $n = 3, 4, 5$ modes causes a change in the gradient of the particle density in the vicinity of the $n = 2$ mode, initially stable due to the large damping rate. If the drive is not too large, the modes all stabilize and fluctuate with amplitudes below the critical values producing large scale resonance overlap.

14.3.3   *Mode resonance overlap*

With a 10 percent increase in drive these modes grow to a magnitude allowing strong destabilization of the n=2 mode. Additional further destabilization of the n=5 mode, the only one with resonances located closer to the plasma boundary and the appearance of significant higher order Fibonaci sequence resonances (see Fig. 14.5), leads to a path for particle flux to the plasma edge. The particle evolution cannot be followed during the whole event in these simulations because the modification of the distribution quickly exceeds the limits of the δf formalism, but it is clear that large scale particle loss can be involved. It is also possible that the particle loss lowers the gradients driving the modes, stabilizing them and halting

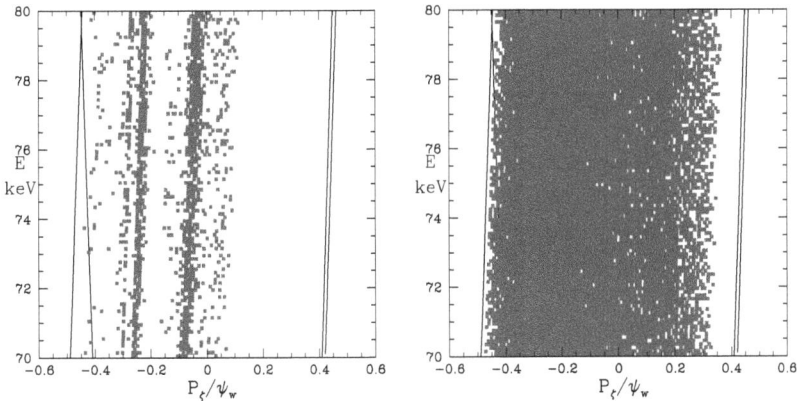

Fig. 14.8 Domain with broken KAM surfaces for the mode amplitudes observed with normal drive (left) and with drive multiplied by 1.1 (right). With normal drive there are good KAM surfaces preventing significant loss, allowing only local flattening of the distribution, but a ten percent increase in the drive leads to global stochastic loss.

the flow of particles to the plasma edge before total loss occurs. It is not possible to distinguish these two cases with a δf simulation.

In Fig. 14.8 are shown the stochastic domains produced for the final mode amplitudes in the simulations in the two cases. With the drive set to produce the nominal growth rates only isolated domains of broken KAM surfaces are present, corresponding to the resonances of the modes. The modes do not overlap, there are large domains of good KAM surfaces preventing significant particle loss. If the drive is increased by 10 percent the resonances overlap, producing a continuous stochastic domain leading to global particle loss.

A plausible scenario for the existence of an Alfvén mode avalanche in NSTX has been simulated, using modes observed to be present in the NSTX discharge shot 141711 at t = 470 msec, shortly before the occurance of an avalanche. Values of mode growth rate and damping for the modes given by theoretical analysis are used. The four modes observed are shown to produce resonances that can provide a path for large scale stochastic loss of particles with only a small increase in the high energy particle gradients producing mode drive. If the mode drive is below this critical value, the mode amplitudes oscillate around saturation levels which do not produce significant modification of the high energy particle distribution.

14.4 References

- R.B. White, V.N. Duarte, N.N. Gorelenkov, E.D. Fredrickson, M. Podesta, Phys Plasmas 27 (2020)
- S. M. Kaye, M.G. Bell, R.E. Bell, S. Bernabei, J. Bialek, T. Biewer, W. Blanchard, J. Boedo, C. Bush, M.D. Carter et al Nuclear Fusion 45 S168-S180 (2005)
- R. B. White and M. S. Chance Phys. Plasmas [6], 226 (1999)
- N. N. Gorelenkov, C. Z. Cheng, and G Fu, Phys Plasmas 6, 2802 (1999)
- A. Pankin, D. McCune, R. Andre, G. Bateman, and A. Kritz, Comp.Phys.Comm. 159,157 (2004)
- R. B. White, N. Gorelenkov, M. Gorelenkova, M. Podesta, S. Ethier, Y. Chen, Plasma Physics and Controlled Fusion, 58, 2016
- M. Zhou and Roscoe White, Plasma Phys. Control. Fusion xx xx (2016)
- Chen, Y., R. B. White, Guo-Yong Fu, Raffi Nazikian, Phys. Plasmas [6], 226 (1999)
- A. N. Kolmogorov, Proc. Int. Congr. Mathematicians, Amsterdam, Vol 1 315 (1957), V. I. Arnold, Russ. Math. Surv. 18(5):9, 1963, J. Moser, Math. Phys. Kl. II 1,1 Kl(1):1, 1962.

Chapter 15

The Lithium Wall Fusion Concept

15.1 THE IDEA OF MAGNETIC FUSION

A fusion plasma is required to have a temperature of 10–30 keV, while the temperature of the plasma facing material (wall) is limited to about 0.1 eV. The plasma density is as low as $0.5 - 2 \cdot 10^{-6}$ g/cm^3, while the density of the wall material is about $\simeq 6$ g/cm^3. Because of the large difference in densities, the potential damage to the wall by the plasma is reduced to sputtering or melting of the surface layer of the wall. At the same time, even a minimal amount of wall material penetrating the plasma can extinguish it by direct cooling or through radiation.

The idea of magnetic fusion is to use the magnetic field in order to reduce the plasma-wall interaction. The wall itself cannot and, in fact, should not be insulated from the plasma: plasma particles leaving the reaction zone should be removed by the wall structure. But the plasma should certainly be insulated from the wall.

Particles coming to the plasma from the wall consist of wall material and plasma particles, which hit the wall and return to the plasma edge as neutral atoms. Both enter the plasma edge and then contribute to the plasma flux to the wall, creating the so-called recycling.

Wall erosion can be reduced by lowering the plasma energy flux and by the choice of appropriate materials. Even better would be to arrange a situation of a "pumping wall", which absorbs particles, and prevents their return to the plasma edge as cold neutrals. Combined with the core fueling of the reaction zone by energetic Neutral Beam Injection (NBI), a pumping wall would be an implementation of plasma insulation from the wall.

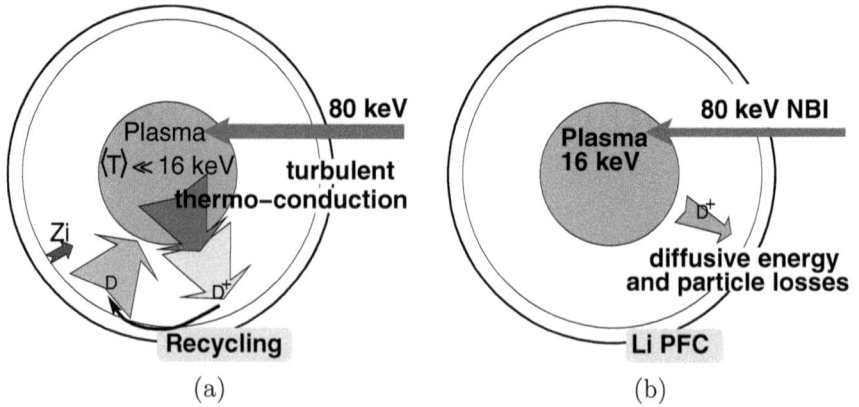

Fig. 15.1 Plasma column heated by energetic NBI. (a) The case of high recycling as in present fusion devices with a strong effect of the wall on the plasma. (b) The case of a pumping wall, utilizing pumping properties of liquid lithium and making the wall invisible to the plasma.

Unfortunately, this original idea of magnetic fusion has not been implemented in practice. Fig. 15.1 shows the cross-section of a plasma column in a tokamak, heated by NBI, which delivers energetic particles (60–120 keV) and energy to the plasma core. In present toroidal fusion research devices, as illustrated by Fig. 15.1a, particles, which leave the hot plasma and hit the wall, are returned back to the plasma edge as cold neutrals. These cold neutrals are converted into cold ions by the charge exchange process, typically dominating at the edge. Hot plasma ions go directly to the wall where they are neutralized. As a result, the plasma edge is cooled very efficiently.

The situation is totally different in the case of a pumping wall, shown in Fig. 15.1b. In the idealized case of perfect pumping, there are no cold particles in the system – energetic plasma particles leaving the confinement zone are absorbed by the wall. The plasma temperature, resulting from thermalization of the heating beam, is determined exclusively by the energy of the beam,

$$E^{NBI} = \left(\frac{3}{2} + 1\right)(T_e^{edge} + T_i^{edge}). \tag{15.1}$$

On the right-hand side, "3/2" results from the definition of the temperature in terms of energy, while "1", exact for a Maxwellian distribution function,

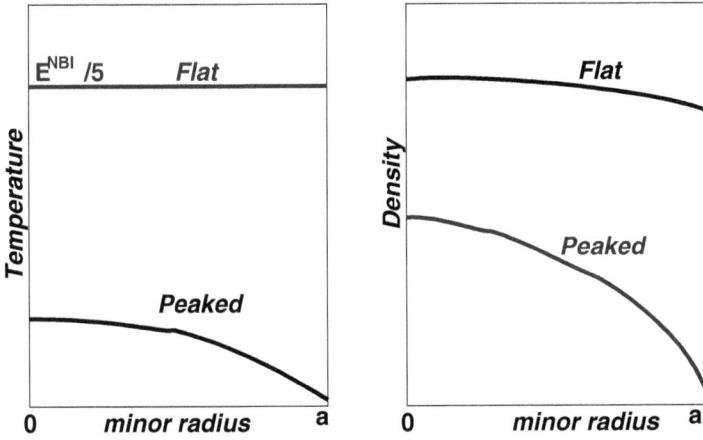

Fig. 15.2 Plasma (a) temperature and (b) density profiles in the case of conventional (black) and a pumping (blue) wall.

is related to the particle diffusion. The characteristic value $E^{NBI}/5$ of the plasma temperature (Fig. 15.2a) is predictable and is independent of the configuration confinement properties. On the other hand, the density is low at the edge (Fig. 15.2b) because of pumping. In the conventional wall situation the temperature profile is peaked. Its central value is determined by thermal conduction and the heat source, rather than by beam energy, and is unpredictable because of turbulence associated with the temperature gradient. Because of large energy losses, the beam energy is much higher than the plasma temperature and NBI cannot be used for fueling. In this case, the density is supplied mostly from the edge and has a flat profile.

Between these two cases there are a number of fundamental differences in plasma regimes and approaches to fusion.

15.2 CONFINEMENT REGIME CONTROLLED BY THERMAL CONDUCTION

The total heat flux Q^{heat} through a magnetic surface consists of thermal conduction and diffusion

$$Q^{heat}(r) = -n(r)\chi\frac{dT(r)}{dr} + \frac{5}{2}T(r)\Gamma = P^{heat}, \tag{15.2}$$

where χ is the thermal conduction coefficient, n is plasma density, and Γ is the particle flux, and P^{heat} is the total heating power. In the case of a conventional wall, the heating of the core leads to a peaked plasma temperature profile. The temperature gradient generates turbulence and large energy losses in the form of anomalous heat conduction, primarily through the electron channel, *i.e.* through the first term in the heat flux. In this regime, a low level of plasma temperature is determined by the heating power and thermal conduction, rather than by the energy E^{NBI} of the beam.

As a consequence, the current approach to the design of a fusion reactor has a number of problems associated with it:

1. Very large size devices are necessary even for the mere demonstration of fusion power production.

2. In order to compensate for turbulent energy losses, the large size must be combined with large plasma current, which creates numerous problems associated with disruptions.

3. In its turn, the high plasma current requires a strong magnetic field, whose amplitude is limited by the super-conducting materials and by strength of the structures.

4. High heating power is required to maintain high plasma temperature.

5. High heat flux, associated with poor confinement and directed in a narrow scrape off layer to the divertor target plates, creates significant, or even unresolvable, problems with power extraction.

6. Strong plasma-wall interaction makes questionable the stationary, or even stable plasma regime, required for a power reactor.

7. The physics of such a plasma is so complicated that exa-scale high performance computing is necessary for simulation of plasma control and performance.

8. The expected utilization of the plasma volume for fusion due to the peaked temperature is very poor – only 25–30 percent.

15.3   DIFFUSION BASED CONFINEMENT REGIME

Utilization of the pumping wall changes the approach in all significant aspects based on the fact that for magnetically confined plasma it is much more efficient to prevent plasma cooling by pumping plasma particles and suppressing the recycling, rather than relying on excessive heating power to compensate for large energy losses.

It has been taken for granted for many years that a toroidally confined plasma must have a very low temperature at the boundary with the confinement vessel. In fact, the plasma edge temperature is determined simply by

$$\gamma_{i,e} \Gamma^{edge \to wall} T_{i,e}^{edge} = P_{i,e}^{heat}, \tag{15.3}$$

where $\gamma_{i,e}$ is a coefficient, $5/2$ in a Maxwellian plasma, and $\Gamma^{edge \to wall}$ is the particle flux to the wall.

The edge should be understood as the end of the confinement zone. It can be different for electrons and ions, typically deeper into the plasma core for electrons. Beyond the edge, a significant fraction of plasma particles move freely to the material surface. Eq. 15.3 plays the role of a boundary condition for transport equations inside the confinement zone for ions and electrons.

The particle flux $\Gamma^{edge \to wall}$ depends on the recycling coefficient R^{ecycle}

$$\Gamma^{edge \to wall} = \frac{\Gamma^{core \to edge}}{1 - R^{ecycle}}, \tag{15.4}$$

where $\Gamma^{core \to edge}$ is the internal particle flux from the core to the plasma edge. In the case of a conventional wall, R^{ecycle} is close to one, especially with large $\Gamma^{core \to edge}$ and carbon walls. As a consequence, also $\Gamma^{edge \to wall}$ is large, and the edge temperature is small. The rate of recycling R^{ecycle} can be reduced by the design of the wall. But at low edge density, expected for the regime of a pumping wall, exhaust pumps are inefficient. A practical implementation exists in the utilization of the unique properties of liquid lithium to absorb hydrogen isotopes. In this case R^{ecycle} can be made significantly less than one. Any level of $R^{ecycle} < 0.5$, achievable with liquid lithium, would dramatically change the plasma regime.

In the case of NBI fueling, the edge temperature will be simply $T_{i,e}^{edge} \simeq (1 - R^{ecycle}) E^{NBI}/5$, thus directly related to the externally controlled beam energy. Even with the worst assumption of unlimited electron anomalous thermal conduction, the core temperature cannot be lower than this boundary temperature.

In this regime, called the LiWall Fusion (LiWF) regime, thermal conduction plays no role in energy confinement. All energy losses are determined by the particle losses, which are limited by the best confined component, *i.e.* by ion diffusion.

If the concept of a liquid lithium wall can be practically realized in the future, overcoming possible technological difficulties, the diffusion based confinement regime is superior to the conventional one:

1. Energy confinement is determined by the ions, the best confined plasma component, with an expected order of magnitude improvement in energy confinement time.

2. The temperature profiles, important for global stability, are well controlled by the NBI energy and recycling, rather than by uncontrolled turbulence in the core. Plasma turbulence, if any, or other anomalies in the particle behavior, do not affect the plasma temperature regime. Only the level of plasma density can be affect ed.

3. Plasma stability is enhanced: (a) sawtooth oscillations are absent in the absence of peaked current density profile (possible only with a peaked temperature); (b) the Greenwald density limit is absent due to the automatically low edge density; (c) the plasma edge is stable to the edge localized modes (ELM), as predicted and confirmed experimentally (Zakharov *et al.*, 2007).

4. The thermo-force, which drives impurities from the target plates to the plasma, is absent in the high-temperature, collisionless scrape off layer of the LiWF regime.

5. The plasma-wall interaction is intrinsically stationary, impossible with the conventional divertor target plates.

6. Low edge density in combination with high edge temperature provides the best conditions for plasma coupling with non-inductive radio-frequency current drive systems.

7. A new burning plasma regime with high plasma temperature and moderate density is possible. The undesirable heating of electrons (not producing fusion energy) by the fusion α particles is avoided, the energy is channeled to synchrotron radiation, thus keeping $T_e < T_i$ and reducing the power deposition to the divertor.

8. The entire plasma volume, rather than a small fraction of it, is used for fusion energy production.

15.4   IMPLEMENTATION

Neutral beam injection (Fig. 15.3a) with the necessary beam energy $E^{NBI} = 80$–120 keV has been routinely used in tokamaks for several

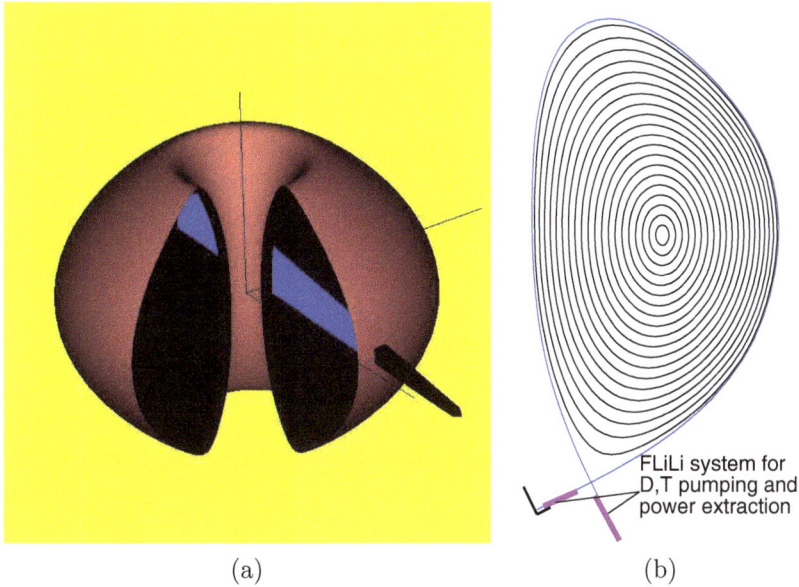

(a)   (b)

Fig. 15.3   Two key components of LiWF: (a) NBI as the heating/core-fueling method, and (b) FLiLi layer on the plasma facing surface of the target plates.

decades. The implementation of the pumping wall is also realistic. It requires a very small flow rate of flowing liquid lithium (FLiLi), *i.e.* 1 g/s or 2 cm^3/s of FLiLi to pump 10^{21}-10^{22} plasma particles/s. This can be achieved with a slow (less than 1 cm/s) moving thin (about 0.1 mm) FLiLi layer driven by gravity. Recycling can be significantly reduced by lithium covered divertors as shown in Fig. 15.3, technologically much easier than covering the entire wall.

The very small necessary flow rate makes practical stationary FLiLi systems for experimental devices, resolving the major technology problem of using liquid lithium in tokamaks, *i.e.* elimination of the crust which develops on a lithium surface exposed to a small but finite water, oxygen, etc., outflow from the in-vessel components.

Small FLiLi velocity and thickness make the layer insensitive to MHD effects, providing an easy scaling of system designs from the test level to operational systems in existing and future tokamaks.

At the same time, the heat flux from the plasma to the FLiLi surface cannot be removed by a flowing liquid metal: the required high velocities are forbidden by the strong magnetic field of the tokamaks. Instead, good thermal contact of the FLiLi layer with the guide plate, serving as a heat sink (actively cooled) is necessary for the heat removal.

While it is not possible to rely solely on material development for the power extraction from the burning plasma, the reduction of the power flux, possible only in the LiWF regime, provides a practical approach for solving the difficult problem of the power extraction in magnetic fusion.

15.5   ALPHA PARTICLE CONFINEMENT

As seen in the discussions of plasma stability in Chap. 6 and Chap. 7, and induced particle loss in Chap. 12 and 14, operation of a tokamak with large toroidal current producing values of q near to one or below it, along with significant plasma pressure or β, leads to destabilization of the internal kink, the sawtooth, and the fishbone modes. A large value of poloidal current has been assumed necessary because of the need to confine alpha particles. It is worth exploring the idea of giving up alpha particle confinement, and instead directly heating the plasma with neutral beams, as has been done with experimental devices all along. Without confining alpha particles, the device could be made small enough so that neutral beam heating is possible for the whole plasma. Alpha particles can be allowed to escape directly to the first wall, giving a uniform high energy particle distribution at the wall rather than a localized one, greatly reducing the problems associated with localized power extraction. In addition this means that there would be no helium ash accumulating in the plasma center: the ash appears at the plasma edge where it can be extracted through the separatrix and pumped out.

Even more importantly, this would remove the heating of the plasma ions by the alpha particles through the electron channel. Almost all tokamak plasmas studied to date have been hot ion plasmas, with the ion temperature higher or equal to the electron temperature. A hot electron plasma can be produced artificially, and is inferior in many ways (Petty *et al.*, 1999) because the electrons are much more subject to turbulence. Relying on ion heating through the electrons, as is currently planned with conventional tokamaks, could prove to be very unfortunate.

Removing the requirement of confining alpha particles relaxes considerably the design parameters, including poloidal field strength (see Sec. 3.8) and therefore plasma current, and ripple magnitude limitations (see Sec. 12.7). This gives improved possibility of avoiding plasma disruption. It would also mean that an ignition device could be made much smaller than what has been so far envisioned. In addition this would eliminate the problem of burn control, Sec. 12.13. The fusion rate would be completely in control of the beam injection.

15.6 DIVERTOR PLATES

Experiments, especially the last experiments on TFTR before it was unfortunately decommissioned in 1997, showed that the use of lithium deposited on a tokamak wall leads to much higher and flatter temperature profiles (Mansfield *et al.*, 2001). The highest parameters obtained with discharges in TFTR, the so called supershots, were through the use of lithium deposition. Some additional work has been done in this direction, but so far no experiments have been undertaken using a full surface covering. However, the experiments with partial wall covering that have been carried out indicate that lithium does indeed produce flatter themperature profiles and higher internal temperatures. To cover the whole wall it is not necessary to engineer a liquid lithium surface; a solid or molten layer of lithium deposited on the surface is sufficient to test absorption for the duration of a discharge, and a flowing lithium layer on a divertor may also be sufficient to greatly flatten temperature profiles. The situation is complicated by the fact that most devices possess carbon tiles, which rapidly absorb a lithium coating, exposing the carbon and reintroducing large recycling. Experiments using a fully covered liquid lithium divertor are underway in China.

15.7 References

- Antar, G.Y., R. P. Doerner, R. Kaita, R. Majeski *et al.*, Fusion Engineering and Design 60, 157 (2002).
- Evtikhin, V., I. E. Lyublinski, A. V. Vertkov, S. V. Mirnov *et al.*, Fusion Engineering and Design 56, 363 (2001).
- Kaita, R., R. Majeski, M. Boaz, P. Efthimion *et al.*, Fusion Engineering and Design 61, 217 (2002).
- Lazarev, V.B., A. G. Alekseev, A. M. Belov, S. V. Mirnov *et al.*, Plasma Phys. Rep. 28, 802 (2002).
- Majeski, R., M. Boaz, D. Hoffman, R. Kaita *et al.*, J. Nuclear Mater. 313, 625 (2003).
- Majeski, R. R. Doerner, T. Gray, T., R. Kaita *et al.*, Phys. Rev. Lett. 97, 075002, (2006).
- Majeski, R., L. Berzak, T. Gray, R. Kaita *et al.*, Nucl. Fusion, 49, 055014, (2009).
- Mansfield, D.K., D. W. Johnson, B. Grek, H. W. Kugel *et al.*, Nucl. Fusion 41, 1827 (2001).
- Petty, C. C., M. R. Wade, J. E. Kinsey, R. J. Groebner *et al.*, Phys. Rev. Lett. 83, 3661 (1999).
- Zakharov, L.E., N.N. Gorelenkov, R. B. White, S. I. Krashininikov *et al.*, Fusion Engineering and Design 72, 149 (2004).
- Zakharov, L.E., W. Blanchard, R. Kaita, H. Kugel *et al.*, J. Nuclear Mater. 363, 453 (2007).

Chapter 16

Phase Integral Methods

16.1　INTRODUCTION

The power and simplicity of phase integral methods (Heading, 1962) for the approximate solution of differential equations make them a common tool in many branches of physics, and a particularly useful one in plasma physics, where the equations are often too cumbersome to solve by standard exact methods. Many of the differential equations of interest can be put in the form

$$\frac{d^2\psi}{dz^2} + Q(z,\omega)\psi = 0. \tag{16.1}$$

In these cases, the existence of solutions and the approximate complex eigenfrequencies ω can often be determined by phase integral methods. This technique has been used to examine unstable modes in plasmas in the areas of ballooning modes, drift waves, and other microinstabilities, as well as to investigate parametric instabilities associated with incident wave energy both in tokamak heating and laser pellet fusion problems. It is not possible to cover all aspects of the method of phase integrals here. We restrict ourselves to deriving the most essential results, which should permit the solution of most problems arising in plasma physics that can be addressed by these methods.

　We take Eq. 16.1 as standard form for the differential equation to be examined. The complex frequency ω generally plays the role of an unknown eigenvalue, and $z = x + iy$ is a complex variable. The physical problem is initially defined on the real axis and the equation has been analytically continued into the complex plane. The physical problems include searching for the existence of an instability, in which case outgoing wave boundary conditions are imposed for a growing mode and ω becomes an eigenvalue to be determined, or finding the amplitudes and phases of reflected and transmitted waves given an incoming wave of frequency ω.

Briefly, the WKBJ approximate solutions of Eq. 16.1, so named after Wentzel (1926), Kramers (1926), Brillouin (1926), and Jeffreys (1923), take the form

$$\psi_\pm = Q^{-1/4} e^{\pm i \int^z Q^{1/2} dz}, \tag{16.2}$$

and provided that

$$\left| \frac{dQ}{dz} Q^{-3/2} \right| \ll 1 \tag{16.3}$$

a general solution of Eq. 16.1 can be approximated by

$$\psi = a_+ \psi_+ + a_- \psi_-. \tag{16.4}$$

The solutions $\psi_\pm$ are local, not global solutions of Eq. 16.1. Clearly, inequality (16.3) is not valid in the vicinity of a zero of $Q(z, \omega)$, commonly called a turning point. Aside from this, however, $\psi_\pm$ are not approximations of a continuous solution of Eq. 16.1 in the whole z plane; *i.e.* if ψ is to approximate a continuous solution of Eq. 16.1, then the coefficients $a_\pm$ are not fixed over the whole z plane. The method of phase integrals consists in relating, for a given solution of Eq. 16.1, the WKBJ approximation in one region of the z plane to that in another.

These regions are separated by the so-called Stokes and anti-Stokes lines (Stokes, 1899) associated with $Q(z, \omega)$, and thus the qualitative properties of the solution are determined once these lines are known. The Stokes (anti-Stokes) lines associated with $Q(z, \omega)$ are paths in the z plane, emanating from zeros or singularities of $Q(z, \omega)$, along which $\int Q^{1/2}(z, \omega) dz$ is imaginary (real). We review first the characteristic properties of these lines and then the way in which they determine the global nature of a WKBJ solution.

Define a local anti-Stokes line to be, for any z_0 an infinitesimal path dz emanating from z_0 along which $Q^{l/2} dz$ is real. Along this path $|\psi_\pm|$ are essentially constant, *i.e.* the solutions are oscillatory. If $Q(z_0, \omega)$ is finite and well behaved, the local anti-Stokes line is given by setting dz equal to a real number times $\pm Q(z_0)^{-1/2}$, *i.e.*, from z_0 there issue two oppositely directed lines. Points at which $Q(z_0)$ is zero or infinity must be analyzed with more

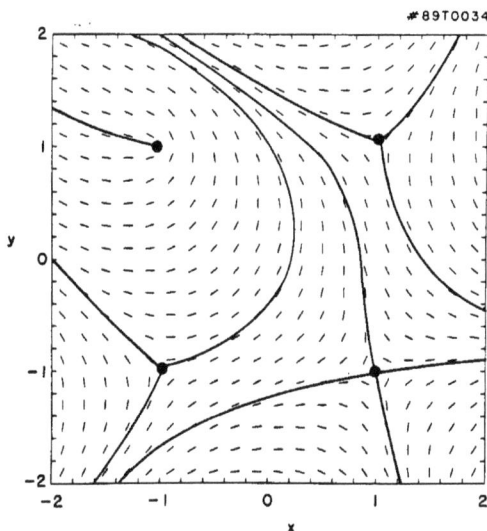

Fig. 16.1   Stokes diagram for $Q = (z - z_1)(z - z_2)(z - z_3)^2/(z - z4)$ with $z_1 = l + i$, $z_2 = -1 - i$, $z_3 = 1 - i$, and $z_4 = -1 + i$.

care. When the zero is first order, consider an infinitesimal line emanating from z_0, with $dz = (z - z_0)$. Then write $Q(z) \simeq Q'(z_0)dz$, and require that dz be a local anti-Stokes line, *i.e.* that $Q(z)^{1/2}dz = Q'(z_0)^{1/2}dz^{3/2}$ be real. Since dz is then proportional to $Q'(z_0)^{-1/3}e^{i(2n\pi)/3}$ with n integer, we find that three anti-Stokes lines emanate from z_0. Similarly, one finds that from a double root there issue four anti-Stokes lines, from a simple pole a single line, etc.. It is thus quite easy to read the locations of zeros, poles, etc., of a function from a plot of the z plane upon which are displayed the local anti-Stokes lines, which we will refer to as a Stokes diagram. An example is shown in Fig. 16.1 for a function Q, which possesses simple zeros in the first and third quadrants, a second order zero in the fourth quadrant, a pole in the second quadrant, and no other zeros or singularities. In refering to Stokes diagrams, we will refer to both zeros and singularities of Q(z) as singular points, since it is the function $Q^{1/2}$ which is relevant in this diagram.

A display of this nature allows a qualitative survey of the analytic structure of a function, without the numerical complication of an actual search for roots (White, 1979, 2000).

Using the local anti-Stokes lines as guides, we can form global, continuous anti-Stokes lines for those particular lines which emerge from the singular points of the Stokes plot, and these lines have been added to Fig. 16.1. Along the global anti-Stokes lines the functions $\psi_\pm$ are, within the validity of the WKBJ approximation, of constant amplitude, *i.e.* oscillatory. We similarly define local and global Stokes lines to be lines emerging from the singular points for which the integral $\int Q^{1/2} dz$ is imaginary. Along the Stokes lines the WKBJ solutions are exponentially increasing or decreasing with fixed phase. Except at singular points, the Stokes and anti-Stokes lines are orthogonal. The global anti-Stokes and Stokes lines which are attached to the singular points of the Stokes diagram, along with the Riemann cut lines, determine the global properties of the WKBJ solutions.

In the notation of Heading, including the slow $Q^{-1/4}$ dependence, a WKBJ solution is denoted by

$$(a, z)_s = Q^{-1/4} e^{i \int_a^z Q^{1/2} dz} \tag{16.5}$$

where the subscript s(d) indicates that the solution is subdominant (dominant); *i.e.* exponentially decreasing (increasing) for increasing $|z - a|$ in a particular region of the z plane, bounded by Stokes and anti-Stokes lines. The point a is taken to be a nearby singular point to which the dominancy or subdominancy refers. The two independent local WKBJ approximate solutions of Eq. 16.1 in this notation are given by (z,a) and (a,z). Clearly, if (z,a) is subdominant, then (a,z) is dominant. It is readily verified that upon crossing an anti-Stokes line these two solutions reverse character. Thus we find that upon crossing an anti-Stokes line we must make the change $(a, z)_d \to (a, z)_s$ and $(z, a)_s \to (z, a)_d$. This is the first of the connection formulae, a collection of rules for continuing a solution through the z plane in the presence of cuts, Stokes lines, and anti-Stokes lines.

16.2   CONNECTION FORMULAE

We first consider the continuation of a solution about an isolated turning point, located at $z = 0$. Later we show how one can pass from one turning point to another, allowing continuation through the entire z plane. The connection formulae depend on the nature of the turning point, and for simplicity we first consider a first order turning point, $Q \sim -z$, the associated Stokes diagram shown in Fig. 16.2.

First consider crossing a cut. Analytically continuing the solution, Eq. 16.5, counter-clockwise around the turning point a we find that $\psi_\pm \to -i\psi_\mp$. Thus in crossing the cut in a clockwise sense, in order to ensure continuity of our continued solution, we must make the changes

$$(0, z) \to i(z, 0)$$
$$(z, 0) \to i(0, z). \qquad (16.6)$$

Dominancy (or subdominancy) is not changed.

Now consider the process of crossing an anti-Stokes line, where dominant and subdominant solutions exchange character. Begin in the vicinity of a nearby Stokes line, and suppose the solution to Eq. 16.1 is approximated by a dominant expression $(z, 0)_d$, given by Eq. 16.5. A small subdominant part could also be present, so to speak, lost in the noise of the WKBJ approximation.

Trying to continue the solution past an anti-Stokes line creates a problem, because the previously small subdominant part, with an unknown coefficient, becomes dominant, making our solution totally inaccurate. To correct this, one must, in the vicinity of the Stokes line, choose the coefficient of the subdominant solution so that the continuation to a nearby anti-Stokes line will give the correct solution. The necessary coefficient of the subdominant solution is called the Stokes constant.

For a first order turning point, the Stokes constant can be derived simply by requiring that the solution be single valued upon continuation about the turning point. We know this is true because for $Q = -z$ the point $z = 0$ is a regular point of the differential equation and the solution is representable by a Taylor series with infinite radius of convergence. This is not the case for all forms of Q, in some cases the solution itself may possess cuts originating at zeros or singularities of Q. Begin with a subdominant solution, $(0, z)_s$ along the positive real axis of Fig. 16.2, a Stokes line. We deliberately begin with a subdominant solution so that the solution is small and cannot contain any dominant part due to the approximate nature of the WKBJ solution. Now continue this solution in both directions about the turning point.

Passing upward into domain 3, or downward into domain 7, an anti-Stokes line is passed and the solution becomes dominant, reaching maximal dominancy along the Stokes lines separating domains 3, 4 and domains 6, 7. On crossing these Stokes lines we must add subdominant parts to make up for any lost by the WKBJ approximation. Thus on passing into

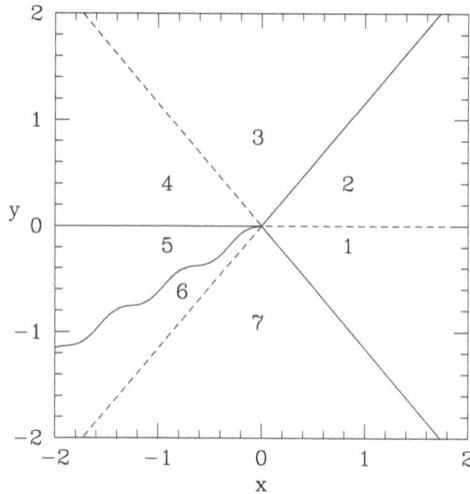

Fig. 16.2   Stokes diagram for a first order turning point.

domain 4 the solution becomes $(0, z)_d + T_1(z, 0)_s$ and in domain 6 it becomes $(0, z)_d - T_2(z, 0)_s$ where T_1 and T_2 are Stokes constants, and the signs are chosen to reflect the fact that the Stokes line is crossed clockwise in one case and counterclockwise in the other. Now continue both solutions into domain 5. Coming from domain 6 we cross the cut clockwise, giving $i(z, 0)_d - iT_2(0, z)_s$. From domain 4 we cross an anti-Stokes line, giving $(0, z)_s + T_1(z, 0)_d$. Equating these determines the Stokes constants to be $T_1 = T_2 = i$.

We can now use this result to relate the asymptotic behavior at $+\infty$ to that at $-\infty$. For large positive x the solution has the form $(0, z)_s \sim e^{-\frac{2}{3}x^{3/2}}/x^{1/4}$. Note that the branch of the square root is determined by the fact that this solution must be subdominant, and that we have multiplied by a constant phase to choose the solution to be real. Writing $z = re^{i\pi}$, a phase choice determined by our placement of the cut in Fig. 16.2, we find $(0, z) = e^{i(\frac{2}{3}r^{3/2} - \pi/4)}/r^{1/4}$ and $(z, 0) = e^{-i(\frac{2}{3}r^{3/2} + \pi/4)}/r^{1/4}$. Placing the cut in another position will modify some expressions, but not final results. The solution in domain 4 is $(0, z)_d + i(z, 0)_s = 2\sin(2r^{3/2}/3 + \pi/4)$. Choose overall normalization for positive x to be $Ai(x) \simeq e^{-\frac{2}{3}x^{3/2}}/(2\sqrt{\pi}x^{1/4})$, giving for large negative x $Ai(x) \simeq \sin(2r^{3/2}/3 + \pi/4)/(\sqrt{\pi}r^{1/4})$, with $r = |x|$, which is of course the correct asymptotic expression for the Airy function. The continuation of the second function, $Bi(x)$, is slightly more subtle, since the solution involves dominant and subdominant solutions on a Stokes line.

To obtain the correct expression for a solution on a Stokes line, we must consider flux conservation. Multiply the differential equation $d^2\psi/dz^2 + Q\psi = 0$ by ψ^* and its complex conjugate by ψ and subtract the two equations, giving, for Q real,

$$\frac{d}{dz}[\psi^*\psi' - \psi'^*\psi] = 0. \tag{16.7}$$

Thus $S = Im\psi^*\psi'$, referred to as the flux, is conserved if Q is real. This is not an approximate WKBJ result, it is exact. Now consider the solution $(0, z) = e^{i\frac{2}{3}z^{3/2}}/z^{1/4}$ for z negative and real in Fig. 16.2. We have $(0, z)' = ie^{i\frac{2}{3}z^{3/2}}z^{1/4}$ plus a higher order term in $1/z$, giving for the flux $S = 1$. Now continue to large positive x. In domain 4 this solution is exponentially decreasing, so we have $(0, z)_s$, and the same in domain 3. Finally we have in domain 2 $(0, z)_d$. Now continue in the lower half plane. In 5 we have $(0, z)_d$, and in 6 $-i(z, 0)_d$, giving in 7 $-i(z, 0)_d + (0, z)_s$, and in 1 $-i(z, 0)_s + (0, z)_d$. Now we recognize a problem, since the solutions in domains 1 and 2 disagree by the presence of the subdominant term. The correct expression can be obtained by considering the flux, since the flux for the correct solution must be one, and this is an exact result. Write the solution as $(0, z)_d - T(z, 0)_s$. The flux is then $T - T^*$ and thus the correct value for T is $i/2$, not i. Thus the correct value on the real line, a Stokes line, is just the average of the values obtained in domains 1 and 2, above and below the line. We thus obtain a special rule regarding Stokes lines lying on the real axis. Use half the Stokes constant to step on to the Stokes line, and again half to step off. The value exactly on the line is given by the mean of the values above and below the line.

Return to the Airy function $Bi(x)$. Choose for large negative x the asymptotic form $Bi(x) \simeq cos(\frac{2}{3}r^{3/2} + \pi/4)/(\sqrt{\pi}r^{1/4})$, express this in terms of $(0, z)$ and $(z, 0)$ and continue to the right, giving for the solution exactly on the real line, $Bi(x) \simeq e^{\frac{2}{3}x^{3/2}}/(\sqrt{\pi}x^{1/4})$. It is often stated that the choice of an asymptotically dominant form does not uniquely determine the solution because of possible subdominant parts, but if the "half Stokes constant" rule is used, the solution is unique.

It is possible to find the Stokes constant for a turning point of arbitrary order by analytically solving the exact differential equation in the vicinity of the turning point. Consider the vicinity of a turning point of order n,

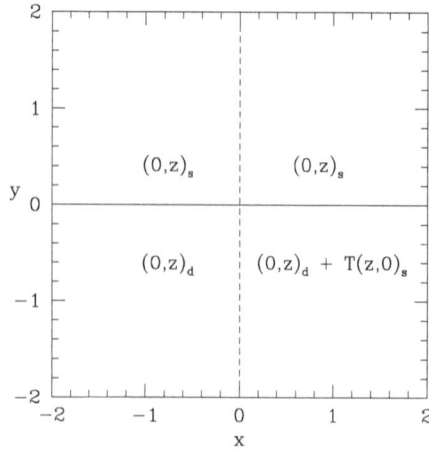

Fig. 16.3   Stokes diagram for the Bessel function.

where the differential equation has the form

$$\frac{d^2\phi}{dw^2} + w^n\phi = 0. \tag{16.8}$$

Substitute $\phi = w^{1/2}u$, $z = 2w^{(n+2)/2}/(n+2)$, giving Bessel's equation of order $1/(n+2)$,

$$u'' + \frac{1}{z}u' + \left(1 - \frac{\nu^2}{z^2}\right)u = 0, \tag{16.9}$$

with $\nu = 1/(n+2)$. To find the Stokes constants, we analytically continue the solution to the Bessel equation. The canonical form, Eq. 16.1, can be obtained in the variable z by letting $u = \psi/z^{1/2}$, giving $\psi'' + Q(z)\psi = 0$, with $Q(z) = 1 + (1/4 - \nu^2)/z^2$. For large $|z|$ the solutions have the form

$$u \simeq \frac{1}{\sqrt{z}}e^{\pm iz}. \tag{16.10}$$

The Stokes diagram is shown in Fig. 16.3. The origin is a second order pole of Q, the positive and negative real axes are anti-Stokes lines, and the positive and negative imaginary axes are Stokes lines.

Also shown are the WKBJ approximations to the solution, where we have chosen the form along the positive real axis to be $u \sim \frac{1}{\sqrt{x}}e^{ix}$, subdominant in the upper half plane. This solution has been continued counterclockwise around the origin to obtain the expressions shown in each quadrant. Note that the origin is a regular singular point of the differential equation, so the expression in quadrant 4 cannot be continued across the real axis and equated to the expression in quadrant 1, because the solution itself possesses a cut.

A general solution of the Bessel equation can be written

$$u = Az^{\nu} P_1(z^2) + Bz^{-\nu} P_2(z^2) \tag{16.11}$$

where P_1 and P_2 are convergent series in z^2. See, for example, Whittaker and Watson (1962). We choose A and B so that the asymptotic form of u matches the expressions in Fig. 16.3. Along the real axis in the upper half plane we have for large x

$$Ax^{\nu} P_1(x^2) + Bx^{-\nu} P_2(x^2) = \frac{1}{\sqrt{x}} e^{ix}. \tag{16.12}$$

Continuing counter clockwise by taking $z = e^{i\pi} r$ and matching to the subdominant form we find for large r

$$Ae^{i\nu\pi} r^{\nu} P_1(r^2) + Be^{-i\nu\pi} r^{-\nu} P_2(r^2) = \frac{e^{-i\pi/2}}{\sqrt{r}} e^{-ir}. \tag{16.13}$$

Similarly letting $z = re^{i2\pi}$ and matching to the form in the lower right quadrant we find

$$Ae^{i2m\pi} r^{\nu} P_1(r^2) + Be^{-i2\nu\pi} r^{-\nu} P_2(r^2) = \frac{e^{-i\pi}}{\sqrt{r}} e^{ir} + T\frac{e^{-i\pi}}{\sqrt{r}} e^{-ir}. \tag{16.14}$$

These three equations can be written as a matrix equation for the vector $(Ar^{\nu} P_1(r^2), Br^{\nu} P_2(r^2), r^{-1/2} e^{ir})$. There is a solution only if

$$\det \begin{pmatrix} 1 & 1 & -1 \\ e^{i\nu\pi} & e^{-i\nu\pi} & e^{-i2r} \\ e^{i2\nu\pi} & e^{-i2\nu\pi} & 1+Te^{-i2r} \end{pmatrix} = 0 \tag{16.15}$$

which gives $T = 2i\cos\nu\pi$ for the Stokes constant. Returning to Eq. 16.8 we find the Stokes constant for a turning point of order n to be given by

$$T = 2i\cos\left(\frac{\pi}{n+2}\right). \tag{16.16}$$

The form of the WKBJ solutions in the w plane is $\phi \sim e^{\pm i \int w^{n/2}} dw$, and in the z plane $u \sim e^{\pm i \int dz}$. Since $dz \sim w^{n/2} dw$ the Stokes and anti-Stokes lines in the two planes correspond as they must. The Stokes constant for crossing Stokes lines in the complex w plane of Eq. 16.8 is thus given by Eq. 16.16.

These results can be simply summarized. They prescribe a set of rules, first given by Heading, for obtaining a globally defined WKBJ solution which corresponds to the approximation of a single solution of the differential equation.

Begin with a particular solution in one region of the z plane, choosing that combination of subdominant and dominant solutions which gives the desired boundary conditions in this region. The global solution is obtained by continuing this solution through the whole z plane effecting the following changes:

1. If a_d and a_s are respectively the coefficients of the dominant and subdominant terms of a solution, then upon crossing a Stokes line in a counterclockwise sense a_s must be replaced by $a_s + T a_d$ where T is called the Stokes constant. When the Stokes line originates at an isolated zero of order n, $T = 2i\cos(\pi/(n+2))$.

2. Upon crossing a cut in a counterclockwise sense, the cut originating from a first order zero of Q at the point a, we have

$$(a, z) \rightarrow -i(z, a)$$
$$(z, a) \rightarrow -i(a, z). \tag{16.17}$$

Properties of dominancy or subdominancy are preserved in this process.

3. Upon crossing an anti-Stokes line, subdominant solutions become dominant and vice versa.

4. Reconnect from singularity a to singularity b using $(z, a) = (z, b)[b, a]$ with $[b, a] = e^{i \int_b^a Q^{1/2} dz}$. If a, b are joined by a Stokes line, reconnect while on the line, using $1/2$ the usual Stokes constant to step on the line, and again $1/2$ to step off.

Using these rules we can pass from region to region across the cuts, Stokes and anti-Stokes lines emanating from a turning point. Beginning with any combination of dominant and subdominant solutions in one region, this process leads to a globally defined single valued approximate solution of Eq. 16.1. Although it would appear that the first rule gives rise to a discontinuous solution, this is not the case. At the Stokes line, in the presence of a dominant solution, the discontinuity produced is small compared to the error due to the WKBJ approximation itself. As one continues further away from the Stokes line, however, the subdominant term will begin to be important, and the modified coefficient is the correct one.

A Stokes structure consisting of more than an isolated singularity is more complicated, in that the Stokes constants are modified by the proximity of the other singularities. However, the modification is normally exponentially small, and one can, in most cases, use the values of the Stokes constants obtained for isolated singularities, also for complex structures. In the following, for the sake of completeness, we will derive the Stokes constants for structures involving two singularities to explicitly display this behaviour. The complete expressions are essential to understand, for example, the behaviour of the Stokes phenomena when two singularities approach one another, in which case the mutual influence of the singularities on one another is not negligible.

However, as will be seen, the values of the Stokes constants for the case of a bound state, the most common problem encountered in searching for instabilities, are exactly those given by the isolated singularities, and for scattering problems the use of the values for isolated singularities produces only exponentially small errors. Thus for essentially all practical calculations, the values given by the isolated singularities can be used.

16.3 CAUSALITY

To properly understand the choice of boundary conditions in solving differential equations, it is necessary to invoke causality, namely to require an outward group velocity in directions from which there is assumed to be no incoming wave. Assume for simplicity that Q tends to a constant k^2 for large $|x|$. The WKBJ solutions take the form $\Psi_\pm = e^{\pm ikx}$. In a physics problem the time dependence is also relevant, and the wave has a frequency $\omega(k)$ with the k dependence given by properties of the physical system. Form a wave packet

$$\Psi_+(x,t) = \int dk f(k) e^{i(kx - \omega t)} \tag{16.18}$$

where $f(k)$ is localized about k_0. Expand $\omega(k) \simeq \omega_0 + d\omega/dk(k-k_0)$, giving

$$\Psi_+(x,t) = e^{i(k_0 x - \omega_0 t)} \int dk f(k) e^{i(k-k_0)(x-td\omega/dk)}. \qquad (16.19)$$

The integral is very small unless $x \simeq (d\omega/dk)t$, giving the usual identification of $d\omega/dk$ as group velocity, and thus Ψ_+ is outgoing for $x > 0$ if $d\omega/dk > 0$. Now consider the spatial dependence of Ψ_+ with a complex frequency. If the mode frequency is $\omega = \omega_r + i\gamma$ expanding $e^{ik(\omega)x}$ gives

$$\Psi_+ \simeq e^{(ik_0 x - \gamma x dk/d\omega)}. \qquad (16.20)$$

We then obtain the rules for the asymptotic behaviour of the solution. If $\gamma > 0$ and $d\omega/dk > 0$ the solution is decreasing (subdominant) for large x. Physically the spatial behaviour can be simply understood in terms of information propagation. If the mode is growing the news of its growth propagates outward at the group velocity, so the mode is largest for small $|x|$, *i.e.* it is damped in the direction of propagation. In the following we will use the introduction of a small dissipation to clarify the choice of the solution.[*]

16.4 BOUND STATES – INSTABILITIES

The bound state problem, or the search for instability, is generically given by a function $Q(z)$ which is real on the real axis, with two first order zeros at points a and b, and Q is positive between a and b and negative otherwise. The Stokes diagram is shown in Fig. 16.4. Denote the Stokes constants as S_k, where k refers to a line bordering domain k. We will find that the boundary conditions immediately give the Bohr–Sommerfeld condition, which determines the energy of the bound state, or equivalently, the growth rate of the instability, independent of the values of the Stokes constants. We further find the six Stokes constants can be represented by a single magnitude and one phase δ, but with a sign which must be distributed as shown in Fig. 16.4. In addition, the bound state boundary conditions fix the Stokes constants to be equal to the value for isolated singularities.

Begin at large positive x with a subdominant solution. Define $[a,b] = e^{iW}$. Along the real axis between a and b, Q is real and positive and thus W

---

[*]Note that the convention $e^{-i\omega t}$ is opposite from that used by Heading, which switches upper and lower half planes in some arguments.

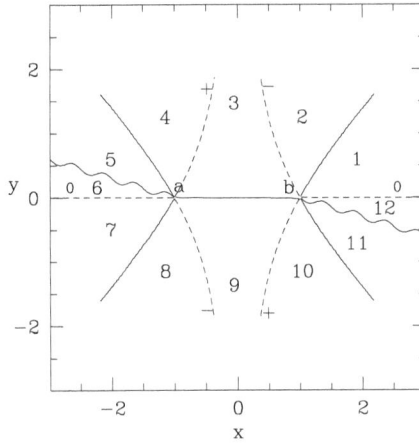

Fig. 16.4   Stokes plot for the bound state problem.

is real and positive. With the choice of cuts as shown, and using Q positive and real for $a < x < b$ we have for $x > b$ and for $x < a$ that $Q^{1/2} \sim e^{-i\pi/2}$, $Q^{1/4} \sim e^{-i\pi/4}$. Thus for $b < z$ the solution (b, z) is dominant. Begin with a subdominant solution at large positive x and continue:

(1) $(z, b)_s$

(2) $(z, b)_d$

(3) $(z, b)_d + S_2(b, z)_s = e^{iW}(z, a)_d + e^{-iW}S_2(a, z)_s$

(4) $e^{iW}(z, a)_d + [e^{iW}S_4 + e^{-iW}S_2](a, z)_s$

(5) $e^{iW}(z, a)_s + [e^{iW}S_4 + e^{-iW}S_2](a, z)_d$

(6) $-ie^{iW}(a, z)_s - i[e^{iW}S_4 + e^{-iW}S_2](z, a)_d$.

The differential equation is real, and thus the complex phase of the solution is constant along the real axis. $(z, a)_d$ in domain 6 (z real) has the same complex phase as $(z, b)_s$ in domain 1 (z real), thus the coefficient of the dominant solution must be real for any W, giving $S_4 = iSe^{i\delta}$, $= S_2 = iSe^{-i\delta}$, with S, δ real. We then have

(6) $-ie^{iW}(a, z)_s + 2S\cos(W + \delta)(z, a)_d$.

If the dominant part is nonzero, the subdominant correction is not relevant and we cannot require it to be real. Requiring the dominant solution to be zero gives $W + \delta = (n + 1/2)\pi$, in which case the subdominant part must be real, giving $W = (n + 1/2)\pi$, and therefore $\delta = 0$. S remains undetermined. Note, however, that bound state boundary conditions, namely the requirement that the dominant solution vanish to the left, gives $W = (n + 1/2)\pi$, the Bohr–Sommerfeld quantization condition,

$$\int_a^b Q^{1/2} dx = (n + 1/2)\pi, \tag{16.21}$$

independent of the value of the Stokes constant. In the following we continue the solution completely around the complex plane in an attempt to determine S. This turns out to be only possible if the Bohr–Sommerfeld condition is imposed.

Continuing we find

(7) $2S\cos(W + \delta)(z, a)_d + [-ie^{iW} + 2S\cos(W + \delta)S_6](a, z)_s$

(8) $2S\cos(W + \delta)(z, a)_s + [-ie^{iW} + 2S\cos(W + \delta)S_6](a, z)_d$.

Beginning again in domain 1 and continuing in the lower half plane we find

(12) $(z, b)_s$

(11) $i(b, z)_s$

(10) $i(b, z)_d$

(9) $i(b, z)_d - iS_{10}(z, b)_s = ie^{-iW}(a, z)_d - iS_{10}e^{iW}(z, a)_s$

(8) $ie^{-iW}(a, z)_d - i[S_{10}e^{iW} + S_8 e^{-iW}](z, a)_s$

(7) $ie^{-iW}(a, z)_s - i[S_{10}e^{iW} + S_8 e^{-iW}](z, a)_d$.

Now compare the solutions obtained by continuing in each direction. The matching can be carrried out in both domains 7 and 8, where dominant and subdominant terms switch roles. Comparing, we find $S_{10} = iSe^{i\delta}$, $S_8 = iSe^{-i\delta}$, and $S_6 = iS$,

$$S = \sqrt{\frac{\cos(W)}{\cos(W + \delta)}}. \tag{16.22}$$

The relative distribution of positive, negative, and zero phases δ are thus determined and shown in Fig. 16.4, but matching determines neither the phase δ nor the magnitude S. However, at each bound state solution $\delta = 0$ and thus $S = 1$, *i.e.* the Stokes constant is equal to its value for an isolated first order zero. The general value is of interest only for the limit $a \to b$, in which case the Stokes constant differs significantly from the isolated singularity value.

The limit $a \to b$ is singular in that each pair of anti-Stokes lines extending vertically merges to become one, and thus the Stokes constants associated with two lines become that of one. The Stokes diagram obtained in this limit is shown in Fig. 16.5. For $a \to b$ the continuation is

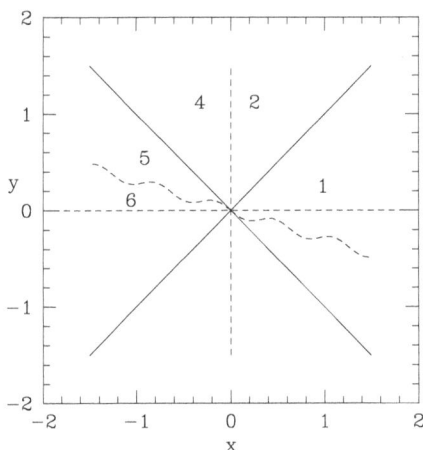

Fig. 16.5   Stokes plot for the bound state, $a \to b$.

given by the above with $W \to 0$.

(1) $(z,0)_s$

(2) $(z,0)_d$

To pass to domain 4 we cross two Stokes lines. Substituting the values for S_2 and S_4 we find

(4) $(z,0)_d + 2i\sqrt{cos(\delta)}(0,z)_s$

(5) $(z,0)_s + 2i\sqrt{cos(\delta)}(0,z)_d$.

Whereas in the singular limit, $a = b$, the Stokes constant is $\sqrt{2}i$ and the continuation is

(1) $(z,0)_s$

(2) $(z,0)_d$

(4) $(z,0)_d + \sqrt{2}i(0,z)_s$

(5) $(z,0)_s + \sqrt{2}i(0,z)_d$.

Comparing we find $\delta = \pi/3$. The Stokes constants for the two cases are not identical, with $S_k = e^{i\pi/3}\sqrt{2}i,\ e^{-i\pi/3}\sqrt{2}i,\ \sqrt{2}i$ in the case $a \to b$ (six Stokes lines) and $S_k = \sqrt{2}i$ for $a = b$ (four Stokes lines), with the two Stokes constants associated with the merging vertical lines, $S_k = e^{\pm i\pi/3}\sqrt{2}i$ adding to give $\sqrt{2}i$.

A numerical determination of the phase δ and an analytic fit as a function of W for the case $Q = b^2 - z^2$ are shown in Fig. 16.6. The two curves

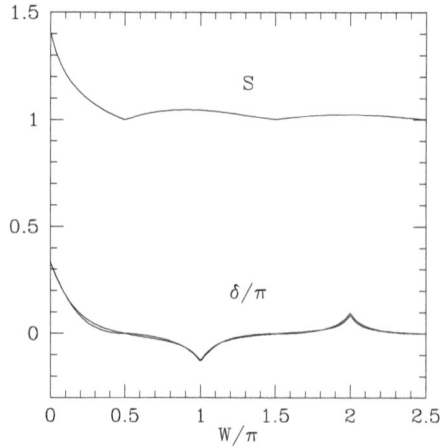

Fig. 16.6 δ and S for the bound state.

are almost indistinguishable. Note that at all bound states $W = (n + \frac{1}{2})\pi$ the phase is zero and $S = 1$, and at $W = 0$ $S = \sqrt{2}$. The approximate fit to S satisfying these conditions and giving a very good fit to the numerical δ is given by $S = \sqrt{1 + |cos(W)|/(1 + 3W)}$ and shown in the figure, with of course S and δ related by Eq. 16.22.

16.5   OVERDENSE BARRIER – SCATTERING

The overdense barrier is given by a function $Q(z)$ which is real on the real axis, with two first order zeros at real points a and b, and Q is negative between a and b and positive otherwise. The Stokes diagram is shown in Fig. 16.7, and propagating oscillatory solutions exist for large positive and negative x.

We consider an incident wave from the left, the problem being to determine the reflected and transmitted waves. In classical physics problems, the absolute square of these coefficients gives the reflected and transmitted power, and in quantum mechanical problems, the probability of reflection and transmission. Thus causality requires outgoing boundary conditions at the far right. Define W through $[a, b] = e^{-W}$. Take Q to be real and positive for $b < x$, then with the choice of the cuts as shown in Fig. 16.7 along the real axis between a and b, Q has phase $Q \sim e^{i\pi}$ and W is real

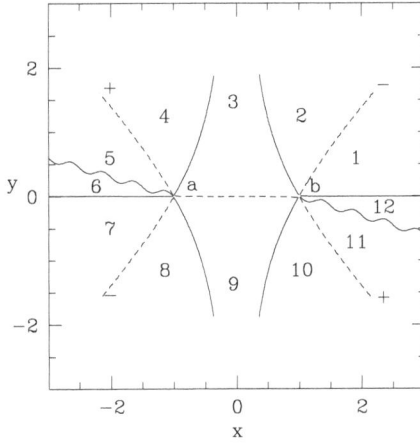

Fig. 16.7 Stokes plot for the overdense barrier.

and positive, and for $x < a$, Q is again real and positive. Requiring outgoing wave conditions for large positive x gives boundary conditions of a subdominant solution in domain 1. Continuation through the upper half plane gives:

(1) $(b, z)_s$

(2) $(b, z)_s$

(3) $(b, z)_d = e^W (a, z)_s$

(4) $e^W (a, z)_d$

(5) $e^W (a, z)_d + S_4 e^W (z, a)_s$

(6) $-i e^W (z, a)_d - i S_4 e^W (a, z)_s$.

Now consider the flux. In domain 1 we have $flux = Im(\psi^* \psi') = 1$. In domain 6 we find $flux = S_4 S_4^* e^{2W} - e^{2W}$. Without loss of generality write

$$S_4 = i\sqrt{1 + e^{-2W}} e^{i\delta}, \qquad (16.23)$$

with δ undetermined.

The solution in domain (6) becomes $\sqrt{1 + e^{-2W}} e^{i\delta} e^W (a, z)_s - i e^W (z, a)_d$. In domain (6) the subdominant solution is right-moving, and the dominant solution is thus the reflected wave. Using the edges of the propagation domains, a and b as the reference points for phase changes, we find reflection

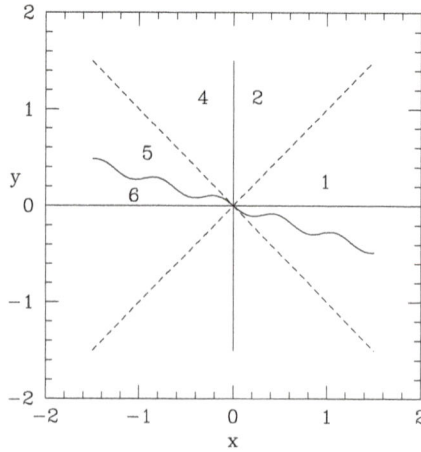

Fig. 16.8 $Q = z^2$, Stokes plot.

and transmission coefficients of

$$r = -\frac{ie^{-i\delta}}{\sqrt{1 + e^{-2W}}}, \tag{16.24}$$

$$t = \frac{e^{-W-i\delta}}{\sqrt{1 + e^{-2W}}}. \tag{16.25}$$

We thus find the probabilities for reflection and transmission to be

$$|r|^2 = \frac{1}{1 + e^{-2W}}, \tag{16.26}$$

$$|t|^2 = \frac{e^{-2W}}{1 + e^{-2W}}. \tag{16.27}$$

The phase δ is not given in general by WKBJ analysis, but can be determined in two limits. First examine the limit $a \to b$. The Stokes diagram collapses to a single second order zero, and the Stokes plot is shown in Fig. 16.8. We retain the cuts to make the results coincide exactly with those resulting from the limit $a \to b$ in Fig. 16.7. The Stokes constants have a continuous limit, all lines remaining distinct. In this case the Stokes constant is $\sqrt{2}i$ and the continuation is given by the above with $W = 0$,

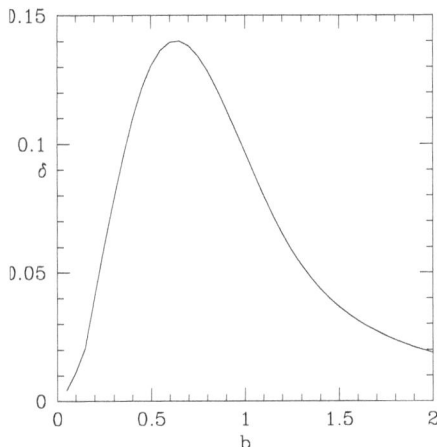

Fig. 16.9 Phase δ for overdense scattering, $Q = z^2 - b^2$.

$\delta = 0$, giving for the reflection and transmission coefficients $r = -i/\sqrt{2}$, $t = -1/\sqrt{2}$, and $|r|^2 = |t|^2 = 1/2$. In the limit $|a - b| \to \infty$ the singularities are isolated and $\delta = 0$, $S = 1$. Note that the analytic continuation carried out for large $|z|$ is valid for all W, large W is not a requirement for the WKBJ analysis.

Similarly by continuation through the lower half plane it can be shown that all Stokes constants can be written in the form $S_k = iSe^{\pm i\delta}$, $S = \sqrt{1 + e^{-2W}}$, for the line bordering domain k, with the sign of the phase δ distributed as shown in Fig. 16.7.

The phase can be determined for a particular functional form of $Q(z)$ by an analytical continuation of the solution (Bender–Orszag, 1978; Ford, 1959a, 1959b; Soop, 1965), a numerical integration of the differential equation, or by a matching of the solution to a numerical evaluation of the two Taylor series solutions. Shown in Figs. 16.9, 16.10 is the phase δ and the Stokes constant for the overdense barrier scattering problem with $Q = z^2 - b^2$, determined using the Taylor series to find the reflection and transmission coefficients. Note that the Stokes constant deviates significantly from the value given by the isolated singularity analysis, $S = i$ only for $b < 1$.

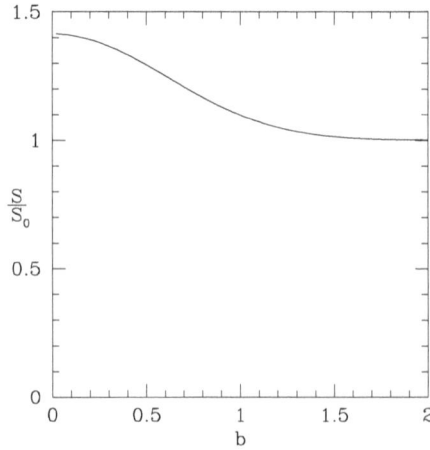

Fig. 16.10   Stokes constant for overdense scattering, $Q = z^2 - b^2$.

16.6   UNDERDENSE BARRIER – SCATTERING

The underdense barrier problem is given by a function $Q(z)$ which is real on the real axis, with two first order zeros at points a and b which are pure imaginary, and Q is positive everywhere on the real axis. The Stokes diagram is shown in Fig. 16.11. Again define W through $[a, b] = e^{-W}$. Along the imaginary axis between a and b Q is real and positive and W is real and positive. Consider an incident wave from the left, giving again outgoing boundary conditions at the far right. For this problem, Heading discusses two methods which he discards as approximate. The first method is continuation in the lower half plane alone. This gives a transmission coefficient of one and zero reflection. It is used by Berry (1990) in an analysis of the birth point of reflected waves.

Continuation in the upper half plane gives a transmission coefficient of one and a reflection coefficient of $r = -ie^{-W}$. This continuation involves the singular point $z = b$ only, and the correct Stokes constant is that associated with an isolated first order zero, $S = i$. This is because, as far as the continuation is concerned, the second singularity at $z = a$ does not exist, and the Stokes diagram consists only of the structure in the upper half plane.

However, neither of these continuations give solutions which conserve flux and neither can be considered correct. A third method is given by

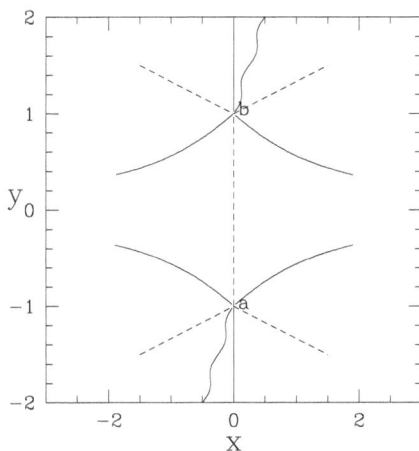

Fig. 16.11   Stokes plot for the underdense barrier.

Heading, who describes it as more accurate, but does not give a justification for its use. A justification for its use comes from the consideration of causality. Not only is it necessary to consider outgoing wave boundary conditions, a small dissipation must be considered so as to cause waves to damp in the direction of propagation. This is done by multiplying $Q(z)$ by $e^{i\nu}$ with ν small and positive. Since Stokes lines are given by $dz \sim 1/\sqrt{Q(z)}$, this results in a rotation of the Stokes plot. Shown in Fig. 16.12 is the resulting plot (the rotation has been exaggerated for clarity). Now the only possible continuation from large real positive z to large real negative z is clear. It is necessary to begin in domain 1 and continue to domain 7, a process which necessarily involves both singular points.

Continuation then gives:

(1) $(b, z)_s$

(2) $(b, z)_s$

(3) $i(z, b)_s$

(4) $i(z, b)_d$

(5) $i(z, b)_d + iS_-(b, z)_s$

(6) $i(z, b)_s + iS_-(b, z)_s$

(6) $ie^{-W}(z, a)_d + iS_- e^{2W}(a, z)_s$

(7) $i(z, a) + iS_-(0, z)$.

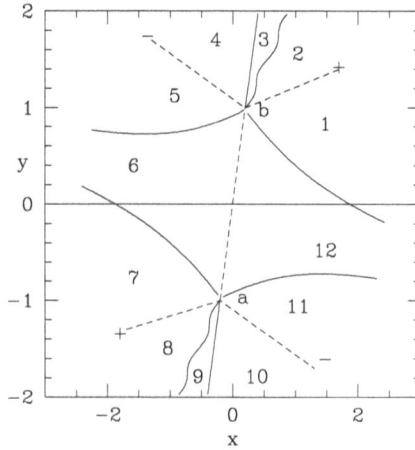

Fig. 16.12   Underdense barrier, rotated Stokes plot.

The incoming solution is $(a, z) = e^{-W}(b, z)$, and the reflected wave is $(z, a) = e^{-W}(z, b)$. We then have

$$r = \frac{e^{-W - i\delta}}{S}, \tag{16.28}$$

$$t = \frac{-ie^{-i\delta}}{S} \tag{16.29}$$

and equating flux for $x \to -\infty$ $x \to \infty$ gives again $S = i\sqrt{1 + e^{-2W}}$, but again δ is undetermined. This gives

$$|r|^2 = \frac{e^{-2W}}{1 + e^{-2W}}, \tag{16.30}$$

$$|t|^2 = \frac{1}{1 + e^{-2W}}. \tag{16.31}$$

A comparison with numerical integration for the case $Q = z^2 + b^2$ is shown in Fig 16.13, obtained numerically at $|x| = 5$ using 100 terms in the Taylor series, with the constants in the series fixed by requiring outgoing wave conditions at $x = 5$. Solid points show the data from the Taylor

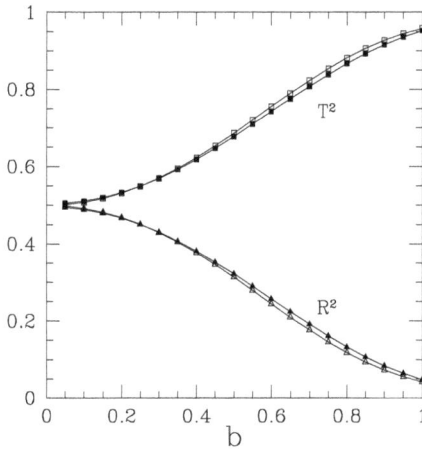

Fig. 16.13 $Q = z^2 + b^2$, transmission and reflection vs b.

expansion and open points the data from the WKBJ continuation, using Eqs. 16.30 and 16.31.

The phase δ can again be determined in two limits. First examine the limit $a \to b$. The Stokes diagram collapses to a single second order zero, and the Stokes plot is shown in Fig. 16.14, the same as in the case of the overdense barrier Fig. 16.8, except for cut placement. We retain the cuts to make the results coincide exactly with those resulting from the limit $a \to b$ in Fig. 16.12. Again the Stokes constants have a continuous limit. In this case the Stokes constant is $\sqrt{2}i$ and the continuation is given by the above with $W = 0$, giving for the reflection and transmission coefficients $r = -i/\sqrt{2}$, $t = -1/\sqrt{2}$, and $|r|^2 = |t|^2 = 1/2$.

The phase δ is found numerically to be the same as in the overdense case, which is not surprising since the two Stokes diagrams are equivalent under rotation.

16.7   EIGENVALUE PROBLEMS

Normally in plasma physics problems, the function Q is a complex function of the frequency of the mode ω. For an arbitrary guess for ω the Stokes plot will be quite complicated. To find an unstable mode one must identify

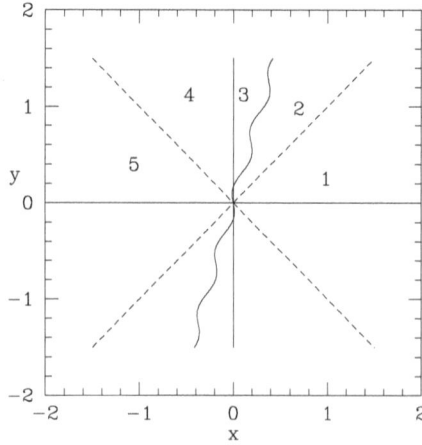

Fig. 16.14 $Q = z^2$, Stokes plot.

turning points and check whether there exists a value of ω giving a solution which satisfies the required boundary conditions. This is normally best accomplished numerically (White, 1979). Unless ω is chosen to be the eigenvalue, pairs of turning points will not exhibit the canonical attractive well Stokes structure shown in Fig. 16.4. A typical example of what is seen is shown in Fig. 16.15. In addition there may be other nearby singularities and zeros of $Q(z, \omega)$, making it difficult to recognize possible unstable mode structures. This is especially true if the physical problem possesses several parameters which must be explored.

As an example we give a problem which arose in the theory of the collisionless drift wave. The differential equation for this problem is given on the real axis by

$$Q = -\frac{\omega L}{2}\left(K^2 - 1 - \frac{1}{\omega}\right) + \frac{x^2}{4} + \frac{L}{2}(1 - \omega)AZ(A) \qquad (16.32)$$

where L, K, R are real parameters, $A = (\omega RL)^{1/2}/|x|$ and Z is the plasma dispersion function. A more complete description of the derivation and solutions of this equation have been reported elsewhere, Chen *et al.* (1978); Pearlstein and Berk (1969); Tsang *et al.* (1978); Ross and Mahajan (1978). The expression $|x|$ arises in a derivation valid on the real axis, and the

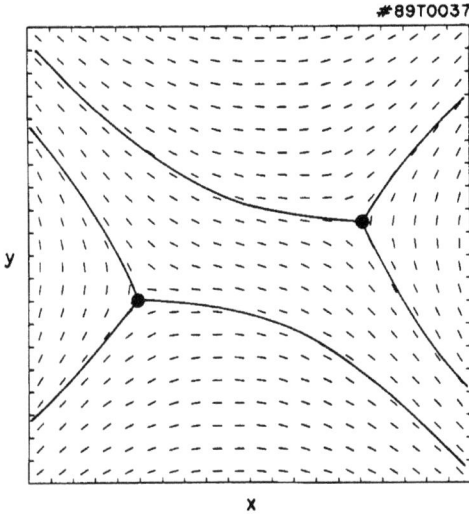

Fig. 16.15   Stokes diagram showing two turning points.

analytic continuation into the complex plane is provided by $|x| = (z^2)^{1/2}$ with an appropriate choice of cuts.

The physical real axis is determined by $|x| \geq 0$, and thus the physical plane is divided into two parts by branch cuts which can be taken along the imaginary axis. Thus in the physical plane for $Re z > 0$, $|x|$ is replaced by z, and for $Re z < 0$, $|x|$ is replaced by $-z$. The continuation of these functions through the cuts defines a second plane, which we will refer to as the nonphysical plane. It is also divided into two distinct parts. For values of K much less than a critical value K_c which depends on R and L (for $R = 1/1837$ and $L = 50$, $K_c = 0.36$), there are two turning points located approximately along the line $x = -y$ in the physical plane. As K increases, these turning points begin to migrate toward the imaginary axis. In Fig. 16.16 is shown the location of the turning points (v and $-v$) for a mode with $L = 50$, $K = 0.26$, and $R = 1/1837$. Note the presence of another pair of turning points (g and $-g$), located in the nonphysical plane. Assume a decaying mode, in which case the solution for $x \to \infty$ must be dominant, $\psi(z) = (z, v)_d$. Continuing in toward $x = 0$ we cross a Stokes line associated with v and thus the solution becomes $\psi(z) = (z, v)_d + i(v, z)_s$.

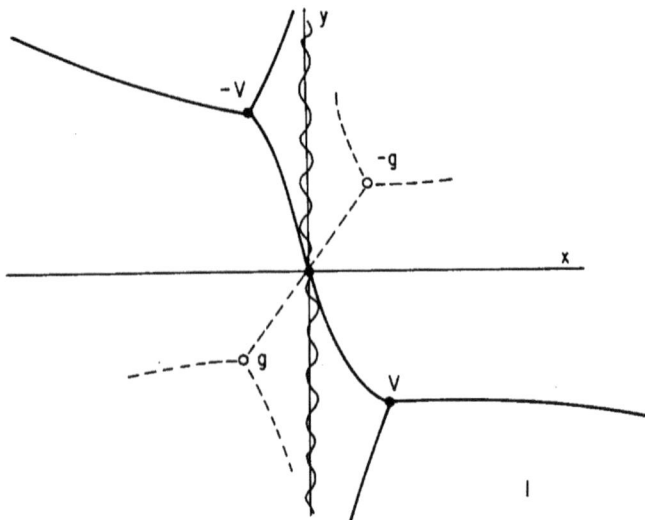

Fig. 16.16   Stokes structure for $K < K_c$.

The differential equation and the turning points under consideration are symmetric about $z = 0$, thus we can choose a solution with a particular parity with respect to z. For an odd solution we require $\psi(0) = 0$, or $(0, v) + i(v, 0) = 0$. This has the solution $2 \int_0^v Q^{1/2}dz = [n + 1/2]\pi$, n odd. For the even solutions we require that $d\psi/dz$ vanish at $z = 0$, or $i(0, v) + (v, 0) = 0$. This has the solution $2 \int_0^v Q^{1/2}dz = [n + 1/2]\pi$, n even. Thus we obtain the standard connection formula between $v, -v$ which for the parameters given above gives $\omega = 0.908 - 0.008i$. The mode is decaying in agreement with our initial assumption of a purely dominant solution for large x. It is readily verified that the anti-Stokes line emanating from v toward positive x does not cross the real axis.

As the parameter K increases, provided $L > 3R^{-1/4}$, the turning points $v, -v$ as well as the turning points $g, -g$ approach the imaginary axis and coalesce for $K = K_c$ after which the Stokes diagram takes the form of Fig. 16.17. The role of turning point for the determination of the solution has been passed on from v to g. Once again the analysis presented above carries through, only now the connection formula is determined by performing the integral from g to $-g$. We then discover that the growth rate $Im(\omega)$ is zero.

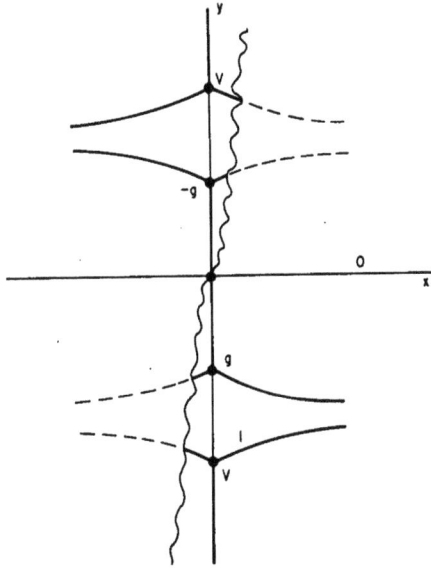

Fig. 16.17   Stokes structure for $K > K_c$.

If $L < 3R^{-1/4}$ this coalescense does not occur for any value of K, and the mode continues to be determined by the turning point v.

For K less than the critical value, *i.e.* with Stokes structure as shown in Fig. 16.16, imposing the connection formula between g and $-g$ gives rise to a growing mode $Im(w) > 0$. However, this ghost mode remains in the nonphysical plane, interpretable as an outwardly propagating solution only in this plane. The critical value of K is seen to be the coalescense of the turning points of a nonphysical growing mode and a physical damped mode. The nonphysical turning point then dominates to produce a marginally stable mode for all larger values of K.

16.8   DETERMINING STOKES CONSTANTS

For an isolated sungularity, the Stokes constant is determined to be i by requiring the solution to be single valued in the plane. Interestingly, the determination of the energies for the simplest bound state problem, the Weber equation, does not depend on the value of the Stokes constant, something not true for more complicated potentials. It is of interest to try to determine Stokes constants for other more complicated problems. See R. B. White and A.G. Kutlin, Archive 1704.01170v5. In the following we examine determination of the Stokes constants for the Weber equation, the Budden problem, fourth and sixth order potentials, and a non Hermitian potential.

16.8.1 The Weber Equation

A general solution of the Weber equation can be written with use of parabolic cylinder functions:

$$y(z) = C_1 U(-E/2, \sqrt{2}z) + C_2 U(E/2, i\sqrt{2}z). \tag{16.33}$$

To find Stokes constants in the upper half-plane of the complex z-plane let us choose C_1 and C_2 such that $y(z) \sim (a, z)$ for $-\pi/4 < Arg(z) < \pi/4$. According to Eq. 16.5

$$(a, z) = (E - z^2)^{-1/4} e^{i \int_a^z (E - z^2)^{1/2} dz}$$
$$= (E - z^2)^{-1/4} e^{-z^2/2 + E/4 + (E/2) \ln(2z/\sqrt{E}) + O(1/z)}, \tag{16.34}$$

and in a limit of large z we can write

$$(a, z) \sim e^{i\pi/4} \left(2\sqrt{\frac{e}{E}} \right)^{E/2} z^{-1/2 + E/2} e^{-z^2/2}, \tag{16.35}$$

$$(z, a) \sim e^{i\pi/4} \left(2\sqrt{\frac{e}{E}} \right)^{-E/2} z^{-1/2 - E/2} e^{z^2/2}. \tag{16.36}$$

The common factor $e^{i\pi/4}$ can be omitted for further calculations due to the linearity of the Weber equation.

Using asymptotic expansions for the parabolic cylinder functions we find that the required values for the arbitrary constants are

$$C_1 = 2^{-1/4} (2e/E)^{-E/4}, C_2 = 0. \tag{16.37}$$

Once we have determined the values of the arbitrary constants in Eq.16.33, we can write an asymptotic expansion of our solution in any domain. Particularly, for $\pi/4 < Arg(z) < 3\pi/4$ it has the form

$$y(x) \sim (a, z)_d + i \frac{\sqrt{2\pi} e^{i\pi E/2}}{\Gamma(\frac{1-E}{2})} (2e/E)^{E/2} (z, a)_s. \tag{16.38}$$

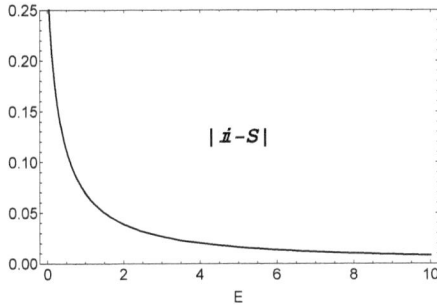

Fig. 16.18   The difference between i and the actual value of the Stokes constant for the Weber equation

Now compare this relation with the one from the WKB analysis performed above, line 4. One might think that the coefficient of the subdominant function is equal to $S(e^{-iW} + e^{iW})$ but this is not quite true. Since Stokes constants depend on the lower limit of integration used in the WKB-approximation we should at first write everything in a unique base and only then compare rigorous result with the WKB one. And since $(a, z) = [a, -a](-a, z)$ and $(z, a) = (z, -a)[-a, a]$ we can write

$$S(1 + e^{2iW}) = i\frac{\sqrt{2\pi}e^{i\pi E/2}}{\Gamma(\frac{1-E}{2})}(2e/E)^{E/2}. \tag{16.39}$$

As long as we chose a branch of the square root such that (a, z) is subdominant for large positive z, $W = \pi E/2$ and

$$S = i\frac{\sqrt{2\pi}e^{i\pi E/2}}{\Gamma(\frac{1-E}{2})}(2e/E)^{E/2}(1 + e^{i\pi E})^{-1}. \tag{16.40}$$

Now one can easily see that S approaches i for large energies.

16.8.2   The Budden equation

Another example of a potential which allows exact calculation of Stokes constants is the Budden potential with $Q(z) = 1+c/z$. We will see that the Stokes constants approach i for large values of c exactly as in the previous chapter.

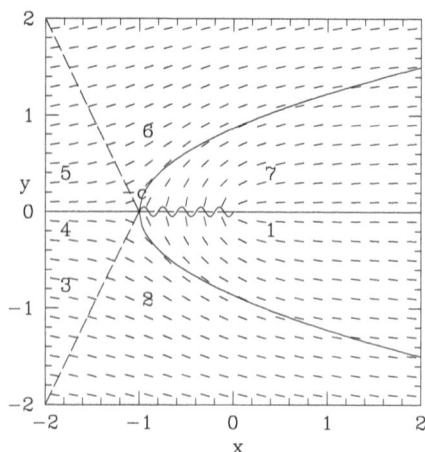

Fig. 16.19   Stokes plot for the Budden problem $Q = 1 + c/z$.

Before we calculate Stokes constants let us perform a WKB analysis. Choosing zero as a lower limit of integration we have

$$i \int_0^z \sqrt{1 + \frac{c}{x}} \, dx = iz + \frac{ic}{2} - \frac{ic}{2} \ln\left(\frac{c}{4z}\right) + O\left(\frac{1}{\sqrt{z}}\right) \tag{16.41}$$

and in a limit of large z

$$(0, z) \sim \left(\frac{4e}{c}\right)^{ic/2} z^{ic/2} e^{iz}, \tag{16.42}$$

$$(z, 0) \sim \left(\frac{4e}{c}\right)^{-ic/2} z^{-ic/2} e^{-iz}. \tag{16.43}$$

To perform an analytical continuation around the origin, place the cut between $z = 0$ and $z = -c$ along the real axis as shown in Fig.16.19. Starting with $y(z) \sim (0, z)$ at large negative z and using the rules of a continuation, one finds that the continuation is

3. $(0, z)_d = [0, -c](-c, z)_d$

2. $[0, -c](-c, z)_d + S[0, -c](z, -c)_s$

1. $[0, -c](-c, z)_s + S[0, -c](z, -c)_d = (0, z)_d + S[0, -c]^2(z, 0)_s$

All integrals here were evaluated below the cut and give $[0, -c] = e^{\frac{\pi c}{2}}$.

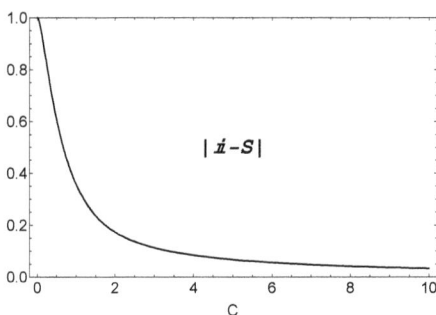

Fig. 16.20   The difference between i and actual value of Stokes constant for the Budden equation

Now we can find an exact value of S. A general solution for the Budden equation is

$$y(z) = ze^{-iz}[C_1 U(1 + \frac{ic}{2}, 2, 2iz) + C_2 M(1 + ic/2, 2, 2iz)], \quad (16.44)$$

where $U(a, b, z)$ and $M(a, b, z)$ are Kummer functions. Requiring the solution $y(z)$ to be asymptotic to $(0, z)$ for large negative z we find that

$$C_1 = \frac{\Gamma(1 + \frac{ic}{2})}{\Gamma(1 - \frac{ic}{2})} e^{3\pi c/4} \left(\frac{2e}{c}\right)^{ic/2}, \quad (16.45)$$

$$C_2 = \Gamma(1 + \frac{ic}{2}) e^{\pi c/4} \left(\frac{2e}{c}\right)^{ic/2}. \quad (16.46)$$

Now, using an asymptotic expansion of this solution for large positive z we have

$$y(z) \sim (0, z) + \frac{\Gamma(1 + \frac{ic}{2})}{\Gamma(1 - \frac{ic}{2})} \left(\frac{2e}{c}\right)^{ic} (e^{\pi c} - 1)(z, 0). \quad (16.47)$$

Finally, comparing this asymptotic relation with the one from our WKB analysis we deduce that

$$S = \frac{\Gamma(1 + \frac{ic}{2})}{\Gamma(1 - \frac{ic}{2})} \frac{2e}{c})^{ic} (1 - e^{-\pi c}). \quad (16.48)$$

It can be easily verified that this Stokes constant approaches i for large values of c, as shown in Fig. 16.20.

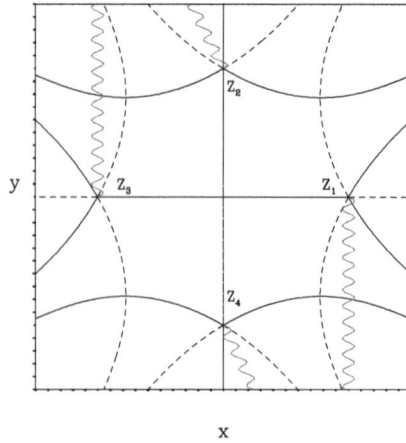

Fig. 16.21   Stokes plot for the bound state problem $Q = E - z^4$.

16.8.3   *A fourth order potential*

Now consider a more complicated potential, that of an anharmonic oscillator, with more turning points in the complex plane. As an example we take $Q(z) = E - z^4$. The Stokes structure is shown in Fig. 16.21. The cuts have been chosen to give symmetry in the continuation between the upper and lower half planes. We assume the Stokes constants to be the same at all singular points. This is probably not true for general separation, but we will be restricted to an asymptotic evaluation for large separation, and this turns out to be true to leading order.

In order to do the connections, we need the expressions $[k, l] = e^{\int_{z_k}^{z_l} i\sqrt{Q(z)}dz}$. Note that the sheet of $i\sqrt{Q(z)}$ is defined by the cut locations, with the initial sheet determined by the fact that (z_1, z) is subdominant for $x \to +\infty$, meaning that $i\sqrt{Q}(z) = -real$ in this domain.

Carrying out the integrals then gives

$$[1, 2] = e^{W/2}e^{iW/2}, \quad [1, 3] = e^{iW}, \quad [2, 3] = e^{-W/2}e^{iW/2} \quad (16.49)$$

where $W = E^{3/4} \int_{-1}^{1} \sqrt{1 - u^4}du$, and $\int_{-1}^{1} \sqrt{1 - u^4}du = 1.74804$.

Begin with a subdominant solution $\psi(z) = (Z_1, z)_s$ at large positive x and continue through the upper half plane above the singularity at $iE^{1/4}$

to large negative x. Using the symmetry of the potential we have

$$Se^{-W} + Se^{iW} = -(-1)^n. \tag{16.50}$$

Also we find a condition for the vanishing of the dominant solution

$$e^W[1 + S^2] + 2S^2 cos(W) + e^{-W}S^2 = 0. \tag{16.51}$$

For large turning point separation W is large and we have a solution given by the isolated turning point values, $S = i$ and $W_n = (n + 1/2)\pi$, the usual approximate WKB solution. There is a natural perturbation expansion parameter given by the existence of the exponential term e^{-W}, present because of the additional singular points not on the real axis. Even for the lowest bound state as given by the WKB approximation $e^{-W_0} \simeq 0.2$, and for the next level $e^{-W_1} \simeq 0.009$.

Perturbing about the WKB value $W_n = (n + 1/2)\pi$ gives the solution

$$S \simeq i\left[1 - cos(W_n)e^{-W_n} - \frac{e^{-2W_n}}{2}\right], \qquad cos(W) = -e^{-W_n}. \tag{16.52}$$

Thus $S \simeq i(1 + e^{-2W}/2)$ and $sin(W) = (-1)^n\sqrt{1 - cos^2(W)} \simeq (-1)^n(1 - e^{-2W}/2)$, where we have dropped terms of order e^{-4W}, amounting to a correction to the ground state energy of a fraction of a percent.

Values of the exact energy levels, the WKB approximation, and the Phase Integral evaluation are shown in table I. For the ground state E_{PI} has a 4 percent error, E_{WKB} has a 20 percent error.

This problem has also been approached using higher order phase integral approximations. A third order solution using the phase integral series due to Froman is given in table 2.1 of Child. The third order approximation to the ground state eigenvalue is given as 0.98076, with an error over 7 percent.

Table I. Energy Levels $Q = E - z^4$

n	E_{exact}	E_{wkb}	$cos(W)$	E_{PI}
0	1.0604	0.8671	-0.207879	1.0246
1	3.7964	3.7519	-8.9833×10^{-3}	3.7424
2	7.45567	7.4139	-3.8820×10^{-4}	7.4144
3	11.6374	11.6114	-1.6776×10^{-5}	11.6114
4	16.2618	16.2335	-7.2495×10^{-7}	16.2335

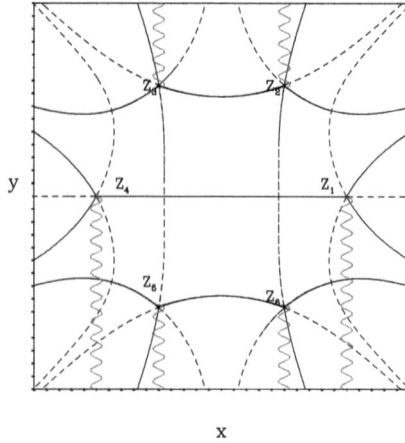

Fig. 16.22   Stokes plot for the bound state problem $Q = E - z^6$.

16.8.4   A sixth order potential

Now consider an anharmonic oscillator with more singular points in the complex plane, $Q(z) = E - z^6$. The Stokes structure is shown in Fig. 16.22 with first order zeros located at $Z_k = E^{1/6} e^{i(k-1)\pi/3}$ with $k = 1, 2, 3, 4, 5, 6$. We assume the Stokes constants to be the same at all singular points.

In order to do the connections, we need the expressions $[k, l] = e^{\int_{z_k}^{z_l} i\sqrt{Q(z)}dz}$. Note that the sheet of $i\sqrt{Q(z)}$ is defined by the cut locations, with the initial sheet determined by the fact that (z_1, z) is subdominant for $x \to +\infty$, meaning that $i\sqrt{Q(z)} = -real$ in this domain. Carrying out the integrals then gives

$$[1, 2] = e^{\sqrt{3}W/4} e^{iW/4}, \qquad [3, 2] = e^{-iW/2}, \qquad [3, 4] = e^{-\sqrt{3}W/4} e^{iW/4} \quad (16.53)$$

where $W = E^{2/3} \int_{-1}^{1} \sqrt{1 - u^6} du$, and $\int_{-1}^{1} \sqrt{1 - u^6} du = 1.821488$.

Begin with a subdominant solution $\psi(z) = (Z_1, z)_s$ at large positive x and continue through the upper half plane above all singularities to large negative x. Choosing the solution to be real for $x \to \infty$ and using the symmetry of the potential, but also noting that with the choice of cuts we have $Q^{1/4} = e^{i\pi/2}$ for large positive x and $Q^{1/4} = e^{-i\pi/2}$ for large negative x we find

$$(-1)^n i = 1 + S^2[1 + cos(W) + isin(W) + 2e^{-\sqrt{3}W/2} cos(W/2)]. \quad (16.54)$$

Also we find a condition for the vanishing of the dominant solution

$$1 + e^{\sqrt{3}W/2}cos(W/2) + S^2[e^{\sqrt{3}W/2} + 2cos(W/2) + e^{-\sqrt{3}W/2}]cos(W/2) = 0, \tag{16.55}$$

giving $S^2 = -1 + O(e^{-\sqrt{3}W})$.

For large turning point separation W is large and we have a solution given by the isolated turning point values, $S = i$ and $W_n = (n + 1/2)\pi$, the usual approximate WKB solution. A first order perturbation about the WKB value W_n gives the solution

$$S = i, \qquad cos(W) = -2e^{-\sqrt{3}W_n/2}cos(W_n/2). \tag{16.56}$$

It is interesting to note that these solutions break down at the second order in $\epsilon = e^{-\sqrt{3}W/2}$, meaning that one or more of the Stokes constants has a second order correction not given by Eq. 16.55. Note that Eq. 16.55 is real, but expanding $sin(W)$ in Eq. 16.54 through $sin(W) \simeq (-1)^n(1 - cos^2(W)/2)$ gives an additional second order imaginary term, but there is no second order term to balance it, and we conclude that S^2 must possess an imaginary second order term, not given by Eq. 16.55, so this equation can be trusted only to first order.

Values of the exact energy levels, the WKB approximation, and the Phase Integral evaluation are shown in table II. For the ground state E_{PI} has a 4 percent error, E_{WKB} has a 30 percent error.

Table II. Energy Levels $Q = E - z^6$

n	E_{exact}	E_{wkb}	$cos(W)$	E_{PI}
0	1.1448	0.8008	-0.36206	1.1009
1	4.3332	4.1612	2.3888×10^{-2}	4.1929
2	9.0731	8.9535	1.5727×10^{-3}	8.9508
3	14.9195	14.8316	-1.0354×10^{-4}	14.8314
4	21.7140	21.6224	-6.81617×10^{-6}	21.6224

16.8.5 *A non Hermitian Hamiltonian*

Non-Hermitian Hamiltonians having PT symnmetry have been shown to have real spectra, following a conjecture by D. Bessis that the spectrum of the Hamiltonian $H = p^2 + x^2 + ix^3$ is real and positive. A non-Hermitian Hamiltonian problem studied by Bender and Boettcher is given by the function $Q(z) = E + (iz)^N$. The energy spectrum is positive because of

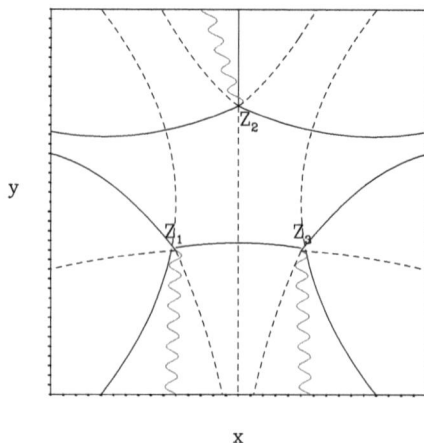

Fig. 16.23   Stokes plot for the bound state problem, $Q = E + (iz)^3$.

symmetry under the product of parity and time reversal. As an example we take $N = 3$.

The Stokes diagram is shown in Fig. 16.23, with three singular points located at $E^{1/3}e^{i\pi/2}$, $E^{1/3}e^{-i\pi/6}$, and $E^{1/3}e^{-i5\pi/6}$. Subdominant regions include the positive and negative real axis for $|x| \to \infty$. We carry out the continuation in the upper half complex plane in order to take account of the singular point at Z_2.

In order to do the connections, we need the expressions $[k, l] = e^{\int_{z_k}^{z_l} i\sqrt{Q(z)}dz}$. Note that the sheet of $i\sqrt{Q(z)}$ is defined by the cut locations, with the initial sheet determined by the fact that (z_3, z) is subdominant for $x \to +\infty$, meaning that $i\sqrt{Q}(z) = -real$ in this domain.

Carrying out the integrals then gives

$$[1, 2] = e^{\sqrt{3}W/2}e^{-iW/2}, \quad [1, 3] = e^{-iW}, \quad [2, 3] = e^{-\sqrt{3}W/2}e^{-iW/2} \quad (16.57)$$

where $W = E^{5/6}cos(\pi/6) \int_{-1}^{1} \sqrt{1 - u^3}du$, and $\int_{-1}^{1} \sqrt{1 - u^3}du = 1.68262$.

Begin for positive real x with $\psi(z) = (Z_3, z)_s$, and continue to large negative x, giving a subdominant and a dominant term, which we set to zero, giving

$$0 = e^{\sqrt{3}W}(1 + S^2) + 2S^2cos(W) + e^{-\sqrt{3}W}S^2 \quad (16.58)$$

leaving

$$\psi(z) = -(Z_1, z)_s(Se^{-\sqrt{3}W} + Se^{iW}), \qquad (16.59)$$

very similar to the equations found in section 16.8.3, since the Stokes structure in the upper half plane is the same. In this case the differential equation is not real, but choosing symmetric integration paths from $x = 0$ asymptotically along the Stokes lines to the right and to the left, and using the symmetry of $Q(z)$ we again conclude that the phases of the solutions for large positive x and large negative x are equal within a sign.

The cut locations give the fact that whereas $Q^{1/4} = e^{i\pi/4}$ for large positive x, $Q^{1/4} = e^{i3\pi/4}$ for large negative x, so in fact choosing the eigenfunction to be real for $x \to +\infty$ and requiring that it be real for large negative x gives from Eq. 16.59 $cos(W) + e^{-\sqrt{3}W} = 0$. Again in this case the asymptotic value of the Stokes constant is given by $S = i + O(\epsilon^2)$ with $\epsilon = e^{-\sqrt{3}W}$.

The first few energy levels for $N = 3$, with WKB and numerical values given by Bender, and values from this Phase Integral analysis are given in Table III. The WKB ground state energy has an error of 6 percent, and the Phase Integral value an error of 0.6 percent.

Table III. Energy Levels $Q = E - iz^3$

n	E_{exact}	E_{wkb}	$cos(W)$	E_{PI}
0	1.1562	1.09427	-6.5834×10^{-2}	1.1496
1	4.1092	4.08949	-2.8533×10^{-4}	4.0892
2	7.5621	7.54898	-1.2366×10^{-6}	7.54898
3	11.3143	11.3043	-5.3598×10^{-9}	11.3043
4	15.2916	15.2832	-2.3228×10^{-11}	15.2832

16.8.6 *Potentials*

The four potentials for the bound state cases are plotted in Fig. 16.24. The harmonic oscillator potential (a) $V(x) = z^2 - E$ and the two anharmonic oscillator potentials (c) $V(x) = z^4 - E$ and (d) $V(x) = z^6 - E$ are real on the real axis, $z = x$. The potential associated with $Q = E + (iz)^3$ (b) is real giving subdominant solutions along the lines $y = -|x|/\sqrt{3}$, with $V(x) = |x|^3(\sqrt{3}-1/3) - E$. The energies used in the plot are the ground state values. It is seen that the degree of distortion from the harmonic oscillator potential

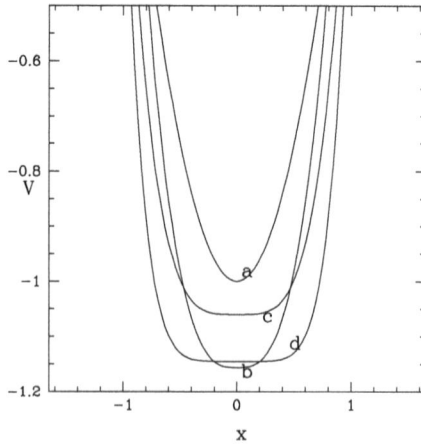

Fig. 16.24   Potentials for the four bound state cases discussed, the harmonic oscillator associated with the Weber equation, $V = x^2 - E$, (a), the non-Hermitian potential $Q = E + (iz)^3$ (b), and the Hermitian anharmonic oscillators $V = x^4 - E$ (c), and $V = x^6 - E$ (d). The exact ground state energies were used for each plot.

shape is inceasingly larger for the $(iz)^3$ and the z^4 and z^6 cases, associated with the larger number of singularities in $Q(z)$ in addition to the two turning point singularities. We see that the error in the WKB energy levels increases with the deviation from the harmonic oscillator potential shape, along with a corresponding improvement of the Phase Integral evaluation over the WKB value.

16.8.7   *Conclusion*

A proper use of Phase Integral methods can improve the eigenvalue determination for bound states significantly compared to a simple WKB evaluation. This improvement increases along with the increasing deviation of the potential shape from that of a harmonic oscillator. For all potentials possessing zeros or singularities in the complex plane in addition to the principal turning points, a small parameter is defined by $exp \int (i\sqrt{Q}) dz$, with the integration taken from them to the principle turning points, allowing a perturbation expansion. No such parameter exists for the Weber equation or the Budden problem. However in these cases the Stokes constants can be calculated analytically. It is remarkable that the asymptotic value of the Stokes constant in each case is $S = i$, the value for an isolated first order zero, and that in the cases examined in perturbation theory the corrections to S are second order or higher in the small parameter given by the separation of the singular points. We have had to make the simplifying assumption that the Stokes constants are all equal, undoubtedly not true to higher order. The two complex equations resulting from the vanishing of the dominant solution and the symmetry or anti-symmetry of the subdominant solution do not allow proceeding to higher order, additional information is needed. It is an open question whether such relations exist, and whether the resulting equations lead to a convergent series giving the exact bound state energy. Of course the bound state energies do not form an open set, so S is not determined as an analytic function of E, only the values at the bound eigenstates are fixed. However, we conjecture that it is a common feature that all Stokes constants associated with lines emerging from a simple zero approach i for large separation between singularities.

16.9   Problems

1. Sketch the anti-Stokes lines associated with $d^2\psi/dz^2 + Q\psi = 0$ for $|z| < 4$,

$$Q = \frac{z - i\pi}{z + i\pi} sinz.$$

2. Sketch the anti-Stokes and Stokes lines and find the Stokes constant T for a second order zero, $Q = z^2$, by directly continuing the solution around the turning point and requiring the solution to be single valued. Check your result using Eq. 16.16.

3. Consider the eigenvalue problem $y'' + Ecos(x)y = 0$ with $y(\pi) = 0$, and $y'(0) = 0$, $E > 0$. Sketch the Stokes plot for y.

   a. Use WKBJ to find the three smallest eigenvalues E, approximating Stokes constants as those from isolated singularities.

   Hint 1: The bounday condition $y(\pi) = 0$ is satisfied ON a Stokes line.

   Hint 2: Continue the WKBJ solution to $x = 0$ and require $y'(0) = 0$, giving a transcendental equation for $w = \int_0^{\pi/2} \sqrt{Ecos(x)}dx$. Solve this iteratively. Note

$$\int_0^{\pi/2} \sqrt{cos(x)}dx = \int_{\pi/2}^{\pi} \sqrt{-cos(x)}dx \simeq 1.19814025.$$

   b. Integrate the equation numerically using the WKBJ values as first approximations to E. Adjust E to give $y(\pi) = 0$. What is the accuracy of the WKBJ eigenvalues?

4. Consider the differential equation

$$\frac{d^2y}{dx^2} + \left(\omega^2 - \frac{x^2}{4}\right)y = 0.$$

   a. Find the asymptotic behaviour of the two solutions for $x \to \pm\infty$.

   b. Write the solutions as $y \sim e^S W(x)$ and find a series representation for $W(x)$ for the solution tending to zero at ∞. Find the radius of convergence of the series.

c. Draw the Stokes diagram. Use WKBJ analysis to find ω such that the solution tends to zero at $\pm\infty$. These are the eigenvalues given by the WKBJ analysis.

d. What happens to the series for $W(x)$ for these values of ω?

e. What is the accuracy of the eigenvalues and eigenfunctions given by the WKBJ method?

5. Consider the differential equation

$$\frac{d^2\psi}{dx^2} + B(x^2 - x^4 - 1/4 + E)\psi = 0,$$

boundary conditions $\psi(-\infty) = 0, \psi(\infty) = 0$, with $B = 10^4, 0 < E << 1$.

a. Sketch the Stokes plot for ψ.
Neglecting exponentially small tunneling effects there are two states with the smallest value of E (ground states), one symmetric and one anti-symmetric in x.

b. Sketch the ground state eigenfunctions and write an expression determining the energy. Use $E \ll 1$ to expand the potential to second order in x around the minima and calculate the ground state energy.

6. Use WKB theory and require that the two ground states be exactly symmetric and antisymmetric to find an expression for the changes in the energy from the degenerate value of problem 5. Use Stokes constants for isolated singularities. Hint: It is not necessary to follow the solution through the whole Stokes diagram, for symmetry use $\psi'(0) = 0$ and for antisymmetry use $\psi(0) = 0$.

7. A normally incident plane polarized electromagnetic wave enters a magnetized plasma $\vec{B}_0 = B_0\hat{z}$ where $\omega_p(x)$ increases with x. If $\vec{k}$ and $\vec{E}$ of the wave are perpendicular to $\vec{B}_0$, $k_y = k\hat{y}$, $\vec{E} = E\hat{y}$ the y component of $\vec{E}$ satisfies

$$\frac{d^2 E_y}{dx^2} + k_0^2\epsilon(x)E_y = 0.$$

The dielectric function has a cutoff, $\epsilon = 0$, and resonance $\epsilon = \infty$. Model this dielectric through $k^2\epsilon(x) = 1 + a/x$. Draw the Stokes diagram and find the reflection and transmission coefficients for a wave incident from the left using phase integral methods. This is called the Budden problem, see Budden (1979), White and Chen, Plasma Physics 16, 565 (1974). Note that $|R|^2 + |T|^2 \neq 1$. Why?

16.10 References

- Asymptotic Analysis of Differential Equations, R. B. White, World Scientific Press, 2005
- Bender, C. M., and S. A. Orszag, Advanced Mathematical Methods for Scientists and Engineers (McGraw-Hill, New York, 1978) p. 531.
- Berk, H. L., Nevins, W. M., and Roberts, K. V., J. of Math. Phys. 23, 988 (1982).
- Berry, M. V., in Proc. of the Royal Society 427, 265 (1990).
- Bohm, D., Quantum Theory (Prentice-Hall, New Jersey 1951).
- Brillouin, L., C.R. Acad. Sci. Paris 183, 24 (1926).
- Budden, K. G., Phil. Trans. Royal Soc. London 290, 405 (1979).
- Chen, L., P. K. Kaw, C. R. Oberman, P. Guzdar, *et al.*, Phys. Rev. Lett. 41, 649 (1978).
- Copson, E. T., Introduction to the Theory of Function of a Complex Variable (Oxford Press, London, 1935) p. 118.
- Dewar, R. L., and B. Davies, J. Plasma Phys. 32, 443 (1984).
- Furry, W. H., Phys. Rev. 71, 360 (1947).
- Ford, An Phys. NY. 7, 287 (1959).
- Heading, J., An Introduction to Phase Integral Methods (Wiley, New York, 1962).
- Jeffries, H., Proc. Lond. Math. Soc. 23, 428 (1923).
- Johnston, T. W., and P. Picard, Phys. Rev. Lett. 51, 574 (1983).
- Johnston, T. W., and P. Picard, Phys. Fluids 28, 859 (1985).
- Kramers, H. A., Zeit. f. Phys. 39, 828 (1926).
- Pearlstein, L. D., and H. L. Berk, Phys. Rev. Lett. 23, 220 (1969).
- Ross, D. W., and S. M. Mahajan, Phys. Rev. Lett. 40, 324 (1978).
- Stokes, G. G., Proc. Camb. Phil. Soc. 6, 362 (1889).
- Soop, M., Ark. Fys. 30, 217 (1965).
- Tsang, K. T., P. J. Catto, J. C. Whitson, and J. Smith, Phys. Rev. Lett. 40, 327 (1978).
- Wentzel, G., Zeit. f. Phys. 38, 518 (1926).
- White, R. B., J. of Comput. Phys. 31, 409 (1979).
- White, R. B., (2000) – The code WKB, written in fortran 99, is available upon request.
- R.B. Dingle *Asymmpotic Expansions: Their Derivation and Interpretation* Academic Press, London and New York (1973)
- C. M. Bender and S. Boettcher, Phys Rev Lett 80, 5243 (1998)

- M.S. Child *Semiclassical Mechanics with Molecular Applications,* second edition, Oxford University Press (2014)
- M.V. Berry, Asymmpotics, Superasymptotics, Hyperasymptotics, in *Asymptotics Beyond All Orders*, H. Segur et al (eds.) Plenum Press NY (1991)
- C. M. Bender, K. Olausson, P.S. Wang, Phys. Rev. D 16, 1740 (1977)
- Whittaker, E. T., and G. N. Watson, A Course on Modern Analysis (University Press, Cambridge, 1962).

Index

www.ingramcontent.com/pod-product-compliance
Ingram Content Group UK Ltd.
Pitfield, Milton Keynes, MK11 3LW, UK
UKHW052219130325
456149UK00002B/4